HANDBOOK OF ASSAY DEVELOPMENT IN DRUG DISCOVERY

Drug Discovery Series

Series Editor

Andrew Carmen

Johnson & Johnson PRD, LLC
San Diego, California, U.S.A.

Drug Discovery Series/5

HANDBOOK
OF ASSAY
DEVELOPMENT
IN DRUG
DISCOVERY

EDITED BY LISA K. MINOR

Taylor & Francis
Taylor & Francis Group
Boca Raton   London   New York

A CRC title, part of the Taylor & Francis imprint, a member of the Taylor & Francis Group, the academic division of T&F Informa plc.

Published in 2006 by
CRC Press
Taylor & Francis Group
6000 Broken Sound Parkway NW, Suite 300
Boca Raton, FL 33487-2742

International Standard Book Number-10: 1-57444-471-9 (Hardcover)
International Standard Book Number-13: 978-1-57444-471-1 (Hardcover)
Library of Congress Card Number 2005053104

Library of Congress Cataloging-in-Publication Data

Handbook of assay development in drug discovery / edited by Lisa Minor.
    p. cm. -- (Drug discovery series ; 5)
  Includes bibliographical references and index.
    ISBN 1-57444-471-9 (alk. paper)
    1. Drug development--Handbooks, manuals, etc. 2. Drugs--Research--Handbooks, manuals, etc. 3.
Drugs--Design--Handbooks, manuals, etc. I. Minor, Lisa. II. Series.

RM301.25.H34 2006
615'.19--dc22   2005053104

Taylor & Francis Group
is the Academic Division of Informa plc.

Visit the Taylor & Francis Web site at
http://www.taylorandfrancis.com

and the CRC Press Web site at
http://www.crcpress.com

Dedication

This book is dedicated to my husband, Keith Demarest, who has been so understanding and supportive of me throughout my science career.

Preface

Multiwell assays for drug discovery have become a staple of the scientific community. This has been led by the need to screen targets more rapidly and with more efficiency than in previous years. This has also been led by advances in chemistry allowing parallel or multiplex chemical synthesis to provide more compounds for screening. These multiwell assay formats are both *in vitro* and cell based and include a mixture of simple enzyme assays as well as those more complex, such as cell imaging. This handbook is designed to serve as a reference guide to all of those in discovery research who are developing assays, as well as those in the academic world who wish to know what is available to perform the assay they want as well as where to find those tools. This book will provide general guidelines as well as specific examples of each assay system. These include radiometric assays, fluorescent-based assays, reporter assays, and others, for such targets as kinases, proteases, nuclear receptors, and G-protein–coupled receptors (GPCRs). The book includes chapters from assay developers in the pharmaceutical as well as in the vendor community. It is meant as a general guidebook that one can take to the bench. It includes descriptions of methods, exact protocols used to perform such methods, and troubleshooting tools. I hope that you will find it beneficial and that use of it will provide the information and guidance that you seek.

Editor

Lisa Minor received her Ph.D. from Pennsylvania State University, where she studied enzyme regulation in primary cultures of rat hepatocytes. She received her postdoctoral training at the Medical College of Pennsylvania, where she studied the metabolism of lipid that had accumulated in smooth muscle foam cells. She joined Johnson & Johnson immediately after her postdoctoral studies. At Johnson and Johnson she has been a member of therapeutic area teams and a high-throughput screening group. She has been responsible for developing, optimizing and implementing enzyme and cell culture assays for target identification and high-throughput screening and has been instrumental in developing new technologies for cell-based assays. Included are assays to measure the translocation of G-protein–coupled receptors from the membrane to the cytoplasm using cell-based image analysis as well as developing a high-throughput screening mRNA detection assay using branched DNA and enzyme-linked immunosorbent assay (ELISA)-based platforms for measuring phosphorylation in a cell-based system. Currently, Dr. Minor is a member of the Vascular Research Team and is on the governing board of the Society for Biomolecular Screening.

Contributors

Elaine J. Adie
GE Healthcare
The Maynard Centre
Whitchurch, Cardiff, U.K.

Jenny A. Berry
GE Healthcare
The Maynard Centre
Whitchurch, Cardiff, U.K.

Wayne P. Bocchinfuso
Lilly Research Laboratories
Eli Lilly & Co.
Indianapolis, Indiana

Kristen M. Borchert
Becton-Dickinson Technologies
Research Triangle Park, North Carolina

Roger Bossé
PerkinElmer Life and Analytical Sciences
PerkinElmerBioSignal, Inc.
Montreal, Quebec, Canada

Nathalie Bouchard
PerkinElmer Life and Analytical Sciences
PerkinElmerBioSignal, Inc.
Montreal, Quebec, Canada

Philip E. Brandish
Merck & Co.
North Wales, Pennsylvania

Bob Bulleit
Research and Development
Promega Corporation
Madison, Wisconsin

Susan M. Catalano
Drug Discovery Imaging
Hayward, California

Cailin Chen
Drug Discovery
Johnson & Johnson Pharmaceutical Research
 and Development, LLC
Spring House, Pennsylvania

Jeffrey R. Cook
GE Healthcare–Biosciences
Piscataway, New Jersey

Conrad L. Cowan
Carad Therapeutics
Tarrytown, New York

Bruce Damiano
Drug Discovery
Johnson & Johnson Pharmaceutical Research
 and Development, LLC
Spring House, Pennsylvania

Michele S. Dowless
Lilly Research Laboratories
Eli Lilly & Co.
Indianapolis, Indiana

John Dunlop
Neuroscience Discovery Research
Wyeth Research
Princeton, New Jersey

James P. Edwards
Immunology, Drug Discovery
Johnson & Johnson Pharmaceutical Research
 and Development, LLC
San Diego, California

Richard M. Eglen
DiscoveRx Corp.
Fremont, California

Stuart Emanuel
Cancer Therapeutics Research
Johnson & Johnson Pharmaceutical Research
 & Development, LLC
Raritan, New Jersey

Michael J. Francis
GE Healthcare
The Maynard Centre
Whitchurch, Cardiff, U.K.

Peter Fung
DiscoveRx Corp.
Fremont, California

Ralph J. Garippa
Hoffmann-La Roche, Inc.
Nutley, New Jersey

Tina Garyantes
Lead Discovery Technology
Sanofi Aventis
Bridgewater, New Jersey

Said A. Goueli
Research and Development
Promega Corporation and
Department of Pathology and Lab Medicine
University of Wisconsin Medical School
Madison, Wisconsin

Robert Graves
GE Healthcare
Piscataway, New Jersey

Harry Harney
PerkinElmer Life and Analytical Sciences
Boston, Massachusetts

Erika M. Hawkins
Research and Development
Promega Corporation
Madison, Wisconsin

Ilkka Hemmilä
PerkinElmer Life and Analytical Sciences
Turku, Finland

Ann F. Hoffman
Hoffmann-La Roche, Inc.
Nutley, New Jersey

Keith A. Houck
Lilly Research Laboratories
Eli Lilly & Co.
Indianapolis, Indiana

Kevin Hsiao
Research and Development
Promega Corporation
Madison, Wisconsin

Christine C. Hudson
Xsira Pharmaceuticals
Morrisville, North Carolina

Kelvin T. Hughes
GE Healthcare–Biosciences
Whitchurch, Cardiff, U.K.

Pertti Hurskainen
PerkinElmer Life and Analytical Sciences
Turku, Finland

Chantal Illy
PerkinElmer Life and Analytical Sciences
PerkinElmerBioSignal, Inc.
Montreal, Quebec, Canada

Dana L. Johnson
Oncology Research
Johnson & Johnson Pharmaceutical Research
  and Development, LLC
Spring House, Pennsylvania

Patricia Kasila
PerkinElmer Life and Analytical Sciences
Boston, Massachusetts

Lindy Kauffman
DiscoveRx Corp.
Fremont, California

Dianne Kowal
Neuroscience Discovery Research
Wyeth Research
Princeton, New Jersey

Priya Kunapuli
Department of Automated Biotechnology
Merck Research Laboratories
North Wales, Pennsylvania

Zhuyin Li
Lead Discovery Technology
Sanofi Aventis
Bridgewater, New Jersey

Anice Lim
DiscoveRx Corp.
Fremont, California

Stephen Lin
Neuroscience Discovery Research
Wyeth Research
Princeton, New Jersey

Carson R. Loomis
National Institute of Health
Bethesda, Maryland

Qiang Lü
Neuroscience Discovery Research
Wyeth Research
Princeton, New Jersey

Lisa Minor
Drug Discovery
Johnson & Johnson Pharmaceutical Research
  and Development, LLC
Spring House, Pennsylvania

Richard A. Moravec
Research and Development
Promega Corporation
Madison, Wisconsin

Tabassum Naqvi
DiscoveRx Corp.
Fremont, California

Andrew Niles
Research and Development
Promega Corporation
Madison, Wisconsin

Martha A. O'Brien
Promega Corporation
Madison, Wisconsin

Robert H. Oakley
Xsira Pharmaceuticals
Morrisville, North Carolina

Patricia D. Pelton
Johnson & Johnson Pharmaceutical Research
  and Development, LLC
Raritan, New Jersey

Dominique Perrin
Serono Pharmaceutical Research Institute
Geneva, Switzerland

David Powell
GE Healthcare–Biosciences
The Maynard Centre
Whitchurch, Cardiff, U.K.

Molly J. Price-Jones
Molecular Light Technology Research Ltd.
Whitchurch, Cardiff, U.K.

Robert Ring
Neuroscience Discovery Research
Wyeth Research
Princeton, New Jersey

Terry L. Riss
Research and Development
Promega Corporation
Madison, Wisconsin

Riaz Rouhani
DiscoveRx Corp.
Fremont, California

Nathalie Rouleau
PerkinElmer Life and Analytical Sciences
PerkinElmerBioSignal, Inc.
Montreal, Quebec, Canada

Jan L. Sechler
Oncology Research
Johnson & Johnson Pharmaceutical Research
  and Development, LLC
Spring House, Pennsylvania

Rajendra Singh
DiscoveRx Corp.
Fremont, California

Charles Smith
Drug Discovery
Johnson & Johnson Pharmaceutical Research
  and Development, LLC
Spring House, Pennsylvania

Lynne Smith
GE Healthcare
The Maynard Centre
Whitchurch, Cardiff, U.K.

Richard Sportsman
Molecular Devices Corp.
Sunnyvale, California

Jacques Andre St. Pierre
PerkinElmer Life and Analytical Sciences
PerkinElmerBioSignal, Inc.
Montreal, Quebec, Canada

Robin L. Thurmond
Immunology, Drug Discovery
Johnson & Johnson Pharmaceutical Research
  and Development, LLC
San Diego, California

Inna Vainshtein
DiscoveRx Corp.
Fremont, California

Rob Hooft van Huijsduijnen
Serono Pharmaceutical Research Institute
Geneva, Switzerland

Dean Wenham
Insymbiosis
Montreal, Canada

Stephen K.-F. Wong
Pfizer Inc.
Department of Neuroscience
Groton, Connecticut

Yingxin Zhang
Neuroscience Discovery Research
Wyeth Research
Princeton, New Jersey

Wei Zheng
National Institute of Health
National Human Genome Research Institute
NIH Chemical Genomics Center
Bethesda, Maryland

Table of Contents

1 Protein Kinases in Drug Discovery: Rationale, Success, and Challenge

Dana L. Johnson and Jan L. Sechler

CONTENTS

1.1 OVERVIEW

The central role that protein phosphorylation plays in almost all facets of cellular and ultimately systemic function has become evident since the advent of molecular biology in the last three decades. The phosphorylation of various cellular substrates is accomplished by the broad superfamily of enzymes known as protein kinases. Due to their essential role in a variety of regulatory processes, protein kinases have been a subject of intensive focus in the pharmaceutical industry. The critical action of this enzyme family cannot be overestimated, and with the emergence of targeted drug discovery approaches, new therapeutic agents that target kinases have achieved regulatory approval. Here we will provide an overview of the functional importance of this class of proteins and review some therapeutic success stories that clinically validate protein kinases as drug discovery targets. This chapter only serves as a preface to the topic since thousands of citations concerning protein kinases, including their potential as therapeutic targets, exist, and therefore a comprehensive review is beyond the scope of this introduction.

1.2 HUMAN PROTEIN KINASES

With the advent of the Human Genome Project and subsequent complete sequence determination and analysis, it is now thought that a genomic complement of at least 518 protein kinases exists. Fully 244 of these kinases map to chromosomal loci associated with pathological conditions, most often those associated with tumors [1]. This is a rather astounding finding suggesting that disregulations of kinases are critical drivers of diverse pathological conditions and that their potential as targets for therapeutic intervention will only increase as knowledge of the proteome, and more specifically the kinome, grows. Given this information, the focus upon these enzymes as therapeutic targets in oncology will likely increase. However, oncology is not the only therapeutic area in which kinases appear to be value added targets; considerable efforts on various kinases are under way in inflammation and diabetes drug discovery.

Protein kinases exist in both membrane-bound and soluble forms and can be broadly classified in a number of ways. For those kinases that principally target protein substrates, two general classes, tyrosine and serine/threonine kinases, are acknowledged based on the site of terminal substrate phosphorylation. Further classification is based on the primary sequence of the catalytic domain and has been extensively annotated [1].

In cancer research, the receptor and soluble tyrosine kinases appear particularly attractive as therapeutic targets [2–4]. They have been directly implicated in oncogenic transformations and can function as oncogenes; the role played by HER2 is an example [5]. Further, they functionally mediate signaling events related to growth factor–initiated proliferative pathways [6]. These pathways can be relevant in both transformed and normal cells because of the role they play in angiogenesis, the process of new blood vessel formation, as well as tumor growth [7]. Normal angiogenesis in adults appears to be fairly restricted to endometrial turnover and wound healing, with pathological angiogenesis associated with cancer, rheumatoid arthritis, and ocular vascular disorders such as diabetic retinopathy and macular degeneration [7,8]. The angiogenic process appears to be induced by locally hypoxic environments and leads to the increased expression of a number of growth factors including variants of vascular endothelial growth factor (VEGF), platelet-derived growth factor (PDGF), and fibroblast growth factor (FGF), among others [7]. Therefore, the various cognate receptor tyrosine kinases for these growth factors have become important drug discovery targets [9]. However, to date no small molecule inhibitors that target these three kinases have received regulatory approval, and only recently was the antiangiogenic agent Avastin® approved for therapeutic use [10]. This agent does not bind the VEGF receptor but rather the VEGF ligand. Although this does not provide clinical validation of the VEGF receptor as a target, it is consistent with the idea that the pathway is a valid target.

Another class of oncology-centric kinases is the cyclin-dependent kinases (cdk), whose coordinated and controlled sequential activation is critical to the replicating cell's ability to transverse the cell cycle and generate a copy of itself. As indicated by their name, the cdks require association with a cognate cyclin for their activity [11,12]. There is considerable evidence suggesting that in neoplastic cells, the expression and/or control of various cdks or critical substrate molecules can be altered or aberrant. Since uncontrolled cellular proliferation is perhaps the single most important characteristic of the cancer cell phenotype, the regulation of cellular proliferation has become a central theme in targeted drug discovery. Clinical trials of first generation cdk inhibitors are ongoing, but no agents targeting a cdk have yet been approved for therapeutic use [13].

Interest in this class of enzymes is not limited to oncology as at least one cdk, cdk5, has also been suggested to have therapeutic target potential in neurodegenerative disorders such as Alzheimer's disease, Parkinson's disease, and amyotrophic lateral sclerosis. The activity of cdk5 appears to be somewhat restricted to the nervous system and is thought to be implicitly involved in several processes leading to neuronal pathology [14]. It is interesting to speculate that inhibitors broadly active against cdks may have therapeutic potential in this poorly served patient population.

The above description covers just a few representatives of the many possible kinase target classes in drug discovery. Biological research has not yet taught us the cellular roles of all the kinome members or which kinases can be therapeutically manipulated without overt toxicity or unacceptable risk. However, considerable progress has been made on certain pathways or molecular targets, and new therapeutic agents have begun to emerge.

1.3 THERAPEUTIC EXAMPLES

The first therapeutic agent targeting a kinase was not a small molecule inhibitor of the kinase enzyme activity but rather trastuzumab (Herceptin®), the antibody to the epidermal growth factor family oncogene HER2 [15]. This agent appears to principally function by effectively liganding the oncogenic protein on the cell surface. This binding event ultimately results in the internalization and removal of the receptor from the cell surface, effectively silencing the proliferative signal of the oncogenic kinase, although the cellular effects of the agent seem to be multifaceted and are still under active investigation [5].

The first small molecule kinase inhibitor to reach the market was imatinib mesylate (also called Gleevec®, Glivec®, STI-571, CGP 57418B) [16]. This remarkable agent has transformed the treatment paradigm for chronic myelogenous leukemia (CML) by virtue of its ability to inhibit the Bcr-Abl kinase fusion gene product that is the hallmark of CML. CML is characterized by a chromosomal translocation that results in 9:22 reciprocal exchange producing a constitutively active form of the Abl kinase gene; this is termed the Philadelphia chromosome. The uncontrolled activity of the kinase fusion product appears to be adequate to drive the replication of the CML cell [16].

Imatinib mesylate is an orally available Bcr-Abl kinase inhibitor with activity against the PDGF receptor as well as the c-kit tyrosine kinase. In early clinical trials, response rates of 40 to 60% were observed, depending on the disease classification. While minimal adverse events were reported, there clearly was some toxicity associated with the clinical administration of the agent; however, it appeared to be manageable [16]. Subsequently, the activity of Gleevec against the c-kit tyrosine kinase led to the approval and use of the agent in the treatment of gastrointestinal stromal tumors (GIST) [17,18].

Another tyrosine kinase receptor that has been of intense therapeutic interest is the epidermal growth factor receptor (EGFR). EGFR is overexpressed in many solid tumors of epithelial origin, making it an attractive target for anticancer therapy. The EGFR kinase inhibitor, gefitinib (Iressa®), has recently been approved in the U.S. and Japan for the treatment of nonsmall cell lung carcinoma (NSCLC). Initial clinical response to gefitinib was limited to only 10 to 19 percent of patients with advanced NSCLC. However, those patients who did respond showed remarkable and rapid improvement upon treatment [19]. The observation that there was no correlation between immunohistochemical staining of the tumor for EGFR and response to the drug led investigators to look for EGFR mutations in the subset of responding patients [19,20]. These patients were found to harbor gain-of-function heterozygous mutations in the EGFR gene that clustered around the ATP-binding pocket and enhanced kinase activity in response to EGF. The mutations are believed to stabilize the interaction between the drug and the kinase, thereby enhancing the effect of inhibition. EGFR gene alterations were not identified in tumor samples from the patients who did not respond to gefitinib [19]. Since the presence of mutations appears to be an indicator of clinical responsiveness, sequence analysis of EGFR from lung cancer patients should aid clinicians in the choice of treatment.

As previously mentioned, drug discovery efforts toward the development of kinase inhibitors have not been limited to the field of oncology. The p38 MAPK signaling pathway plays an important role in regulating proinflammatory pathways, including the production of IL-6, making it a popular target for the development of inhibitors for the treatment of inflammatory conditions. While there are no marketed drugs in this class, several p38-targeted agents are in development for the treatment of rheumatoid arthritis and inflammatory bowel disease [21]. These include SB242235, which has demonstrated efficacy in animal models of antigen-induced arthritis and is currently being developed by Glaxo Smithkline (GSK) [21]. Another promising compound is CNI-1493, currently in preparation for Phase II/III trials, with which significant clinical benefit has been observed in Crohn's disease patients [21,22]. Finally, preliminary results from a clinical trial centering on RWJ67657 have demonstrated the ability of this compound to suppress proinflammatory cytokine response in a dose-dependent manner upon exposure to endotoxin without observed toxicities [23]. Although much work lies ahead for these compounds, significant progress has been made toward therapeutic application of this group of agents.

1.4 FUTURE CHALLENGES

Despite considerable efforts across both industry and academia to exploit protein kinases as drug targets, relatively few effective, approved, well-tolerated medicines have emerged. To date, most of the reported agents appear to target the catalytic site of the kinase and act as ATP-competitive inhibitors [24,25]. Given that this structural motif is fairly well conserved, this strategy typically results in a kinase inhibitor that exhibits less-than-desirable selectivity. While this selectivity (or lack thereof) can lead to additional therapeutic applications, as has been the case for Gleevec with its action against c-kit and Bcr-Abl kinases, it may also result in additional toxicities if the activity is undesirable. Further, there is growing evidence that mutation of this binding pocket can also lead to clinical resistance, at least in the case of Gleevec [26]. This is a significant concern for ATP-competitive agents, and clinical experience has not yet revealed whether this plasticity will hold true for all agents or whether it will be restricted to agents with certain binding modes.

It has been postulated that exploitation of the principles of structure-based drug design (SBDD) can be used to create more selective kinase inhibitors [25]. However, this approach can be limited by the inability to crystallize a target protein. Ideally, the use of SBDD creates direct knowledge of critical interactions of a candidate inhibitor with the amino acids of the target protein, allowing optimization of binding affinity and selectivity by exploiting interactions with amino acid residues adjacent to the ATP binding domain but unique to the target enzyme.

Further, as was found for Gleevec, both activated and inactive forms of many target kinases exist, and any given drug's ability to modulate a disease many be dependent on the nature of this interaction. In the case of Gleevec, it appears that the compound stabilizes the inactive conformation of the protein, thereby gaining effective high affinity [27]. The choice of utilizing an activated or an inactive form of a kinase in drug screening and optimization could potentially alter the structural chemotypes capable of interaction with the target and might potentially lead one astray. Careful consideration of these issues is certainly warranted when conducting a kinase-targeted drug discovery program and can save considerable time and effort as a therapeutic program matures. This is highlighted by the recent finding that Iressa appears to be especially active on certain EGFR mutants, as described above [19,20].

Finally, additional kinase-active chemotypes are needed because of the depth and breath of potential kinase targets. Through additional chemotypes, it may be possible to improve selectivity, solubility, and activity of kinase inhibitors by expanding the known chemical subtypes with reasonable drug properties. Numerous biochemical and cell-based screening methods for profiling kinase inhibitors exist. One reasonable way of obtaining these new chemical structures is through the application of high-throughput and ultra high-throughput screening paradigms. Recent advances in the development of new screening technologies should aid in accelerating progress toward the identification of protein kinase inhibitors with novel chemotypes.

REFERENCES

1. Manning C, Whyte DB, Martinez R, Hunter T, Sudarsanam S. The protein complement of the human genome. *Science* 2002;298:1912–1918.
2. Nam NH, Parang, K. Current targets for anticancer drug discovery. *Curr Drug Targ* 2003;4:159–179.
3. Fabbro D, Garcia-Echeverria CG. Targeting protein kinases in cancer therapy. *Curr Opin Drug Discovery Devel* 2004;5:701–712.
4. Fabbro D, Ruetz S, Buchdunger E, Cowan-Jacob SW, Fendrich G, Liebetanz J, Mestan J, O'Reilly T, Traxler P, Chaudhuri B, Fretz H, Zimmermann J, Meyer T, Caravatti G, Furet P, Manley PW. Protein kinases as targets for anticancer agents: from inhibitors to useful drugs. *Pharmacol Ther* 2002;93:79–98.
5. Menard S, Pupam SM, Campliglio M, Tagliabu E. Biologic and therapeutic role of HER2 in cancer. *Oncogene* 2003;22:6570–6578.
6. Gschwind A, Fischer OM, Ullrich A. The discovery of receptor tyrosine kinases: targets for cancer therapy. *Nature Rev* 2004;4:361–370.
7. Tonini T, Rossi F, Claudio P. Molecular basis of angiogenesis and cancer. *Oncogene* 2003;22:6549–6556.

8. Witmer A, Vrensen G, Van Noorden C, Schlingemann R. Vascular endothelial growth factors and angiogenesis in eye disease. *Prog Retinal Eye Res* 2003;22:1–29.

9. Drevs J, Medinger M, Schmidt-Gersbach C, Weber R, Unger C. Receptor tyrosine kinases: the main targets for anticancer therapy. *Curr Drug Targ* 2003;4:113–121.

10. Ferrara N, Hillan K, Gerber H, Novotny W. Discovery and development of bevacizumab, an anti-VEGF antibody for treating cancer. *Nature Rev Drug Discovery* 2004;3:391–400.

11. Senderowicz AM. Cyclin-dependent kinases as targets for cancer therapy. *Cancer Chemother Biol Response Mod* 2002;20:169–196.

12. Lee MH, Yang HY. Molecular targets for cell cycle inhibition and cancer therapy. *Expert Opin Ther Patents* 2003;13:329–346.

13. Dai Y, Grant S. Small molecule inhibitors targeting cyclin-dependent kinases as anticancer agents. *Curr Oncol Rep* 2004;6:123–130.

14. Kesavapany S, Li B-S, Amin N, Zheng Y-L, Grant P, Pant HC. Neuronal cyclin-dependent kinase 5: role in nervous system function and its specific inhibition by the Cdk5 inhibitory peptide. *Biochim Biophys Acta* 2003;1967:143–153.

15. Abou-Jawde R, Choueiri T, Aleman C, Mekhail T. An overview of targeted treatments in cancer. *Clin Ther* 2003;25:2121–2137.

16. Cohen MH, Williams G, Johnson JR, Duan J, Gobburu J, Rahman A, Benson K, Leighton J, Kim SK, Wood R, Rothmann M, Che G, Maung KU, Staten AM, Pazdur R. Approval summary for imatinib mesylate capsules in the treatment of chronic myelogenous leukemia, *Clin Cancer Res* 2002;8:935–942.

17. Dagher R, Cohen M, Williams G, Rothmann M, Gobburu J, Robbie G, Rahman A, Chen G, Staten A, Griebel D, Pazdur R. Approval summary: imatinib mesylate in the treatment of metastatic and/or unresectable malignant gastrointestinal stromal tumors. *Clin Cancer Res* 2002;8:3034–3038.

18. Demetri GD, Imatinib (STI571, Gleevec) as therapy to target the molecular pathophysiology of gastrointestinal stromal tumors: a model for rational drug development. *Prog Oncol* 2002;95–112.

19. Lynch TJ, Bell DW, Sordella R, Gurubhafavatula S, Okimoto RA, Brannigan BW, Harris PL, Haserlat SM, Supko JG, Haluska FG, Louis DN, Christiani DC, Settleman J, Haber DA. Activating mutations in the epidermal growth factor receptor underlying responsiveness of non-small-cell lung cancer to gefitinib. *N Engl J Med* 2004;350:2129–2139.

20. Paez JG, Janne PA, Lee JC, Tracy S, Greulich H, Gabriel S, Herman P, Kaye FJ, Lindeman N, Boggon TJ, Naoki K, Sasaki H, Fujii Y, Eck MJ, Sellers WR, Johnson BE, Meyerson M. EGFR mutations in lung cancer: correlation with clinical response to gefitinib therapy. *Science* 2004;304:1497–1500.

21. Boldt S, Kolch W. Targeting MAPK signaling: Prometheus' fire or Pandora's box? *Curr Pharm Design* 2004;10:1885–1905.

22. Hommes D, van den Blink B, Plasse T, Bartelsman J, Xu C, Macpherson B, Tytgat G, Peppelenbosch M, Van Deventer S. Inhibition of stress-activated MAP kinases induces clinical improvement in moderate to severe Crohn's disease. *Gastroenterology* 2002;122:7–14.

23. Fijen JW, Zijlstra JG, De Boer P, Spanjersberg R, Tervaert JW, Van Der Werf TS, Ligtenberg JJ, Tulleken JE. Suppression of the clinical and cytokine response to endotoxin by RWJ-67657, a p38 mitogen-activated protein-kinase inhibitor, in healthy human volunteers. *Clin Exp Immunol* 2001;124:16–20.

24. Al-Obeidi FA, Wu JJ, Lam KS. Protein tyrosine kinases: structure, substrate specificity, and drug discovery. *Biopolymers* 1998;47:197–223.

25. Noble MEM, Endicott JA, Johnson LN. Protein kinase inhibitors: insights into drug design from structure. *Science* 2004;303:1800–1805.

26. Weisberg E, Griffin JD. Resistance to imatinib (Glivec): update on clinical mechanisms. *Drug Resistance Updates* 2003;6:231–238.

27. Schinder T, Bornmann W, Pellicena P, Miller WT, Clarson B, Kuriyan J. Structural mechanism for STI-571 inhibition of Ableson tyrosine kinase. *Science* 2000;289:1938–1942.

2 An Introduction to the Protein Tyrosine Phosphatase Gene Family and Screening Assay Development

Dominique Perrin and Rob Hooft van Huijsduijnen

CONTENTS

2.1 ABSTRACT

The protein tyrosine phosphatase (PTP) family comprises some 90 genes, all of which have been discovered in the last 15 years. It has become clear that these enzymes play a highly dynamic role in intracellular processes, and several of these enzymes have emerged as clear drug targets. This introduction will list these PTPs and summarize the evidence for their potential role in disease, as well as discussing assays that are available to screen these enzymes.

2.2 INTRODUCTION

Protein tyrosine phosphatases are rapidly gaining interest in both the scientific community and the pharmaceutical industry, with the growing realization that these enzymes play key roles in many intracellular processes often associated with health and disease. PTP1B has recently emerged as a major drug target for the treatment of diabetes and other diseases associated with insulin resistance, such as polycystic ovarian syndrome [1]. Within 4 years after the description of PTP1B knockout mice [2,3], which suggested that PTP1B regulates insulin receptor signaling, two PTP1B drugs have been tested in the clinic. In addition, several other phosphatases are being pursued as potential drug targets [reviewed in 4–6].

2.3 WHAT ARE PHOSPHATASES?

Kinases are intracellular enzymes that transfer phosphate groups from simple precursors (adenosine triphosphate or guanosine triphosphate) to other proteins, lipids, or carbohydrates. Phosphatases catalyze the reverse reaction and produce free phosphate. Although cellular isolates with phosphatase activity have been described since the 1960s, the first phosphatase cDNAs were cloned only in the late 1980s (see [7–14] for reviews).

Tyrosine phosphatases can be subdivided into "classical PTPs" (which are phosphotyrosine specific) and "dual-specific" or DS-PTPs, which in addition dephosphorylate serine- and threonine-phosphate residues. Classical and DS-PTPs are characterized by the PTP signature motif ([H/V]C[X]$_5$R[S/T]). Weaker sequence similarity extends to a ~250–amino acid domain [15,16]. These PTPs form the bulk of the tyrosine phosphatases, but the signature motif alone is also found in other, smaller subfamilies whose overall sequence similarity with these two main classes is not obvious. These include cdc25A-C, LMW-PTPs (low molecular weight PTPs), and the MTM (myotubularin-) related and tensin (TPTE-like) phosphatases. Very recently, a previously unsuspected, new family of tyrosine phosphatases was identified that lacks the "PTP signature motif" and shows no primary sequence similarity with known PTPs [17]. Serine/threonine protein phsophatases (the PP family) acquire diversity mostly through combination with various regulatory subunits [18]. Finally, a single protein histidine phosphatase has been identified [19].

2.3.1 A Character Sketch of the Main Tyrosine Phosphatases

PTPs have one or two catalytic domains and are expressed in different cellular locations. PTPs can be located in the nucleus, and in many cases, different splice forms that derive from the same PTP gene result in different compartmentalization [20]. Many PTPs are type I transmembrane proteins ("receptor-PTPs"). An interesting characteristic of many (but not all) membrane PTPs is that they have two regions with homology to catalytic domains. However, in most cases only the plasma membrane-proximal (D1) domain has catalytic activity. It has been hypothesized that the inactive D2 domain can be involved in repression of PTP activity [21–23]. In a mechanism that is the exact opposite from how receptor kinases are activated, receptor PTPs would be *inactive* in the dimerized state and are active as monomers. Although evidence for this mechanism has been obtained for a number of PTPs, including CD45, PTP-α, PTP-ζ, and PTP-μ, it is by no means clear whether this activation mode describes all receptor PTPs. These experiments also lead to the question of what natural ligands may exist for these receptor PTPs. So far, only a single bona fide ligand, midkine, has been identified for a receptor PTP (PTP-ζ) [24]. Many other receptor PTPs have adhesion-like extracellular domains and are likely to be involved in homo- or heterotypic intercellular binding [25–29]. Alternatively, covalent modification of the receptor may result in signaling: for CD45, dimers that form among the products of different splice forms have been shown to result in proteins whose phosphatase activity correlates with *O*-glycosylation of the extracellular domain [30].

Understanding the physiological role of a PTP requires knowledge of its physiological substrates. A useful discovery was that mutation of a specific PTP amino acid in the catalytic domain results in so-called "substrate-trapping mutants." Such PTP mutants recognize their substrates with the same specificity as their wild-type equivalents do, yet they cannot complete enzymatic conversion and remain tightly associated with their substrates [31]. The technology has been used often to "pull out" substrates from cells that overexpress such a PTP mutant [e.g., 31–38]. We have used this technology in a systematic manner to find PTPs with specificity for the autophosphorylated insulin receptor [39]. However, it has not been possible to find sharply defined phosphopeptide preferences for tyrosine phosphatases [40]. Overexpression of any active PTP, or *in vitro* incubation with almost any protein substrate, will usually result in effective dephosphorylation. By contrast, cellular activity results from a subtle convergence of events that unite enzyme and substrate at the right time and in the right place, a context and concentration that are practically impossible to reproduce artificially. An interesting illustration of these difficulties is the long gestation period of PTP1B as a drug target for diabetes. Since the 19th century, vanadate has been known to have a therapeutic benefit in diabetes [41]. In 1988, when the first PTP, PTP1B, was purified to homogeneity [42], it was understood that vanadate was an inhibitor of an unknown PTP, resulting in enhanced insulin receptor phosphorylation and a therapeutic benefit for diabetes [43,44]. Remarkably, PTP1B was one of the first PTPs that was sequenced, an accomplishment that occurred in 1989 [45], yet it would take 10 more years before its critical involvement in diabetes was recognized, and this only because of PTP1B knockout animals' striking phenotype. In the meantime, it had been shown for over half a dozen different PTPs that their overexpression inhibits insulin receptor–initiated signaling [reviewed in 4]. This illustrates that overexpression of wild-type or dominant negative PTPs compromises both natural PTP substrate specificity and regulation. Much more so than for other signaling targets, the predictive power of such experiments for PTP target identification is very limited. The recent discovery that siRNA can potently and selectively inhibit mammalian gene expression opens the door to high-throughput analysis of gene function in cellular assays [46–50]. Even though such approaches cannot distinguish between enzymatic and structural functions of PTPs (e.g., their role as a scaffold), they require less effort to perform than constructing knockout animals does and produce much more information than overexpression experiments with either wild-type or dominant negative mutants do.

From a functional point of view, it is important to realize that phosphorylation and dephosphorylation are highly dynamic processes that involve multiple feedback loops. For instance, insulin stimulation generates a burst of intracellular H_2O_2 that is associated with reversible oxidative inhibition of up to 62% of overall cellular PTPase activity [51].

Interestingly, a number of PTPs have single catalytic domains that carry "mutations" at critical amino acid positions, as a result of which they are devoid of catalytic activity. Such PTPs have been named "STYX" proteins [52]. In some cases, a single point mutation can reconvert a STYX back into an active PTP [53], but in most cases it is clear that these proteins are completely devoid of enzymatic activity. It is possible that they play a role in regulating the activity of other PTPs.

A number of PTP mutant mice have now been constructed. Their phenotypes, as gathered from various sources, are summarized in Table 2.1. Interpretation of knockout phenotypes from a drug discovery perspective is not straightforward. Animals born without a gene may compensate developmentally and end up with a different appearance than they would have if the gene had been inhibited at the adult stage, resulting in either a milder or a more severe phenotype. The details on how the gene was knocked out are also relevant. Was it only mutated in its catalytic function, or were only certain splice variants affected? In many cases, knockout phenotypes depend on the genetic background of the animals (*intra*species variation) and do not correspond to the equivalent situation where the gene is mutated in humans (*inter*species variation). Finally, if "no phenotype" is observed, gene function may still be highly relevant in disease situations, but discovering in which disease the gene activity needs to be modified for a beneficial effect is far from trivial. These questions are obviously crucial from a drug discovery perspective. For instance, targeting a gene

TABLE 2.1
Mouse and Human PTP Mutation Phenotypes

Gene Name	Viability	Brief Description of Phenotype	References
CD45	+	Lack T cells, immature B cells	106,107
Cdc25b	+[a]	Meiotic arrest	108
Cdc25c	+	No obvious phenotype	109
DEP/CD148/PTPβ 2	–	Die at midgestation with severe defects in vascular organization	87
		Frequently deleted in human cancers	110
EPMA2/Lafora	–	Progressive myoclonus epilepsy	111,112
GLEPP1	+	Reduced renal filtration surface area	70
JSP/JKAP		No JNK activation by TNF-α or TGF-β (EC only, not animals)	113
LAR	+	Mammary gland defect [98]; reduced glycaemia and insulinaemia [97]	97,98
LC-PTP (He-PTP)	+	No obvious phenotype	114
MKP-1	+	No obvious phenotype	115
MTM1	+[b]	Human mutations: X-linked severe hypotonia and generalized muscle weakness	116
MTMR2	+	Human mutations: Charcot-Marie-Tooth type 4B disease (CMT4B), progressive demyelination	117
PEZ/PTP36	+	Androgenization of female mice, no mammary gland tissue	patent appl. WO 02/45500 A2
PTEN	–	Early embryonic lethality; frequently mutated in human cancers	118,119
PTP1B	+	Enhanced leptin and insulin sensitivity	2,3
PTP-PEST	–	Die at embryonal stage d 8.5, defective neural fold closure, lack primary hepatocytes	120; patent appl. WO 02/079421 A2
PTP-α	+	No obvious phenotype, learning deficit, decreased anxiety	121,122
PTP-β	–	Die E9.5 to10.5, reduced vascular development, heterozygotes normal	Patent appl. Deltagen 10/005,220
PTP-δ	–	Growth retardation, early mortality, posture and motor defects	123
PTP-ε	+	Hypomyelination, reduced src activity	124,125
PTP-IA-2	+	Abnormal glucose-stimulated insulin secretion	126
PTP-κ	+	No obvious phenotype; tumor suppressor for non-Hodgkin lymphoma (?)	127,128
PTP-μ	+	No obvious phenotype	129
PTP-σ	–	Pituitary dysplasia, defects in olfactory lobes, reduction in CNS size and cell number	130
PTP-ζ	+	"Suggestion of a fragility of myelin," resistant to *Helicobacter pylori*	131,132
SHP-1	–	Motheaten mice (me) lethal, hemopoietic dysregulation, splenomegaly, runting, autoimmune disease	133–135
SHP-2	–	Lethal at E8.5 to 10, defect in mesoderm patterning, often mutated in Noonan syndrome	136–138
TC-PTP	–	Born normal, die 3 to 5 wk p.p.; runting, splenomegaly, lymphadenopathy	139; patent appl. U.S.2003/0101470A1

Note: Phenotypes of PTP knock out mice and genetic mutations. +, healthy, overtly normal, breed normally. EAE: Experimental allergic encephalitis (an animal model for multiple sclerosis). Updated from 5.

[a] Females sterile.
[b] Human mutation.

TABLE 2.2
Phosphatases Currently Targeted for Drug Discovery

Phosphatase	Disease Indication	Company	Highest Development Status	Compound Name
PTP1B	Diabetes	ISIS	Phase II	ISIS-113715
PTP1B	Diabetes	Wyeth	Phase II (discontinued)	Ertiprotafib
PTP1B	Diabetes	Various[a]	Discovery	Not named
TC-PTP	Cancer	Kinetek	Discovery	KP-304 series
PTP-PEST	Cancer	Kinetek	Discovery	KP-305 series
HC-PTP	Wound healing	Ontogen Corp.	Discovery	OC420-873
SHP-2	Cancer	Ontogen Corp.	Discovery	Not named
CD45	Inflammation	Taisho/Astra Zenica	Discovery	TU-572
Cdc25a-c	Cancer	Pfizer	Discovery	PNU-108937
Cdc25	Cancer	Maxia	Discovery	MX-7214
Cdc25b	Cancer	Pfizer	Discovery	PNU-108937
Cdc25	Cancer	Taiho	Discovery	TYP-835, QX-101
Cdc25	Cancer	U. of Pittburgh	Discovery	NSC-321206
Cdc25	Cancer	GPC BioTech	Discovery	Sulfircin analogs
Prl-3	Cancer	Genzyme	Discovery	Not named
Undisclosed PTP	Inflammation/cancer	Cerylid	Discovery	CBL-113 and CBL-114
Undisclosed PTP	Inflammation	Ceptyr	Discovery	Not named

[a] A number of companies, including Serono, Novo-Nordisk, Roche, Abbott, Merck-Frosst, Array Biopharma, Molecumetics, and Cengent, are developing PTP1B inhibitors.

Source: The Investigational Drugs Database (IDdb3) http://www.iddb3.com/.

that is "only" required during development and that produces an embryonically lethal phenotype may still be an acceptable rationale for a drug that is to be prescribed for adults. Many marketed drugs, such as tetracycline, are simply not prescribed to children or pregnant women. The recent discovery that small, double-stranded RNAs ("RNAi" or "siRNA") can disrupt gene expression in mammalian cells does not resolve all these issues, but it offers at least the potential to do high throughput, opportunistic target validation, in particular for signaling components such as phosphatases.

Table 2.2 lists a number of tyrosine phosphatases that have been or are still being pursued as drug targets and their indications.

Finally, we will now highlight a few of the PTPs in more detail and discuss their characteristics.

2.4 DETAILED CHARACTERISTICS OF SEVERAL PTPs

2.4.1 PTP1B

PTP1B was one of the first PTPs whose cDNA was cloned. Only after analysis of two independent PTP1B knockout lines was it apparent that this PTP is an important antagonist of insulin and leptin receptor signaling. PTP1B knockout mice displayed reduced plasma glucose and insulin levels, obesity resistance, and increased insulin receptor and IRS-1 tyrosine phosphorylation [2,3]. Later it became apparent that PTP1B is also involved in leptin signaling [54–56].These findings have prompted a number of companies to start PTP1B drug screening projects. The development of an early drug candidate (Ertiprotafib from Wyeth) was discontinued following a diabetes Phase II trial because of efficacy and side effect issues. However, the prospects for PTP1B-targeting treatments are still bright, especially following impressive results from once-weekly injections of antisense reagents by Isis Pharmaceuticals [57]. An antisense drug candidate entered clinical trials in mid-2003.

As is typical for well-studied PTPs, a host of other potential PTP1B substrates have been identified, such as the EGFR, p130^{cas}, p210 bcr-abl, STAT5, src, β-catenin, Jak2, and Tyk2 [58], but such a wide role for this enzyme is somewhat unlikely in the light of the modest knockout phenotypes.

2.4.2 SHP1 AND SHP2

Both of these PTPs are characterized by domains containing *N*-terminal SH-2 (src-homology 2). Ancestry for the genes can be traced to *Caenorhabditis elegans* and *Drosophila* [59]. They have been studied intensively [reviewed in 60–65]. SHP-1 is also called PTP-1c, SHPTP1, HCP, or PTPN6. This proliferation of names is unfortunately typical for members of the phosphatase gene family and wholly unnecessary since sets of orthologs from mammalian species are very sharply demarcated. Consequently, reading the PTP literature requires the reader to have an updated list of synonyms at hand. SHP1 is mostly expressed in cells from the hematopoietic lineage. Its general role seems to be the inactivation of Jak kinases, which are associated with cytokine receptors — Jak1, which is a second messenger for the IFN-a receptor, and Jak2, which is associated with the EPO receptor. Furthermore, this PTP has been linked with p56lck, SLP76, p120 catenin, myosin (after mIgM crosslinking), the IL3R, FcγRIIB (BCR) via ITIM (downregulating), and CD22.

The related SHP2 is also known as SHPTP2, syp, PTP1D, PTP2C or SH-PTP3, PTP-SHβ, PTP-L1, and PTPN11, and is ubiquitously expressed. It is believed to regulate PDGFR, GHR, EGFR, IRS-1 and -2, STAT5, and gp130.

2.4.3 GLEPP1 (GLOMERULAR EPITHELIAL PROTEIN 1)

This is another example of a PTP that received a different name whenever another (obvious) ortholog was cloned from mouse (GLEPP-1) [66], human (PTP-Φ) [67], and rabbit (PTP-oc) [68]. In addition, partial cDNAs are known as PTP-U2, PTP-U2L, and PTP-U2S [66]. This is a receptor PTP that has only a single catalytic domain and eight extracellular FNIII domains. It is predominantly expressed in kidney, brain, and macrophages. Interestingly, there are three lines of thought about how this PTP may function. One group has examined the role of GLEPP1 in renal function and discovered that knockout mice show a somewhat reduced filtration rate [70]. Another group has demonstrated that GLEPP-1 dephosphorylates paxillin, which plays an important role in CSF-1–induced macrophage responses [67]. A role for GLEPP-1 in the immune response is also suggested by the observation that its mRNA is up-regulated by 12-O-tetradecanoyl phorbal-13-acetate (TPA) [69]. Finally, another group has looked at osteoclast bone resorption and used antisense to show that GLEPP-1 is required for this [71].

2.4.4 CD45

Also named leukocyte antigen (LCA), PTPRC, T200, B220, or Ly-5, CD45 is required for lymphocyte signaling in both B and T cells [see 72–79 for reviews]. Its knockout phenotype is well correlated with a lack of T- and B-cell antigen receptor signaling, and human mutations in the gene have been associated with immunodeficiency or multiple sclerosis (an autoimmune disease).

2.4.5 IA2-β/PTPIA2

Synonyms are ICAAR, IA-2, PTP-π, PTPRP, PTP NE6, phogrin (in the rat), PTP-IAR, PTP-NP, and R-PTP-X. This and PTPIA-2 (also known as PTPRN or R-PTP-N) are "STYX" PTPs that have no detectable enzyme activity. IA2-β is expressed in brain and pancreas (insuloma) and is a major autoantigen in insulin-dependent diabetes.

2.4.6 SAP-1

SAP-1 (PTPRH) stands for stomach cancer–associated PTP. It is a receptor PTP with a single catalytic domain that is mainly expressed in brain and liver and to lesser extent in heart and stomach as a 4.2-kilobase mRNA [80], but it has not been detected in pancreas or colon. However, it is (over)expressed in gastrointestinal cancers [81]. It has been associated with the substrates p130^{cas}, FAK, and p62 (dok).

2.4.7 FAP-1

FAP-1 (fas-associated phosphatase-1) is also known as hPTP1E, PTPL1, PTP-BL (rat), PTP-BAS, or RIP (mouse). It reportedly antagonizes Fas signaling [82–84].

2.4.8 DEP-1

DEP-1 (density-enhanced phosphatase) is also known as hPTP-η, PTPRJ, PTP-U1, PTP β2, Byp, or CD148. It has a single catalytic domain and is involved in signal transduction in lymphocytes and in the differentiation of erythrocytes. DEP-1 has been found to be widely expressed on B and T cells, granulocytes, macrophages, and certain dendritic cells as well as mature thymocytes [85]. It can be induced by LPS and CSF-1 on macrophages [86] and is required for vascular development [87]. Overexpression inhibits the growth of breast cancer cells *in vitro*. In one of the few examples where a receptor PTP has been shown to function as a bona fide receptor it was shown that on eosinophils, cross-linking results in degranulation [88].

2.4.9 LAR

LAR (leukocyte antigen–related PTP), PTPRF RPTP is the prototype of a family of typical receptor PTPs that is characterized by homo- or heterotypic interactions. This PTP has long been suspected to play a role in insulin resistance. Thus, its expression has been associated with insulin resistance [89], and a voluminous literature exists on the ability of LAR to regulate insulin receptor signaling in cells [89–96]. However, LAR KO mice do not confirm a convincing role for LAR in plasma glucose homeostasis.

2.4.10   PTP-δ

(R)PTP-δ (PTPRD, PTPD, or HPTP) is expressed predominantly in specific brain areas [99] and is a typical RPTP with two catalytic domains and a transmembrane region. It is part of the LAR family member and heterodimerizes with PTP-σ.

2.5 PRACTICAL CONSIDERATIONS FOR PTP ASSAY DEVELOPMENT

2.5.1 cDNA Cloning and Enzyme Production

Many PTPs have, in addition to the catalytic domain, extra domains that reversibly inhibit catalytic activity (such as SHP-1 and SHP-2) or that target the PTP to its substrate. Therefore, there is justification for developing a PTP enzymatic screen that involves only the catalytic domain. We have extensive experience with cloning and expressing the catalytic domains of the majority of PTP catalytic domains as GST-fusion proteins in *Escherichia coli* [39]. In the majority of cases, the resulting fusion protein was soluble and highly active; a few milligrams of protein obtained from a medium-scale fermentation provides ample material for a large screen. Cloning of cDNA can be conveniently done from EST clones obtained for a nominal fee from commercial suppliers

(ATTC or Invitrogen). Since receptor phosphatases are type I receptors (with a single intracellular domain), the catalytic domains are encoded at the 3′ end of the corresponding mRNAs, for which, in general, much EST sequence data is available. Note that many receptor PTPs have two domains with sequence similarity with the PTP catalytic domain. In such cases it is the membrane-proximal domain, the domain located at the 5′ side in the mRNA, that needs to be subcloned for expression.

2.5.2 ASSAY DEVELOPMENT/SUBSTRATE OPTIONS

When considering screening PTPs, two basic options are available: using a generic small molecule substrate or a more specific peptidic substrate. Interestingly, out of 35 recently published papers reporting inhibitors for the enzymatic activity of PTP-1B, 23 (or 67%) reported using *para*-nitrophenyl phosphate (*p*NPP), and six (or 17%) used fluorescein diphosphate (FDP) as generic substrates, while six (or 17%) mentioned the use of malachite green together with a specific phosphopeptide.

The most widely used generic substrate is *p*NPP. The assay is based on the hydrolysis of colorless *p*NPP to *p*-nitrophenol, which is yellow. The activity of the enzyme is therefore measured by light absorbance. The *p*NPP method is cheap and robust. However, the Km value of the PTPs for *p*NPP is very high, namely in the millimolar range. Therefore one weakness of this method is its low sensitivity to competititive inhibitors when using high concentrations of substrate.

Alternatively, quenched fluorescent generic substrates such as FDP and 6,8-difluoro-4-methy-lumbelliferyl phosphate (DiFMUP) have been designed; the substrate is transformed into a fluorogenic product upon dephosphorylation [97]. Therefore, enzyme activity is associated with an increase in fluorescence intensity. These substrates are characterized by Km values in the μM range and a high dynamic signal between substrate and product, allowing the setup of assays using low μM concentrations of substrate and therefore increasing the sensitivity of the assay toward inhibitors. However, the limitation of these generic substrates is the fact that they are small molecules as compared to physiological protein substrates. Consequently, some concerns have been raised that results obtained with inhibitors against these generic substrates might not be applicable to native substrates.

The second option is to use a specific peptidic substrate and measure its dephosphorylation. One popular method to do so is to measure the amount of free phosphate liberated upon dephosphorylation with malachite green [98]. This method, though inexpensive, proved to be relatively insensitive and prone to interference by free phosphate contaminants and some compounds. Therefore, alternative techniques have recently been developed that make use of fluorescent substrates and are based on various fluorescence readouts, notably fluorescence polarization (FP) and fluorescence intensity. The common feature of these methods is the fact that the phosphopeptide substrate is labeled with a fluorochrome. In the FP methods, the phosphopeptide is stabilized by interactions with a phosphate-binding high-molecular-weight group, either an antiphosphotyrosine antibody or a trivalent metal coordination complex coupled to a nanoparticle (IMAP from Molecular Probes, Fremont, CA).

Therefore, the polarization signal of the stabilized phosphopeptide is high in the absence of phosphatase activity and decreases upon dephosphorylation of the substrate. However, one of the limitations of using antiphosphopeptides antibodies is cost, making generic binders such as the IMAP beads competitive. Alternatively, other technologies are grounded on the measurement of fluorescence intensity. One method is based on the quenching of the fluorescent peptide through the binding of the phosphate group by an iron ion coupled to a proprietary quencher (IQ technology from Pierce). Therefore, removal of phosphate from the substrate by a phosphatase is associated with an increase in fluorescence intensity. Another assay is based on a quenched fluorogenic phosphotyrosine-containing peptide. Upon dephosphorylation, the peptide is susceptible to cleavage by chymotrypsin [99], releasing a fluorogenic peptide fragment. The main limitation of this method is the need to incorporate a chymotrypsin cleavage site in the peptide, which might alter its efficacy as a phosphatase substrate.

Finally, a very promising technology, also relying on fluorescence intensity, is the Caliper's mobility shift assay (Caliper Life Sciences, Hopkinton, MA), a nanofluidics-based technology that involves electrophoretic separation of the fluorescently labeled substrate and the phosphorylated product of the reaction on a microchip. Fluorescence intensity of the peaks is the parameter that is measured. Reactions can be run on-chip, with the reactions taking place in 1 minute in a reaction chamber, or off-chip, with the reaction taking place in a microplate well and the microchip used solely as a separation device between substrate and product. Interestingly, though, Caliper's technology has been used with success using DiFMUP substrate, too [100].

A main limitation of the peptide-based substrate approach is the need to determine an efficient substrate. Several approaches are possible, including synthesizing specific peptide arrays based on known or putative sequences or the use of a set of generic peptides corresponding to known phosphorylation sites. Peptide arrays can be synthesized in-house on cellulose membrane [101,102] or can be bought from suppliers (e.g., Jerini, Berlin, FRG) either as off-the-shelf libraries or as custom-designed ones. Efficacy of the dephosphorylation process can be assessed by autoradiography using an antiphosphotyrosine antibody coupled to peroxidase and a luminescent substrate. However, one specific constraint is affecting FP- and Caliper-based methods, namely the size (up to 20 aa) and charge (between −3 and +3) of the peptide.

Another more general concern about fluorescence-based methods is their well-known sensitivity to both quenching and fluorogenic compounds. The mobility shift assay technology bypasses elegantly this problem by elecrophoretically separating substrate and product, with the fluorescent compounds appearing as single peaks.

2.5.3 SCREENING

Colorimetry- or fluorescence-based phosphatase assays using a generic substrate are particularly easy to set up and to automate. Establishing the assay only requires determining the Km value of the enzyme for the substrate and the kinetics of the reaction as a function of both time and enzyme concentration. Similarly, automation is simple as a single substrate dispenser is involved, and protocols can be adapted from one phosphatase to another; the only difference is the identity of the enzyme and the concentration of substrate used, as we are screening at a concentration of substrate corresponding to the Km value. In our hands, the assays have been easily run on a Tecan (Tecan, Hombrechtikon, Switzerland), as well as on a Biomek FX (Beckman Coulter) workstation in a 384-well format. Robustness of the assays proved excellent, with Z′ factor values over 0.70 (Figure 2.1). In terms of readers, Victor V as well as Fusion (Perkin-Elmer Life Sciences, Boston, MA) proved suitable. One added advantage of this screening format is the very low cost of fluorescent substrates such as DiFMUP and FDP, in the range of 0.2 cents/well. On the other hand, malachite green–based assays using a specific phosphopeptide substrate were evaluated and quickly dropped due to the high sensitivity of the assay to contaminants, notably free phosphate. Moreover, Caliper's mobility shift assay technology has been validated and put in place for screening kinase inhibitors as well (Perrin et al., manuscript in preparation) and evaluation for phosphatases is currently under way. If the outcome is positive, this technology will become our method of choice for screening phosphatases, as is presently already the case for screening kinases.

2.6 SUMMARY AND CONCLUSIONS

In the short time that they have been investigated, protein tyrosine phosphatases have emerged as a group of diverse and active participants in cell signaling. The group includes one very clearcut drug target (PTP1B), and it is likely that several other PTP members will follow as drug candidates. Future PTP target validation is likely to benefit from RNAi and careful mouse knockout approaches,

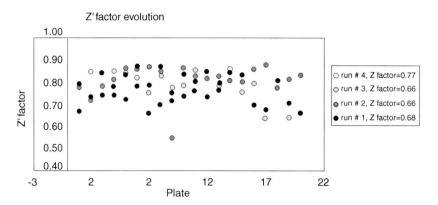

FIGURE 2.1 Evolution of the Z′ factor during screening runs. To evaluate the quality of the DiFMUP-based assay for use in a high-throughput phosphatase screening (HTS) campaign, performance was assessed by calculating the Z′ value using the following equation: $Z' = 1 - [(3SD \text{ of positive control}) + (3SD \text{ of negative control})]/ABS (\text{mean of positive control} - \text{mean of negative control})$. The data for four individual runs are shown.

as opposed to earlier "sledgehammer" methods that involved the overexpression of mutated or wild-type enzymes to deduce their physiological role. The current focus on PTP1B for inhibitor discovery has led to the identification of many interesting chemical series that may form the basis for structure-based drug design or testing focused libraries of chemicals on other members of the family. From a practical point of view, most phosphatases are relatively easy to produce and screen, which will facilitate the search for better inhibitors.

REFERENCES

1. Park, K. H., Kim, J. Y., Ahn, C. W., Song, Y. D., Lim, S. K., and Lee, H. C. (2001) Polycystic ovarian syndrome (PCOS) and insulin resistance. *Int J Gynaecol Obstet* 74, 261–267.
2. Elchebly, M., Payette, P., Michaliszyn, E., Cromlish, W., Collins, S., Loy, A. L., Normandin, D., Cheng, A., Himms-Hagen, J., Chan, C. C., Ramachandran, C., Gresser, M. J., Tremblay, M. L., and Kennedy, B. P. (1999) Increased insulin sensitivity and obesity resistance in mice lacking the protein tyrosine phosphatase-1B gene. *Science* 283, 1544–1548.
3. Klaman, L. D., Boss, O., Peroni, O. D., Kim, J. K., Martino, J. L., Zabolotny, J. M., Moghal, N., Lubkin, M., Kim, Y. B., Sharpe, A. H., Stricker-Krongrad, A., Shulman, G. I., Neel, B. G., and Kahn, B. B. (2000) Increased energy expenditure, decreased adiposity, and tissue-specific insulin sensitivity in protein-tyrosine phosphatase 1B-deficient mice. *Mol Cell Biol* 20, 5479–5489.
4. Espanel, X., Wälchli, S., Pescini Gobert, R., El Alama, M., Curchod, M.-L., Gullu-Isler, N., and Hooft van Huijsduijnen, R. (2001) Pulling strings below the surface: hormone receptor signaling through inhibition of protein tyrosine phosphatases. *Endocrine* 15, 19–28.
5. Hooft van Huijsduijnen, R., Wälchli, S., Ibberson, M., and Harrenga, A. (2002) Protein tyrosine phosphatases as drug targets: PTP1B and beyond. *Expert Opin Ther Targets* 6, 637–647.
6. Hooft van Huijsduijnen, R., Bombrun, A., and Swinnen, D. (2002) Selecting protein tyrosine phosphatases as drug targets. *Drug Discovery Today* 7, 1013–1019.
7. Wang, W. Q., Sun, J. P., and Zhang, Z. Y. (2003) An overview of the protein tyrosine phosphatase superfamily. *Curr Top Med Chem* 3, 739–748.
8. Mustelin, T. and Tasken, K. (2003) Positive and negative regulation of T-cell activation through kinases and phosphatases. *Biochem J* 371, 15–27.
9. Johnson, K. G. and Van Vactor, D. (2003) Receptor protein tyrosine phosphatases in nervous system development. *Phys Rev* 83, 1–24.
10. Asante-Appiah, E. and Kennedy, B. P. (2003) Protein tyrosine phosphatases: the quest for negative regulators of insulin action. *Am J Physiol Endocrinol Metab* 284, E663–E670.

11. Zhang, Z. Y. (2002) Protein tyrosine phosphatases: structure and function, substrate specificity, and inhibitor development. *Annu Rev Pharmacol Toxicol* 42, 209–234.
12. Zhang, Z. Y., Zhou, B., and Xie, L. (2002) Modulation of protein kinase signaling by protein phosphatases and inhibitors. *Pharmacol Ther* 93, 307–317.
13. Mustelin, T., Feng, G. S., Bottini, N., Alonso, A., Kholod, N., Birle, D., Merlo, J., and Huynh, H. (2002) Protein tyrosine phosphatases. *Frontiers Biosci* 7, d85–d142.
14. Mustelin, T., Abraham, R. T., Rudd, C. E., Alonso, A., and Merlo, J. J. (2002) Protein tyrosine phosphorylation in T cell signaling. *Frontiers Biosci* 7, d918–d969.
15. Hooft van Huijsduijnen, R. (1998) Protein tyrosine phosphatases: counting the trees in the forest. *Gene* 225, 1–8.
16. Andersen, J. N., Mortensen, O. H., Peters, G. H., Drake, P. G., Iversen, L. F., Olsen, O. H., Jansen, P. G., Andersen, H. S., Tonks, N. K., and Moller, N. P. (2001) Structural and evolutionary relationships among protein tyrosine phosphatase domains. *Mol Cell Biol* 21, 7117–7136.
17. Rayapureddi, J. P., Kattamuri, C., Steinmetz, B. D., Frankfort, B. J., Ostrin, E. J., Mardon, G., and Hegde, R. S. (2003) Eyes absent represents a class of protein tyrosine phosphatases. *Nature* 426, 295–298.
18. Barford, D. (1996) Molecular mechanisms of the protein serine/threonine phosphatases. *Trends Biochem Sci* 21, 407–412.
19. Klumpp, S., Hermesmeier, J., Selke, D., Baumeister, R., Kellner, R., and Krieglstein, J. (2002) Protein histidine phosphatase: a novel enzyme with potency for neuronal signaling. *J Cereb Blood Flow Metab* 22, 1420–1424.
20. Kamatkar, S., Radha, V., Nambirajan, S., Reddy, R. S., and Swarup, G. (1996) Two splice variants of a tyrosine phosphatase differ in substrate specificity, DNA binding, and subcellular location. *J Biol Chem* 271, 26,755–26,761.
21. Takeda, A., Wu, J. J., and Maizel, A. L. (1992) Evidence for monomeric and dimeric forms of CD45 associated with a 30-kDa phosphorylated protein. *J Biol Chem* 267, 16,651–16,659.
22. Bilwes, A. M., den Hertog, J., Hunter, T., and Noel, J. P. (1996) Structural basis for inhibition of receptor protein-tyrosine phosphatase-alpha by dimerization. *Nature* 382, 555–559.
23. Majeti, R., Bilwes, A. M., Noel, J. P., Hunter, T., and Weiss, A. (1998) Dimerization-induced inhibition of receptor protein tyrosine phosphatase function through an inhibitory wedge. *Science* 279, 88–91.
24. Maeda, N., Ichihara-Tanaka, K., Kimura, T., Kadomatsu, K., Muramatsu, T., and Noda, M. (1999) A receptor-like protein-tyrosine phosphatase PTPzeta/RPTPbeta binds a heparin-binding growth factor midkine. Involvement of arginine 78 of midkine in the high affinity binding to PTPzeta. *J Biol Chem* 274, 12,474–12,479.
25. Spertini, F., Wang, A. V., Chatila, T., and Geha, R. S. (1994) Engagement of the common leukocyte antigen CD45 induces homotypic adhesion of activated human T cells. *J Immunol* 153, 1593–1602.
26. Avraham, S., London, R., Tulloch, G. A., Ellis, M., Fu, Y., Jiang, S., White, R. A., Painter, C., Steinberger, A. A., and Avraham, H. (1997) Characterization and chromosomal localization of PTPRO, a novel receptor protein tyrosine phosphatase, expressed in hematopoietic stem cells. *Gene* 204, 5–16.
27. Feiken, E., van Etten, I., Gebbink, M. F., Moolenaar, W. H., and Zondag, G. C. (2000) Intramolecular interactions between the juxtamembrane domain and phosphatase domains of receptor protein-tyrosine phosphatase RPTPmu. Regulation of catalytic activity. *J Biol Chem* 275, 15,350–15,356.
28. Haj, F., McKinnell, I., and Stoker, A. (1999) Retinotectal ligands for the receptor tyrosine phosphatase CRYPalpha. *Mol Cell Neurosci* 14, 225–240.
29. Juneja, H. S., Schmalstieg, F. C., Lee, S., and Chen, J. (1998) CD45 partially mediates heterotypic adhesion between murine leukemia/lymphoma cell line L5178Y and marrow stromal cells. *Leukemia Res* 22, 805–815.
30. Xu, Z., and Weiss, A. (2002) Negative regulation of CD45 by differential homodimerization of the alternatively spliced isoforms. *Nat Immunol* 3, 764–771.
31. Flint, A. J., Tiganis, T., Barford, D., and Tonks, N. K. (1997) Development of "substrate-trapping" mutants to identify physiological substrates of protein tyrosine phosphatases. *Proc Natl Acad Sci USA* 94, 1680–1685.
32. Agazie, Y. M., and Hayman, M. J. (2003) Development of an efficient "substrate-trapping" mutant of Src homology phosphotyrosine phosphatase 2 and identification of the epidermal growth factor receptor, Gab1, and three other proteins as target substrates. *J Biol Chem* 278, 13,952–13,958.

33. Kawachi, H., Fujikawa, A., Maeda, N., and Noda, M. (2001) Identification of GIT1/Cat-1 as a substrate molecule of protein tyrosine phosphatase zeta/beta by the yeast substrate-trapping system. *Proc Natl Acad Sci USA* 98, 6593–6598.

34. Xie, L., Zhang, Y. L., and Zhang, Z. Y. (2002) Design and characterization of an improved protein tyrosine phosphatase substrate-trapping mutant. *Biochemistry* 41, 4032–4039.

35. Cote, J. F., Charest, A., Wagner, J., and Tremblay, M. L. (1998) Combination of gene targeting and substrate trapping to identify substrates of protein tyrosine phosphatases using PTP-PEST as a model. *Biochemistry* 37, 13,128–13,137.

36. Nakai, Y., Irie, S., and Sato, T. A. (2000) Identification of IkappaBalpha as a substrate of Fas-associated phosphatase-1. *Eur J Biochem* 267, 7170–7175.

37. Tiganis, T., Bennett, A. M., Ravichandran, K. S., and Tonks, N. K. (1998) Epidermal growth factor receptor and the adaptor protein p52Shc are specific substrates of T-cell protein tyrosine phosphatase. *Mol Cell Biol* 18, 1622–1634.

38. Timms, J. F., Carlberg, K., Gu, H., Chen, H., Kamatkar, S., Nadler, M. J., Rohrschneider, L. R., and Neel, B. G. (1998) Identification of major binding proteins and substrates for the SH2-containing protein tyrosine phosphatase SHP-1 in macrophages. *Mol Cell Biol* 18, 3838–3850.

39. Wälchli, S., Curchod, M.-L., Pescini Gobert, R., Arkinstall, S., and Hooft van Huijsduijnen, R. (2000) Identification of tyrosine phosphatases that dephosphorylate the insulin receptor: a brute-force approach based on "substrate-trapping" mutants. *J Biol Chem* 275, 9792–9796.

40. Wälchli, S., Espanel, X., Harrenga, A., Rossi, M., Cesareni, G., and Hooft van Huijsduijnen, R. (2004) Probing protein tyrosine phosphatase substrate specificity using a phosphotyrosine-containing phage library. *J Biol Chem* 279, 311–318.

41. Lyonnet, B., Martz, M. E., and Martin, E. (1899) L'emploi thérapeutique des dérivés du vanadium. *La Presse Med* 1, 191–192.

42. Tonks, N. K., Charbonneau, H., Diltz, C. D., Fischer, E. H., and Walsh, K. A. (1988) Demonstration that the leukocyte common antigen CD45 is a protein tyrosine phosphatase. *Biochemistry* 27, 8695–8701.

43. Fantus, I. G., Kadota, S., Deragon, G., Foster, B., and Posner, B. I. (1989) Pervanadate [peroxide(s) of vanadate] mimics insulin action in rat adipocytes via activation of the insulin receptor tyrosine kinase. *Biochemistry* 28, 8864–8871.

44. Jackson, T. K., Salhanick, A. I., Sparks, J. D., Sparks, C. E., Bolognino, M., and Amatruda, J. M. (1988) Insulin-mimetic effects of vanadate in primary cultures of rat hepatocytes. *Diabetes* 37, 1234–1240.

45. Charbonneau, H., Tonks, N. K., Kumar, S., Diltz, C. D., Harrylock, M., Cool, D. E., Krebs, E. G., Fischer, E. H., and Walsh, K. A. (1989) Human placenta protein-tyrosine-phosphatase: amino acid sequence and relationship to a family of receptor-like proteins. *Proc Natl Acad Sci USA* 86, 5252–5256.

46. Elbashir, S. M., Lendeckel, W., and Tuschl, T. (2001) RNA interference is mediated by 21- and 22-nucleotide RNAs. *Genes Dev* 15, 188–200.

47. Elbashir, S. M., Harborth, J., Lendeckel, W., Yalcin, A., Weber, K., and Tuschl, T. (2001) Duplexes of 21-nucleotide RNAs mediate RNA interference in cultured mammalian cells. *Nature* 411, 494–498.

48. Xia, H., Mao, Q., Paulson, H. L., and Davidson, B. L. (2002) siRNA-mediated gene silencing *in vitro* and *in vivo*. *Nat Biotechnol* 20, 1006–1010.

49. Yu, J. Y., DeRuiter, S. L., and Turner, D. L. (2002) RNA interference by expression of short-interfering RNAs and hairpin RNAs in mammalian cells. *Proc Natl Acad Sci USA* 99, 6047–6052.

50. Paul, C. P., Good, P. D., Winer, I., and Engelke, D. R. (2002) Effective expression of small interfering RNA in human cells. *Nat Biotechnol* 20, 505–508.

51. Mahadev, K., Zilbering, A., Zhu, L., and Goldstein, B. J. (2001) Insulin-stimulated hydrogen peroxide reversibly inhibits protein-tyrosine phosphatase 1B *in vivo* and enhances the early insulin action cascade. *J Biol Chem* 276, 21,938–21,942.

52. Wishart, M. J., and Dixon, J. E. (1998) Gathering STYX: phosphatase-like form predicts functions for unique protein-interaction domains. *Trends Biochem Sci* 23, 301–306.

53. Wishart, M. J., Denu, J. M., Williams, J. A., and Dixon, J. E. (1995) A single mutation converts a novel phosphotyrosine binding domain into a dual-specificity phosphatase. *J Biol Chem* 270, 26,782–26,785.

54. Cook, W. S., and Unger, R. H. (2002) Protein tyrosine phosphatase 1B: a potential leptin resistance factor of obesity. *Dev Cell* 2, 385–387.

55. Zabolotny, J. M., Bence-Hanulec, K. K., Stricker-Krongrad, A., Haj, F., Wang, Y., Minokoshi, Y., Kim, Y. B., Elmquist, J. K., Tartaglia, L. A., Kahn, B. B., and Neel, B. G. (2002) PTP1B regulates leptin signal transduction *in vivo*. *Dev Cell* 2, 489–495.

56. Kaszubska, W., Falls, H. D., Schaefer, V. G., Haasch, D., Frost, L., Hessler, P., Kroeger, P. E., White, D. W., Jirousek, M. R., and Trevillyan, J. M. (2002) Protein tyrosine phosphatase 1B negatively regulates leptin signaling in a hypothalamic cell line. *Mol Cell Endocrinol* 195, 109–118.

57. Zinker, B. A., Rondinone, C. M., Trevillyan, J. M., Gum, R. J., Clampit, J. E., Waring, J. F., Xie, N., Wilcox, D., Jacobson, P., Frost, L., Kroeger, P. E., Reilly, R. M., Koterski, S., Opgenorth, T. J., Ulrich, R. G., Crosby, S., Butler, M., Murray, S. F., McKay, R. A., Bhanot, S., Monia, B. P., and Jirousek, M. R. (2002) PTP1B antisense oligonucleotide lowers PTP1B protein, normalizes blood glucose, and improves insulin sensitivity in diabetic mice. *Proc Natl Acad Sci USA* 99, 11,357–11,362.

58. Myers, M. P., Andersen, J. N., Cheng, A., Tremblay, M. L., Horvath, C. M., Parisien, J. P., Salmeen, A., Barford, D., and Tonks, N. K. (2001) TYK2 and JAK2 are substrates of protein-tyrosine phosphatase 1B. *J Biol Chem* 276, 47,771–47,774.

59. Wälchli, S., Colinge, J., and Hooft van Huijsduijnen, R. (2000) MetaBlasts: tracing protein tyrosine phosphatase gene family roots from man to *Drosophila melanogaster* and *Caenorhabditis elegans* genomes. *Gene* 253, 137–143.

60. Wu, C., Sun, M., Liu, L., and Zhou, G. W. (2003) The function of the protein tyrosine phosphatase SHP-1 in cancer. *Gene* 306, 1–12.

61. Feng, G. S. (1999) Shp-2 tyrosine phosphatase: signaling one cell or many. *Exp Cell Res* 253, 47–54.

62. Siminovitch, K. A. and Neel, B. G. (1998) Regulation of B cell signal transduction by SH2-containing protein-tyrosine phosphatases. *Semin Immunol* 10, 329–347.

63. Shultz, L. D., Rajan, T. V., and Greiner, D. L. (1997) Severe defects in immunity and hematopoiesis caused by SHP-1 protein-tyrosine-phosphatase deficiency. *Trends Biotechnol* 15, 302–307.

64. Siminovitch, K. A., Lamhonwah, A. M., Somani, A. K., Cardiff, R., and Mills, G. B. (1999) Involvement of the SHP-1 tyrosine phosphatase in regulating B lymphocyte antigen receptor signaling, proliferation and transformation. *Curr Top Microbiol Immunol* 246, 291–297; discussion 298.

65. Tamir, I., Dal Porto, J. M., and Cambier, J. C. (2000) Cytoplasmic protein tyrosine phosphatases SHP-1 and SHP-2: regulators of B cell signal transduction. *Curr Opin Immunol* 12, 307–315.

66. Thomas, P. E., Wharram, B. L., Goyal, M., Wiggins, J. E., Holzman, L. B., and Wiggins, R. C. (1994) GLEPP1, a renal glomerular epithelial cell (podocyte) membrane protein-tyrosine phosphatase. Identification, molecular cloning, and characterization in rabbit. *J of Biol Chem* 269, 19953–19962.

67. Pixley, F. J., Lee, P. S., Condeelis, J. S., and Stanley, E. R. (2001) Protein tyrosine phosphatase phi regulates paxillin tyrosine phosphorylation and mediates colony-stimulating factor 1-induced morphological changes in macrophages. *Mol Cell Biol* 21, 1795–1809.

68. Wu, L. W., Baylink, D. J., and Lau, K. H. (1996) Molecular cloning and expression of a unique rabbit osteroclastic phosphotyrosyl phosphatase. *Biochem J* 315 (Pt 2), 515–523.

69. Seimiya, H., Sawabe, T., Inazawa, J., and Tsuruo, T. (1995) Cloning, expression, and chromosomal localization of a novel gene for protein tyrosine phosphatase (PTP-U2) induced by various differentiation-inducing agents. *Oncogene* 10, 1731–1738.

70. Wharram, B. L., Goyal, M., Gillespie, P. J., Wiggins, J. E., Kershaw, D. B., Holzman, L. B., Dysko, R. C., Saunders, T. L., Samuelson, L. C., and Wiggins, R. C. (2000) Altered podocyte structure in GLEPP1 (Ptpro)-deficient mice associated with hypertension and low glomerular filtration rate. *J Clin Invest* 106, 1281–1290.

71. Suhr, S. M., Pamula, S., Baylink, D. J., and Lau, K. H. (2001) Antisense oligodeoxynucleotide evidence that a unique osteoclastic protein-tyrosine phosphatase is essential for osteoclastic resorption. *J Bone Miner Res* 16, 1795–1803.

72. Lee, K., and Burke, T. R., Jr. (2003) CD45 protein-tyrosine phosphatase inhibitor development. *Curr Top Med Chem* 3, 797–807.

73. Irie-Sasaki, J., Sasaki, T., and Penninger, J. M. (2003) CD45 regulated signaling pathways. *Curr Top Med Chem* 3, 783–796.

74. Tchilian, E. Z., and Beverley, P. C. (2002) CD45 in memory and disease. *Arch Immunol Ther Exp* 50, 85–93.

75. Penninger, J. M., Irie-Sasaki, J., Sasaki, T., and Oliveira-dos-Santos, A. J. (2001) CD45: new jobs for an old acquaintance. *Nature Immunol* 2, 389–396.

76. Alexander, D. R. (2000) The CD45 tyrosine phosphatase: a positive and negative regulator of immune cell function. *Semin Immunol* 12, 349–359.

77. Altin, J. G., and Sloan, E. K. (1997) The role of CD45 and CD45-associated molecules in T cell activation. *Immunol Cell Biol* 75, 430–445.

78. Poppema, S., Lai, R., Visser, L., and Yan, X. J. (1996) CD45 (leucocyte common antigen) expression in T and B lymphocyte subsets. *Leukemia Lymphoma* 20, 217–222.

79. Trowbridge, I. S., and Thomas, M. L. (1994) CD45: an emerging role as a protein tyrosine phosphatase required for lymphocyte activation and development. *Annu Rev Immunol* 12, 85–116.

80. Matozaki, T., Suzuki, T., Uchida, T., Inazawa, J., Ariyama, T., Matsuda, K., Horita, K., Noguchi, H., Mizuno, H., Sakamoto, C., and Kasuga, M. (1994) Molecular cloning of a human transmembrane-type protein tyrosine phosphatase and its expression in gastrointestinal cancers. *J Biol Chem* 269, 2075–2081.

81. Seo, Y., Matozaki, T., Tsuda, M., Hayashi, Y., Itoh, H., and Kasuga, M. (1997) Overexpression of SAP-1, a transmembrane-type protein tyrosine phosphatase, in human colorectal cancers. *Biochem Biophys Res Commun* 231, 705–711.

82. Sato, T., Irie, S., Kitada, S., and Reed, J. C. (1995) FAP-1: a protein tyrosine phosphatase that associates with Fas. *Science* 268, 411–415.

83. Yanagisawa, J., Takahashi, M., Kanki, H., Yano-Yanagisawa, H., Tazunoki, T., Sawa, E., Nishitoba, T., Kamishohara, M., Kobayashi, E., Kataoka, S., and Sato, T. (1997) The molecular interaction of Fas and FAP-1. A tripeptide blocker of human Fas interaction with FAP-1 promotes Fas-induced apoptosis. *J Biol Chem* 272, 8539–8545.

84. Arai, M., Kannagi, M., Matsuoka, M., Sato, T., Yamamoto, N., and Fujii, M. (1998) Expression of FAP-1 (Fas-associated phosphatase) and resistance to Fas-mediated apoptosis in T cell lines derived from human T cell leukemia virus type 1-associated myelopathy/tropical spastic paraparesis patients. *AIDS Res Hum Retroviruses* 14, 261–267.

85. Autschbach, F., Palou, E., Mechtersheimer, G., Rohr, C., Pirotto, F., Gassler, N., Otto, H. F., Schraven, B., and Gaya, A. (1999) Expression of the membrane protein tyrosine phosphatase CD148 in human tissues. *Tissue Antigens* 54, 485–498.

86. Osborne, J. M., den Elzen, N., Lichanska, A. M., Costelloe, E. O., Yamada, T., Cassady, A. I., and Hume, D. A. (1998) Murine DEP-1, a receptor protein tyrosine phosphatase, is expressed in macrophages and is regulated by CSF-1 and LPS. *J Leukoc Biol* 64, 692–701.

87. Takahashi, T., Takahashi, K., St. John, P. L., Fleming, P. A., Tomemori, T., Watanabe, T., Abrahamson, D. R., Drake, C. J., Shirasawa, T., and Daniel, T. O. (2003) A mutant receptor tyrosine phosphatase, CD148, causes defects in vascular development. *Mol Cell Biol* 23, 1817–1831.

88. del Pozo, V., Pirotto, F., Cardaba, B., Cortegano, I., Gallardo, S., Rojo, M., Arrieta, I., Aceituno, E., Palomino, P., Gaya, A., and Lahoz, C. (2000) Expression on human eosinophils of CD148: a membrane tyrosine phosphatase. Implications in the effector function of eosinophils. *J Leukoc Biol* 68, 31–37.

89. Ahmad, F., Azevedo, J. L., Cortright, R., Dohm, G. L., and Goldstein, B. J. (1997) Alterations in skeletal muscle protein-tyrosine phosphatase activity and expression in insulin-resistant human obesity and diabetes. *J Clin Invest* 100, 449–458.

90. Mooney, R. A., Kulas, D. T., Bleyle, L. A., and Novak, J. S. (1997) The protein tyrosine phosphatase LAR has a major impact on insulin receptor dephosphorylation. *Biochem Biophys Res Commun* 235, 709–712.

91. Ahmad, F., and Goldstein, B. J. (1997) Functional association between the insulin receptor and the transmembrane protein-tyrosine phosphatase LAR in intact cells. *J Biol Chem* 272, 448–457.

92. Hashimoto, N., Feener, E. P., Zhang, W. R., and Goldstein, B. J. (1992) Insulin receptor protein-tyrosine phosphatases. Leukocyte common antigen-related phosphatase rapidly deactivates the insulin receptor kinase by preferential dephosphorylation of the receptor regulatory domain. *J Biol Chem* 267, 13,811–13,814.

93. Kulas, D. T., Zhang, W. R., Goldstein, B. J., Furlanetto, R. W., and Mooney, R. A. (1995) Insulin receptor signaling is augmented by antisense inhibition of the protein tyrosine phosphatase LAR. *J Biol Chem* 270, 2435–2438.

94. Ahmad, F., Considine, R. V., and Goldstein, B. J. (1995) Increased abundance of the receptor-type protein-tyrosine phosphatase LAR accounts for the elevated insulin receptor dephosphorylating activity in adipose tissue of obese human subjects. *J Clin Invest* 95, 2806–2812.
95. Zhang, W. R., Li, P. M., Oswald, M. A., and Goldstein, B. J. (1996) Modulation of insulin signal transduction by eutopic overexpression of the receptor-type protein-tyrosine phosphatase LAR. *Mol Endocrinol* 10, 575–584.
96. Li, P. M., Zhang, W. R., and Goldstein, B. J. (1996) Suppression of insulin receptor activation by overexpression of the protein-tyrosine phosphatase LAR in hepatoma cells. *Cell Signal* 8, 467–473.
97. Ren, J. M., Li, P. M., Zhang, W. R., Sweet, L. J., Cline, G., Shulman, G. I., Livingston, J. N., and Goldstein, B. J. (1998) Transgenic mice deficient in the LAR protein-tyrosine phosphatase exhibit profound defects in glucose homeostasis. *Diabetes* 47, 493–497.
98. Schaapveld, R. Q., Schepens, J. T., Robinson, G. W., Attema, J., Oerlemans, F. T., Fransen, J. A., Streuli, M., Wieringa, B., Hennighausen, L., and Hendriks, W. J. (1997) Impaired mammary gland development and function in mice lacking LAR receptor-like tyrosine phosphatase activity. *Dev Biol* 188, 134–146.
99. Mizuno, K., Hasegawa, K., Katagiri, T., Ogimoto, M., Ichikawa, T., and Yakura, H. (1993) MPTP delta, a putative murine homolog of HPTP delta, is expressed in specialized regions of the brain and in the B-cell lineage. *Mol Cell Biol* 13, 5513–5523.
100. Gee, K. R., Sun, W. C., Bhalgat, M. K., Upson, R. H., Klaubert, D. H., Latham, K. A., and Haugland, R. P. (1999) Fluorogenic substrates based on fluorinated umbelliferones for continuous assays of phosphatases and beta-galactosidases. *Anal Biochem* 273, 41–48.
101. Lanzetta, P. A., Alvarez, L. J., Reinach, P. S., and Candia, O. A. (1979) An improved assay for nanomole amounts of inorganic phosphate. *Anal Biochem* 100, 95–97.
102. Nishikata, M., Suzuki, K., Yoshimura, Y., Deyama, Y., and Matsumoto, A. (1999) A phosphotyrosine-containing quenched fluorogenic peptide as a novel substrate for protein tyrosine phosphatases. *Biochem J* 343 Pt 2, 385–391.
103. Kerby, M., and Chien, R. L. (2001) A fluorogenic assay using pressure-driven flow on a microchip. *Electrophoresis* 22, 3916–3923.
104. Espanel, X., and Sudol, M. (2001) Yes-associated protein and p53-binding protein-2 interact through their WW and SH3 domains. *J Biol Chem* 276, 14,514–14,523.
105. Espanel, X., Huguenin-Reggiani, M., and Hooft van Huijsduijnen, R. (2002) The SPOT technique as a tool for studying protein tyrosine phosphatase substrate specificities. *Protein Sci* 11, 2326–2334.
106. Byth, K. F., Conroy, L. A., Howlett, S., Smith, A. J., May, J., Alexander, D. R., and Holmes, N. (1996) CD45-null transgenic mice reveal a positive regulatory role for CD45 in early thymocyte development, in the selection of CD4+CD8+ thymocytes, and B cell maturation. *J Exp Med* 183, 1707–1718.
107. Kishihara, K., Penninger, J., Wallace, V. A., Kundig, T. M., Kawai, K., Wakeham, A., Timms, E., Pfeffer, K., Ohashi, P. S., Thomas, M. L. et al. (1993) Normal B lymphocyte development but impaired T cell maturation in CD45-exon6 protein tyrosine phosphatase-deficient mice. *Cell* 74, 143–156.
108. Lincoln, A. J., Wickramasinghe, D., Stein, P., Schultz, R. M., Palko, M. E., De Miguel, M. P., Tessarollo, L., and Donovan, P. J. (2002) Cdc25b phosphatase is required for resumption of meiosis during oocyte maturation. *Nat Genet* 30, 446–449.
109. Chen, M. S., Hurov, J., White, L. S., Woodford-Thomas, T., and Piwnica-Worms, H. (2001) Absence of apparent phenotype in mice lacking Cdc25C protein phosphatase. *Mol Cell Biol* 21, 3853–3861.
110. Ruivenkamp, C. A., Van Wezel, T., Zanon, C., Stassen, A. P., Vlcek, C., Csikos, T., Klous, A. M., Tripodis, N., Perrakis, A., Boerrigter, L., Groot, P. C., Lindeman, J., Mooi, W. J., Meijjer, G. A., Scholten, G., Dauwerse, H., Paces, V., Van Zandwijk, N., Van Ommen, G. J., and Demant, P. (2002) Ptprj is a candidate for the mouse colon-cancer susceptibility locus Scc1 and is frequently deleted in human cancers. *Nat Genet* 31, 295–300.
111. Minassian, B. A., Lee, J. R., Herbrick, J. A., Huizenga, J., Soder, S., Mungall, A. J., Dunham, I., Gardner, R., Fong, C. Y., Carpenter, S., Jardim, L., Satishchandra, P., Andermann, E., Snead, O. C., 3rd, Lopes-Cendes, I., Tsui, L. C., Delgado-Escueta, A. V., Rouleau, G. A., and Scherer, S. W. (1998) Mutations in a gene encoding a novel protein tyrosine phosphatase cause progressive myoclonus epilepsy. *Nat Genet* 20, 171–174.

112. Ganesh, S., Delgado-Escueta, A. V., Sakamoto, T., Avila, M. R., Machado-Salas, J., Hoshii, Y., Akagi, T., Gomi, H., Suzuki, T., Amano, K., Agarwala, K. L., Hasegawa, Y., Bai, D. S., Ishihara, T., Hashikawa, T., Itohara, S., Cornford, E. M., Niki, H., and Yamakawa, K. (2002) Targeted disruption of the Epm2a gene causes formation of Lafora inclusion bodies, neurodegeneration, ataxia, myoclonus epilepsy and impaired behavioral response in mice. *Hum Mol Genet* 11, 1251–1262.

113. Chen, A. J., Zhou, G., Juan, T., Colicos, S. M., Cannon, J. P., Cabriera-Hansen, M., Meyer, C. F., Jurecic, R., Copeland, N. G., Gilbert, D. J., Jenkins, N. A., Fletcher, F., Tan, T. H., and Belmont, J. W. (2002) The dual specificity JKAP specifically activates the c-Jun N-terminal kinase pathway. *J Biol Chem* 277, 36,592–36,601.

114. Gronda, M., Arab, S., Iafrate, B., Suzuki, H., and Zanke, B. W. (2001) Hematopoietic protein tyrosine phosphatase suppresses extracellular stimulus-regulated kinase activation. *Mol Cell Biol* 21, 6851–6858.

115. Dorfman, K., Carrasco, D., Gruda, M., Ryan, C., Lira, S. A., and Bravo, R. (1996) Disruption of the erp/mkp-1 gene does not affect mouse development: normal MAP kinase activity in ERP/MKP-1-deficient fibroblasts. *Oncogene* 13, 925–931.

116. Laporte, J., Hu, L. J., Kretz, C., Mandel, J. L., Kioschis, P., Coy, J. F., Klauck, S. M., Poustka, A., and Dahl, N. (1996) A gene mutated in X-linked myotubular myopathy defines a new putative tyrosine phosphatase family conserved in yeast. *Nat Genet* 13, 175–182.

117. Bolino, A., Muglia, M., Conforti, F. L., LeGuern, E., Salih, M. A., Georgiou, D. M., Christodoulou, K., Hausmanowa-Petrusewicz, I., Mandich, P., Schenone, A., Gambardella, A., Bono, F., Quattrone, A., Devoto, M., and Monaco, A. P. (2000) Charcot-Marie-Tooth type 4B is caused by mutations in the gene encoding myotubularin-related protein-2. *Nat Genet* 25, 17–19.

118. Di Cristofano, A., Pesce, B., Cordon-Cardo, C., and Pandolfi, P. P. (1998) PTEN is essential for embryonic development and tumour suppression. *Nat Genet* 19, 348–355.

119. Dahia, P. L. (2000) PTEN, a unique tumor suppressor gene. *Endocr Relat Cancer* 7, 115–129.

120. Simoncic, P. D., Lee-Loy, A., Barber, D. L., Tremblay, M. L., and McGlade, C. J. (2002) The T cell protein tyrosine phosphatase is a negative regulator of janus family kinases 1 and 3. *Curr Biol* 12, 446–453.

121. Ponniah, S., Wang, D. Z., Lim, K. L., and Pallen, C. J. (1999) Targeted disruption of the tyrosine phosphatase PTPalpha leads to constitutive downregulation of the kinases Src and Fyn. *Curr Biol* 9, 535–538.

122. Skelton, M. R., Ponniah, S., Wang, D. Z., Doetschman, T., Vorhees, C. V., and Pallen, C. J. (2003) Protein tyrosine phosphatase α (PTPα) knockout mice show deficits in Morris water maze learning, decreased locomotor activity, and decreases in anxiety. *Brain Res* 984, 1–10.

123. Uetani, N., Kato, K., Ogura, H., Mizuno, K., Kawano, K., Mikoshiba, K., Yakura, H., Asano, M., and Iwakura, Y. (2000) Impaired learning with enhanced hippocampal long-term potentiation in PTPδ-deficient mice. *EMBO J* 19, 2775–2785.

124. Peretz, A., Gil-Henn, H., Sobko, A., Shinder, V., Attali, B., and Elson, A. (2000) Hypomyelination and increased activity of voltage-gated K(+) channels in mice lacking protein tyrosine phosphatase epsilon. *Embo J* 19, 4036–4045.

125. Dombrádi, V., Krieglstein, J., and Klumpp, S. (2002) Regulating the regulators. Conference on protein phosphorylation and protein phosphatases. *EMBO Rep* 3, 120–124.

126. Saeki, K., Zhu, M., Kubosaki, A., Xie, J., Lan, M. S., and Notkins, A. L. (2002) Targeted disruption of the protein tyrosine phosphatase-like molecule IA-2 results in alterations in glucose tolerance tests and insulin secretion. *Diabetes* 51, 1842–1850.

127. Skarnes, W. C., Moss, J. E., Hurtley, S. M., and Beddington, R. S. (1995) Capturing genes encoding membrane and secreted proteins important for mouse development. *Proc Natl Acad Sci USA* 92, 6592–6596.

128. Nakamura, M., Kishi, M., Sakaki, T., Hashimoto, H., Nakase, H., Shimada, K., Ishida, E., and Konishi, N. (2003) Novel tumor suppressor loci on 6q22-23 in primary central nervous system lymphomas. *Cancer Res* 63, 737–741.

129. Koop, E. A., Lopes, S. M., Feiken, E., Bluyssen, H. A., van der Valk, M., Voest, E. E., Mummery, C. L., Moolenaar, W. H., and Gebbink, M. F. (2003) Receptor protein tyrosine phosphatase mu expression as a marker for endothelial cell heterogeneity; analysis of RPTPmu gene expression using LacZ knock-in mice. *Int J Dev Biol* 47, 345–354.

130. Elchebly, M., Wagner, J., Kennedy, T. E., Lanctot, C., Michaliszyn, E., Itie, A., Drouin, J., and Tremblay, M. L. (1999) Neuroendocrine dysplasia in mice lacking protein tyrosine phosphatase sigma. *Nat Genet* 21, 330–333.

131. Harroch, S., Palmeri, M., Rosenbluth, J., Custer, A., Okigaki, M., Shrager, P., Blum, M., Buxbaum, J. D., and Schlessinger, J. (2000) No obvious abnormality in mice deficient in receptor protein tyrosine phosphatase β. *Mol Cell Biol* 20, 7706–7715.

132. Fujikawa, A., Shirasaka, D., Yamamoto, S., Ota, H., Yahiro, K., Fukada, M., Shintani, T., Wada, A., Aoyama, N., Hirayama, T., Fukamachi, H., and Noda, M. (2003) Mice deficient in protein tyrosine phosphatase receptor type Z are resistant to gastric ulcer induction by VacA of *Helicobacter pylori*. *Nat Genet* 33, 375–381.

133. Tsui, H. W., Siminovitch, K. A., de Souza, L., and Tsui, F. W. (1993) Motheaten and viable motheaten mice have mutations in the haematopoietic cell phosphatase gene. *Nat Genet* 4, 124–129.

134. Kozlowski, M., Mlinaric-Rascan, I., Feng, G. S., Shen, R., Pawson, T., and Siminovitch, K. A. (1993) Expression and catalytic activity of the tyrosine phosphatase PTP1C is severely impaired in motheaten and viable motheaten mice. *J Exp Med* 178, 2157–2163.

135. Umeda, S., Beamer, W. G., Takagi, K., Naito, M., Hayashi, S., Yonemitsu, H., Yi, T., and Shultz, L. D. (1999) Deficiency of SHP-1 protein-tyrosine phosphatase activity results in heightened osteoclast function and decreased bone density. *Am J Pathol* 155, 223–233.

136. Saxton, T. M., Ciruna, B. G., Holmyard, D., Kulkarni, S., Harpal, K., Rossant, J., and Pawson, T. (2000) The SH2 tyrosine phosphatase SHP2 is required for mammalian limb development. *Nat Genet* 24, 420–423.

137. Yu, D. H., Qu, C. K., Henegariu, O., Lu, X., and Feng, G. S. (1998) Protein-tyrosine phosphatase Shp-2 regulates cell spreading, migration, and focal adhesion. *J Biol Chem* 273, 21,125–21,131.

138. Arrandale, J. M., Gore-Willse, A., Rocks, S., Ren, J. M., Zhu, J., Davis, A., Livingston, J. N., and Rabin, D. U. (1996) Insulin signaling in mice expressing reduced levels of Syp. *J Biol Chem* 271, 21,353–21,358.

139. You-Ten, K. E., Muise, E. S., Itie, A., Michaliszyn, E., Wagner, J., Jothy, S., Lapp, W. S., and Tremblay, M. L. (1997) Impaired bone marrow microenvironment and immune function in T cell protein tyrosine phosphatase-deficient mice. *J Exp Med* 186, 683–693.

3 Time-Resolved Fluorescence Based Assays for Kinases

Zhuyin Li and Tina Garyantes

CONTENTS

3.1 INTRODUCTION

Substrate phosphorylation assays are functional assays for kinases that enable the identification of compounds that are competitive inhibitors at the ATP and phosphate-accepting substrate binding sites, as well as allosteric inhibitors. Fluorescence-based detection methods, such as prompt fluorescence, fluorescence polarization (FP), and fluorescence resonance energy transfer (FRET), in conjunction with antibodies specific to the phosphorylated substrates, have become the assays of choice for kinase studies that require high throughput.

The physical properties of fluorophores, such as excitation and emission wavelengths, fluorescence lifetime, intensity, polarization, reactivity, and environmental sensitivities, when creatively selected, can be used to develop a wide variety of assays. The highly sensitive nature of fluorescence labels enables miniaturization without sacrificing precision and throughput. Most fluorescent assay protocols are compatible with existing liquid handling and detection systems and can be implemented without extensive instrument development. In addition, fluorescence-based assay technologies are considered safer and more environmentally friendly than radioactive assays.

However, most conventional fluorescence measurements are susceptible to interference from background fluorescence, such as autofluorescent compounds, instrument artifacts, light scattering, and fluorescence from microtiter plates. Although these sources of interference can be minimized

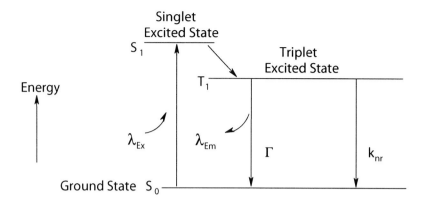

FIGURE 3.1 A simplified Jablonski diagram to illustrate processes that occur between the absorption and emission of light. S_0, ground energy state; S_1 and T_1, excited energy states; λ_{Ex}, absorption wavelength; λ_{Em}, emission wavelength; Γ, radioactive decay rate; and k_{nr}, nonradioactive decay rate.

using ratiometric readout methods and red-shifted fluorophores, high false positive and negative rates coming from interference result in low assay sensitivity, thus confusing results, which delays compound development. Time-resolved fluorescence assays that take advantage of the long fluorescence lifetimes of lanthanide ions can minimize background interference, resulting in high sensitivity and low false positive and negative rates. This assay technology has been widely applied to many biological systems.

3.2 PRINCIPLE OF TIME-RESOLVED FLUORESCENCE AND TIME-RESOLVED FLUORESCENCE RESONANCE ENERGY TRANSFER

3.2.1 FLUORESCENCE IN BIOLOGICAL ASSAYS

When a molecule absorbs energy from a photon (λ_{Ex}), an electron is excited from the ground energy state (S_0) to an excited energy state (S_1). As the electron returns to the ground state, the excess energy is released partly as radioactivity (Γ, radioactive decay rate) with a lower-energy (longer-wavelength, λ_{Em}) photon, and partly as nonradioactive energy (K_{nr}, nonradioactive decay rate) such as heat (Figure 3.1). In biological assays, fluorescent labels are usually linked to biologically interesting molecules and used to monitor structural or environmental changes of the molecule. The fluorescence intensity can be described as:

$$I_{(t)} = I_0 exp(-t/\tau) \tag{3.1}$$

where $I_{(t)}$ is fluorescence intensity at time t after excitation, I_0 is the intensity at time zero, and τ is fluorescence lifetime [1].

3.2.2 TIME-RESOLVED FLUORESCENCE WITH LANTHANIDE CHELATES

The fluorescence lifetime τ is the average amount of time that a fluorophore remains in the excited state following excitation. It is the concept upon which time-resolved fluorescence is based. It can be expressed as the inverse of the total fluorescence decay rate:

$$\tau = \frac{1}{\Gamma + K_{nr}} \tag{3.2}$$

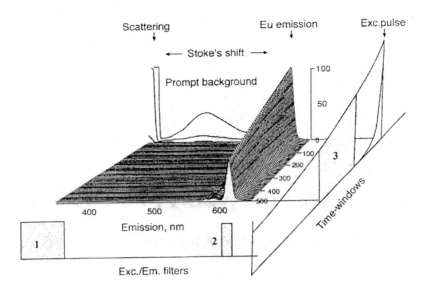

FIGURE 3.2 Emission profile of fluorescent europium chelates showing excitation position (1), emission filter (2), and optimized counting window (3). (Reprinted with permission from Hemmila I. *J Biomol Screen* 1999; 4:303–307.)

Almost all organic fluorophores, such as fluorescein and rhodamine, display lifetimes between 1 and 10 nsec. Europium ion (Eu^{3+}) and other rare earth lanthanide ions are uniquely fluorescent metal ions, which emit photons in aqueous solution with exceptionally long lifetimes of 0.5 to 3 msec. However, their fluorescence emission intensity is very weak, since the direct decays from the excited state to the ground state for lanthanide ions are forbidden transitions [2–4]. Furthermore, because of poor light absorption coefficients, it is unusual for europium ion and other lanthanide ions to be directly excited by light. Usually, they are excited through chelated organic ligands. The ligand absorbs photons and transfers the absorbed energy by intersystem crossing and intramolecular transfer to the chelated lanthanide ion. The lanthanide ion then emits photons with an exceptionally long separation between the excitation and emission wavelengths (Stoke's shift over 250 nm) and a phenomenally long fluorescence lifetime of 0.1 to 1 msec [5,6] (Figure 3.2). For lanthanide chelates to be considered as suitable fluorescence probes for biological assays, the chelates must have thermodynamic and kinetic stability, high fluorescence quantum yields, the ability to be conjugated to a biomolecule of interest, and excitation and emission wavelengths different from those of reagents, such as assay buffers and test compounds, that have the potential to interfere with measurement.

3.2.3 TIME-RESOLVED FLUORESCENCE AND DISSOCIATION-ENHANCED LANTHANIDE FLUORESCENCE IMMUNOASSAY METHOD

Heterogeneous assays require washing steps and are thus more cumbersome than addition-only assay formats, but they are also more sensitive due to the removal of interfering background sources with the wash. Most heterogeneous time-resolved fluorescence-based kinase assays are configured based upon the principle of dissociation-enhanced lanthanide fluorescence immunoassay (DELFIA) [7–10]. This method uses europium chelate–labeled antibodies specific to the phosphorylated substrate.

Europium β-diketone chelates have excellent fluorescence properties [7]. However, binding between the europium ion and the β-diketone ligand is not strong enough to undergo the chelate–antibody conjugation process. This problem was circumvented by labeling the antibody with stable europium chelates, such as europium ion coupled with ethylenediaminetetraacetic acid

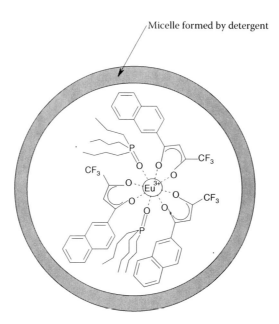

FIGURE 3.3 Europium β-diketone chelate stabilized by micelle.

(EDTA) derivatives, which have low fluorescent emissions. After binding of labeled antibodies to the phosphorylated substrate immobilized on the surface of microtiter plates, unbound antibodies are washed away. Subsequently, bound europium ions are preferentially dissociated from chelates using a low-pH enhancer solution. The low-pH enhancer solution also contains β-diketone or its derivatives to rechelate released europium ions. A Lewis base is used to stabilize the europium β-diketone chelate, and a nonionic detergent (Triton X-100 or Tween 20) that forms micelles is used to solubilize the chelate, as well as to provide a suitable hydrophobic environment favorable to the measurement of fluorescence [8] (Figure 3.3).

After the switch of ligand for europium ion, the wells are excited at 320 nm and emission is measured at 615 nm (Figure 3.4). Typically, samples are excited with nanosecond-long light pulses repeated at a rate of 1000/sec for 1 sec. Because europium chelates have a millisecond fluorescence lifetime, they continue to emit photons long after the scatter of excitation light and emission of background fluorescence (on the order of nanoseconds) have dissipated (Figure 3.2).

After each excitation flash, the fluorescence emission from the sample is allowed to decay for 10 to 100 microseconds before measurement. The fluorescence emission is then collected for several hundred microseconds. The assay result is based on the measurement of integrated fluorescence intensity, not the decay time, and should not be confused with fluorescence lifetime measurement.

The DELFIA-based assay method remarkably reduces the interference from background fluorescence, such that a sensitivity of 0.2 fmol of phosphate transferred per well, or 1 pmol/L concentration change, can be attained with high-affinity antibodies. Sometimes, this level of sensitivity is not achievable due to a lower binding affinity between the antibody and the phosphorylated substrate [11,12]. Unfortunately, the multiple-step washes and solid-phase nature of the reaction make DELFIA a time-consuming assay. Naturally, a washless or homogeneous time-resolved fluorescence assay for kinases would be highly desirable.

3.2.4 Time-Resolved Fluorescence Resonance Energy Transfer

Time-resolved fluorescence resonance energy transfer (TR-FRET) provides an alternative approach for kinase assays. This technique combines the benefits of time-resolved fluorescence (TRF) with fluorescence resonance energy transfer (FRET), where the excited state energy is transferred from

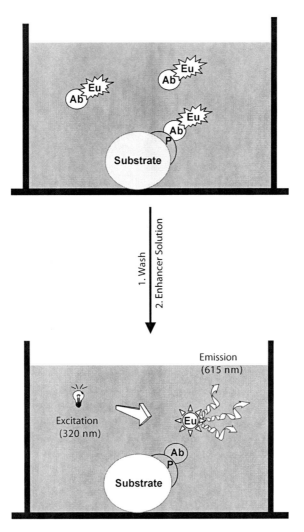

FIGURE 3.4 The principle of DELFIA. Substrate, kinase substrate immobilized on the surface of microtiter plate; P, phosphorylated amino acid in the substrate; Ab, specific antibody against phosphorylated substrate; Eu, europium chelate.

a "donor" fluorophore to a nearby "acceptor" fluorophore, which subsequently emits a photon. Because the efficiency of the transfer between donor and acceptor fluorophores is dependent on the distance and orientation between them, among other factors, this technique is very sensitive for use in monitoring binding or folding events at the molecular level.

Fluorescence resonance energy transfer (FRET) is based on the Förster energy transfer theory [13]. According to Förster's theory, the rate of energy transfer k_{ET} and the energy transfer efficiency E between a donor–acceptor pair can be expressed as

$$k_{ET} = \frac{1}{\tau_d}(\frac{R_0}{R})^6 \tag{3.3}$$

$$E = \frac{R_0^6}{R_0^6 + R^6} \tag{3.4}$$

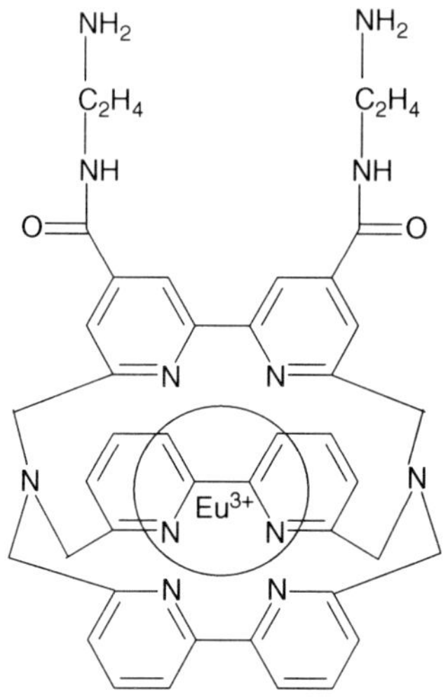

FIGURE 3.5 Molecular structure of TBP(Eu^{3+}) [europium cryptate].

where R_0 is the distance at which the transfer efficiency is 50%.

$$R_0 = (JK^2 \, \Phi_D \, n^{-4})^{1/6} \times 9.78 \times 10^3 \, \text{Å} \qquad (3.5)$$

The spectroscopic factors in the above equations are J, the spectral overlap between donor and acceptor; n, the refractive index of the medium between the donor and acceptor; τ_d, the fluorescence lifetime of the excited state of the donor in the absence of acceptor; and Φ_D, the fluorescence quantum yield of the donor in the absence of acceptor. The geometric factors are R, the distance between the centers of the donor and acceptor fluorophores; and K^2, the orientation factor for a dipole–dipole interaction.

According to the theory, the energy transfer efficiency E has a sixth power dependence in R_0, and R_0 is influenced by J and Φ_D. Therefore, providing suitable spectral overlap (J) and quantum yield (Φ_D) values, FRET can be used as a spectroscopic ruler to reveal the proximity of biological structures. Using commonly available organic fluorescent donor–acceptor pairs, the FRET principle has been applied to design simple biological assays, in which the donor and acceptor can be placed within 10 to 60 Å (1 to 6 nm) of each other. However, for complicated biological assays, such as recognition of phosphorylated kinase substrates using specific antibodies, the immune complex can be large (50 to 100 Å), and the selection of donor–acceptor pairs becomes restricted.

The discovery of lanthanide–pyridine derivative chelates made configuration of TR-FRET-based assays for complicated large biological molecules feasible [6,14–18]. Lanthanide–pyridine derivative chelates are long-lifetime fluorescence donors that pair well with allophycocyanin (APC), which is a highly efficient fluorescence acceptor. Lanthanide–pyridine derivative chelates, such as europium trisbipyridine diamine [TBP(Eu^{3+}), commonly known as europium cryptate] (Figure 3.5), are kinetically stable compounds that can be used to directly label biological molecules, such as antibodies. The pyridine derivatives do not interfere with the fluorescent emission of europium ions. More importantly, they protect the europium ions from fluorescence quenching and are capable of absorbing excitation energy and transferring it to europium ions, thus increasing the quantum yield (Φ_D) of europium ions (Figure 3.6).

APC has a high molar absorption coefficient in the wavelength range covering the multiple emission peaks of TBP(Eu^{3+}) and a very high fluorescence quantum yield (Φ_A) (Figure 3.6 and Figure 3.7), which provide two important advantages. First, the large spectral overlap (J) of the

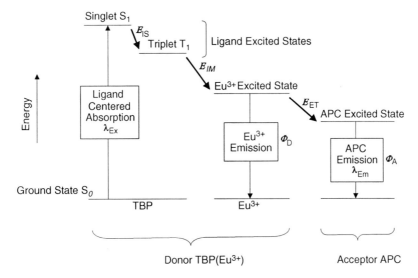

FIGURE 3.6 Simplified intersystem, intramolecular and intermolecular energy transfer pathways in the TR-FRET assay with a europium cryptate as the donor and an APC as the acceptor. E_{IS}, intersystem energy transfer efficiency of the TBP ligand from the singlet excited state to the triplet excited state; E_{IM}, intramolecular energy transfer efficiency between TBP ligand and europium ion; E_{ET}, intermolecular energy transfer efficiency between europium ion and APC; Φ_D, fluorescence quantum yield of europium ion; Φ_A, fluorescence quantum yield of APC; λ_{Ex}, excitation wavelength of TR-FRET assays; λ_{Em}, emission wavelength of TR-FRET assays.

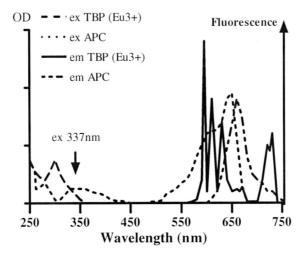

FIGURE 3.7 Emission and excitation spectra of TBP(Eu^{3+}) cryptate and allophycocyanin (APC). (Reproduced with permission from Mathis G. *Clin Chem* 1993; 39:1953–1959.)

donor and acceptor, together with the high quantum yield (Φ_D) of the donor, allow a longer-distance (50 to 100 Å) energy transfer between europium ion and APC than is common with other pairs. Second, the energy transfer from TBP(Eu^{3+}) to APC ($E_{ET}*\Phi_A$) is more efficient than the emission of fluorescence from TBP(Eu^{3+}) alone (Φ_D), thus amplifying the overall fluorescence signal ($E_{ET}*\Phi_A > \Phi_D$, Figure 3.6) [14].

One of the two fluorophores required for a FRET assay is attached to the antibody against phosphorylated substrate. To attach the second fluorophore, a peptide substrate is typically biotinylated; and if it is a protein substrate, it will be tagged with generic protein tags (His, MBP, FLAG,

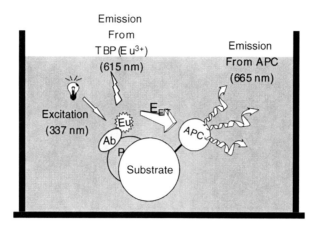

FIGURE 3.8 The principle of TR-HTRF detection. Substrate, biotinylated kinase substrate; P, phosphorylated amino acid in the substrate; Ab, specific antibody against phosphorylated substrate; Eu, europium cryptate [TBP(Eu^{3+})]; APC, streptavidin-labeled allophycocyanin; E$_{ET}$, energy transfer direction.

etc.). To detect phosphorylation of substrates, a detection mixture containing streptavidin-labeled APC or antitag antibody-labeled APC, and europium chelate–labeled antiphosphorylated substrate antibody, is allowed to bind to the phosphorylated substrate in solution. Phosphorylation of the substrate and specific immunoreaction and/or streptavidin-biotin binding will bring the europium chelate and APC into close proximity (Figure 3.8). Following illumination of the bound europium chelate, a highly specific fluorescence emission from the immunocomplex-bound APC is generated, which is proportional to the phosphorylation of substrate.

TR-FRET provides both spectral and temporal selections for the specific fluorescence signal resulting from resonance energy transfer between bound europium chelate and bound APC in the presence of free europium chelate and free APC. When the solution is illuminated at 337 nm, both the europium chelate and APC are excited and fluoresce (Figure 3.7). An emission spectral filter centered at 665 nm passes only the desired emissions from APC bound to the immunocomplex with europium chelate and does not pass the emission from the europium chelate. The fluorescence emission at 665 nm is measured after a delay of 10 to 100 μsec. This temporal delay eliminates the nonspecific fluorescence from free APC that only has a lifetime of a few nanoseconds and avoids the background fluorescence from compounds and instruments, which also tends to be short-lived. The TR-FRET technology is highly sensitive and can detect picomoles of phosphorylated substrate per well, or nanomole/liter concentration changes [6,14,15]. This sensitivity compares well with radioisotope labeling-based heterogeneous assays, but the technique is not as sensitive as the heterogeneous DELFIA assay.

3.3 DEVELOPMENT OF TIME-RESOLVED FLUORESCENCE AND TIME-RESOLVED FLUORESCENCE RESONANCE ENERGY TRANSFER–BASED KINASE ASSAYS

3.3.1 CONSIDERATIONS FOR ASSAY DESIGN

Assays that measure the extent of substrate phosphorylation could be considered as the optimal assay format for kinases since they enable one to identify kinase inhibitors with all possible modes of action, including competitive, noncompetitive, and uncompetitive inhibitors. Deciding which detection format, DELFIA, TR-FRET, or radioisotopes, to use to measure the phosphorylation of the substrate can be influenced by many factors. For example: Are special reagents, such as antibody against phosphorylated substrate that is required for the fluorescent assays, available? Does fluorescence labeling change the pharmacological characterization of the enzymatic reaction? Is it an autophosphorylation or a substrate phosphorylation assay? How much protein is required? Are reagents affordable? What is the Michaelis constant (K_m) of ATP and peptide substrate? What is the specific activity of the kinase enzyme? Is the assay format automation friendly?

During assay design, independent of the final assay format, we would also assess the suitable sequences for enzyme (catalytic domain vs. full length protein, tags, etc.) and substrate (peptide vs. protein), applicable species (human, mouse, rat, etc.), whether to target activated or inactivated kinases, and the desired inhibition modes (competitive vs. noncompetitive and uncompetitive).

3.3.2 REAGENTS AND INSTRUMENTS

Antibodies that can distinguish a phosphorylated substrate from an unphosphorylated one are a prerequisite for developing DELFIA and TR-FRET assays. If detection sensitivity is a concern due to low enzyme specific activity or a high substrate K_m value, then DELFIA is preferred, since it is difficult to configure TR-FRET–based assays under these conditions. DELFIA reagents that include europium labeling kits and enhancer solution are available from PerkinElmer Life Sciences (Boston, MA; http://las.perkinelmer.com). Many ready-to-use europium-labeled antibodies against phosphorylated substrates are available as well.

If the specific activity of an enzyme and the K_m values of the substrates are reasonable, then the TR-FRET assay format is preferred due to its homogeneous nature. Both CIS bio international (Ceze Cedex, France; www.htrf-assays.com) and PerkinElmer provide reagents for TR-FRET–based assays. CIS bio supplies HTRF (homogeneous time-resolved fluorescence) reagents that contain europium cryptates and XL665, a modified APC. PerkinElmer offers LANCE (LANthanide Chelate Excitation) reagents that include europium or terbium chelates and a number of fluorescence acceptors with similar optical properties to those of APC. Labeling kits, generic reagents (such as prelabeled streptavidin or antiphosphorylated tyrosine antibodies), and labeling services are provided by both companies.

Many instruments with time-resolved fluorescence detection features are commercially available. These include, but are not limited to, ViewLux and EnVision by PerkinElmer, RUBYstar by BMG (Offenburg, Germany; www.bmglabtech.com), Analyst by Molecular Devices (Sunnyvale, CA; www.moleculardevices.com), and LEADseeker by GE Healthcare (formerly Amersham Biosciences, Piscataway, NJ; www.gehealthcare.com). ViewLux and LEADseeker were built using a temperature-controlled charge-coupled device (CCD) camera as the detection system. Since the CCD images the entire plate at once, it provides very fast detection speeds. Other instruments are constructed using photomultiplier tubes (PMTs) that read one well per PMT at a time. Theoretically, this slows the rate at which a plate can be read. Since PMTs typically have higher signal-to-noise ratios than CCD cameras do, the speed difference is not pronounced. Most of these instruments are capable of dual-wavelength detection that can be useful for ratiometric readouts. However, for time-resolved fluorescence, there is an advantage for instruments that have two separate simultaneous detection systems, as opposed to instruments that switch detection between the two wavelengths.

3.3.3 ASSAY DEVELOPMENT

Kinases differ in specific activity, cofactor requirements, and K_m values for ATP and phosphate-accepting substrates. Therefore, the optimized assay conditions and final assay qualities will vary. Let us examine a typical assay development for a tyrosine kinase with a biotinylated peptide substrate.

For a DELFIA-based assay, streptavidin precoated microtiter plates could be used to capture the biotinylated substrate. Typically, a kinase reaction is initiated by placing kinase enzyme in the microtiter plates, followed by addition of test compounds, allowing enzyme and compound to interact for a minimum of 30 minutes to catch both fast- and slow-on inhibitors. Subsequently, a mixture of ATP and biotinylated peptide substrate is added. Often, all reagents are diluted in an assay solution that might include cofactors (magnesium ion, manganese ion), detergents (Triton X-100, Tween-20), carrier proteins (γ-globulin), antioxidants (DTT), and phosphatase inhibitors (β-glycerophosphate, sodium fluoride, sodium orthovanadate). HEPES and Tris are commonly used to buffer the assay solution. Phosphate buffer should not be used for any kinase assay since it interferes with the enzymatic reaction.

After incubation at room temperature or 37°C for an appropriate amount of time (20 to 120 min), the plates are washed using HEPES or Tris buffer containing detergent and EDTA to quench metal cofactors and stop the kinase reaction. High reaction temperatures could accelerate the enzymatic reaction. Once the plates are blocked with carrier proteins such as bovine serum albumin (BSA) to prevent nonspecific binding, europium chelate–labeled antiphosphorylated tyrosine antibody in DELFIA assay buffer (PerkinElmer) is added, which binds to the phosphorylated tyrosine in the substrate. After 1 h of immunoreaction, unbound antibodies are removed by washing, typically with HEPES or Tris buffer containing detergent. Subsequently, DELFIA enhancer solution is added, and the mixture is incubated for 30 min. Enhancer solution will free europium ion from its chelate with EDTA derivate and recapture the ion with β-diketone ligands. The plates are then read with a series of nanosecond excitation pulses (320 nm) at a frequency of 0.5 to 2 kilohertz (500 to 2000 pulses per second), and light is collected at 615 nm for a few hundred microseconds, starting about 100 μsec after each pulse.

If precoated streptavidin plates interfere with the enzymatic reaction, the reaction should be performed on a separate low binding plate before being transferred to the streptavidin precoated plates for capture, wash, and detection. Concentrations of all reagents, including enzyme, substrates, cofactors, additives, and enhancer solution, as well as pH values, should be carefully titrated to obtain optimal conditions. Reaction kinetics should also be carefully monitored to ensure that at the end of the reaction, substrate phosphorylation is still in the linear range, which is typically less than 5% phosphorylated.

The homogeneous TR-FRET assay format is highly amenable to automation. As with the DELFIA assay, the enzymatic reaction is carried out by the addition of an ATP and biotinylated substrate mixture onto nonbinding microtiter plates containing enzyme and compound in an appropriate assay buffer. After incubation at room temperature or 37°C for an appropriate amount of time, typically until 5 to 10% of the substrate is phosphorylated, a detection solution is added. The detection solution contains antiphosphotyrosine monoclonal antibody labeled with europium chelates, such as europium cryptates, and a streptavidin labeled with a suitable acceptor fluorophore, such as XL665. The reaction mixture is allowed to incubate for an appropriate length of time in order to catch the phosphorylated tyrosine in the substrate. Subsequently, the plates are subjected to time-resolved fluorescence readings with an excitation wavelength of 337 nm and an emission wavelength of 665 nm (Figure 3.9).

As with other assay formats, all reagents and reaction times should be carefully optimized for TR-FRET–based assays (Figure 3.10). Particular attention should be paid to the concentration ratio among biotinylated substrates, europium chelate–labeled antibody and APC- or XL-665-labeled streptavidin. TR-FRET occurs only when an antiphosphorylated substrate antibody-conjugated europium chelate is excited at 337 nm and is in close proximity to transfer energy to an appropriate acceptor, such as APC or XL-665 bound to streptavidin, which results in a fluorescence increase at 665 nm. Consequently, a "donor-phosphorylated substrate-acceptor" immunocomplex must be created to facilitate this energy transfer. An accurate balance of donor, acceptor, and substrate concentrations is critical to maximize sensitivity since an excess of free donor or acceptor will result in high background levels, but a dearth will reduce assay sensitivity.

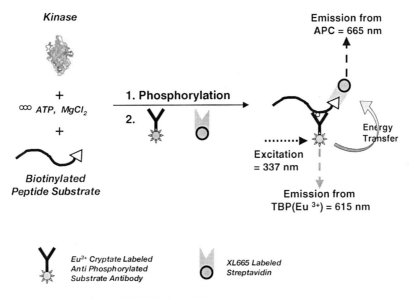

FIGURE 3.9 The principle of the TR-FRET–based kinase assay.

3.4 LIMITATIONS OF TIME-RESOLVED FLUORESCENCE AND TIME-RESOLVED FLUORESCENCE RESONANCE ENERGY TRANSFER

Time-resolved fluorescence technologies provide many advantages relative to prompt fluorescence technologies and have replaced conventional radioisotope-based detection for screening of kinase inhibitors. The spectral and temporal resolution allows one to develop robust assays for the identification of kinase inhibitors and permits the study of inhibition mechanisms of compounds. Nevertheless, there are trade-offs for both the DELFIA and TR-FRET assay formats. DELFIA assays have exceptional sensitivity and a large dynamic range of detection and are less prone to interference from background fluorescence. However, multiple steps and washings make them susceptible to liquid handling errors and consequently they are difficult to automate.

In both homogeneous TR-FRET assays, LANCE and HTRF, the measurement of emission is conducted in the presence of test compounds, which may interfere in ways that are not corrected by simple time resolution. These mechanisms include attenuation of the excitation wavelength, inner filter effects on donor or acceptor emission wavelengths (compounds absorbing at donor or acceptor emission wavelengths), and nonradioactive deactivation of the long-lived excitation state of europium ions through collisional interaction. Excitation attenuation and inner filter effects could be corrected using dual wavelength (615 and 665 nm) detection with a ratiometric readout. HTRF technology applies potassium fluoride (KF) to protect europium ions from molecular collision, thus stabilizing the fluorescence signal, since fluoride ions and cryptate form better cages for europium ions than does cryptate alone [18]. LANCE technology utilizes EDTA to remove manganese and zinc ions before the addition of the detection mixture since these two ions are known to cause nonradioactive deactivation of europium–pyridine chelates by collision with the europium ions [19]. Due to the presence of free acceptor and donor, and the potential spectral cross-talk between europium chelate emission and APC emission, the detection limit and dynamic range of both TR-FRET methods are compromised, compared with the DELFIA method. Consequently, it is relatively difficult to develop assays for kinases with high K_m values for phosphate-accepting substrates. Finally, both TR-FRET methods are suboptimal for determining kinase autophosphorylation because of the difficulty in balancing enzyme, substrate, donor, and acceptor concentrations.

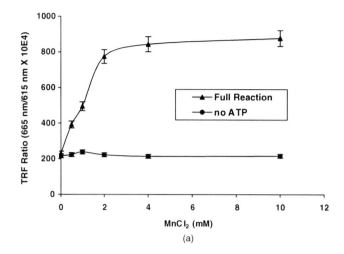

(a)

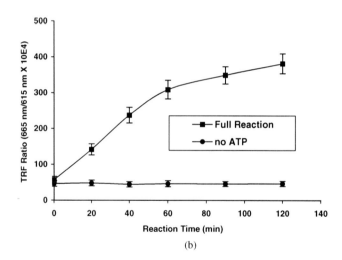

(b)

FIGURE 3.10 Examples of reagent titration for the development of a tyrosine kinase assay using LANCE reagents. (a) Requirement of manganese ion (Mn^{++}) for kinase activity. (b) Reaction kinetics. (c) Measurement of Michaelis constant (K_m) for peptide substrate. Concentration of ATP used was 50 μM. (d) Measurement of Michaelis constant (K_m) for ATP. Peptide substrate concentration used was 320 nM (n = 4 for all data points).

3.5 CONCLUSION

There are many *in vitro* kinase assay formats from which to choose when developing an assay. Often the choice is based on very practical issues that are specific to the laboratory or researcher, such as the availability of an appropriate detector or radioactivity license, access to labeling facilities, automation equipment, cost, or experience with support from a local vendor. In addition, the particular characteristics of the enzyme will dictate which assay format to use. For instance, it may be difficult to label a protein substrate with the second fluorophore at the proper distance for efficient energy transfer in a TR-FRET assay, or the labels may interfere directly with the phosphorylation reaction. An enzyme with a high substrate K_m might require so much labeled substrate in the reaction mixture that TR-FRET methods are unsuitable. For other kinases, there might not be phosphopeptide-specific antibody available, or the available antibodies may be so weak that an excessively high concentration would be required for practicality.

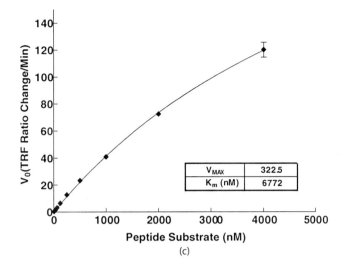

(c)

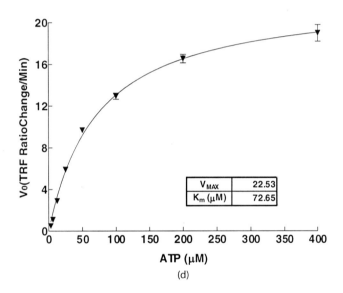

(d)

FIGURE 3.10 (CONTINUED)

After all practical limitations have been identified, it is important that the assay developer consider the type of compound that is desired and how the assay will be used when selecting an assay format. For instance, if the assay is for library screening, then high throughput, homogeneous, nonradioactive formats like TR-FRET should be favored. If the library to be screened consists only of ATP-like compounds, almost any assay format will work. If the library is more structurally diverse and the goal is to identify substrate-competitive inhibitors, then the assay conditions should be set to disfavor the identification of ATP-competitive compounds, such as utilizing very high ATP concentrations. For compounds that stabilize the inactivated form of the enzyme (which often have very slow on rates, presumably because they only act on a rare conformation of the enzyme), the assay should be performed with the enzyme preincubated with the compounds. To identify molecules that only inhibit the inactivated form of the protein, a cascade assay where the upstream activating kinase is incubated with the inactivated targeted kinase and its substrate might be of interest.

Frequently, assays are developed at later stages of the drug discovery process after a chemical series has been identified for optimization. These assays may be used for specificity profiling, selectivity profiling, toxicity identification, and the study of the mechanism of inhibitions. Generally, throughput and simplicity are secondary concerns to reproducibility and sensitivity at this stage. Therefore, the high sensitivity of a heterogeneous assay, such as DELFIA or other TRF methods, is often favored, even though the washing steps can be onerous.

Barring any specific limitations, the choice of assay format for phosphate transfer enzymes is almost limitless. Time-resolved fluorescence-based methods are generally highly sensitive, nonradioactive, functional assays that should be in the toolbox of every assay developer. Homogeneous TR-FRET–based methods should be considered strongly for high throughput applications. Heterogeneous formats, such as DELFIA, should be considered when sensitivity is critical.

REFERENCES

1. Lakowicz JR. *Principles of Fluorescence Spectroscopy.* New York, NY. Kluwer Academic/Plenum Publishers, 1999:96–97.
2. Richardson FS. Terbium (III) and europium (III) ions as luminescent probes and stains for biomelecular systems. *Chem Rev* 1982; 82:541–552.
3. Sabbatini N, Guardigli M. Luminescent lanthanide complexes as photochemical supramolecular devices. *Coord Chem Rev* 1993; 123:201–228.
4. Balzani V, Ballardini R. New trends in the design of luminescent metal complexes. *Photochem Photobiol* 1990; 52:409–416.
5. Alpha B, Lehn JM, Mathis G. Energy transfer luminescence of europium (III) and terbium (III) crypates of macrobicyclic polypyridine ligands. *Angew Chem Int Ed Engl* 1987; 26:266–267.
6. Hemmila I. LANCE: homogeneous assay platform for HTS. *J Biomol Screen* 1999; 4:303–307.
7. Hemmila I, Dakubu S, Mukkala VM, Siitari H, Lovgren T. Europium as a label in time-resolved immunofluorometric assays. *Anal Biochem* 1984; 137:335–345.
8. Soini E. Pulsed light, time-resolved fluorometric immunoassay. In: Bizollon CA ed. *Monoclonal Antibodies and New Trends in Immunoassays.* New York, NY. Elsevier, 1984:197–208.
9. Lovgre T, Pettersson K. Time-resolved fluoroimmunoassay, advantages and limitations. In: Van Dyke, K, Van Dyke R, eds. *Luminescence Immunoassay and Molecular Applications,* Boca Raton, FL. CRC Press, 1990:234–250.
10. Suonpaa M, Markcla E, Stahlberg T, Hemmila I. Europium-labeled streptavidin as a highly sensitive universal label. *J Immun Meth* 1992; 149:247–253.
11. Braunwalder A, Yarwood DR, Sills MA, and Lipson KE. Measurement of the protein tyrosine kinase activity of c-src using time-resolved fluorometry of europium chelates. *Anal Biochem* 1996; 238:159–163.
12. Sadler TM, Achilleos M, Ragunathan S, Pitkin A, LaRocque J, Morin J, Annable R, Greenberger LM, Frost P, Zhang Y. Development and comparison of two nonradioactive kinase assays for Ikappa B kinase. *Anal Biochem* 2004; 326:106–113.
13. Förster T. Zwischen molekare energiewanderung und fluoreszenz. *Ann Physik* 1848; 2:55–75.
14. Mathis G. Rare earth cryptates and homogeneous fluoroimmunoassays with human sera. *Clin Chem* 1993; 39:1953–1959.
15. Morrison LE. Time-resolved detection of energy transfer: theory and application to immunoassay. *Anal Biochem* 1988; 174:101–120.
16. Pope A. LANCE vs. HTRF technologies (or vice versa). *J Biomol Screen* 1999; 4:301–302.
17. Mathis G. HTRF technology. *J Biomol Screen* 1999; 4:309–313.
18. Kolb AJ, Kaplita PV, Hayes DJ, Park YW, Pernell C, Major JS, Mathis G. Tyrosine kinase assays adapted to homogeneous time-resolved fluorescence. *DDT* 1998; 3:333–342.
19. Application notes: Miniaturization of LANCE kinase assays. http://las.PerkinElmer.com.

4 Development of High-Throughput Screening Assays in Scintillating Microplates (FlashPlates) to Identify Inhibitors of Kinase Activity

Stuart Emanuel

CONTENTS

4.1 INTRODUCTION

Kinases represent one of the most numerous groups of proteins in the human genome and play an important role in regulating cell growth and cellular processes. It is estimated that as many as 518 protein kinases exist in the human genome, corresponding to nearly 1.7% of all human genes [1]. Protein kinases typically contain a conserved catalytic domain characteristic of the eukaryotic protein kinase superfamily and are classified into seven major groups, which can be further subdivided into families and subfamilies based on the sequence of their catalytic domains. Because they control processes related to growth, cell movement, cytoskeletal organization, and differentiation, loss of normal regulation can disrupt signaling cascades leading to diseases such as cancer and diabetes [2], which makes kinases logical targets for small-molecule therapeutic intervention. As many as 400 diseases may be associated with dysregulated kinase signaling [3].

Some kinase enzymes specifically catalyze phosphate addition on serine/threonine residues, others phosphorylate only on tyrosine residues, and some possess dual specificity acting on both tyrosine and serine/threonine. Most serine/threonine kinases are located in the soluble cytoplasmic compartment of the cell, while the receptor tyrosine kinases (RTK) are membrane bound. RTKs serve as growth factor receptors controlling cell proliferation, development, and differentiation, and these genes are often mutated in cancer. This class of receptors has been classified into at least 19 unique subfamilies [4], many of which act as oncogenes. A typical RTK consists of an extracellular ligand-binding domain, a transmembrane domain, and an intracellular domain containing the kinase activity. The extracellular domain is stimulated following ligand binding, a conformational change is induced, and typically, receptor dimerization and catalytic activation follow [5]. Autophosphorylation occurs when adjacent receptors phosphorylate each other on tyrosine residues, which then serve as binding sites for molecules containing Src homology-2 (SH2) and protein tyrosine-binding (PTB) domains that transduce signals into the cell [6].

Therapeutic validation for targeting protein kinases began with the approval in Japan of Fasudil (HA1077, AT877) in 1995 for the treatment of cerebral vasospasm, which may result from this compound's inhibition of Rho kinase [7], and ultimately, with the Food and Drug Administration (FDA) approval of Gleevec (STI-571, Imatinib) in 2001 for treatment of chronic myelogenous leukemia. Gleevec blocks uncontrolled signaling by the Abl tyrosine kinase and has shown remarkable efficacy with minimal side effects [8]. Due to the appeal of protein kinases as drug targets, there is a need to develop assays which monitor the activity of these enzymes suitable for high-throughput screening (HTS). We have made extensive use of the FlashPlate to develop assays for identification of inhibitors of protein kinase activity. The format described here has been successfully adapted to develop more than 25 HTS assays.

The FlashPlate (PerkinElmer, Boston, MA) is a white 96-well polystyrene microplate that contains a scintillant coated onto the bottom of each well. The scintillant plate surface can be coated with a number of reagents to capture radiolabeled product, such as streptavidin, antibodies, protein-A, nickel chelate, glutathione, wheat germ agglutinin, or cAMP. When the radioactive label is brought into close proximity with the scintillant, light is emitted, which can be detected on a scintillation counter [9,10]. A nonradioactive substrate can be coated onto the bottom of the plates, for example myelin basic protein, which serves as a phosphate acceptor; an enzymatic reaction transfers a radiolabeled phosphate to the substrate producing a radioactive product that excites the scintillant and generates a signal. Many radioisotopes can be used to produce a signal in FlashPlates, including ^{32}P, ^{33}P, ^{3}H, ^{14}C, ^{35}S, ^{125}I, and ^{45}Ca. Although the FlashPlate can be used to develop homogeneous radioactive assays where separation of bound and free isotope is not required with low-energy beta emitters such as ^{3}H, ^{14}C, and ^{35}S, we have found that removing the unbound reactants results in a more robust assay with a much greater signal-to-noise ratio. When higher-energy beta emitters are required, such as ^{32}P and ^{33}P-γ-ATP used in kinase assays, unreacted isotope produces a higher background due to the excitation of the scintillant in the well by unincorporated label. The kinase assays described here all utilize removal of the unreacted components and washing steps and can be used for high-throughput screening with the aid of automated plate washers. Aspiration of reaction components from the microplate also serves to remove any colored compounds, which may produce color quench and interfere with counting. Regardless of the assay format, FlashPlates can be read on various instruments such as TopCount (Packard), MicroBeta (Wallac), or Envision (NEN). These instruments are able to read plates in both the 96- and 384-well formats.

4.2 METHODS

Some reaction variables to consider when setting up a kinase assay include amount of enzyme, ratio of labeled to unlabeled ATP and total amount of ATP, choice and amount of substrate, buffer conditions, and reaction time and temperature. The source of kinase enzymes can be extracts abundant in a particular protein. For example, active protein kinase-A has been reported to be purified from bovine heart [11] or skeletal muscle [12]. Cell lines known to overexpress the kinase of interest can also serve as the source of protein; for example, A431 cells have been reported to contain high levels of epidermal growth factor receptor kinase [13]. Protein can also be produced recombinantly in mammalian cells, yeast, bacteria, or insect cells; this method has additional advantages in that it allows introduction of an affinity tag to facilitate purification, such as poly-histidine (6HIS), glutathione-*S*-transferase (GST), and others. The affinity tag can also serve to immobilize the kinase enzyme in the FlashPlate during the assay. For some kinase enzymes it may not be necessary to produce a clone of the full-length protein. Soluble cytoplasmic kinases are usually expressed as full-length proteins; however, for membrane-bound proteins possessing intra-cellular domains with kinase activity, expression of the hydrophobic membrane-spanning sequences would lead to improper folding and aggregation. The extracellular domains of transmembrane receptor tyrosine kinases are not required to obtain an active kinase, as activity is induced by addition of magnesium or manganese in the presence of ATP. For these proteins it is sufficient to express the intracellular portion of the enzyme containing the kinase domain and often, only the kinase domain itself need be expressed.

4.2.1 SUBSTRATE SELECTION

Selecting an appropriate substrate can determine how the enzyme performs in the assay and influence specificity of the kinase. Peptide substrates are commonly used; however, these short sequences do not reproduce the three-dimensional structure surrounding the phosphoacceptor res-idues found *in vivo*. Full length *in vivo* substrates may more closely recapitulate the phosphorylation sites encountered in cells, but they can contain numerous sequences capable of being modified by the kinase, lack specificity, and be difficult to immobilize. Artificial substrates such as Poly G:T can also be used but do not represent phosphoacceptor molecules likely to be encountered *in vivo* and may yield artefactual results. In general, short peptide substrates derived from actual cellular substrates that contain consensus sequences or recognition elements specific to the enzyme under investigation are the most useful for protein kinase activity assays (Table 4.1) [14]. Peptides that include residues known to be phosphorylated *in vivo* provide a more physiologically relevant measure of enzyme function. We evaluated several fragments derived from phospholipase-C as possible substrates for vascular endothelial growth factor receptor (VEGF-R) and platelet derived growth factor receptor (PDGF-R) assays. The most suitable fragment contained residues reported to be acted upon by PDGF-R and other RTKs, which are functionally important for signaling and essential for biological activity [15].

TABLE 4.1
Substrate Peptides Utilized for Kinase Activity Assays

Kinase	Substrate Name	Peptide Sequence	Reference
VEGF-R2, PDGF-R, RET	PLCγ1	Biotin-AEPDYGALYEGRNPGFYVEANP-amide	15
CDK1, CDK2	Histone-H1	Biotin-KTPKKAKKPKTPKKAKKL-amide	18
EGF-R	Angiotensin 2	Biotin-DRVYIHPF-amide	13
Protein kinase A, Aurora-A	BetterThanKemptide	Biotin-GRTGRRNSI-amide	19
PKC	PKC pseudopeptide	Biotin-RFARKGSLRQKNV-NH$_2$	20
Casein kinase 1	Glycogen synthase	Biotin-KRRRALS(phospho)VASLPGL-amide	21,22
Casein kinase 2	Nef	Biotin-RREEETEEE-amide	23
Calmodulin kinase	AutoCamtide II	Biotin-KKALRRQETVDAL-amide	24
GSK-3	CREB	Biotin-KRREILSRRP(phospho)SYR-amide	25
MAP kinase ERK-2	MBP	Biotin-APRTPGGRR-amide	26
Insulin receptor kinase	Insulin kinase domain	Biotin-TRDIYETDYYRK-amide	27, 28
HER2, FGF-R2	Poly(GT) 4:1	Biotin-poly(GT) 4:1	29
c-kit	None	Autophosphorylation	

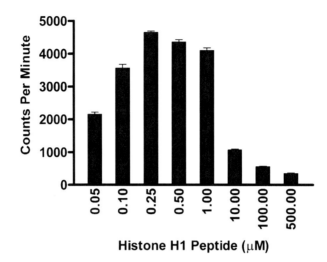

FIGURE 4.1 Peptide substrate titration. The CDK1 kinase assay was carried out as described in Table 4.2 with 5 units (2.5 ng) of recombinant CDK1 enzyme and increasing amounts of biotinylated histone H1 peptide substrate in a streptavidin FlashPlate. Each peptide concentration was tested in triplicate. Bars indicate mean ± SD of total counts per minute ^{33}P-γ-ATP incorporation into immobilized peptide.

4.2.2 Optimization of Substrate Concentration

For FlashPlates, which are available precoated with substrate, the amount of substrate is predetermined as with the myelin basic protein–coated plates that can be phosphorylated by mitogen activated protein (MAP) kinase isoforms ERK1/2. For plates that come precoated with a tethering reagent such as streptavidin, protein-A, or nickel chelate, it is necessary to determine the optimum amount of substrate to use in the reaction. Streptavidin-coated plates are frequently used to immobilize a biotinylated peptide substrate, which can be phosphorylated by an appropriate enzyme. The surface area in each well of the plate has a limited binding capacity for substrate, which can vary from lot to lot. One option is to precoat the entire plate with a saturating amount of substrate, allow binding to occur, and wash away the unbound material. Although the precoated plates would be stable and could be stored for several weeks or more, this requires an extra step and is wasteful of substrate [16]. Our preferred method is to add the peptide substrate in the reaction mix and allow binding and capture to proceed during the course of the reaction. By keeping the amount of enzyme constant and varying the amount of peptide, the optimum amount of peptide to use can be determined. With this technique, the signal will increase until saturating levels of substrate are reached, at which point the signal plateaus. Above this level, the signal will decrease as excess substrate competes for the radiolabel and dilutes out the signal (Figure 4.1). In a kinase reaction containing 5 μM ATP, the majority of the ATP transferred to the peptide is cold and does not contribute to the signal. To maximize the radiolabel incorporated into the substrate, a saturating amount of peptide is used in the reaction while avoiding excess peptide. In the example shown, CDK1 kinase enzyme was used to phosphorylate an 18-residue peptide containing a repeat of the histone H1 phosphorylation site. A 1 mM stock solution of the peptide was prepared in 50 mM Tris-HCl pH 8.0. The molecular weight of the peptide was 1723 daltons, so a 1 mM concentration corresponds to 1.723 mg/mL. From Figure 4.1, saturating levels of peptide were achieved at 0.25 μM peptide in a total volume of 100 μL or 43 ng peptide per well of a streptavidin FlashPlate. In general, for peptides of similar length and molecular weight, saturating amounts will be obtained in the range of 0.25 to 0.5 μM peptide. The actual amount can vary based on the sequence of the peptide and the lot of FlashPlates. Peptide stock solutions can also be prepared in tris or hepes buffers at acidic pH or in 100% dimethylsulfoxide (DMSO) if solubility is low in 50 mM Tris-HCl pH 8.0.

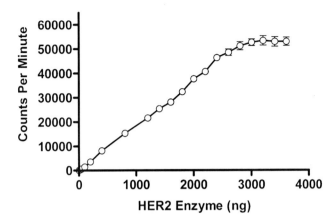

FIGURE 4.2 Enzyme titration for HER2 assay. A kinase assay was carried out in which all components of the reaction were held constant and increasing amounts of a 6HIS-tagged HER2 kinase enzyme were added to separate wells of a nickel chelate FlashPlate. Enzyme autophosphorylation at each concentration was measured in triplicate. Bars indicate mean ± SD of total counts per minute ^{33}P-γ-ATP incorporation into immobilized enzyme.

4.2.3 OPTIMIZATION OF ENZYME CONCENTRATION AND STABILITY PARAMETERS

The amount of enzyme used in the reaction should also be determined empirically by keeping the amount of peptide and isotope and the reaction time constant and adding increasing amounts of enzyme. As increasing amounts of enzyme are added, the signal obtained will increase and eventually plateau. An amount of enzyme in the linear range should be chosen. The enzyme titration is carried out in the same way when an autophosphorylation assay is developed, but in this case the kinase protein itself serves as the substrate in the absence of peptide. The example shown in Figure 4.2 is for a HER2 autophosphorylation assay performed in a nickel chelate FlashPlate in which the HER2 enzyme binds via an N-terminal 6HIS tag. In this case, the binding capacity of the well has been reached at around 3000 ng of enzyme. In situations where the signal is low, the reaction time can be increased to allow for further processivity of the enzyme toward the substrate without increasing the amount of enzyme in the reaction. A reaction timecourse will reveal whether extended incubation will allow additional incorporation of label into the substrate; however, background may also increase in proportion to incubation time, and this should be taken into account as well. When signal increases in a linear manner and background remains low, improved signal-to noise-ratios can be realized. Although the maximum assay signal increases in the example shown, the minimum signal also increases slightly over time, so the signal-to-noise ratio remains about 20-fold throughout the time course of the reaction (Figure 4.3).

Enzyme stability should also be evaluated, as loss of enzyme activity may result in nonlinear behavior, especially when reaction time is increased. For HTS considerations, enzyme stability is also an issue, as enzyme reservoirs may need to be incubated on ice or diluted enzyme may need to be replenished at regular intervals to avoid fluctuations in signal throughout a run. When diluted to working concentrations, the enzyme may be more sensitive to temperature effects. Figure 4.4 shows a loss of activity over time for VEGF-R kinase enzyme diluted in 50 mM Tris containing 1% bovine serum albumin (BSA). In this case, incubation of the diluted enzyme stock at 4°C prior to addition to the assay minimized degradation of the activity that occured when the enzyme stock was maintained at room temperature during the HTS run. A gradual decrease in maximum signal as an HTS run progresses may be a sign that the enzyme is experiencing some degradation (Figure 4.5). The incubation temperature of the reaction can be varied within a narrow range, and selection of a suitable temperature may be influenced by the stability of the enzyme. Temperatures

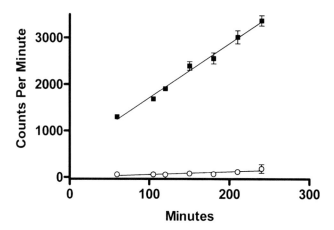

FIGURE 4.3 Time course of VEGF-R kinase assay. A kinase reaction was initiated with the VEGF-R enzyme and a biotinylated peptide substrate corresponding to a fragment of phospholipase Cγ1 (PLC-γ1). All components of the reaction were held constant and total activity was measured over a period of 1 to 4 h (■). Background was determined by addition of an inhibitor of enzyme activity (○). Enzyme reaction proceeded in a linear manner from 60 to 240 min. Data reflect mean ± SD of total counts per minute ^{33}P-γ-ATP incorporation into immobilized peptide (n = 3).

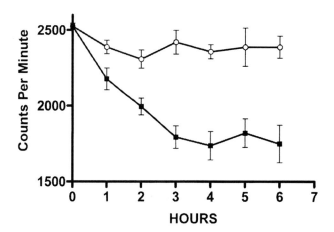

FIGURE 4.4 Stability of VEGF-R enzyme. The percent loss of activity of VEGF-R kinase enzyme diluted in 50 mM Tris HCl pH = 8 containing 1% BSA was evaluated over a 6-h period when incubated at room temperature (■) and on ice (○). The mean ± SD (n = 5) of the total counts per minute for the maximum signal in the assay are shown.

from 4 to 37°C are compatible with biological activity of kinase enzymes but 25 to 30°C is generally preferred. Reaction temperature can be decreased for sensitive enzymes that exhibit a rapid loss of activity or increased to 37°C to accelerate the rate of reaction and shorten incubation time. As long as the temperature remains constant during the course of the reaction and the reaction time course remains linear throughout the incubation period, the results will be consistent from assay to assay. Once an incubation temperature and time are selected, the conditions should be standardized, as values determined under different conditions may vary considerably. Reactions should not incubated on the bench at room temperature as this value may vary depending on environmental conditions in the building and seasonal changes. A temperature-controlled incubator is critical to maintaining a constant temperature and rate of reaction.

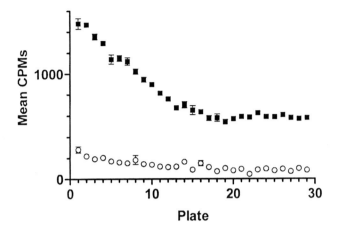

FIGURE 4.5 Stability of CDK1 enzyme at room temperature during an HTS run. Average maximum (■) and minimum (○) signal for first 25 plates of a CDK1 HTS kinase assay. Maximum signal was calculated in column 12 rows A to D by addition of DMSO vehicle only. The minimum signal was obtained by addition of an inhibitor of enzyme activity to column 12 rows E to H. Data reflects mean ± SD of total counts per minute ^{33}P-γ-ATP incorporation into immobilized histone H1 peptide (n = 4).

4.2.4 CONCENTRATIONS OF RADIOLABELED AND NONRADIOLABELED ATP

Another important factor to consider in the development of a kinase assay is the amount of radiolabel present in the reaction (usually ^{33}P-γ-ATP) relative to the concentration of cold ATP. The majority of ATP present in the reaction will be cold (nonradioactive), and transfer of a phosphate group that does not contain ^{33}P to the substrate will not contribute to the signal. However, a certain amount of cold ATP must be added to achieve concentrations of ATP near the K_m for the enzyme. If the total concentration of hot plus cold ATP in the reaction is too low, the enzyme will not be able to function. The K_m can be determined experimentally by keeping reaction conditions constant and varying the amount of ATP present in the reaction. Total ATP concentrations between 1 and 15 μM are usually sufficient to allow most kinases to catalyze transfer of phosphate groups to their cognate substrates. Although much higher ATP concentrations are encountered in the cell, on the order of 1 mM, the amount of cold ATP in the reaction is usually much lower, to maintain a high ratio of radiolabeled ATP to cold ATP and maximize the signal. Because the amount of radiolabeled ATP in the reaction is very small compared to the unlabeled ATP, it is possible to increase the assay signal by raising the concentration of radiolabeled ^{33}P-γ-ATP, but this will result in higher cost and generation of a larger amount of radioactive waste. Because the total ATP concentration in most kinase assays is lower than that found *in vivo*, the IC_{50} of an ATP-competitive inhibitor in cells will be much higher than the value calculated in the *in vitro* kinase assay. As increasing amounts of total ATP are added to the *in vitro* kinase assay, the observed IC_{50} value of an ATP competitive inhibitor will increase. Figure 4.6 shows an example where the quantity of radiolabel is held constant and the amount of cold ATP is increased in the reaction as the IC_{50} of an inhibitor is measured.

4.2.5 GENERAL KINASE ASSAY PROTOCOL AND PERFORMANCE

The general procedure we have developed to assay for kinase activity is as follows: A kinase reaction mix is prepared in 50 mM Tris-HCl pH = 8, 10 mM MgCl$_2$, 0.1 mM Na$_3$VO$_4$, 1 mM DTT, 10 μM ATP, 0.25 to 1 μM biotinylated peptide substrate, 0.2 to 0.8 μCi per well ^{33}P-γ-ATP [2000 to 3000 Ci/mmol]. Assay conditions vary slightly for each protein kinase; for example, insulin receptor kinase requires 10 mM MnCl$_2$ for activity and calmodulin-dependent protein kinase

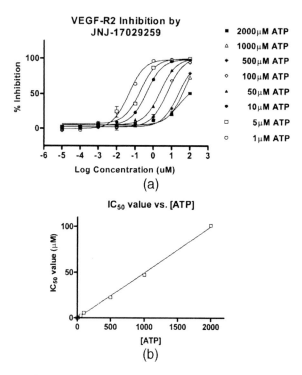

FIGURE 4.6 Inhibition as a function of ATP concentration. (a) The IC_{50} value of an inhibitor of the VEGF-R2 kinase activity was calculated at various concentrations of ATP ranging from 1 to 2000 μM in an *in vitro* kinase assay by plotting the percent inhibition vs. the log concentration of compound. (b) The IC_{50} values were plotted against the ATP concentration used in the assay to show that a linear relationship exists between inhibition and the amount of ATP present in the assay, as would be expected for an ATP-competitive inhibitor of kinase activity.

requires calmodulin and 10 m*M* CaCl$_2$. The reaction mix containing all components except enzyme is dispensed into the wells of a streptavidin-coated FlashPlate, and 1 μL drug stock in 100% DMSO is added to a 100-μL reaction volume, resulting in a final concentration of 1% DMSO in the reaction. Enzyme is diluted in 50 m*M* Tris-HCl pH = 8.0, 0.1% BSA, and added to each well. The reaction is incubated for 1 h at 30°C in the presence or absence of test compound. After 1 h, the reaction mix is aspirated from the plate, and the plate is washed with phosphate-buffered saline (PBS) containing 100 m*M* ethylenediaminetetraacetic acid (EDTA). The plate is read on a scintillation counter to determine the amount of ^{33}P-γ-ATP incorporated into the immobilized peptide. Test compounds are typically assayed at eight concentrations [100, 10, and 1 μM; 100, 10, and 1 n*M*; 100 and 10 p*M*]. A maximum and minimum signal for the assay is determined on each plate. We routinely use column 12 rows A to D to determine the maximum signal and column 12 rows E to H to establish the minimum (background) signal. The IC_{50} is calculated from the dose-response curve of the percent inhibition of the maximum signal in the assay according to the formula [100 − ([max signal − background/test compound signal − background] (100)] = % inhibition] by graphing the percent inhibition against the log concentration of test compound. The value used to calculate the background can be determined by adding a known inhibitor, which will block 100% of activity or by adding all components to the reaction except the enzyme. Known inhibitor compounds appropriate for the kinase being assayed are also included on each plate. We reserve column 11 of each plate for a single-column IC_{50} determination with a known inhibitor of the kinase enzyme as an intraplate control. The remainder of the plate, columns 1 to 10, are used for single-point inhibitor screening or eight-point IC_{50} determinations. A detailed procedure for performing the kinase assay is presented in Table 4.2.

TABLE 4.2
VEGF-Receptor Kinase Assay Procedure

Reagents
1. 10× Kinase buffer [500 mM Tris-HCl pH = 8, 100 mM MgCl$_2$, 1 mM Na$_3$VO$_4$]
2. 10 mM DTT [final concentration 1 mM]
3. 10 mM ATP [final concentration 5 μM]
4. ^{33}P-γ-ATP [2000 to 3000 Ci/mmol] 0.8 μCi per well at 10 μCi/μL = 0.08 μL/well
5. VEGF-R enzyme at 2.5 mg/mL; use 200 ng/well = 0.08 μL/well
6. Enzyme dilution buffer [50 mM Tris-HCl pH = 8.0, 0.1% BSA]
7. Wash/stop buffer [PBS +100 mM EDTA]
8. NEN Streptavidin FlashPlates (NEN #SMP-103).
9. PLCγ1 peptide substrate at 1 mM in 50 mM Tris-HCL pH = 8.0 (Use at 0.25 μM).

Master Mix Chart

Reagent	Per Well	One Plate (µL)
10× Kinase buffer	10	1100
10 mM DTT	10	1100
10 mM cold ATP	0.05	5.5
1 mM PLC1 peptide	0.025	2.75
^{33}P-γ-ATP at 10 μCi/μL	0.08	8.8
H$_2$O	49.77	5475

Protocol
1. Prepare Master Mix according to chart above. The indicated volumes will prepare enough reaction mix for one plate including an extra 10% to account for dead volume.
2. Dispense 70 μL of Master Mix into each well of a FlashPlate.
3. Add 1 μL of drug stock in 100% DMSO to appropriate wells (final DMSO concentration = 1%).
4. Prepare Enzyme Mix as follows: Enzyme dilution buffer   3291 μL
                                       VEGF-R enzyme   8.8 μL
5. Start reaction by adding 30 μL of diluted Enzyme Mix to each well except column 12 rows E to H. These wells are used to calculate the minimum signal in the assay (background). Add 1 μL of DMSO containing no inhibitor to column 12 rows A to D; these wells are used to calculate the maximum signal in the assay. Swirl plate to mix.
6. Incubate plate at 30°C for 60 min.
7. Dump reaction mix and wash plate 3× with 200 μL per well Wash/stop buffer.
8. Fill each well with 200 μL Wash/stop buffer.
9. Seal plate and count for 1 min per well on Packard TopCount.

4.2.6 ASSAY VALIDATION

Performance of the kinase assay should be evaluated to select conditions that result in a reliable, robust, and reproducible assay. Identification of inhibitors requires a sufficient window for reduction of activity to be detected by a test compound, which is the signal-to-noise (S/N) ratio of the assay calculated by (S/N = average maximum signal/average minimum signal). In general, a S/N ratio of at least 10:1 is required to implement an automatable HTS screen. The kinase assay protocol we describe here results in a S/N level of greater than 20:1 depending on the processivity of the enzyme/substrate combination. For the VEGF-R assay, a S/N of 30:1 was obtained. The coefficient of variability (CV) across the rows and columns of an assay plate can be determined by processing

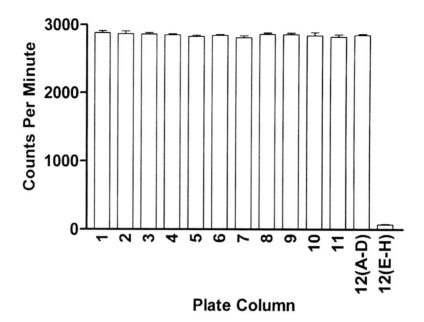

FIGURE 4.7 Average column signal across a plate. Mean counts per minute values for the maximum assay signal show very little variability in the binding of phosphorylated biotinylated peptide substrate across a 96-well streptavidin FlashPlate. Each bar represents one plate column where values were calculated from (n = 8), except for column 12, where rows A to D are normally used to determine the maximum and rows E to H are used to determine the minimum signal in the assay (n = 4) for HTS.

several plates that contain all maximum signal points and several plates that contain all minimum signal points. The columns and rows can be analyzed in Excel to calculate the CV across each row and up and down each column and ensure statistical accuracy over the entire plate. For the VEGF-R kinase assay, row and column CVs did not exceed 3% (data not shown). Figure 4.7 shows the mean signal produced in each column of a 96-well plate for the VEGF-R kinase assay. This type of analysis will reveal any inconsistencies in the signal across the plate and guarantees that the FlashPlates show consistent binding in all columns with no edge effects. Where a control inhibitor is available, it can be utilized to demonstrate that IC_{50} values calculated in each row of a 96-well plate will be consistent. The z'-factor parameter takes into account the S/N and the degree of variation in the data from an enzymatic assay [17]. A z'-factor value of 1.0 indicates an ideal assay with no variation and wide dynamic range; a z' value between 0.5 and 1.0 is considered excellent with low variation in the minimum and maximum signals and a good S/N ratio; values lower than 0.5 indicate a poor assay with high variability and a small S/N ratio. Table 4.3 shows some statistics calculated for the VEGF-R kinase assay, which had a z'-factor of 0.92. All statistical parameters should be measured on several plates and on different days to evaluate day-to-day assay variability and assess reproducibility.

4.2.7 DEFINITION AND SOURCE OF KINASE ENZYMES

The vascular endothelial growth factor receptor-2 (VEGF-R2) is a fusion protein containing a polyhistidine tag at the *N*-terminus followed by amino acids 786-1343 of the rat VEGF-R2 kinase domain (GenBank accession #U93306). The platelet-derived growth factor receptor beta (PDGF-Rβ) is a fusion protein containing a polyhistidine tag at the *N*-terminus followed by nucleotides 1874-3507 of the human PDGF-R subunit kinase domain (accession #M21616). The intracellular domain of the rearranged during transformation (RET) tyrosine kinase (accession #X12949) containing an *N*-terminal histidine tag was expressed and purified from Hi5 insect cells. Cyclin

TABLE 4.3
VEGF-Receptor Kinase Assay
Performance

Assay Parameter	Value
Mean maximum signal (CPM)	2837
SD maximum signal (CPM)	57
Mean minimum signal (CPM)	91
SD minimum signal (CPM)	13
Maximum signal % CV	2.0
Z'-factor	0.92

dependent kinase 1 (CDK1) was isolated from insect cells expressing both the human CDK1 catalytic subunit and its positive regulatory subunit cyclin B (New England Biolabs, Beverly, MA). CDK2 in complex with cyclin A consists of a C-terminal His-tagged CDK2 and an N-terminal GST-tagged cyclin A produced in Sf21 cells (Upstate Biotech, Lake Placid, NY). The epidermal growth factor receptor (EGFR) is purified from human A431 cell membranes (Sigma, St. Louis, MO). Protein kinase-A (PKA) is the catalytic subunit of cAMP-dependent protein kinase-A purified from bovine heart (Upstate Biotech, Lake Placid, NY). Aurora-A is the mouse protein containing an N-terminal polyhistidine tag and was expressed and purified in insect cells (Accession #BC014711). Protein kinase-C (PKC) is the gamma or beta isoform of the human protein produced in insect cells (BIOMOL, Plymouth Meeting, PA). Casein Kinase 1 is a truncation at amino acid 318 of the C-terminal portion of the rat delta isoform produced in E. coli (New England Biolabs, Beverly, MA). Casein Kinase 2 includes the alpha and beta subunits of the human protein produced in E. coli (New England Biolabs, Beverly, MA). Calmodulin-dependent protein kinase 2 is a truncated version of the alpha subunit of the rat protein produced in insect cells (New England Biolabs, Beverly, MA). Glycogen synthase kinase-3β (GSK3β) is the beta isoform of the rabbit enzyme produced in E. coli (New England Biolabs, Beverly, MA). Mitogen activated protein (MAP) kinase is the rat ERK-2 isoform containing a polyhistidine tag at the N-terminus produced in E. coli and activated by phosphorylation with MEK1 prior to purification (BIOMOL, Plymouth Meeting, PA). Insulin receptor kinase consists of residues 941-1313 of the cytoplasmic domain of the beta-subunit of the human insulin receptor (BIOMOL, Plymouth Meeting, PA). c-kit consists of the intracellular domain of the human receptor with an N-terminal polyhistidine tag (PanVera, Madison, WI). The human epidermal growth factor receptor-2 (HER2) construct contains a polyhistidine tag at the N-terminus followed by 24 additional amino acids and begins at amino acid 676 (accession #M11730), followed by the remainder of the HER2 cytoplasmic domain. The human fibroblast growth factor receptor-2 (FGF-R2) contains residues 453-765 of the cytoplasmic domain (accession #P21802) and was purified from baculovirus-infected Sf9 cells.

4.3 SUMMARY

We have described a rapid and robust FlashPlate scintillation proximity assay that has been developed for high-throughput screening of large compound libraries to identify inhibitors of kinase activity. The assay is modular in format and can be used to measure activity of tyrosine or serine/threonine kinases by replacing the enzyme with an enzyme of interest and a biotinylated peptide substrate capable of being phosphorylated by the enzyme. This procedure has also been used with GST and 6HIS tagged protein kinases in the absence of substrate to measure autophosphorylation of the kinase domain. The assay is sensitive and reproducible enough to determine accurate IC_{50} values and has been successfully applied to over 20 protein kinases. The adaptability and simplicity of this procedure make it amenable to automated screening and should aid in the identification of small-molecule inhibitors of this important class of drug targets.

REFERENCES

1. Manning, G., Whyte, D.B., Martinez, R., Hunter, T., and Sudarsanam, S. The protein kinase complement of the human genome. *Science* 298:1912–1934, 2002.
2. Aaronson, S.A. Growth factors and cancer. *Science* 254:1146–1153, 1991.
3. Plowman, G.D., Ullrich, A., and Shawver, L.K. Receptor tyrosine kinases as targets for drug intervention. *Drug News and Perspectives* 7:334–339, 1994.
4. Dunn, D. Mining the human kinome. *Drug Discovery Today* 7:1121–1222, 2002.
5. Heldin, C.-H. Dimerization of cell surface receptors in signal transduction. *Cell* 80:213–223, 1995.
6. Ullrich, A. and Schlessinger, J., Signal transduction by receptors with tyrosine kinase activity. *Cell* 61:203–212, 1990.
7. Cohen, P. Protein kinases — the major drug targets of the twenty-first century. *Nature Rev Drug Disc* 1:309–315, 2002.
8. Drucker, B.J., Talpaz, M., Resta, D.J., Peng, B., Buchdunger, E., Ford, J.M., Lydon, N.B., Kantarjian, H., Capdeville, R., Ohno-Jones, S., and Sawyers, C.L. Efficacy and safety of a specific inhibitor of the BCR-ABL tyrosine kinase in chronic myeloid leukemia. *New Engl J Med* 344:1031–1037, 2001.
9. Hart, H.E. and Greenwald, E.B. Scintillation proximity assay (SPA)-method of immunoassay. *Molec Immunol* 16:265–267, 1979.
10. Pocius, D. and Amrein, K. Detection of the activities of tyrosine(Y) kinase and tyrosine phosphatase (PTPs) that regulate the phosphorylation status of SRC family tyrosine kinases utilizing a novel scintillation proximity assay (SPA) system. *FASEB J* 10:1458, 1996.
11. Gilman, A.G. A protein binding assay for adenosine 3′:5′-cyclic monophosphate. *Proc Natl Acad Sci USA* 67:305–312, 1970.
12. Beavo, J.A., Bechtel, P.J., and Krebs, E.G. Preparation of homogeneous cyclic AMP-dependent protein kinase(s) and its subunits from rabbit skeletal muscle. *Methods Enzymol* 38:299–308, 1974.
13. Weber, W., Bertics, P.J., and Gill, G.N. Immunoaffinity purification of the epidermal growth factor receptor. Stoichiometry of binding and kinetics of self-phosphorylation. *J Biol Chem* 259:14,631–14,636, 1984.
14. Kennelly, P.J. and Krebs, E.G. Consensus sequences as substrate specificity determinants for protein kinases and protein phosphatases. *J Biol Chem* 266:15,555–15,558, 1991.
15. Emanuel, S.L., Gruninger, R.H., Fuentes-Pesquera, A., Connolly, P.J., Seamon, J.A., Hazel, S., Tominovich, R., Hollister, B., Napier, C., Reuman, M., Bignan, G., Tuman, R., Johnson, D., Moffatt, D., Batchelor, M., Foley, A., O'Connell, J., Allen, R., Perry, M., Jolliffe, L., and Middleton, S.A. A VEGF-R2 kinase inhibitor potentiates the activity of the conventional chemotherapeutic agents paclitaxel and doxorubicin in tumor xenograft models. *Mol Pharm* 66:3,635–647, 2004.
16. Turlais, F., Hardcastle, A., Rowlands, M., Newbatt, Y., Bannistaer, A., Kouzarides, T., Workman, P., and Wynne Aherne, G. High-throughput screening for identification of small-molecule inhibitors of histone acetyltransferases using scintillating microplates (FlashPlate). *Anal Biochem* 298:62–68, 2001.
17. Zhang, J.-H., Chung, T.D.Y., and Oldenberg, K.R. A simple statistical parameter for use in evaluation and validation of high-throughput screening assays. *J Biomol Screen* 4:67–73, 1999.
18. Lew, J., Beaudette, K., Litwin, C.M.E., and Wang, J.H. Purification and characterization of a novel proline-directed protein kinase from bovine brain. *J Biol Chem* 267:13,383–13,390, 1992.
19. Glass, D.B., Cheng, H.-C., Mende-Mueller, L., Reed, J., and Walsh, D.A. Primary structural determinants essential for potent inhibition of cAMP-dependent protein kinase by inhibitory peptides corresponding to the active portion of the heat-stable inhibitor protein. *J Biol Chem* 264:8802–8810, 1989.
20. House, C. and Kemp, B.E. Protein kinase C pseudosubstrate prototope: structure–function relationships. *Cell Signal* 2:187–190, 1990.
21. Flotow, H. and Roach, P.J. Role of acidic residues as substrate determinants for casein kinase I. *J Biol Chem* 266:3724–3727, 1991.

22. Flotow, H., Graves, P.R., Wang, A., Fiol, C.J., Roeske, R.W., and Roach, P.J. Phosphate groups as substrate determinants for casein kinase I. *J Biol Chem* 265:14,624–14,629, 1990.

23. Klarlund, J.K. and Czech, M.P. Insulin-like growth factor I and insulin rapidly increase casein kinase II activity in BALB/c 3T3 fibroblasts. *J Biol Chem* 263:15,872–15,875, 1988.

24. Hanson, P.I., Kapiloff, M.S., Lou, L.L., Rosenfeld, M.G., and Schulman, H. Expression of a multi-functional Ca^{2+}/calmodulin-dependent protein kinase and mutational analysis of its autoregulation. *Neuron* 3:59–70, 1989.

25. Hoeffler, J.P., Meyer, T.E., Yun, Y., Jameson, J.L., and Habener, J.F. Cyclic AMP–responsive DNA-binding protein: structure based on a cloned placental cDNA. *Science* 242:1430–1433, 1988.

26. Clark-Lewis, I., Sanghera, J.S., and Pelech, S.L. Definition of a consensus sequence for peptide substrate recognition by p44[mpk], the meiosis-activated myelin basic protein kinase. *J Biol Chem* 266:15,180–15,184, 1991.

27. Cho, H., Krishnaraj, R., Itoh, M., Kitas, E., Bannwarth, W., Saito, H., and Walsh, C.T. Substrate specificities of catalytic fragments of protein tyrosine phosphatases (HPTP, LAR, and CD45) toward phosphotyrosylpeptide substrates and thiophosphotyrosylated peptides as inhibitors. *Prot Sci* 2:977–984, 1993.

28. Tornqvist, H.E., Pierce, M.W., Frackelton, A.R., Nemenoff, R.A., and Avruch, J., Identification of insulin receptor tyrosine residues autophosphorylated *in vitro*. *J Biol Chem* 262:10,212–10,219, 1987.

29. Durocher, Y., Chapdelaine, A., and Chevalier, S. Tyrosine protein kinase activity of human hyperplastic prostate and carcinoma cell lines PC3 and DU145. *Cancer Res* 49:4818–4823, 1989.

5 Development of High-Throughput Screening Assays for Kinase Drug Targets Using AlphaScreen™ Technology

Dean Wenham, Chantal Illy, Jacques Andre St.Pierre, and Nathalie Bouchard

CONTENTS

5.1 INTRODUCTION

Protein phosphorylation is recognized to play a critical role in the regulation of cell growth and development. Protein kinases covalently modify protein substrates by attaching phosphate groups from ATP to serine, threonine, and/or tyrosine residues, leading to modifications in the functional properties of the substrate. Activation of the protein kinase cascades serves to amplify signals generated at the cell surface into complex biological responses, including the activation of transcription factors and gene expression.

The pathophysiological dysfunction of protein kinase signaling pathways underlies the molecular basis of many cancers (Kohno and Pouyssegur, 2003; Ghosh et al., 2003) and of several manifestations of cardiovascular (Behr et al., 2003; Jones et al., 2003), inflammatory, and immunological diseases (English and Cobb, 2002; Kumar et al., 2003). Given their role in such a wide variety of disease states, protein kinases are becoming extremely attractive targets for drug discovery, where biopharmaceutical companies are in search of enzyme inhibitors (English and Cobb, 2002). Indeed, protein kinases, together with G-protein coupled receptors (GPCRs), currently represent the most prevalently used therapeutic target type in pharmaceutical and large biotechnology companies.

An increasing number of assay technologies (including radioactive and nonradioactive platforms) is routinely utilized to develop and perform high-throughput screening (HTS) assays for kinases. At the present time, AlphaScreen represents an attractive technology for protein kinase screening, not only because it offers a nonradioactive and homogeneous platform, but also because it presents the advantages of good sensitivity and signal-to-background ratios, low cost, and short screening time due to a reduced number of addition steps, as well as being simple in nature and therefore providing rapid assay development.

In the present chapter, the process of developing *in vitro* AlphaScreen kinase assays that are amenable to HTS is described by illustrating the process with some key examples. These examples include two kinases of the phosphoinositide 3-kinases (PI 3-K) pathway (PI 3-K lipid kinase and Akt serine/threonine kinase); the serine/threonine ERK1 kinase, a member of the MAPK (mitogen-activated protein kinase) family; and the soluble β-insulin receptor tyrosine kinase domain (IRKD), an example of a tyrosine kinase. Other examples of AlphaScreen kinase assays have been described (Warner et al., 2004).

5.2 ALPHASCREEN: GENERAL PRINCIPLES

AlphaScreen is a bead-based chemistry used to study biomolecular interactions. The acronym ALPHA given to the technology by PerkinElmerBioSignal Inc. represents Amplified Luminescent Proximity Homogeneous Assay, but the technology was originally called LOCI (Luminescent Oxygen Channeling Immunoassay; Ullman et al., 1994) when it was developed by Syva Company (now called Dade Behring). AlphaScreen relies on the transfer of energy between an acceptor bead and a donor bead brought into proximity via a biological interaction. The donor beads are embedded with a photosensitizer (phtalocyanine), which converts ambient oxygen to an excited state singlet oxygen molecule upon illumination at 680 nm. Within its 4 μsec half-life, the excited oxygen diffuses approximately 200 nm in solution. If a biomolecular interaction positions the acceptor bead into close proximity with a donor bead, energy will be transferred from the singlet oxygen to the acceptor bead. Different fluorophores are present in the acceptor beads (thioxene, anthracene, and rubrene). Upon excitation by singlet oxygen, thioxene will emit light and induce a cascade of fluorescence resonance energy transfer leading to the emission of light at 520 to 620 nm. In the absence of any biological interaction, the singlet oxygen will return to ground state, and no signal will be produced (Figure 5.1).

Each donor bead contains a concentration of photosensitizer capable of generating up to 60,000 singlet oxygen particles per second resulting in substantial signal amplification. The chemical

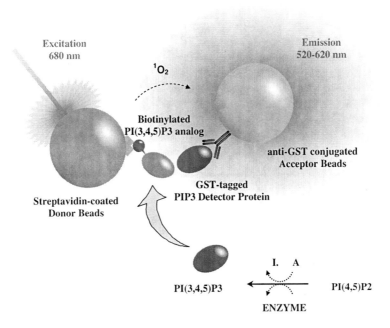

FIGURE 5.1 Schematic diagram of AlphaScreen general principle illustrated using phosphatidylinositol 3-kinases (PI 3-K) assay as an example. Binding of biological partners [in this case GST-PIP$_3$ detector protein and biotinylated-PI(3,4,5)P$_3$] will bring donor and acceptor beads into close proximity (200 nm) and generate fluorescent emission between 520 and 620 nm upon excitation at 680 nm. The generation of phosphatidyl-inositol 3′,4′,5′-trisphosphate [PI(3,4,5)P$_3$] as a result of PI 3-K activity will result in a competition event due to its binding to the GST-PIP$_3$ detector protein and therefore reduce the AlphaScreen signal.

energy transfer between acceptor and donor beads results in an emission at a shorter wavelength (520 to 620 nm) than excitation (680 nm), which reduces background. The singlet oxygen produced has a half-life of around 0.3 sec, allowing measurements in a time-resolved mode.

5.3 ALPHASCREEN: KINASE ASSAY DESIGN

An AlphaScreen kinase assay is comprised of two major steps:

- The kinase reaction, in which the target kinase is incubated with the substrate and ATP for a certain period of time
- The detection reaction, where the kinase reaction is stopped and the phosphorylated product detected with AlphaScreen reagents

Thus, for example, an increase in AlphaScreen signal is observed as the substrate becomes increasingly phosphorylated by the kinase where, for example, a substrate specific antiphospho antibody is used in the detection stage. When screening for kinase inhibitors, the test compounds are preincubated with the enzyme before addition of the substrate and ATP.

Kinase assays for HTS require qualities of sensitivity, linearity, and robustness. Assay robustness involves stability of reagents over time, low sensitivity to dimethylsulfoxide (DMSO) or other organic solvents used to prepare chemical libraries, low coefficient of variation (CV%) across microplates, predictability of known pharmacological inhibitors, good signal-to-background ratios, and Z′ factors above 0.5 (Zhang, 1999). Optimization and validation of AlphaScreen kinase assays involve several steps that address all these requirements.

One of the obstacles in developing *in vitro* kinase assays lies in the difficulty of obtaining sufficient amounts of pure, active enzymes at an acceptable cost. Because of its high sensitivity of detection, AlphaScreen offers the possibility of using lower amounts of enzyme as well as measuring kinase activity present in cell extracts.

When developing kinase screening assays, the choice of the substrate is a crucial step since it tunes the sensitivity of an assay to detect desired classes of inhibitors. The activity of protein kinases can be traditionally assayed by measuring phosphorylation of protein nonspecific substrates, such as myelin basic protein (MBP), or of broad-spectrum generic polypeptide substrates such as poly[Glu:Tyr] (4:1) and poly[Glu:Ala:Tyr] (1:1:1). As protein kinases have been characterized, more specific peptide substrates have been designed. These peptide substrates are typically based on substrate protein sequences or autophosphorylation sites. Because of the lack of specificity of these peptide substrate consensus sequences, protein substrates are preferably used to assay some protein kinases. In particular, multiple serine/threonine kinases are not active on small peptidic phosphorylation sequences and thus require more complete substrate domains. To allow for the detection of phosphorylated products by AlphaScreen beads, kinase substrates have to be either biotinylated or tagged with specific tags, such as glutathione-*S*-transferase (GST) and 6-histidine (His). Examples of different types of substrate used in AlphaScreen assays are presented in the following sections.

5.3.1 THE KINASE REACTION

As a general rule, kinase reactions should be conducted under initial rate conditions, the product formation should be linear with respect to enzyme concentration and incubation time, and the total consumption of the substrate should be less than 20%. One of the primary tasks to perform during a typical kinase assay development is to optimize the composition of the assay buffer and its pH. This buffer should also contain any reducing chemicals (DTT) or cofactors ($MgCl_2$, $MnCl_2$) required for the optimal functionality of the kinase. In addition, when working with crude enzyme preparations, addition of specific phosphatase inhibitors (such as orthovanadate for tyrosine phosphatases and okadaic acid for some serine/threonine phosphatases) to the buffer is recommended.

The optimal working concentration of substrate and enzyme should be carefully determined. The selection of the enzyme concentration should be based upon a balance of the signal-to-background ratio with the cost of reagent and the incubation time. If the enzyme is precious and limited, a minimum concentration should be used that still produces a linear response and an acceptable signal-to-background ratio. If the enzyme is available in larger quantities, a higher concentration can be added to increase the signal and decrease the incubation time. For HTS compatibility, kinase reactions are generally performed at room temperature (23°C +/– 2°C) with incubation times varying from 15 to 120 min. To detect substrate-competitive inhibitors, a substrate concentration should be selected that is close to or below the apparent K_m.

For kinase reactions, ATP provides the source of phosphate for substrate phosphorylation and is an important parameter in the assay. AlphaScreen kinase assays are typically performed near the K_m concentration for ATP. Indeed, kinase screening campaigns often identify ATP-competitive hits that bind to the ATP binding site on the kinase. Therefore, if it is a strategy to avoid compounds that could potentially compete with ATP, the kinase reaction could be configured with an ATP concentration that is several-fold higher than its K_m. In addition, it is possible to discriminate between ATP-competitive hits and those that inhibit via other mechanisms by constructing competitive curves to the inhibitor at different ATP concentrations.

When developing HTS kinase assays, it is important to study the sensitivity of the assay to DMSO and to show that, under the assay conditions, the enzyme can be inhibited by known inhibitors. Most inhibitors function by preventing the binding of a substrate to the enzyme or by blocking the binding of ATP or essential cofactors to the kinase. Broad-spectrum kinase inhibitors (such as the ATP-competitive staurosporine compound) or class-specific inhibitors (such as genistein for tyrosine kinases) can be used to demonstrate the ability of the assay to detect kinase inhibitors. The inhibition constant (IC_{50} value) of reference inhibitors can be determined from competition inhibition curves. Examples of enzyme inhibition are presented in the body of this chapter.

5.3.2 THE DETECTION REACTION

The flexibility allowed by AlphaScreen technology for detection configuration allows kinase assays to be developed using various different detection strategies. Tyrosine kinase assays are usually performed with a direct format involving generic antiphosphotyrosine antibodies (namely P-Tyr-100, PT66, or PY20) conjugated to the acceptor beads. Biotinylated sequence-specific or generic substrates are preferably used in such assays. After completion of the kinase reaction, the biotinylated phosphorylated product is captured by streptavidin-coated donor beads and antiphosphotyrosine antibodies conjugated to acceptor beads. These binding events lead to the generation of an AlphaScreen signal proportional to the amount of product present.

Recognition of phosphorylated serine or threonine residues is known to be greatly influenced by the nature of the vicinal amino acids and often requires very specific and costly antibodies. For this reason, serine/threonine kinase assays are often performed using an indirect format using, for example, protein A conjugated acceptor beads. This approach takes advantage of the high affinity of protein A for numerous antibodies of rabbit or mouse origin; thus, the quantity of the antibody needed in the assay can be significantly reduced to subnanomolar concentrations. Another indirect strategy involves the capture of specific antiphosphoserine/threonine antibodies by antispecies IgGs conjugated to acceptor beads. Similarly to direct format assays, biotinylated synthetic peptide substrates can be used for serine/threonine kinase activity measurement in indirect configuration and in this case, will be captured by streptavidin-coated donor beads. When GST- or His-tagged substrate domains are used, it is preferable to exploit a sandwich assay format where a biotinylated anti-GST or biotinylated anti-His antibody are used to capture the substrate onto the donor beads. It is also worth noting that kinase assays can be configured where the phosphorylated product competes with the binding between a biotinylated product and an antibody or binding protein. Indeed, this configuration is described in Section 5.5.

Generally, the AlphaScreen beads and antibodies are added after the kinase reaction. The antibody/beads mixture is thus prepared in the detection buffer containing the metal chelator ethylenediaminetetraacetic acid (EDTA) to stop the kinase reaction upon addition of this detection mixture. In other cases, it may be beneficial to include the antiphospho antibody in the kinase reaction to allow for the detection of the phosphorylated product while it is being produced. The detection time may vary from one to many hours depending on the affinity of the antibodies and the level of signal desired.

The process of developing AlphaScreen assays is illustrated by examples in Section 5.5 and Section 5.6, where a detailed description of the optimization of a phosphatidylinositol 3-kinase assay is described specifically in Section 5.5.

5.4 KEY CONSIDERATIONS FOR ALPHASCREEN ASSAYS

5.4.1 ENVIRONMENTAL CONDITIONS AND REAGENT HANDLING

When developing and performing any assay it is important to tightly control the environmental and reagent handling conditions in order to generate superior assay reproducibility. For AlphaScreen assays, the environment for the readout system (i.e., the instrument) should be in an area that is least prone to dramatic temperature fluctuation. It is optimal to avoid manipulating reagents as well as reading the plates under bright light, for example direct sun exposure. If possible, choose a location with dim light (around 100 lux) and/or apply green filters to the light fixtures. By controlling these basic parameters, it is possible to minimize assay-to-assay and day-to-day variability.

AlphaScreen reagents should be stored at 4°C unless specified otherwise and protected from exposure to light (the plastic container in which the beads are provided is quite adequate for this purpose). The beads have a 250 nm diameter and form a stable suspension. They are supplied in Proclin 300 (preservative) to prevent microbial growth. To avoid loss of material in the tube following shipping or longer-term storage, it is recommended to centrifuge (pulse) down the beads and resuspend them by pipetting before use. Further, the beads should be kept away from any source of heat. This precaution does not concern the beads themselves since the dyes and hydrogel coating them are all very stable even up to 95°C (Beaudet et al., 2001) but it instead concerns the protein coating the beads, which may be sensitive to heat denaturation.

5.4.2 INSTRUMENTS

The detection of the AlphaScreen signal requires an instrument equipped with a laser that is able to generate photons at a wavelength of 680 nm in order to excite the donor beads; further, it requires the ability to collect emitted light at 520 to 620 nm from the acceptor beads. AlphaScreen needs a dedicated reader such as the AlphaQuest™ from PerkinElmer Life and Analytical Sciences (www.las.perkinelmer.com) or a multimode instrument installed with a plug-in module such as the Fusion™, EnVision/Alpha™ (again, both from PerkinElmer Life and Analytical Sciences), or the Mithras LB940 (Berthold Technologies; www.bertholdtech.com).

5.4.3 BEAD BINDING CAPACITY AND THE "HOOK" EFFECT

A phenomenon often referred to as the "hook" effect is common to most homogeneous proximity assays and is a direct consequence of the binding capacity of each of the acceptor and donor molecules. In an indirect detection assay using protein A capture, for example, at working concentrations of beads (20 μg/mL), bead saturation with antibodies will occur at approximately 1 to 3 nM of antibody. Donor beads coated with streptavidin will become saturated at about 10 to 30 nM of biotinylated molecules. Upon AlphaScreen beads saturation, nonproductive interactions are generated resulting in loss of signal (Figure 5.2).

5.4.4 INTERFERING COMPOUNDS

A number of agents should be avoided for use in an AlphaScreen assay. Detergents such as Triton X-100 and CHAPS should not exceed 0.1% (v/v). Preparations containing azide as a preservative agent are not recommended since azide is a potent scavenger of singlet oxygen. Eliminating the use of the following transition metal ions is strongly recommended: Fe^{2+}, Fe^{3+}, Cu^{2+}, Ni^{2+}, and Zn^{2+}. These metals have been shown to be potent singlet oxygen quenchers in the millimolar and micromolar ranges. For example, 100 μM of Fe^{2+} can be deleterious to an AlphaScreen assay.

5.5 ASSAY DEVELOPMENT: PI 3-KINASE AS A MODEL

Phosphatidylinositol 3-kinases (PI 3-K) are ubiquitously expressed lipid kinases that phosphorylate phosphatidylinositols at the 3-hydroxyl of the inositol ring. These phosphorylated products act as second messengers and play fundamental roles in cellular responses such as proliferation, survival, adhesion, cell motility, and carbohydrate metabolism (Franke et al., 1997; Shepherd et al., 1998; Wymann et al., 2000). Therefore, PI 3-Ks have become popular targets for drug development, especially in the areas of inflammation and cancer. In this section, the development of a PI 3-K AlphaScreen assay is used to describe some of the general steps that can be undertaken during the development of AlphaScreen kinase assays.

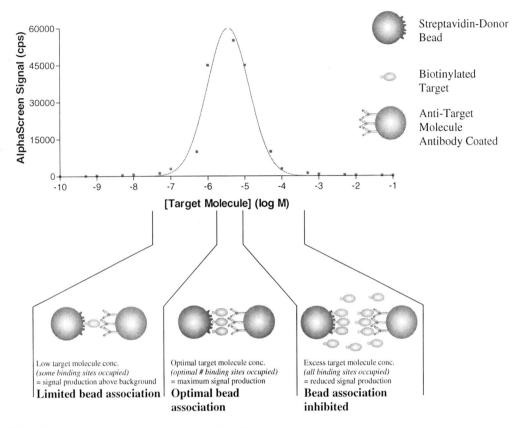

FIGURE 5.2 Graphic representation of the "hook" effect.

5.5.1 ASSAY PRINCIPLE

In this application of AlphaScreen technology, the assay has been configured to detect the phosphorylation and conversion of phosphatidylinositol 4',5'-bisphosphate [PI(4,5)P$_2$; the substrate] into phosphatidylinositol 3',4',5'-trisphosphate [PI(3,4,5)P$_3$; the product] by PI 3-K. A proprietary detector protein was used that is able to bind PI(3,4,5)P$_3$ with high affinity. This detector protein [Det-PI(3,4,5)P$_3$] was tagged with GST. Another key part of the detection system is biotinylated-PI(3,4,5)P$_3$, which binds with high affinity to Det-PI(3,4,5)P$_3$. Both Det-PI(3,4,5)P$_3$ and biotinylated-PI(3,4,5)P$_3$ were obtained from Echelon Biosciences Inc. (Salt Lake City, UT; www.echelon-inc.com) through a collaboration. The binding of Det-PI(3,4,5)P$_3$ to biotinylated-PI(3,4,5)P$_3$ was detected using streptavidin-coated donor beads and acceptor beads conjugated with anti-GST antibody. These complex biomolecular interactions result in the generation of AlphaScreen signal. The presence of nonbiotinylated PI(3,4,5)P$_3$ produced as a consequence of PI 3-K activity acts to compete with the Det-PI(3,4,5)P$_3$ and biotinylated-PI(3,4,5)P$_3$ interaction, thereby reducing AlphaScreen signal. Figure 5.1 shows a diagrammatic representation of this assay.

5.5.2 REAGENTS AND PROTOCOL

In this section, a general description of the optimized assay protocol is provided as a guideline, although details of some of the optimization steps are described in the proceeding sections:

Buffers
- Kinase buffer for PI 3-Kα: Hepes 5 mM (pH 7.4), MgCl$_2$ 2.5 mM, ATP 25 μM
- Detection buffer: Tris 10 mM (pH 7.4), NaCl 150 mM, 0.1% Tween-20, 1 mM DTT, and 7.5 mM EDTA

Protocol
- Assays were performed in a 25 μL final volume in white opaque 384-well plates. Kinase buffer of kinase buffer or test compounds (2.5 μL) were added to the plate followed by 5 μL of enzyme (PI 3-Kα), and the kinase reaction was initiated by addition of 2.5 μL of PI(4,5)P$_2$ substrate. The reaction was incubated at room temperature for a predetermined time (usually 60 to 90 min). Next, 5 μL of biotinylated-PI(3,4,5)P$_3$ (10 nM final) and 5 μL of Det-PI(3,4,5)P$_3$ (10 nM final) prepared in detection buffer were added followed by 5 μL of a mixture of donor and acceptor beads (20 mg/mL final for each). This final mixture was incubated at room temperature for 120 min in darkness followed by the reading of the AlphaScreen signal on an AlphaQuest microplate analyzer.

5.5.3 Titration of Binding Partners [Det-PI(3,4,5)P$_3$ and Biotinylated-PI(3,4,5)P$_3$]

A key first step in the optimization of this assay was the titration of the binding partners used in the detection reaction in order to optimize the signal:background ratio while keeping the reagent concentration as low as possible for cost implications. Figure 5.3 shows the titration of both Det-PI(3,4,5)P$_3$ and biotinylated-PI(3,4,5)P$_3$ over a concentration range of 0.1 to 100 nM. As can be seen, the signal increased with increasing concentrations of the reagents up to 50 nM; after this concentration the signal reduced, probably as a consequence of the "hook effect." From these data, it was decided to choose a subhook concentration of binding partners, thus 10 nM of each partner was selected, which yielded approximately 60,000 counts.

5.5.4 Specificity of the Detection System

The ability of the detection system to differentiate product from substrate is of paramount importance. Therefore, competition curves were constructed to test this specificity as well as the binding of other phosphatidylinositols. As can be seen in Figure 5.4, PI(3,4,5)P$_3$ selectively inhibited the AlphaScreen signal with an IC$_{50}$ = 26 nM and a signal-to-background ratio of 16, thus demonstrating

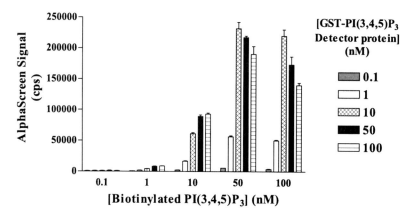

FIGURE 5.3 Binding partner titration for PI 3-K assay. Biotinylated-PI(3,4,5)P$_3$ and GST-PIP$_3$ detector protein were titrated together in order to identify optimal concentrations of each binding partner. The "hook point" was observed at 50 nM for each of the binding partners.

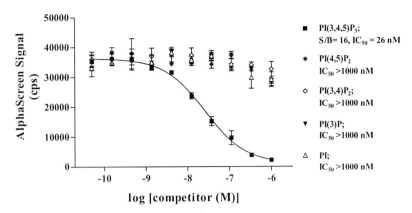

FIGURE 5.4 Specificity of the PI 3-K assay detection system. Increasing concentrations of phosphatidylinositol (PI), phosphatidylinositol 3''-phosphate [PI(3)P], phosphatidylinositol 4',5'-bisphosphate [PI(4,5)P$_2$], phosphatidylinositol 3',4'-bisphosphate [PI(3,4)P'] and phosphatidylinositol 3',4',5'-trisphosphate [PI(3,4,5)P$_3$] were incubated in presence of equimolar concentrations (in this case 5 nM/well) of biotinylated-PI(3,4,5)P$_3$ and GST-PIP$_3$ detector protein. AlphaScreen signal was detected after a 2-h incubation with detection reagents. PI(3,4,5)P$_3$ competed with the interaction between biotinylated-PI(3,4,5)P$_3$ and GST-PIP$_3$ detector protein, although all other phosphatidylinositols tested had no significant effect at the concentrations tested.

the utility of this system for the specific detection of PI(3,4,5)P$_3$ generation. The other phosphatidylinositols tested [phosphatidylinositol (PI), phosphatidylinositol 3'-phosphate (PI(3)P), phosphatidylinositol 4',5'-bisphosphate (PI(4,5)P$_2$) and phosphatidylinositol 3',4'-bisphosphate (PI(3,4)P$_2$)] all demonstrated no significant displacement of signal even up to 1000 nM.

5.5.5 STABILITY OF DETECTION SIGNAL OVER TIME

HTS applications often require that the plate is read minutes to hours after the final detection reagents have been added to the plate, and the stability of this signal often impacts on how the final screen is configured. To this end, it is important to assess the stability of the signal readout of a particular assay during its development in order to design the final screen appropriately. In this case, several standard curves were performed and read at various time points (30 to 300 minutes) postaddition of the final detection reagent. Figure 5.5a shows this series of standard curves, and Figure 5.5b shows a plot of various values derived from these curves. Clearly, the maximal signal (and signal-to-background ratio) increased with time, reaching a maximum at 240 minutes. The pIC$_{50}$ value, however, appeared to be stable from 60 minutes onwards. Thus these data suggest that for screening applications, plates either need to be read at consistently accurate time points when a read time between 30 and 240 min is selected, but for times beyond 240 min, perhaps more flexibility can be built into the accuracy of the final read time for each plate. These data further demonstrate that the final signal-to-background ratio can be indeed optimized by varying the final incubation time.

5.5.6 OPTIMIZATION OF ENZYME AND SUBSTRATE CONCENTRATIONS AND ENZYME REACTION TIME

Enzyme and substrate concentrations were optimized in titration studies in order to maximize the signal window. Figure 5.6a and Figure 5.6b show these titration curves for enzyme (PI 3Kα; obtained from and studies performed at Echelon BioSciences, Salt Lake City, UT) and substrate (PI(4,5)P$_2$). These studies revealed that 25 ng/well of enzyme and between 5 and 10 μM/well of substrate represented optimal concentrations. Using these conditions, a time course study was performed for the enzyme reaction (data not shown). Maximal signal was achieved between 60 and 90 min.

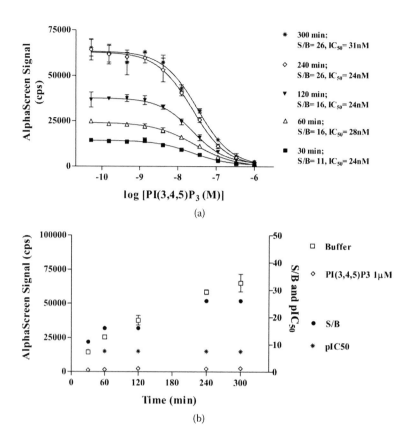

FIGURE 5.5 Stability of PI 3-K detection signal over time. (a) A series of competition curves for PI(3,4,5)P$_3$ was constructed, and the AlphaScreen signal was measured at various time points after the addition of detection reagents. These same data are plotted in (b), where maximal signal (buffer), minimum signal [PI(3,4,5)P$_3$; 1 μM], signal-to-background ratio [S/B = buffer − PI(3,4,5)P$_3$; 1 μM] and the −log IC$_{50}$ (pIC$_{50}$) are represented vs. time. The pIC$_{50}$ can be seen to stabilize rapidly (within 60 min) while the S/B increased over time and appeared to plateau at 240 min.

5.5.7 PHARMACOLOGICAL PROFILE AND TOLERANCE TO DMSO

The ability to identify the activity of known inhibitors of a screening target (where inhibitors are available) is a key step in validating an assay since it allows a screen to be conducted knowing that inhibitors can be detected at pharmacologically relevant concentrations. In this assay, two inhibitors that are known to inhibit PI 3-K activity were tested using the previously optimized assay conditions. Figure 5.7 shows the inhibition curves from this study. Both wortmannin (IC$_{50}$ = 2.5 nM) and LY-294002 (IC$_{50}$ = 4 μM) inhibited PI 3Kα with affinities consistent with those reported previously (Fuchikami et al., 2002). The assay system was further tested for its tolerance to DMSO concentration (data not shown). The assay demonstrated good tolerance up to 1 to 2% DMSO suggesting suitability for compound screening.

5.5.8 ASSAY PRECISION AND SUITABILITY FOR HTS

Once an assay has been optimized, it is important to test the precision of the assay and evaluate whether this precision provides suitable confidence to be able to differentiate kinase inhibitors from

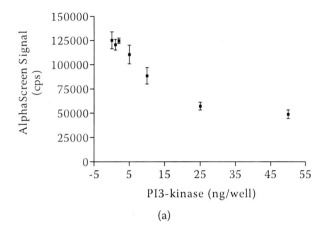

(a)

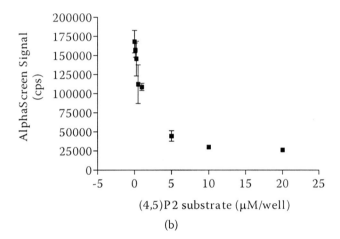

(b)

FIGURE 5.6 Optimization of PI 3-K enzyme and substrate concentrations. (a) PI 3-Kα concentration was titrated from 1 to 50 ng/well while the concentration of PI(4,5)P$_2$ substrate was fixed at 5 μM/well. The enzyme reaction was incubated for 90 min. The maximal response (AlphaScreen signal decrease) was achieved with 25 ng/well of PI 3-Kα. (b) PI(4,5)P$_2$ substrate concentration was titrated from 0.1 to 20 μM/well while the PI 3-Kα enzyme concentration was fixed at 25 ng/well with an enzyme reaction incubation time of 90 min. The maximal response was observed with 10 μM PI(4,5)P$_2$ substrate.

inactive compounds. Precision of screening assays is usually assessed by calculating the Z′ value from an experiment designed to measure signal differences between two defined conditions (e.g., enzyme activity in absence of inhibitor vs. activity in presence of inhibitor). A Z′ value greater than 0.5 in assay development is considered suitable for the transition of the assay to the screening environment. For this assay, kinase assay plates were set up containing 24 replicate wells of the following treatment conditions (all wells containing a 1% final concentration of DMSO) for the kinase reaction:

- Kinase buffer alone
- Wortmannin (30 nM)
- LY-294002 (30 μM)
- As a positive control, the maximal displacement of the AlphaScreen signal was tested in the presence of PI(3,4,5)P$_3$ (1 μM/well)

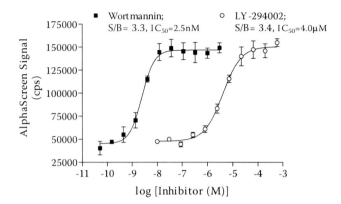

FIGURE 5.7 Pharmacology of PI 3-Kα measured using the PI 3-K AlphaScreen assay. Inhibition curves to known PI 3-K inhibitors were constructed by incubating PI 3-Kα enzyme (25 ng/well) and PI(4,5)P$_2$ substrate (5 μ*M*/well) with increasing concentrations of wortmannin and LY-294002. The incubation time of enzymatic reaction was 60 min, and the detection was performed as described previously. Signal-to-background ratios (S/B) and IC$_{50}$ values were calculated.

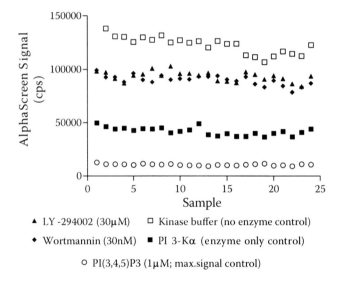

FIGURE 5.8 Assay precision and suitability for HTS. Using the assay conditions described in Figure 5.7, 24 replicate wells of various different treatment conditions (for the enzyme reaction) were tested as described in the graph. A maximal signal control was included by way of incubating with PI(3,4,5)P$_3$ (1 μ*M*/well) in the absence of enzyme or substrate. All wells contained a 1% final concentration of DMSO. Both the inhibitors wortmannin and LY-294002 yielded Z′ = 0.5 vs. the enzyme-only control, and the maximal-signal positive control using PI(3,4,5)P$_3$ produced a Z′ = 0.77.

The data are shown in Figure 5.8, where both inhibitors yielded Z′ = 0.5, and the positive control using PI(3,4,5)P$_3$ produced a Z′ = 0.77. These data suggest that this PI 3-K assay is suitably optimized and ready to transition to compound screening.

5.6 EXAMPLES OF ASSAYS DEVELOPED FOR OTHER KINASES

In this section, examples are described for other kinase assays that have been developed using AlphaScreen technology. In particular, indirect detection strategies are described for the serine/threonine kinases Akt (also known as Protein Kinase B) and ERK1 mitogen-activated protein kinases (ERK1), as well as a direct detection strategy that is described for the insulin receptor kinase domain (IRKD), a tyrosine kinase. In this section, final assay protocols are briefly described, together with some relevant assay development data.

5.6.1 Akt

Akt is a serine/threonine kinase and forms part of the PI 3K pathway. It plays a central role in mediating critical cellular responses including cell growth and survival, angiogenesis, and transcriptional regulation (Blanc et al., 2003; Ghosh et al., 2003). The pathway is activated by insulin as well as various other growth factors. Akt phosphorylates substrates at serine/threonine within a conserved motif characterized by arginine at positions -5 and -3. One such substrate for Akt is GSK-3 (glycogen synthase kinase-3), which is a ubiquitously expressed serine/threonine protein kinase that phosphorylates and inactivates glycogen synthase.

As was described earlier in this chapter, AlphaScreen kinase assays can be configured in either a direct or indirect manner. In this example, the indirect format has been used to develop a sensitive assay for the detection of Akt kinase activity and inhibition. A biotinylated peptide substrate (based on the GSK-3 sequence) was used in this assay (b-GSK-3), which when phosphorylated by Akt, was detected by an antiphosphoserine specific antibody [anti (pS) GSK-3]. This antibody–substrate complex, which accumulates as a consequence of substrate phosphorylation, results in bringing the donor beads and acceptor beads into close proximity via the interaction of b-GSK-3 with the donor beads and anti (pS) GSK-3 with protein A–coated acceptor beads.

The assay itself was performed in a final volume of 25 μL in white opaque 384-well plates. The assay was initiated by adding 5 μL of Akt enzyme and 5 μL inhibitor or assay buffer followed by a 15-min incubation at room temperature. After this inhibitor preincubation, 5 μL b-GSK-3/ATP mix was added, and the kinase reaction was allowed to proceed for 1 h at room temperature. The detection step was initiated by the addition of a 10 μL antibody/acceptor beads/donor beads solution (premixed for 1 h at room temperature) followed by a 1 h incubation. AlphaScreen signal was read on an AlphaQuest microplate analyzer.

During the development of this assay, a number of parameters were optimized. The assay buffers finally selected were as follows: kinase reaction, 25 mM Tris/HCl pH 7.5 containing 10 mM MgCl$_2$, 2 mM DTT, 100 μM Na$_3$VO$_4$, and 0.1% Tween-20; detection reaction, 25 mM Tris/HCl (pH 7.5) containing 200 mM NaCl, 100 mM EDTA, and 0.3% BSA. To determine the ATP concentration, titration curves from 10 μM to 1 mM were performed, and a concentration of 100 μM was selected as optimal. The anti (pS) GSK-3 concentration was selected at 1 nM and b-GSK-3 at 30 nM (although signal-to-background ratios >3 were observed using 0.3 nM). Enzyme amounts were studied over a broad range, where as low as 0.25 U/well enzyme generated 3000 cps (signal-to-background ratio = 5) and 10 U produced 220,000 cps (signal-to-background ratio = 350). Between 0.5 and 2.5 U were used routinely in assays. Figure 5.9a shows a time course for b-GSK-3 phosphorylation performed under these optimized conditions at two different enzyme concentrations. As can be seen, the data demonstrate a linear reaction for at least 2 h at both enzyme concentrations. Using the same two Akt concentrations, inhibition curves were constructed for the nonspecific inhibitor staurosporine (Figure 5.9b). IC$_{50}$ values of 420 and 180 nM were generated for 2.5 and 0.5 U of Akt, respectively. These data suggest that in this example, the magnitude of the assay signal can be "fine tuned" by optimizing enzyme concentration or the incubation time for the kinase reaction, but similar pharmacologies are observed even when, for example, different enzyme concentrations are selected.

5.6.2 MAPK

MAPKs represent important targets for the development of therapeutic agents for a wide spectrum of diseases, including neurodegenerative disorders, inflammation, cancer, and septic shock (Cobb,

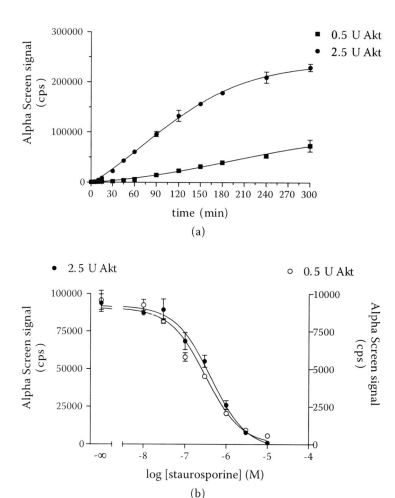

FIGURE 5.9 Characterization of an AlphaScreen assay for Akt kinase activity: Time course and pharmacological study. (a) Using the optimized conditions described in Section 5.6.1, a time course was performed for the phosphorylation of b-GSK-3 at two different Akt concentrations (0.5 and 2.5 U/well). (b) Using the same conditions and Akt concentrations as described in (a), the effect of the inhibitor staurosporine was tested in an inhibition curve.

1999). Mammalian MAPK signaling cascades regulate important processes, including gene expression, cell proliferation, and cell motility, as well as cell survival and death. The cascades include the two stress pathways, JNK and p38, as well as the best-characterized classical MAPK pathway leading to phosphorylation of ERK1 and ERK2 (p44 and p42 MAPK). Here, an AlphaScreen assay is described for the measurement of ERK1 activity.

The AlphaScreen Protein A assay kit was used again in this application, and soluble extracts from insect cells infected with Raf1/MEK1/ERK1 were used as the source of ERK1 active enzyme. A biotinylated peptide derived from myelin basic protein (b-MBP) was used as the substrate, and the resulting phosphorylated peptide was able to interact with the donor beads via the biotin group and with the acceptor beads by using an antiphospho-MBP antibody that was detected by the protein A–coated acceptor beads. Optimized buffers for this assay were as follows: kinase reaction, 8 mM HEPES (pH 7.4) containing 4 mM MgCl$_2$ and 0.24 mM DTT; detection reaction, 100 mM HEPES (pH 7.4) containing 100 mM EDTA and 0.25% BSA. The protocol for this assay was similar to the previous Akt example since both assays are based on indirect detection. Briefly, 5 μL of active

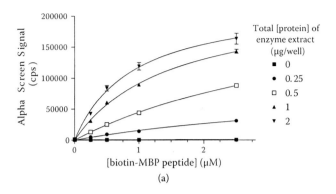

(a)

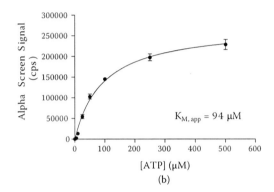

$K_{M, app}$ = 94 µM

[ATP] (µM)

(b)

FIGURE 5.10 Characterization of an AlphaScreen assay for ERK1 kinase activity: Substrate and enzyme titrations. (a) The effect of varying substrate (b-MBP peptide) and ERK1 enzyme (total extract) concentrations on the AlphaScreen signal were tested. Experiments were performed as described in Section 5.6.2. Apparent K_m values for b-MBP were calculated from these data and were 4.8, 4.4, 1.5, and 0.9 µM respectively for 0.25, 0.5, 1, and 2 µg/well of enzyme extract. (b) ATP concentration was titrated using conditions of 0.5 µM of b-MBP substrate and 1 µg/well of ERK1 extract. An apparent K_m = 94 µM was derived.

ERK1 extract (preincubated with inhibitor if required) was incubated with 10 µL b-MBP/ATP mix for 30 min, followed by the addition of 10 µL of detection mix (donor and acceptor beads and antiphospho-MBP [1 nM]) followed by a second incubation of 1 h at room temperature. AlphaScreen signal was measured on an AlphaQuest microplate analyzer.

In this example, data are shown for the optimization of substrate concentrations. Figure 5.10a shows a set of titration curves to the b-MBP substrate at different enzyme extract concentrations. The apparent K_m values for b-MBP that were calculated from these data were 4.8, 4.4, 1.5, and 0.9 µM for enzyme extract (total protein) concentrations of 0.25, 0.5, 1, and 2 µg/well, respectively. For assay sensitivity and linearity, substrate should be used at a concentration equal to or below its K_m, and the amount of enzyme should be as low as possible while maintaining a good signal-to-background ratio. In this example, 0.5 µM of peptide substrate and 1 µg/well of ERK1 extract were determined to be ideal. Under these conditions, a titration to ATP concentration was performed, and a K_m of 94 µM was derived (Figure 5.10b). A concentration of 50 µM of ATP was used for subsequent assays. Here, the data show how simple endpoint determinations of enzyme activity can be used to optimize substrate concentrations for AlphaScreen kinase assays.

5.6.3 INSULIN RECEPTOR KINASE DOMAIN (IRKD)

Protein tyrosine kinases play a key role in signal transduction and normal cell growth and are also involved in numerous disease states including cancer and atherosclerosis, as well as a number of autoimmune diseases. Here, the development of an AlphaScreen assay for the soluble β-insulin receptor kinase domain (IRKD) is described as an example of a protein tyrosine kinase assay. IRKD is derived from the cytoplasmic portion of the human insulin receptor, is constitutively active, and retains its substrate specificities, kinetic constants, and autophosphorylation sites without the requirement for hormone-mediated activation (Cheatham and Kahn, 1995). The assay was configured using the broad-spectrum substrate, poly[Glu:Tyr] (4:1), where the phosphorylated substrate was detected in this case via a direct interaction with acceptor beads that were conjugated with either antiphosphotyrosine P-Tyr-100 or PY20 antibodies.

For this assay, the kinase reaction buffer was composed of 50 mM Tris-HCl (pH 7.5) containing 5 mM MnCl$_2$, 4 mM MgCl$_2$, 2 mM DTT, and 0.01% Tween-20. The detection reaction buffer was composed of 62.5 mM HEPES (pH 7.4) containing 250 mM NaCl, 100 mM EDTA, and 0.25% BSA. The protocol involved the incubation of 5 μL purified IRKD with of 10 μL biotin-poly[Glu:Tyr]/ATP mix, followed by a 30 min incubation. A total of 10 μL detection mix (donor and P-Tyr-100 or PY20 acceptor beads) was then added; the mixture was incubated for 1 h at room temperature and finally read on an AlphaQuest microplate analyzer. In all experiments, substrates poly[Glu:Tyr] and ATP were used at 0.5 nM and 100 μM, respectively.

Here, data are shown demonstrating how the signal can be optimized for AlphaScreen tyrosine kinase assays by varying the enzyme concentration, incubation times, and choice of antibody. Figure 5.11a and Figure 5.11b show that the AlphaScreen signal increases linearly with increasing concentrations of IRKD enzyme and that the choice of the antibody used for detection can make a difference to the signal levels. In this example, P-Tyr-100–coated beads resulted in a 2.4-fold higher signal compared with PY20, when IRKD was used at 50 pg/well. These data also demonstrate that by varying the incubation time of the detection step (in this example, 1 vs. 18 h) allows for signal optimization (panel 5.11a vs. 5.11b). These data are represented as signal:background ratios in panels 5.11c and 5.11d, where it can be seen that signal:background ratios of 200 can be obtained under certain conditions. Overall, these data show that AlphaScreen tyrosine kinase assays can be optimized by varying enzyme concentrations, detection antibodies, and detection times and indeed, in the case of IRKD, a signal:background ratio of 11 could be obtained with a 1 h detection time using enzyme at levels as low as 5 pg/well.

5.7 SUMMARY

The application of AlphaScreen technology to the development of HTS assays for protein kinase targets has been described in this chapter. Clearly, the nonradioactive and homogeneous nature of AlphaScreen confers key advantages to kinase screening, but AlphaScreen also presents other advantages such as good sensitivity and signal-to-background ratios, low cost, and rapid assay development times. Indeed, the generic nature of the AlphaScreen platform allows many different kinase assays to be developed using the same detection strategy (e.g., protein A–coated acceptor beads). Further, the system also confers total flexibility in the final assay design where strategies using direct, indirect, or competition-based detection approaches; specific antibodies or selective binding proteins; specific or generic substrates; and purified recombinant or crude enzyme preparations can all be accommodated by AlphaScreen technology. The "fine tuning" of the final assay methodology has been described here by illustration with examples such as the ubiquitously expressed lipid kinase, PI 3-K, as well as Akt, ERK1, and IRKD, where the importance of the optimization of parameters such as the antibody/binding protein concentrations, signal stability, enzyme and substrate concentrations, and incubation times, as well as the specificity of the system to detect the product of interest has been stressed. Further, factors such as the ability of the final assay to be able to perform pharmacological studies, to demonstrate tolerance to DMSO, and to provide good assay precision are all factors that are key to providing the screening scientist with the confidence to transition the assay from the development laboratory into the screening environment.

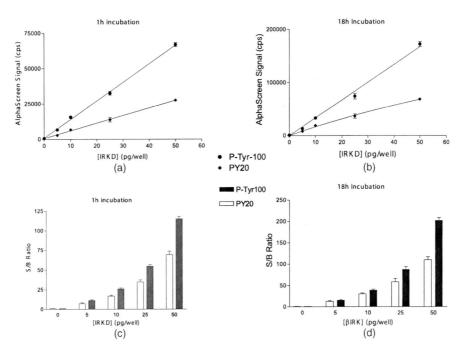

FIGURE 5.11 Characterization of an AlphaScreen assay for IRKD kinase activity: Effect of incubation time and choice of antibody. Titrations to IRKD enzyme concentrations (0 to 50 pg/well) were performed using the protocol described in Section 5.6.3, where detection was performed using acceptor beads coated with either P-Tyr-100 or PY20. The final detection reaction was incubated for either 1 (a) or 18 h (b). These data are represented as signal:background ratios in panels (c) and (d) for 1 and 18 h, respectively.

Thus, the AlphaScreen technology platform has become an established approach for the screening of protein kinases in many laboratories. The use of AlphaScreen for the detection of second messengers (e.g., cAMP) in HTS-based cellular assays for GPCR function is also commonplace, and the development of cellular-based AlphaScreen kinase assays either for readouts of receptor function or specifically for the kinase itself represents the potential next phase in the development of this versatile screening technology.

REFERENCES

Beaudet, L., Bedard, J., Breton, B., Mercuri, R.J., and Budarf, M.L. 2001. Homogeneous assays for single-nucleotide polymorphism typing using AlphaScreen. *Genome Res.* 11: 600–608.

Behr, T.M., Berova, M., Doe, C.P., Ju, H., Angermann, C.E., Boehm, J., and Willette, R.N. 2003. p38 mitogen-activated protein kinase inhibitors for the treatment of chronic cardiovascular disease. *Curr. Opin. Investig. Drugs* 4: 1059–1064.

Blanc, A., Pandey, N.R., and Srivastava, A.K. 2003. Synchronous activation of ERK 1/2, p38 and PKB/Akt signaling by H_2O_2 in vascular snooth muscle cells: potential involvement in vascular disease (review). *Int. J. Mol. Med.* 11: 229–234.

Cheatham, B. and Kahn, C.R. 1995. Insulin action and the insulin signaling network. *Endocr. Rev.* 16: 117–142.

Cobb, M.H. 1999. MAP kinase pathways. *Prog. Biophys. Mol. Biol.* 71: 479–500.

English, J.M. and Cobb, M.H. 2002. Pharmacological inhibitors of MAPK pathways. *Trends Pharmacol. Sci.* 23: 40–45.

Franke, T.F., Kaplan, D.R., and Cantley, L.C. 1997. PI 3K: downstream action blocks apoptosis. *Cell* 88: 435–437.

Fuchikami, K., Togame, H., Sagara, A., Satoh, T., Gantner, F., Bacon, K.B., and Reinemer, P. 2002. A versatile high-throughput screen for inhibitors of lipid kinase activity: development of an immobilized phospholipid plate essay for phosphoinositide 3-kinase. *J. Biomol. Screening* 7: 441–450.

Ghosh, P.M., Malik, S., Bedolla, R., and Kreisberg, J.I. 2003. Akt in prostate cancer: possible role in androgen-independence. *Curr. Drug Metab.* 4: 487–496.

Jones, W.K., Brown, M., Ren, X., He, S., and McGuinness, M. 2003. NF-kappaB as an instigator of diverse signaling pathways: the heart and myocardial signaling? *Cardiovasc. Toxicol.* 3: 229–254.

Kohno, M. and Pouyssegur, J. 2003. Pharmacological inhibitors of the ERK signaling pathways: application of anticancer drugs. *Prog. Cell Cycle Res.* 5: 219–224.

Kumar, S., Boehm, J., and Lee, J.C. 2003. p38 MAP kinases: key signaling molecules as therapeutic targets for inflammatory diseases. *Nat. Rev. Drug Discov.* 2: 717–726.

Shepherd, P.R., Withers, D.J., and Siddle, K. 1998. Phosphoinositide 3-kinase: the key switch mechanism in insulin signaling. *Biochem. J.* 333: 471–490.

Ullman, E.F., Kirakossian, H., Singh, S., Wu, Z.P., Irvin, B.R., Pease, J.S., Switchenko, A.C., Irvine, J.D., Dafforn, A., Skold, C.N., and Wagner, B.B. 1994. Luminescent oxygen channeling immunoassay: Measurement of particle binding kinetics by chemiluminescence. *Proc. Natl. Acad. Sci. USA* 91: 5426–5430.

Warner, G., Illy, C., Pedro, L., Roby, P., and Bosse, R. 2004. AlphaScreen kinase HTS platforms. *Curr. Med. Chem.* 11: 719–728.

Wymann, M.P., Sozzani, S., Altruda, F., Mantovani, A., and Hirsch, E. 2000. Lipids on the move: phosphoinositide 3-kinase in leukocyte function. *Immunol. Today* 21: 260–264.

Zhang J.H. 1999. A simple statistical parameter for use in evaluation and validation of high throughput screening assays. *J. Biomol. Screening* 4: 67–73.

6 Homogeneous High-Throughput Screening (HTS) Assays for Serine/Threonine Kinases Using β-Galactosidase Enzyme Fragment Complementation

Inna Vainshtein, Tabassum Naqvi, Anice Lim, Riaz Rouhani, Lindy Kauffman, Rajendra Singh, and Richard M. Eglen

CONTENTS

6.1   INTRODUCTION

Coordinated protein phosphorylation is a key mechanism in eukaryotic cell function [1]. This process is controlled by kinases, a protein group that represents approximately 2% of the human genome [2]. The kinases collectively form a group of enzymes that catalyze the transfer of the γ phosphate of ATP to amino acid residues on a protein substrate. There are 518 kinases, broadly grouped into 20 classes, based on structural similarities and substrate specificity [2]. Kinases regulate many aspects of cellular function, controlling growth, metabolism, differentiation, and apoptosis. Aberrant expression or function of kinases is associated with all aspects of neoplasia, including metastasis, diabetes, psoriasis, and liver fibrosis [3]. Kinases form two major groups: those that phosphorylate serine and/or threonine residues (Ser/Thr kinases), and those that phosphorylate tyrosine residues (Tyr kinases). Although kinases can be both located in the cytoplasm or on the cell surface, many participate in sequential signaling cascades that mediate cell responses to G-protein–coupled receptors (GPCRs) or receptors of the immune system [3]. Kinases, collectively, are a highly "druggable" class of targets, second in size only to GPCRs.

Inhibitor selectivity is the key issue in kinase drug discovery, as many enzymes are similar in their substrate recognition sites. Like many other cellular proteins, they also functionally bind ATP. Nonetheless, initial concerns relating to the high cellular ATP concentrations and the need to target inhibitors to the kinase ATP enzyme site *per se* appear to have been resolved. Indeed, both therapeutic efficacy and a low incidence of side-effects of several kinase inhibitors have now been clinically demonstrated [4]. Protein tyrosine kinase inhibitors, such as STI-571 (Gleevec) for the treatment of chronic myelogenous leukemia, ZD 1839 (Iressa), Y 27632 (Fasudil), and OSI 774 (Tarceva) for the treatment of solid tumors, are either approved or in development. Ser/Thr kinases inhibitors such as BAY 43-006, LY 333531, and CEP 1347, for the treatment of colon cancers, diabetic neuropathy, and Parkinson's disease, respectively, are also in clinical evaluation [4].

High-throughput assays for measurement of kinase activity are now extensively deployed in drug discovery to identify small molecule inhibitors. Historically, isotopic kinase assays, in which the transfer of radioactive ^{32}P or ^{33}P from ATP to the protein is measured following separation of the product from excess ATP by filtration, were the main approach to compound screening. This approach has the advantage of directly measuring substrate phosphorylation, in contrast to many formats of competitive immunoassays based on phosphopeptide-specific antibodies (see below). However, measuring isotopic phosphate incorporation is a laborious procedure that is difficult to automate. This issue, and the high cost of radioactive waste disposal emanating from high-volume screening, has driven extensive adoption of nonisotopic assays for kinase screening [5].

Early antibody-based assays, developed initially to measure activity of receptor tyrosine kinases, used enzyme-linked immunosorbent assays (ELISAs) [5]. These heterogeneous assays had an advantage in that compounds from a screening library did not directly interfere with the assay signal. However, the various washing and separation steps in the assay protocol meant that they were not easily automated. Consequently, homogeneous assays (i.e., assays in which all reactions are undertaken in the same liquid phase) have been most widely adopted for high-throughput automation systems. Most homogeneous assays employ fluorescent detection methods for kinase inhibitor screening, including time-resolved fluorescence (TRF), time-resolved fluorescence energy transfer (TR-FRET), protease-sensitive FRET, and fluorescence polarization [5]. However, they suffer from the disadvantage of requiring the use of specialized instrumentation for signal detection [5]. A limitation of many fluorescent-based assays, with the exception of TRF-based systems, is that library compounds may optically interfere with the assay signal. Collectively, this has elevated interest in a homogeneous approach to kinase screening in which a chemiluminescent signal is generated, thereby providing an assay technology that is easily automated, has sufficient sensitivity, and is markedly resistant to compound interference [6].

This chapter describes such a novel kinase assay technology that is based on enzyme fragment complementation (EFC), developed and commercialized under the trademark HitHunter™ by DiscoveRx Corp. The technology had been previously developed as an assay technology to measure GPCR function and protease activity [6,7]. Here, the first kinase assay format, described below, measures the kinase activity by assessing peptide substrate phosphorylation. In a similar fashion to other competitive kinase immunoassays, an antibody selective for the phosphorylated, over the nonphosphorylated, substrate is required. The second approach described is a nonantibody-based assay that directly measures the binding of small-molecule inhibitors to the kinase. The first assay format consequently has the advantage that inhibitors are detected by measuring changes in the function of the enzyme; the second provides a direct assessment of an inhibitor's potency at binding sites on the enzyme.

6.2 β-GALACTOSIDASE ENZYME FRAGMENT COMPLEMENTATION

Enzyme fragment complementation (EFC) is the process of restoring enzymatic activity by supplying a fragment of an enzyme to an otherwise inactive mutant. β-D-galactoside galactohydrolase, E.C. 3.2.1.23, or β-galactosidase (β-gal) from *Escherichia coli* is a tetrameric protein consisting of identical 116,000-Da subunits that catalyzes the hydrolysis of D-galactose linked via a β-glycosidic bond to the terminus of a polysaccharide chain. β-gal exhibits intracistronic complementation, i.e., it occurs between fragments encoded by the same gene as in, for example, the *lac Z* gene. As applied to the DiscoveRx assay technology, intracistronic complementation occurs when a peptide containing amino acids 3 to 92 (the α-donor peptide and denoted as "ED"; enzyme donor) restores activity to an otherwise inactive mutant lacking amino acids 11 to 42 (the α-acceptor protein and denoted as "EA"; enzyme acceptor) [6,7]. The use of different β-galactosidase substrates permits the development of homogeneous assays with either chemiluminescent or fluorescent signals that can be detected in simple plate readers found in many screening laboratories. Specifically, a chemiluminescent kinase assay markedly reduces interferences from compound libraries, a feature of the DiscoveRx assay technology that has been extensively exploited in several primary HTS campaigns to date.

The DiscoveRx complementation technology has now been deployed in several homogeneous assays [7–10] designed to measure concentrations of a wide variety of molecules, peptides, and proteins. These assays are designed to be scalable from large to small assay volumes routinely used in high-density microtiter plates. A key feature of assays of this type is that the signal generated by enzyme turnover is proportional to the concentration of labeled analyte to be detected. Consequently, measurement of very low levels of an ED-conjugated analyte is attained even without separation steps, i.e., in a homogeneous assay format.

6.2.1 β-GAL COMPLEMENTATION AND KINASE ASSAY DEVELOPMENT

The HitHunter™ Ser/Thr kinase assay utilizes an ED-phosphopeptide conjugate that is recognized by an antiphosphopeptide peptide antibody; the ED portion also retains its capability to complement to EA to form active β-gal enzyme and thus generate a signal following β-gal substrate hydrolysis. A typical assay is comprised of two stages: in the first the kinase, in the presence or absence of inhibitors, is allowed to phosphorylate the peptide substrate, and in the second the phosphopeptide produced by the kinase and the ED-phospho peptide conjugate compete for a limited number of binding sites on an antiphosphoserine peptide antibody. ED conjugate that is bound to the antibody is sterically hindered from complementing EA, causing a significant decrease in the amount of active β-gal and therefore a decreased signal. Consequently, the concentration of peptide product produced by the kinase is inversely proportional to the amount of ED conjugate bound to the antibody and directly proportional to the signal generated. This reaction scheme is shown in Figure 6.1.

The ED (enzyme donor)–ligand conjugate has the ability to bind either to a binding protein or EA (enzyme acceptor). When ED-ligand binds ("complements") to EA, fully competent β-gal is formed. This results in substrate hydrolysis and signal generation. This can be fluorescent or chemiluminescent, depending upon the β-gal substrate used in the assay. The ED-phosphopeptide binds to a selective antibody that recognizes the ED-phosphopeptide, as well as a nonconjugated peptide. This complex, aided by a suitable secondary antibody, sterically hinders the association of ED to EA, and no β-gal–dependent signal is generated. When the kinase reaction is allowed to proceed, a phosphopeptide is formed that competitively displaces the ED-phosphopeptide from the

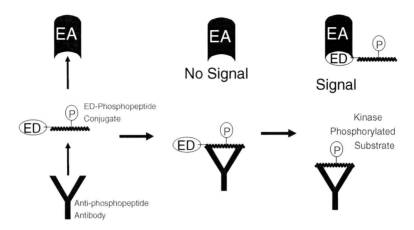

FIGURE 6.1 General principle of HitHunter Ser/Thr kinase assays using β-galactosidase (β-gal) complementation.

antibody. Once liberated, the ED-phosphopeptide then complements to EA and a signal is generated. In this manner, the resultant signal is directly proportional to the concentration of the phosphopeptide produced by the kinase. In applications of the technology to a HitHunter ED-NSIP assay, a similar principle applies, except here the binding protein is the kinase *per se*. ED is conjugated to a small-molecule inhibitor and a competitive binding assay is set up, similar to that described above for the ED-phosphopeptide. Here the term ligand refers to the small-molecule inhibitor. A critical aspect of the ED-NSIP assay is that ED must be attached to the small molecule in such a manner than the affinity of the compound for the kinase is unaffected.

The HitHunter ED-NSIP (ED-non specific inhibitor probe) approach also relies on inhibition of complementation, except in this case, the ED is chemically conjugated to a small-molecule inhibitor of the kinase, such as staurosporine. The conjugation allows the small molecule to retain high affinity for the kinase and the ED to retain its ability to complement EA. Similar to the antibody approach described above, ED conjugate binding to the kinase restricts access of the ED to EA, and thus complementation and assay signal are reduced. In the assay, free inhibitor molecules compete with the ED conjugate for binding sites on the kinase. The assay signal is directly proportional to the amount of free inhibitor bound by the kinase. The following sections describe the methods used in each of the assays and also include some prototypic data to illustrate the use of these assays in kinase screening.

6.3 HITHUNTER SER/THR KINASE ASSAY

6.3.1 ASSAY DEVELOPMENT

Development of a typical HitHunter Ser/Thr kinase assay starts with the determination of optimal concentrations of reagents and their sequence of addition. Figure 6.2 outlines the assay protocols for the 384-well format of the Ser/Thr kinase assay (left panel) and the ED-NSIP assay (right panel). In the Ser/Thr kinase assay, the initial step involves a competitive binding reaction, wherein kinase-produced phosphorylated serine peptide and ED conjugated to phosphorylated serine peptide compete for antibody binding sites. The reagents are incubated for 1 h to allow the establishment of a competitive equilibrium, followed by the addition of signal detection reagents (i.e., EA and substrate), and then a final 1-h incubation. The resulting signals are read using a standard chemiluminescent plate reader [11]. A typical standard curve for the Ser/Thr kinase assay is shown in Figure 6.3, in which there is a concentration-dependent increase in signal over a wide phosphopeptide concentration (2 to 200 nM).

EFC Kinase assay

10 μL **Kinase diluted in Antibody Reagent**
10 μL **Substrate & ATP in kinase reaction buffer**
Incubate 1 hour at RT
20 μL **ED/CL Working Rgt**
Incubate 1 hour at RT
10 μL **EA Rgt**
Incubate at least 1 hour at RT
Read

EFC Kinase binding assay

20 μL **Kinase inhibitor**
10 μL **Kinase**
10 μL **ED probe in kinase reaction buffer**
Incubate 1-3 hours at RT
25 μL **EA and substrate**
Incubate at least 1 hour at RT
Read

FIGURE 6.2 Prototypic assay protocols for HitHunter Ser/Thr kinase functional and ED-NSIP kinase binding assays. In essence the assay protocol consists of a kinase reaction step and an EFC (enzyme fragment complementation) detection step since several reagents are pooled in the assay for simultaneous additions (left panel). Generation of a phosphopeptide standard curve employs antibody binding only and thus includes the phosphopeptide standard, antibody, and ED-phosphopeptide conjugate alone. In a kinase assay, after optimization of the antibody concentration, ATP concentration, etc., the reagents are combined as shown. An analogous protocol is followed in the ED-NSIP assay, which comprises a kinase binding step followed by an EFC detection step (right panel). In each case, the EFC detection step remains the same in that it comprises addition of EA and β-gal substrates (chemiluminescent or fluorescent). The total time for an assay of this nature is generally 2 to 3 h and the assay can be configured for 96-, 384-, or 1536-plate densities.

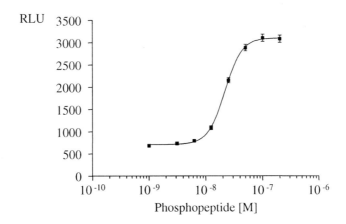

FIGURE 6.3 Prototypic standard curve for a HitHunter Ser/Thr kinase assay using protein kinase Cα (PKCα). Here, serial dilutions of a phosphopeptide standard (20 μL), antiphospho antibody (10 μL), and ED-phosphopeptide (10 μL) were incubated at room temperature for 1 h. Subsequently, EA and chemiluminescent substrate were added (20 μL) to generate signal and the plate read after 1 h using either Victor-2 multireader (PE Life Sciences) or Packard (Packard BioSciences) luminometers. All values shown are mean ± s.e. mean, n = 3, and the experiments were performed in 384 plates.

Typically, the standard curve (containing serial dilutions of phosphopeptide) is used to develop the product detection phase of the kinase assay. As mentioned previously, the product detection phase can be considered in two parts, a competition for antibody-binding sites and signal amplification. During the competition phase, phosphorylated peptide substrate competes with ED labeled phosphosubstrate for antibody binding sites. In assay development, the relative concentrations of ED conjugate and antibody are titered for optimal assay performance, and volume of delivery and incubation time are optimized. The titrations allow the identification of reagent concentrations that

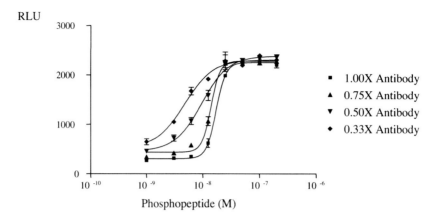

FIGURE 6.4 HitHunter Ser/Thr kinase assay standard curve in the presence of four different concentrations of antibody.

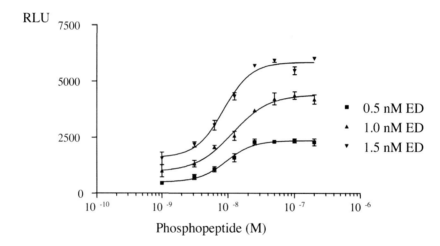

FIGURE 6.5 HitHunter Ser/Thr kinase assay standard curve in the presence of three different concentrations of ED-phosphopeptide conjugate.

provide optimal assay sensitivity and performance. Generally, lower antibody concentrations lead to increased assay sensitivity, but with a concomitant loss in signal-to-background ratio (S/B). Figure 6.4 shows typical antibody titration data. Decreases in ED concentration may increase S/B ratio; however, the decrease must be balanced against the need for sufficient ED for effective complementation. Figure 6.5 contains an example of an ED titration. Antibody binding affinity, the antibody's relative affinity for labeled vs. unlabeled substrate, and the requirement for sufficient ED for complementation mean that optimal concentrations for antibody and ED must be derived experimentally, rather than empirically.

In the final phase of the assay, unbound (free) ED conjugate complements EA, creating active β-gal enzyme, which then hydrolyzes a chemiluminesent substrate. The addition of EA reagent must readjust the assay environment such that conditions are optimal for complementation and substrate hydrolysis. Buffer salts, additives, and pH of the EA reagent are optimized for maximum β-gal activity. In addition, the final EA concentration must be determined. EA concentration is titered to ensure a robust signal, but since EA binds ED with relatively high affinity, EA concentrations cannot be excessive, lest the EA compete for antibody-bound ED and not just the free ED,

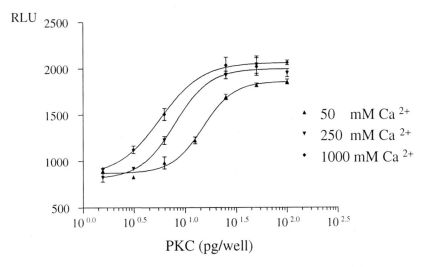

FIGURE 6.6 A PKC-α enzyme concentration–response curve in the presence of three concentrations of calcium.

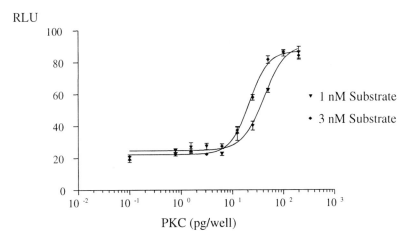

FIGURE 6.7 A PKC-α enzyme concentration–response curve in the presence of two different concentrations of kinase substrate.

as desired. The optimized assay also contains chemiluminescent substrate in excess; substrate should not be limiting. In a well designed EFC assay, antibody-binding sites are limited; all other components are present in optimal concentrations or excess to ensure maximum activity of the kinase, and maximum complementation and hydrolysis of the signal substrate.

Once the product detection phase of the assay is optimized, the kinase reaction conditions must be considered. Kinase reaction conditions are kinase specific, but minimally must include titration of the required components, ATP and substrate. Figure 6.6 shows the effect of substrate titration. In addition, broadly applicable compounds such as magnesium and sodium chloride and kinase-specific co-factors such as phospholipids must be considered. Figure 6.7 shows an example of a calcium titration. All additives must be titered for maximum kinase activity, while ensuring that the assay environment and particularly the final pH and ionic strength are compatible with subsequent assay steps. In general, relatively weak buffering solutions are chosen for the kinase reaction so that subsequent reagent additions return the assay environment to one optimal for EFC.

TABLE 6.1
Precision Characteristics of the HitHunter Ser/Thr Kinase Assay

	Chemiluminescence	Fluorescence
Signal-to-background ratio	5.3	4.9
Z'-factor	0.7	0.6
EC_{50} (nM)	22	21

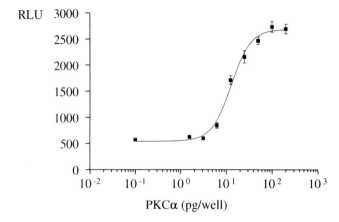

FIGURE 6.8 A PKC-α enzyme concentration–response curve. The kinase reaction and antibody-binding step were combined in the assay, as shown in Figure 6.2. Specifically, the assay buffer, phospholipid activators, antibody, and PKCα peptide substrate were added as a single reagent (20 μL), and serial dilutions of PKCα enzyme, followed by addition of ED-phosphopeptide and ATP, which were used to initiate the kinase reaction. This was allowed to proceed for 90 min in the presence of 4.5 μM kinase peptide substrate and 15 μM ATP. Subsequently, detection reagents were incubated for additional hour and the plates then read as described in Figure 6.3. All values shown are mean ± s.e. mean, n = 3, and the experimentd were performed in 384 plates.

As a final consideration, the assay must also be assessed for its applicability to high-throughput screening (HTS), and performance characteristics such as intra- and inter-plate variation, signal-to-background ratio, replicate imprecision, and Z'-factors must be determined. The HitHunter Ser/Thr assay characteristics are summarized in Table 6.1 and indicate that the assay will provide robust and reproducible performance in the HTS laboratory.

The assay format is flexible. Rather than sequential additions of the kinase reactants, followed by the signal detection reagents, some steps may be combined by pooling compatible assay components. Figure 6.8 shows an example of a typical kinase effector curve, generated using a protocol similar to the one outlined in Figure 6.2. Kinase inhibitors or assay buffer were initially pipetted, followed by a mixture of the kinase substrate peptide, phospholipid kinase activators, and the antibody (reagent 1). This step is followed by serial dilutions of the kinase (reagent 2), incubation, and finally the addition of the signal detection reagents (reagent 3). Figure 6.8 shows that increasing concentrations of PKC-α between 5 and 100 pg/well generated an enzyme concentration–response curve with a signal-to-background ratio of 5 when the reaction was run at room temperature for 90 min. If the assay were used for an inhibitor screen, with a fixed concentration of kinase enzyme (typically 80% of maximal activity, as determined by the kinase titration), the kinase could be combined with reagent 1 and ATP omitted. The phosphorylation reaction would then be initiated by addition of ATP and ED-conjugate (reagent 2).

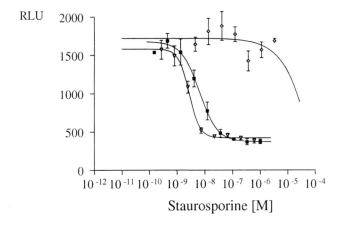

FIGURE 6.9 A staurosporine concentration–effect curve. The PKCα kinase reaction was undertaken using 4.5 μM kinase peptide substrate and 15 μM ATP at room temperature 90 min. Staurosporine were added and pooled compatible reagents then added according to the protocol in Figure 6.2. All values shown are mean ± s.e. mean, n = 3, and the experiments were performed in 384 plates.

TABLE 6.2
Comparison of Inhibitor Potencies (IC$_{50}$, nM) Determined for PKCα in HitHunter Ser/Thr Kinase Assays, HitHunter ED-NSIP Binding Assay with Literature

Inhibitor	HitHunter Ser/Thr Kinase IC$_{50}$	IC$_{50}$ Reported	HitHunter ED-NSIP IC$_{50}$
H-7	10,000	6000	42,000
Staurosporine	13	28	8
Ro 32-0432	28	9	6.3
Go 6976	2.8	2.3	ND

Note: ND, not determined.

The HitHunter Ser/Thr assay was also validated using several known inhibitors of PKCα. Figure 6.9 shows that the inhibitor staurosporine reduced the amount of phosphopeptide product fivefold. Other potent inhibitors, including Go-6976 and Go-0432, as well as weak inhibitors, such as H-7, also inhibited PKCα activity, collectively with potencies in good correspondence with the literature (Table 6.2).

As discussed above, one version of the HitHunter Ser/Thr assay utilizes a PKC-pseudosubstrate. Consequently, it was anticipated that other PKC kinases could be active in the assay. Figure 6.10 shows that a panel of PKC isoforms can indeed be quantified in this way. Furthermore, kinases are promiscuous with regard to substrate specificity and phosphorylate several nonphysiological substrates *in vitro*. Table 6.3 shows that the kinases PKA, Chk1, Chk2, cdk2, and AMPK can also be assayed using the HitHunter approach. Subsequently, it has been found that other kinases, belonging to the CMGC or CaMK families, also possess the ability to phosphorylate the PKC pseudosubstrate and can be assayed using the HitHunter Ser/Thr approach.

Nonetheless, some HTS kinase assays require more "physiological" substrates, i.e., peptide sequences more similar to the endogenous peptide target. The use of these more specific substrates in homogeneous assays requires identification of new "selective" antiphosphoantibodies. For this purpose, two additional substrate–antibody pairs for detection of phosphorylated syntide-2 and Trp^4-kemptide have now been identified at DiscoveRx Corporation for the HitHunter Ser/Thr kinase

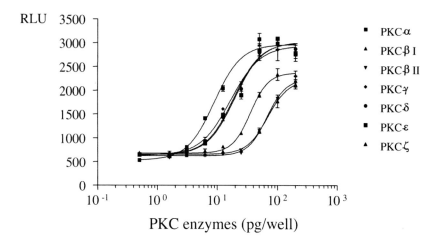

FIGURE 6.10 A panel of PKC kinases screened in the HitHunter Ser/Thr kinase assay. The assays were run in 384-well plates using conditions indicated in Figure 6.4. (Note that the phospholipid activators were not employed for the experiments with PKC-ζ). All values shown are mean ± s.e. mean, n = 3, and the experiments were performed in 384 plates.

TABLE 6.3
Ser/Thr Kinase Potencies Determined in HitHunter Ser/Thr Kinase Assay

	CHK1	CHK2	Cdk2	PKA	AMPK
EC_{50} (*M*)	4.2×10^{-10}	8.6×10^{-10}	4.7×10^{-9}	1.7×10^{-12}	1.5×10^{-9}
Signal-to-background ratio	3.2	3.1	4.1	4.1	4.0

format. A HitHunter assay has been developed using an affinity-purified polyclonal antiphosphosyntide antibody, and its utility has been demonstrated with CaMK II kinase. A second polyclonal antibody against a phosphokemptide substrate was also used to measure phosphorylation of kemptide by both PKA and a novel oncogenic kinase, Aurora A.

6.4   HITHUNTER ED-NSIP (NON-SPECIFIC INHIBITOR PROBE) ASSAYS

The identification of antiphospho antibodies selective for other Ser/Thr kinase substrates is critical for many nonisotopic kinase assays, including those discussed in this chapter. This is particularly relevant for homogeneous kinase assays that require high antibody affinities and specificity. However, relatively few such antibodies are readily available, and the range of kinases now known to be present in the human kinome is large. One solution to this issue has been the development of a fluorescence polarization (FP) method based on affinity capture of phosphorylated substrates using derivatized trivalent metal ions [12]. This technology has the advantage of being nonantibody based; however, it has a disadvantage that high ATP concentrations cannot be used, as other negative ions are captured in addition to phosphopeptides [5]. Moreover, acidic residues in the substrate bind nonspecifically to the metal ion [13], limiting general applicability of the method. DiscoveRx has developed a kinase assay technology that requires neither antibody nor substrate and which employs the kinase *per se* as a binding protein. Direct binding of small molecule inhibitors has been used with radiolabeled compounds, such as staurosporine; however, a nonisotopic method has

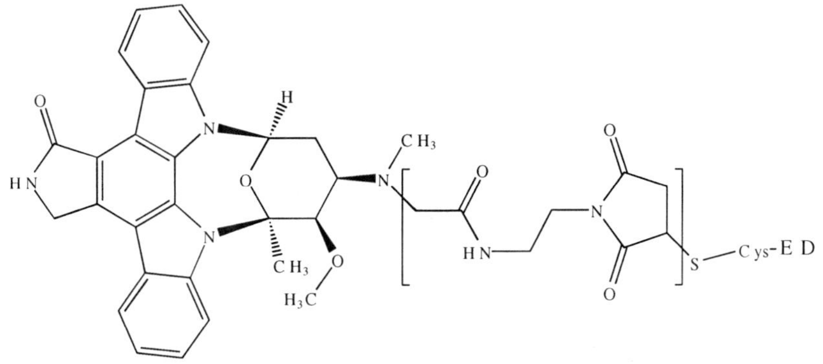

Staurosporine   2-Maleidoethylcarboxymethyl   ED
Linker

FIGURE 6.11 Structure of ED-staurosporine conjugate using a maleidoethylcarboxymethyl linker.

previously been unavailable for HTS applications. This aspect of the technology is extensively covered in a previous publication [10].

Many small molecule inhibitors affect kinase function by directly binding either to the kinase ATP binding site or to the catalytic site [13]. DiscoveRx has designed a proprietary binding probe approach that can be used to directly measure the binding of putative inhibitors. Resolution of a structure for PKA kinase complexed with staurosporine [13] suggests that staurosporine binds to the ATP-binding pocket between amino- and carboxyl-lobes of the enzyme. Other kinases also appear to coordinate staurosporine in the ATP-binding site in close proximity to the exterior of the kinase. These data suggest that staurosporine could also be conjugated to ED via a suitable space linker (Figure 6.11), thus allowing development of a probe for kinase-binding assays. Other kinases that recognize staurosporine may contain an ATP pocket located deeper in the protein. Consequently, staurosporine has been linked to ED via linkers of variable length to allow optimal binding of ED-conjugated staurosporine to these kinase binding sites.

The HitHunter ED-NSIP assay employs ED-staurosporine as binding ligand in a competitive assay to measure inhibitor binding potency. Initially, the approach was validated with PKCα using staurosporine as a free inhibitor. This was a useful example to validate the assay, given the high affinity of the compound for this enzyme [13]. Figure 6.12 shows the displacement curve of ED-staurosporine by unconjugated staurosporine, and Table 6.2 summarizes binding efficiencies of Ro 32-0432 and H-7 compounds. The potencies determined in the assay corresponded to potencies obtained in other kinase assays. The ED-NSIP approach has now been extended to other kinases, including GSK3α kinase [7] and p38 MAP kinase (Figure 6.13). Comparison of the potencies of several GSK3α inhibitors with a functional ^{33}P incorporation assay and the ED-NSIP assay showed a high degree of correspondence [10]. A feature of this assay format is that it allows for measurement of inhibitor potency with either the active or the inactive form of the kinase. Similar binding assays for other enzymes may be established using different small-molecule inhibitors conjugated to ED.

It is unclear at this time how generally the ED-NSIP approach is applicable for other kinases; this will depend on the differing orientations of the ATP-binding pocket and the structure–activity relationships of the ED-conjugated small-molecule inhibitors. At present, experience suggests that the affinity for the small-molecule inhibitor should be 30 nM or higher to achieve a binding assay that requires low amounts of enzyme. DiscoveRx has recently synthesized a variety of linkers conjugated to staurosporine so that Ser/Thr kinases with ATP-binding pockets deeply buried in the protein can be assessed. Beyond simple inhibitor screening, the two HitHunter approaches provide counter screens to allow one to address the mechanism of action of a putative kinase inhibitor, i.e., to determine whether the site of inhibitor action involves binding to the ATP-binding pocket. In

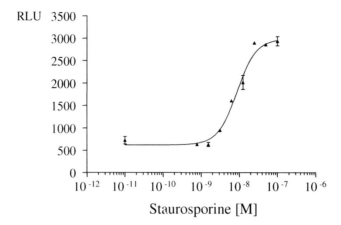

FIGURE 6.12 Competitive displacement of ED-staurosporine from PKCα by unconjugated staurosporine. The assay was undertaken according to the protocol in Figure 6.2. Staurosporine potently displaced ED-staurosporine from PKC kinase with an EC_{50} value of 9 nM, consistent with the literature.

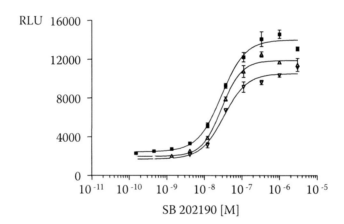

FIGURE 6.13 Competitive displacement of ED-SB 202190 from p38-MAP kinase by unlabeled SB 202190 compound. The assay was undertaken according to the protocol in Figure 6.2. SB 202190 potently displaced ED-SB202190 from P38 MAP kinase with values consistent with the literature.

some cases, this approach may be also used to detect inhibitors acting at sites other than the ATP-binding site, if functional inhibition is seen but lack of binding observed in the DiscoveRx assay.

6.5 HITHUNTER OPTIMIZATION GUIDE

In the HitHunter EFC products, such as the HitHunter Ser/Thr kinase assay, the detection phase is already optimized for sensitivity and signal to background ratio; however, as with all kinase assays, the kinase reaction must be optimized for each new kinase. And since HitHunter is a homogeneous assay, the kinase reaction conditions must be optimized in ways compatible with the subsequent signal detection phase. First, the kinase reaction buffer must be considered. For HitHunter Ser/Thr kinase, a HEPES-based assay buffer is provided, which contains magnesium chloride, ethylene glycol bis(2 amino ethyl ether) N,N,N,N-tetra acetic acid (EGTA), and sodium chloride. These basic additives are usually required for kinase activity; the buffer provides an appropriate starting point for optimization of the kinase environment. Depending on the kinase being studied, additional additives such as calcium and phospholipids may be required and should be spiked into or diluted in the Ser/Thr assay buffer. For fastidious kinases, an entirely different buffer composition may be required, but any alternate buffer should be compared to Ser/Thr assay buffer in a standard curve to ensure that it does not affect detection phase performance. High concentrations of some excipients, such as glycerol and DTT, may have a negative impact on assay performance.

All kinase reactions require the presence of ATP and substrate. Ideally, the substrate concentration should be near its K_m for the kinase, and the ATP in sufficient concentration to support unlimited phosphorylation but not so high as to influence inhibitor binding. The substrate and ATP should be diluted in the assay buffer, and the concentrations titered for optimal kinase activity.

Typical studies to optimize the kinase reaction start with the buffer optimization, followed by testing multiple concentrations of substrate and ATP with serial dilutions of kinase (kinase dilution curve). The selected concentrations of substrate and ATP should be those that give the lowest EC_{50} for the kinase dilution curve. Kinase reaction time and temperature should also be evaluated to ensure best conditions for kinase activity. Once the kinase reaction conditions are optimized, an inhibition curve should be generated (serial dilutions of a known inhibitor, such as staurosporine) with the kinase concentration fixed at approximately the EC_{80} of the kinase dilution curve. The inhibition curve can be used to assess assay "screenability," including such factors as signal to background ratio, IC_{50}, and Z′ factor.

Depending on the results of the inhibitor curve and the ultimate use of the assay, no further optimization of the kinase reaction may be necessary. However, for best assay performance, some fine-tuning may be necessary at this point. If the signal-to-background ratio is lower than desired, one should consider the quality of the kinase preparation, including kinase purity, and whether the kinase has been preactivated (raising the basal level). Slight increases in kinase concentration may improve the signal-to-background ratio. If the inhibitor IC_{50} is higher than expected, decreasing the ATP concentration may help left-shift the curve. Precision may also be improved with attention to order of kinase component addition and volume of component delivery. In general, volumes greater than 5 μL can be delivered more precisely, and larger volumes of delivery provide more effective mixing.

Once the kinase reaction has been optimized, there are several points to consider for overall assay performance. As mentioned previously, the standard curve can be used to verify that any modifications or additions to the assay buffer (made during kinase reaction optimization) do not have a negative affect on the assay performance. A "no-ED control" (as described in the package insert) should be run to verify that none of the reagents or additions has been contaminated with ED or β-galactosidase. Temperature is also important for consistent assay performance. Just as the kinase reaction may vary with changes in temperature, the EFC reagents perform best when used at room temperature consistently. Use of cold reagents or wide swings in ambient temperature can cause day-to-day variations in signal. With just a few well-designed optimization studies, the HitHunter EFC assays, such as HitHunter Ser/Thr kinase, can be used with a wide variety of kinases and provide a rugged and robust screening assay HTS laboratory.

6.6 SUMMARY

Taken together, chemical conjugation of ED to phosphopeptides or small-molecule inhibitors provides a flexible approach to kinase assay development, allowing a range of homogeneous assays to be formulated. A critical feature of this "homogeneous ELISA" approach is that the binding protein (enzyme or antibody) binds both conjugated and unconjugated ligand. Experience with several assay types developed at DiscoveRx indicates that ED is a facile peptide label that minimally perturbs binding to these proteins, thus allowing a great degree of flexibility.

The use of β-gal complementation to generate an amplified signal allows the assay to be conducted in microtiter plate wells, with a high degree of volume scalability. Indeed, several assays have been constructed that have been adapted to 96-, 384-, or 1536-microtiter plate volumes. Since a chemiluminescent signal can be generated in the assay, it provides an option to screen compound libraries that exhibit optical interference with other fluorescent-based approaches. Varying the antibody and ED-phosphosubstrate pairs used in the assay allow several assays to be rapidly configured for different kinases. Thus, assays for tyrosine kinases that use an approach similar to those described above have also been commercialized by DiscoveRx.

ACKNOWLEDGMENTS

The authors wish to acknowledge their colleagues at DiscoveRx Corp. in the development of the kinase assays, including Neil Charter, Peter Fung, Sanga Giertych, Pyare Khanna, and Sharon Zhao. The authors also wish to thank Joy Concepcion for preparation of the figures.

REFERENCES

1. Cohen, P., Timeline: Protein kinases — the major drug targets of the twenty-first century? *Nature Rev Drug Discovery*, 2002, 1, 309
2. Manning, G., Whyte, D.B., Martinez, R., Hunter, T., Sudarsarnam, S., The protein kinase complement of the human genome, *Science*, 2002, 298, 1912.
3. Dancy, J., Sausville, E.A., Issues and progress with protein kinase inhibitors for cancer treatment, *Nature Rev Drug Discovery*, 2003, 2, 296.
4. Levitski, A., Protein kinase inhibitors as a therapeutic modality, *Acc Chem Res*, 2003, 36, 462.
5. Zaman, G.J.R., Garritsen, A., de Boer, T., van Boeckel, C.A.A., Fluorescence assays for high-throughput screening of protein kinases, Combin. Chem. & HTS, 2003, 6, 381.
6. Eglen, R.M. and Singh, R., β-Galactosidase enzyme fragment complementation as a novel technology for high throughput screening, *Combin. Chem. & HTS*, 2003, 6, 313.
7. Coty, W.A., Loor, R., Powell, M., Khanna, P.L., CEDIA® homogeneous immunoassays: current status and future prospects, *J. Clin. Immunoassay*, 1994, 17, 144.
8. Golla, R., Seethala, R., A homogeneous enzyme fragment complementation cyclic AMP screen for GPCR agonists, *J. Biomol. Screen.* 2002, 7, 515.
9. Zhao, X, Vainshtein, V., Gellibolian, R., Shu, Y., Dotimas, H., Wang, X-M.,Peter Fung, P., Horecka, J., Bosano, B.L., Eglen, R.M., Homogeneous assays for cellular protein degradation using β-galactosidase complementation: NF-κ B/I B pathway signaling, *Assay Drug Dev.*, 2003, 6, 823.
10. Vainshtein,I., Silveria S., Kaul,P., Rouhani,R., Eglen,R.M., Wang, J., A high-throughput, nonisotopic, competitive binding assay for kinases using nonselective inhibitor probes (ED-NSIP™), *J. Biomol. Screen.*, 2002, 7, 497.
11. Eglen, R.M., Enzyme fragment complementation: a flexible high-throughput screening assay technology, *Assay Drug Dev.*, 2002, 1, 97.
12. Loomans, E.E.M.G., van Doornmalen, A.M., Wat, J.W.Y., Zaman, G.J.R., High-throughput screening with immobilized metal ion affinity-based fluorescence polarization detection, a homogeneous assay for protein kinases, *Assay Drug Dev.*, 2003, 1, 445.
13. Davies, S.P., Reddy, H., Caivano, M., Cohen, P., Specificity and mechanism of action of some commonly used protein kinase inhibitors, *Biochem. J.*, 2000, 1, 95.

7 IMAP Assays for Protein Kinases

Richard Sportsman

CONTENTS

7.1 INTRODUCTION

IMAP™ is a proprietary assay technology developed by Molecular Devices Corporation (MDC) for high-throughput screening (HTS) of kinases, phosphatases, and phosphodiesterases. IMAP provides a non-radioactive, nonantibody-based alternative for quantitative analysis of protein kinases. The key element is the IMAP "binding reagent," based on nanoparticles derivatized with trivalent metals, M(III), which bind tightly to phosphate groups, e.g., phosphopeptides and phosphoproteins (Figure 7.1). IMAP derives its name from "IMAC" (immobilized metal affinity chromatography), to which it is similar in the sense that it relies on high affinity recognition by immobilized metals. However, the IMAP nanoparticles are used not as a solid phase but rather in a solution-phase homogeneous technique based on fluorescence polarization (FP) for detection [1,2].

IMAP is a "mix-and-read" assay that is highly amenable to HTS applications. It can be scaled to 96-, 384- or 1536-well plate formats. Importantly, IMAP does not rely on antibodies for recognition, giving the technique broad applicability to Ser/Thr and tyrosine kinases and phosphatases. The IMAP kinase assay is a two-step process: enzymatic reaction, followed by binding of the reaction products. As shown in Figure 7.1, by measuring the fluorescence polarization (FP), the amount of phosphopeptide formed by the kinase can be measured and therefore, the kinase activity can be determined.

In this section we will illustrate the use of IMAP in assay development for HTS of kinases and in determination of IC$_{50}$ values for inhibitors. In addition, a method for determination of apparent K$_m$ and K$_i$ values is provided.

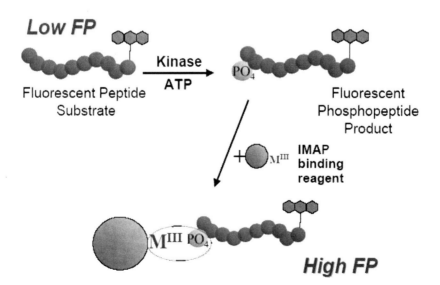

FIGURE 7.1 Principle of the IMAP kinase assay. A fluorescent peptide substrate becomes phosphorylated in the presence of ATP and kinase. The reaction is terminated and quantified by addition of the IMAP binding reagent, a nanoparticulate trivalent metal preparation. The binding of the phosphorylated substrate to the nanoparticles results in an increase in its fluorescence polarization (FP). The FP of the unphosphorylated substrate remains low because it does not bind to the particles.

7.2 ASSAY DEVELOPMENT WITH IMAP

The assay development process is similar to that used for other formats of kinase assays. First, the kinase is titrated against a fixed concentration of substrate and ATP (typically 100 nM and 50 μM, respectively, for IMAP). This determines the optimal range of kinase concentration to be used in subsequent work. The ATP concentration and reaction times can then be studied, and if necessary, further adjustments to kinase concentration can be undertaken. In the following example, we will use the Akt kit from MDC to illustrate titration of kinase and determination of an IC$_{50}$ for an inhibitor. Essentially the same procedure can be used for any substrate and kinase, including those not specifically available as kits from MDC. In the latter case one can use the basic IMAP reagents (buffers and binding reagent) available separately as the Screening Express product or through the IMAP purchase program. A newer form of the IMAP reagents, the Progressive Binding System, provides additional flexibility in assay design [3].

7.2.1 MATERIALS

The following materials are provided with the MDC Akt assay kit. The first three reagents are similar for all IMAP assays and can be obtained separately as mentioned above. The progressive binding system is another option, as mentioned above, and is discussed at the end of this chapter:

- IMAP Binding Reagent (do not freeze the Binding Reagent).
- IMAP Binding Buffer (5×), store at 4°C.
- IMAP Reaction Buffer (5×), store at 4°C. The Complete Reaction Buffer consists of a 1× stock with added DTT (1 mM final reaction concentration).
    - The 1× Reaction Buffer (made from the supplied 5× concentrated stock) contains 10 mM Tris-HCl, 10 mM MgCl$_2$, 0.1% bovine serum albumin (BSA), 0.05% NaN$_3$, pH 7.2. Consult Step 7.2.2.1 below for composition of Complete Reaction Buffer. Other components that can be added without affecting the IMAP system are Mn^{2+}, Ca^{2+}, DTT, 2-mercaptoethanol, certain detergents, and NaCl. The IMAP system can tolerate up to 400 μM EDTA or 40 μM EGTA without significant signal inhibition. Phosphate and structurally related molecules may compete at various affinities with the phosphopeptide for binding to the IMAP Binding Reagent; for this reason their use is not recommended. BSA may bind to and interfere with the action of some inhibitory compounds. If this is the case, phosphate-free bovine gamma globulin (BGG) or detergents such as Tween-20 or Triton X-100 may be substituted for the BSA in the above reaction buffer formulation.
- Akt (Upstate Biotechnology #14-276), store at −70°C. Aliquot if necessary to avoid multiple freeze/thaw cycles. See the Certificate of Analysis enclosed with the kit for specific activity of the supplied lot.
- Fluorescein-labeled Akt Substrate (FAM-Crosstide, Molecular Devices Corp.): (5FAM)-GRPRTSSFAEG. Lyophilized substrate that makes 20 μM when reconstituted in 2.5 mL (store at 4°C).

7.2.2 METHOD FOR TITRATION OF KINASES

7.2.2.1 Prepare the Complete Reaction Buffer

A. Make a 1× solution of Reaction Buffer by adding 40 mL of purified water per 10 mL of the supplied 5× concentrated IMAP Reaction Buffer. When stored at 4°C, the 1× solution of Reaction Buffer is stable for 6 months.

B. To make the Complete Reaction Buffer, add DTT to a final concentration of 1 mM in the 1× solution of Reaction Buffer. Prepare Complete Reaction Buffer fresh for each day of assay. Approximately 15 mL is required per 384 wells; one can scale the volumes up or down as needed.

7.2.2.2 Prepare the Substrate Working Solution*

A. Add 2.5 mL Complete Reaction Buffer to the lyophilized substrate vial to make a 20 μM fluorescent substrate solution.
B. Vortex gently and invert vial to make sure that all of the lyophilized substrate goes into solution.
C. Add 0.1 mL of the 20 μM solution per 4.9 mL of Complete Reaction Buffer to make a 400 nM substrate working solution. This solution is 4× the final reaction concentration of 100 nM substrate.*

7.2.2.3 Prepare the ATP Working Solution*

A. Add 50 μL of a 10 mM stock of ATP to 1200 μL of Complete Reaction Buffer to make a 400 μM ATP stock solution.
B. Add 0.25 mL of this 400 μM stock per 4.75 mL of Complete Reaction Buffer to make a 20 μM working solution. This solution is 4× the final reaction concentration of 5 μM ATP.*

7.2.2.4 Prepare Any Enzyme Inhibitors/Stimulators

A final concentration of up to 7.5% DMSO is tolerated in the 20 μL Akt IMAP reaction.

7.2.2.5 Assay Template

A. An example of a template for setting up an enzyme dilution curve with three replicates is shown in Table 7.1.
B. An example of inhibitor testing with three replicates is shown in Table 7.2. We suggest using a concentration of enzyme that gives 70% of the maximum response for this purpose.

7.2.2.6 Prepare the Enzyme Working Stock Solution

A. Calculate concentration according to the enzyme-specific activity given in the Upstate Certificate of Analysis. For example: 2 μg enzyme in 50 μL with specific activity of 1000 units per mg provides 40 units/mL, where 1 unit is defined as 1 nmol phosphate incorporated into substrate per minute.
B. To prepare an enzyme dilution curve as shown in Figure 7.2 (Step 7.2.2.7B), make a 60 μL stock of approximately 2.40 units/mL in Complete Reaction Buffer, and serially transfer 20 μL of this stock to 40 μL of Complete Reaction Buffer. The resulting dilutions of 2.4, 0.80, 0.27, 0.089, 0.030, and 0.010 units/mL are 4× the final reaction concentrations of 0.60, 0.20, 0.067, 0.022, 0.007, and 0.002 units/mL.

---

* Substrate and ATP may be premixed and added to the reaction as a single aliquot. See Step 7.2.2.7C.

TABLE 7.1
A Template for Kinase Titration

Column →	1	2	3	4	5	6	7
Row ↓							
A B C	0.600 units/mL	0.200 units/mL	0.067 units/mL	0.022 units/mL	0.007 units/mL	0.002 units/mL	Buffer Only control

TABLE 7.2
A Template for Testing of Inhibitor Series

Column →	1	2	3	4	5	6	7
Row ↓							
A B C	Enzyme + inhibitor 1	Enzyme + inhibitor 2	Enzyme + inhibitor 3	Enzyme + inhibitor 4	Enzyme + inhibitor 5	Enzyme + inhibitor 6	Enzyme + inhibitor 7

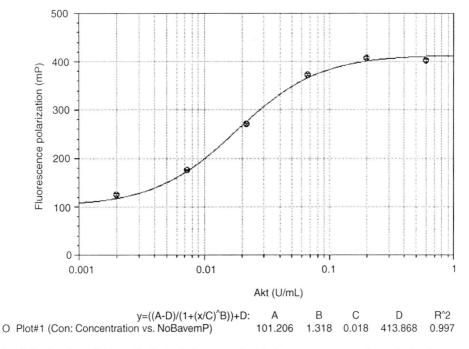

$y=((A-D)/(1+(x/C)^B))+D$:	A	B	C	D	R^2
O Plot#1 (Con: Concentration vs. NoBavemP)	101.206	1.318	0.018	413.868	0.997

FIGURE 7.2 Titration of Akt with the IMAP system. In black 384-well assay plates, 5 μL of enzyme at various concentrations was reacted in a final volume of 20 μL with 5 μM ATP and 100 nM fluorescein-labeled Crosstide substrate. After 45 minutes, 60 μL of IMAP binding reagent was added. Polarization was measured in an Analyst AD multimode reader (Molecular Devices). Enzyme concentrations are those of the 20 μL reaction prior to addition of the binding reagent. Error bars, representing one standard deviation, are less than the circle symbols and are obscured.

7.2.2.7   Add Components to the 384-Well Assay Plate

A. Add 5 μL of any inhibitors/stimulators prepared in Step 7.2.2.4 or add 5 μL Complete Reaction Buffer (for enzyme dilution curve as shown in Step 7.2.2.6B) to the appropriate wells.
B. Add 5 μL of each enzyme dilution prepared in Step 7.2.2.6A or Step 7.2.2.6B to the appropriate wells. For inhibitor testing, use the predetermined EC70 enzyme dilution and incubate at this point as needed at room temperature to allow for interaction with the enzyme.
C. Add 5 μL of the Substrate Solution prepared in Step 7.2.2.2 to the appropriate wells. If ATP and substrate are to be added together, do so in a volume of 10 μL.
D. Add 5 μL of the ATP Working Solution prepared in Step 7.2.2.3 to the appropriate wells.
E. For the Buffer Only background control, add 20 μL of Complete Reaction Buffer to the appropriate wells. Each well of the assay should now have 20 μL volume.

7.2.2.8   Cover the Plate and Protect from Light

Incubate at room temperature for 60 min. You may need to optimize reaction time for your individual needs.

7.2.2.9   Prepare Sufficient IMAP Binding Solution

A. Make a 1× solution of Binding Buffer by adding 40 mL of purified water per 10 mL of the supplied 5× concentrated IMAP Binding Buffer. When stored at 4°C, the 1× solution of Binding Buffer is stable for 6 months.
B. Dilute the IMAP Binding Reagent 1:400 into the 1× solution of Binding Buffer. Store on ice until needed. This solution should be prepared fresh on each day of assay. For example, if you have prepared the enzyme dilution curve in Step 7.2.2.6B, add 4 μL of Binding Reagent per 1.6 mL of 1× Binding Buffer.
C. Add 60 μL of Binding Solution to each assay well, including the Buffer Only wells.

7.2.2.10   Cover the Plate and Protect from Light

Incubate at room temperature for 30 minutes. Longer incubation times can provide a slight increase in response.

7.2.2.11   Measure the Fluorescence Polarization (FP)

On the Analyst AD, HT, or GT, the suggested settings include:

- Continuous lamp
- Excitation fluorescein 485 nm to 20 fwhm
- Emission 530 nm to 25 fwhm
- Fluorescein 505 dichroic
- Z-height 3 mm
- Attenuator out
- SmartRead or Comparator
- Sensitivity 0
- Integration time of 100,000 μsec

7.2.2.12   Analyze Your Results

- Calculate the average background (= Buffer Only wells) for both S and P fluorescent intensity data.
- Subtract the background value from both S and P raw data.
- Calculate FP and plot FP against enzyme concentration.

7.2.3   RESULTS

Figure 7.2 is the titration curve for Akt by the IMAP assay. A large increase in polarization (>300 mP units) occurs over the range of enzyme concentrations, reflecting the conversion of all fluorescent substrate to its phosphorylated form at the highest enzyme concentrations. Considering that the average standard deviation of all points is 2 mP, the S/N for the data of Figure 7.2 is over 150. Another figure of merit is the so called z' factor defined as

$$z' = 1 - \left| \frac{3 * \sigma_{hi} + 3 * \sigma_{lo}}{\hat{\mu}_{hi} - \hat{\mu}_{lo}} \right| \tag{7.1}$$

where $\hat{u}_{hi}$ and $\hat{u}_{lo}$ represent the average values of high and low assay controls, and σ_{hi} and σ_{lo} are the corresponding standard deviations of these values [4]. A value for z' of unity represents a perfect assay, and a value of 0.5 or better is considered to be a "good" assay, one in which active compounds may be expected to be detected with confidence between the limits defined by the assay controls. For the data in Figure 7.2, a fair estimate for z' is 0.92, taking the value at the midpoint of the curve (0.022 U/mL Akt) as the high control and the value at the bottom of the curve (no enzyme) as the low control.

To use the IMAP system to detect or characterize inhibitors, Step 7.2.2.7 is explicitly used to add either known inhibitors (perhaps at various concentrations or as a dilution series, to determine an IC_{50}) or compounds from a screening collection. For the Akt enzyme, we assayed inhibition of kinase activity by a dilution series of the known inhibitor staurosporine. From the enzyme titration data of Figure 7.2, we chose an enzyme concentration of 0.05 U/mL. As shown in Figure 7.3, an

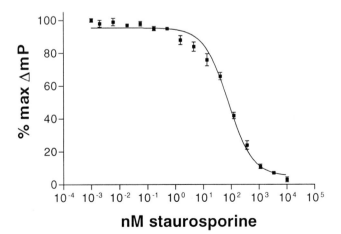

FIGURE 7.3 IMAP assay for determination of the IC_{50} for a known kinase inhibitor acting on the Akt kinase. Akt was assayed as in Figure 7.2, except that its concentration was fixed at 0.05 U/mL, and staurosporine was added in threefold serial dilutions starting at 10,000 n*M*. Max mP (100%) = 364; min mP (0%) = 158.

IC_{50} of 78 nM is determined for staurosporine's activity against Akt in these conditions, close to results determined by other methods [5].

7.2.4 SUMMARY

This section has demonstrated several important advantages of the IMAP platform. IMAP is easy to accommodate in assay development, given the advantages that FP has (ratiometric readout, homogeneous assay) as well as the antibody-free format. IMAP reagents tolerate DMSO to over 30%, so that the limiting concentration of DMSO is always determined by the kinase itself, as here, where Akt showed some inhibition at DMSO concentrations above 7.5%. Other features of IMAP include high signal levels of fluorescence (resistance to fluorescent and colored interferences), direct response to phosphate (coordinate covalent bonding vs. indirect detection through antibody), and very high stability of the end point (25 days at least).

7.3 DETERMINATION OF K_m AND K_i VALUES IN IMAP

In this section we describe the steps needed to determine K_m values for ATP, as well as K_i values for inhibitors for kinases using the IMAP system. These steps are similar to those employed for other techniques for assay of kinases and other enzymes. In the case of protein kinases, the kinetics are complicated by the dual-reactant nature of the system. Typically one will determine apparent K_m values (or $K_{m,app}$ values) for ATP and substrate* at fixed concentrations of the other reactant; a true K_m for the first reactant is considered possible to determine only if the second reactant is held at a concentration well above its K_m.

IMAP generally must be run at peptide substrate concentrations well below K_m because significant fractional substrate conversion must occur in order to get good signal. This complicates the enterprise in two ways. First, it is not possible to set peptide substrate concentration well above K_m, as mentioned above. Second, there is substrate depletion going on, so the simplifying assumptions of Michaelis–Menten kinetics are not strictly applicable. Therefore, the mathematics of the enterprise can be difficult to represent explicitly [6]. Here we present practical solutions to determination of these kinetic parameters.

In order to determine K_i for a competitive inhibitor, one must know the K_m value for the reactant that is competed against.

7.3.1 DETERMINATION OF K_m FOR ATP

The steps for determining the K_m for ATP are similar in IMAP to those for determination by other methods. As noted above, we cannot easily do this under conditions of excess substrate, but we can get a good approximation of the apparent K_m ($K_{m,app}$) by varying ATP concentration in the usual manner:

7.3.1.1 Determine the Kinase's Approximate EC_{50} in the Presence of Relatively High ATP

Follow the above IMAP protocol in Section 7.2 using 100 to 200 nM substrate and 50 to 100 μM ATP, and use a 30- to 60-min reaction time.

---

* In this section we refer to the two substrates for kinases as follows: "substrate" denotes phosphoacceptor (e.g., fluorescent-labeled peptide) and "ATP" refers of course to the phosphodonor (ATP). The term "reactant" will be used to refer to both collectively.

7.3.1.2 Using the Conditions and EC$_{50}$ Kinase Concentration Determined in the Preceding Step, Run Several Reactions in Which ATP Is Varied

For example, it may be varied from 0.25 to 128 μM in doubling increments.

7.3.1.3 Finish the Assay in the Usual Manner with IMAP Binding Solution

Measure the fluorescence polarization in the prescribed fashion.

7.3.1.4 Plot mP vs. ATP

The data can be analyzed by inspection or by curve-fitting programs to determine K_m.

7.3.1.5 Cautions

Plotting mP values vs. [ATP] to determine $K_{m,app}$ by nonlinear least squares analysis (or transforms of these parameters for Lineweaver–Burke plots or similar graphical methods) assumes that the mP response is linear in percent phosphorylation. This is often a good enough approximation, but in some cases there can be deviations from linearity. Phosphopeptide/substrate mixtures can be made to calibrate the mP response, so that data can be expressed as reaction velocities or product concentrations. In that case a more precise method can be followed as below:

A. Convert mP value to fraction substrate reacted, thence to rate, using a calibration curve. Figure 7.4 shows an example of an IMAP calibration curve. This is constructed by measuring the IMAP response of a series of known mixtures of phosphopeptide (i.e., product) and substrate form of a kinase substrate. Here the sum of both species' concentrations is kept constant at 100 nM, as it would be in a kinase reaction, and ATP is added at a typical concentration used in the assay to mimic reaction conditions as closely as possible.

B. With a calibration curve, one can use the equation of the fitted line to interpolate any values from an actual assay using a spreadsheet. Prepare a plot of the polarization data (mP) on the y-axis vs. percent phosphorylation (%P) on the x-axis, as in Figure 7.4. If the data are adequately fit by a linear least squares fit then Equation (7.2) results

$$\%P = (mP - b)/m \tag{7.2}$$

where m and b are slope and y-intercept, respectively, of the linear least squares fitted line. Concentration of product can then be determined as

$$[P] = (\%P/100) * [S]_0 \tag{7.3}$$

where $[S]_0$ is the starting substrate concentration. Finally, rate is computed simply by dividing the product concentration by reaction time.

C. If the calibration curve is not adequately fitted by a simple linear regression, then often a quadratic will do, or one can derive an exact equation from intensity and polarization values of pure substrate and product forms [7,8].

D. Plot rate vs. [S] (or 1/rate vs. 1/[S] for Lineweaver–Burke plots). The data can be analyzed by inspection or by curve-fitting programs to determine $K_{m,app}$ for substrate.

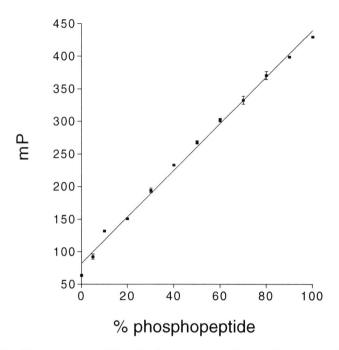

FIGURE 7.4 IMAP calibration curve. IκKβ calibration curve with 100 nM FAM-IκBα-derived peptide/phosphopeptide (products R7254 and R7301, resp., Molecular Devices Corp). Various calibrator solutions representing increasing percent phosphorylation were made by mixing volumes of 100 nM peptide and 100 nM phosphopeptide solutions in appropriate ratios. ATP was added to make 20 μM as in an actual Akt assay. To 20 μL of these calibrator solutions was added 60 μL of binding solution. The binding solution was the Progressive system with 75% A, 25% B, and a 1/600 dilution of the binding reagent. Linear regression was done using Prism 3.0 (Graphpad Software, San Diego, CA).

7.3.1.6 Some Formulations of IMAP-Binding Reagents May Show Inhibition by ATP at Concentrations Greater Than 50 μM

This will be clear if the plot of ATP vs. mP shows a decrease at higher ATP concentrations, but it may be obscured if inhibition closely offsets an actual increase in response. The new Progressive Binding System [3] increases IMAP system capacity and should allow concentrations of ATP over 200 μM without difficulty. However, when possible it is best to check the response of the system artificially by running the corresponding phosphopeptide form of the calibrator in the presence of the maximum intended ATP concentration. If this is not available, run the kinase reaction to completion with excess kinase and evaluate the effect of ATP at the target maximum concentration.

7.3.2 RESULTS

Figure 7.5 is a Lineweaver–Burke representation of data from the determination of the $K_{m,app}$ of ATP for the Akt kinase. In this case a single assay plate was sufficient to analyze reactions at multiple concentrations of peptide substrate. The intersection of the lines at a point below the x-axis (i.e., at x = –0.3), rather than directly on the x-axis, is indicative of a negative *cooperativity* of ATP and substrate that is consistent with a random sequential bisubstrate mechanism frequently seen with protein kinases [9]. In this case $K_{m,app}$ is then –[1/(–0.3)] or 3 μM ATP.

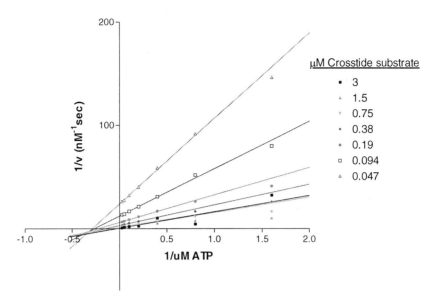

FIGURE 7.5 Determination of $K_{m,app}$ for ATP. Akt at 0.08 U/mL reacted for 20 min with a matrix of different concentrations of ATP and fluorescein-labeled Crosstide substrate. The data were converted from mP to rate (nM Crosstide phosphorylated/sec) as discussed in Section 7.3.1.2 through Section 7.3.1.5 and Figure 7.4, and plotted as Lineweaver–Burke plots. The intersection below the x-axis may be taken as evidence of random bireactant kinetics with negative cooperativity, and K_m in this case is the point on the x-axis above this intersection. (From Segel, IH. *Enzyme Kinetics: Behavior and Analysis of Rapid Equilibrium and Steady-State Enzyme Systems.* Wiley Interscience, New York (1993), pp. 278–281.) Thus, $-1/K_{m,app}$ for ATP is about –0.32, or $K_{m,app} = 3.1 \; \mu M$.

7.3.3 DETERMINATION OF COMPOUND K_i

In principle, once you have established a value for $K_{m,app}$ of a reactant for which your inhibitor competes, you may determine the K_i from IC_{50} data using the Cheng–Prusoff formula for competitive inhibition:

$$IC_{50} = K_i \, (1 + [S]/K_m) \tag{7.4}$$

7.3.3.1 Use the Optimized Conditions and Kinase Concentration Determined in Section 7.3.1.1

Choose an ATP concentration to give the best compromise between assay window and sensitivity of detection of inhibitor. Higher ATP means larger assay window and shorter reaction times. Setting ATP to be less than or equal to its K_m means greater sensitivity to inhibition (lower IC_{50} values, assuming competition with ATP).

7.3.3.2 Preferably, the Maximum Extent of Reaction Should Not Exceed 50%

That is, keep to the original idea of using the enzyme EC_{50}. This is because systematic error in IC_{50} and K_i values increases above the EC_{50}.

7.3.3.3 Assay the Inhibitor at Concentrations Encompassing Its Expected IC_{50} by a Factor of 10 or More

Typically, dilutions of twofold or threefold are sufficient. If the calibration curve is known to be linear, then simply determine IC_{50} by NLLS analysis of a plot of mP vs. concentration of inhibitor. If there is significant nonlinearity of the calibration curve, convert the mP values to percent phosphorylation (as in Section 7.3.1.5) before performing the fit.

7.3.3.4 The K_i Should Then Be Computed from Equation 7.4

Substitute [ATP] for [S] if competitive with ATP.

7.4 CONCLUSION

The IMAP system has grown in popularity since its introduction 3 years ago. Several features of IMAP make it attractive: no antibody, universality of phosphate recognition that is essentially direct, homogeneous assay format, and high fluorescence intensities that produce high precision of mP values and thus high Z' factors [10].

Here we have seen that IMAP is also easy on assay development, and can produce K_m and K_i values. Although IMAP is run at high percent conversion (25 to 75% reaction of substrate), it has been shown to produce values for IC_{50} that are in agreement with other methods [11]. We have assessed the effect of this experimentally and theoretically and determined that, below 70% conversion of substrate, the effect on IC_{50} is less than 50%. This result is in agreement with those reported by others [6].

REFERENCES

1. Huang W, Zhang Y, Sportsman JR. (2002) A fluorescence polarization assay for cyclic nucleotide phosphodiesterases, *J Biomol Screen* 7, 215–222.
2. Gaudet EA, Huang K-S, Zhang Y, Huang W, Mark D, Sportsman JR. (2003) A homogeneous fluorescence polarization assay adaptable for a range of protein serine/threonine and tyrosine kinases, *J Biomol Screen* 8, 164–175.
3. Sportsman JR, Gaudet EA, Boge A. (2004) Immobilized metal ion affinity-based fluorescence polarization (IMAP): advances in kinase screening. *Assay Drug Dev Tech* 2, 205–214.
4. Zhang JH, Chung TD, Oldenburg KR. (1999) A simple statistical parameter for use in evaluation and validation of high throughput screening assays. *J Biomol Screen* 4, 67.
5. Parker GJ, Law TL, Lenoch FJ, Bolger RE. (2000) Development of high throughput screening assays using fluorescence polarization: nuclear receptor-ligand-binding and kinase/phosphatase assays. *J Biomol Screen* 5, 77–88.
6. Wu G, Yuan Y, Hodge CN. (2003) Determining appropriate substrate conversion for enzymatic assays in high-throughput screening. *J Biomol Screen* 8, 694–700.
7. Rajkowski J, Cittanova C. (1981) Corrected equations for the calculation of proteinligand binding results from fluorescence polarization data. *J Theoret Biol* 93, 691–696.
8. Sportsman JR, Boge A, Gaudet EA. (2004) Developing Calibration Curves for IMAP: IMAP app note #4. Molecular Devices Corporation (http://www.moleculardevices.com/).
9. Segel, IH. *Enzyme Kinetics: Behavior and Analysis of Rapid Equilibrium and Steady-State Enzyme Systems*. Wiley Interscience, New York (1993), pp. 274–281.
10. Beasley JR, Dunn DA, Walker TL, Parlato SM, Lehrach JM, Auld DS. (2003) Evaluation of compound interference in immobilized metal ion affinity-based fluorescence polarization detection with a four million member compound collection. *Assay Drug Dev Tech* 1, 455–459.
11. Adams C, Gaudet EA, Boge A. (2003) Assay tutorial: nonradioactive FP-based IMAP assay for kinases, phosphatases, and phosphodiesterases. *Genet Eng News* 23, 38–39.

8 A Homogeneous, Luminescent, High-Throughput, Versatile Assay for a Wide Range of Kinases

Said A. Goueli, Kevin Hsiao, and Bob Bulleit

CONTENTS

8.1 SUMMARY

Phosphotransferases represent a major group of cellular enzymes that utilize ATP as substrate and are implicated in a wide variety of cellular functions such as cellular proliferation, differentiation, and apoptosis. Their substrates encompass a wide range of chemical structures such as proteins, peptides, sugars, and lipids. Because of the significance of these enzymes to cellular functions, they have become valid targets for drug development. The assay we developed (Kinase-Glo® Luminescent Kinase Assay) can be used to monitor the activity of protein kinases, lipid kinases, sugar kinases, sphingosine kinases, etc. The assay is homogeneous, nonradioactive, fast, robust, and amenable to high-throughput applications.

8.2   INTRODUCTION

Protein kinases play a major role in a wide variety of cellular functions and thus represent a very important target for drug discovery [1,2]. The 518 protein kinases in the human genome are involved in phosphorylation of 30% of cellular proteins [3]. In addition, there are many other phosphotransferases that play equally important roles in cellular function that utilize ATP as substrate but do not belong to classical protein kinases. These include inositol phosphate kinases such as phosphoinositide 3-kinases (PI3 kinases) [4], lipid kinases such as sphingosine kinases [5], and sugar kinases such as glucokinase [6]. Screening of kinase inhibitors for the development of new therapeutics has successfully borne fruit with the FDA approval of Gleevec (STI-571) for treatment of chronic myelogenous leukemia (CML), and ZD 1839 (Iressa) for lung cancer. The successful outcomes of the development of those drugs have led many pharmaceutical companies to heighten their search for additional kinase inhibitors that might prove useful in the development of future drugs for a variety of human ailments and diseases [7]. Since the substrates for protein kinases vary from large proteins and small peptides to sugars or lipids, for some kinases that do not recognize protein or peptide substrates, it has become evident that currently known protein kinase assays will not have universal application. Most of the commercially available assays use peptides as substrates with the exception of radioactive assays, which can also use proteins as substrates. Furthermore, few of the current assays are capable of using sugars or lipids as substrates. Here we introduce a new assay that is applicable to a wide variety of substrates such as peptides, proteins, lipids, and sugars. The assay is based on consumption of ATP by the phosphotransferase and monitoring kinase activity by the decrease in ATP using luciferase. Thus, the difference in luminescence output before and after the kinase reaction is a measure of the enzyme activity. The assay is homogeneous, nonradioactive, fast, sensitive, robust, and amenable to high-throughput screening. In addition to its diversity in using any substrate available for the kinase of interest, its strength is manifested by the capability to assay for kinases whose substrates are prephosphorylated such as glycogen synthase 3 kinase (GSK-3) or kinases that phosphorylate their substrates on multiple sites, such as IKKs.

8.3   AN OVERVIEW OF CURRENT KINASE ASSAYS

Protein kinases represent a class of enzymes that transfer inorganic phosphate from their universal substrate, ATP, onto a diverse group of substrates resulting in a phosphorylated product. The first generation of kinase assays utilized radiolabeled [γ-^{32}P]ATP, and the radioactive product was quantified using liquid scintillation counting. The system was developed into high-throughput format using biotinlyated peptide substrates; the phosphorylated biotinlyated peptides were captured on 96-well streptavidin biotin capture membrane (SAM®) plates [8] or SAM membranes in ultrahigh throughput format [9]. Scintillation proximity assay (SPA) is another platform that makes use of the proximity of the radiolabeled peptide substrate to a scintillant-laden matrix, resulting in an enhanced production of photons that can be quantified [10]. Although these technologies are very reliable, sensitive, and robust, they suffer from the hazards of using radioactive material and the cost of their disposal. This led to the development of nonradioactive assays that make use of fluorescently labeled peptides (substrates/products) to monitor enzyme activities by following changes in intensity or polarization of these peptides. The fluorescence polarization strategy relies on the availability of high-affinity, high-selectivity antibodies for the phosphopeptide, and thus its performance is dependent on their availability. The change in polarity of the fluorescently labeled peptide upon phosphorylation and consequent binding to antibodies is a measure of the activity of the kinase, and thus it can be used to monitor the activity of the kinase [11]. A modified version of this approach was introduced recently, where the phosphopeptide product was allowed to bind to a resin instead of an antiphospho antibody, which increased the polarization of the phosphopeptide product upon binding [12] or caused quenching of a fluorophore-laden matrix [13]. More recently, an amplified luminescence proximity homogeneous assay (AlphaScreen) was introduced. It relies on signal amplification of light emitted when phosphorylated peptides on the donor beads are brought in close proximity to antibodies containing acceptor beads upon phosphorylation of the peptides [14]. We have recently introduced a fluorogenic (Profluorescent) kinase assay (ProFluor™ kinase assay) that is sensitive to very low enzyme concentration, is homogeneous and robust, and does not require antibodies [15].

Each of the above-described technologies has limitations: radioactivity-based assays suffer from personnel exposure to radioactive material and the cost of disposal of radioactive isotopes; fluorescence-based assays are hampered by fluorescent compound interference, which can be a problem if the concentration of the substrate is very low (nM), as is the case with fluorescence polarization; phosphobinding resins are limited by the nonspecific binding of peptide substrates that contain acidic amino acids, which are negatively charged and thus interfere with binding of phosphopeptides to the phosphate binding matrices. Most of these assays have issues with the large amount of enzyme required for detection of enzyme activity. In addition optimization of the assays requires very high-affinity, high-selectivity antibodies, and the cost of instrumentation to perform these assays (FP, AlphaScreen) is high.

All of these technologies rely on the detection of the phosphorylated product either by using radioactivity detection systems or by using antiphosphorylated antibodies or matrices to capture the phosphorylated peptide. The technology we present here relies on the detection of kinase activity based on the change in concentration of ATP, which is a universal substrate to all classes of kinases including protein kinases. It is simple, fast, nonradiacative, robust, low-cost, amenable to multiwell and HTS-formatted, and workable with any kinase substrate.

8.4 MATERIALS AND METHODS

- *Materials*: Opaque-walled Costar multiwell plates were purchased from Corning Inc., Corning, NY. Multichannel pipettes or automated pipetting station were purchased from Rainin Instrument, LLC, Woburn, MA. A multiwell luminometer (BMG Labtech, Durham, NC) was used.
- *Reagents*: Kinase-Glo™ Reagent was purchased from Promega Corp. Cat # V671 (Madison, WI) and was used as recommended by the manufacturer (see text below). Protein Kinases, Protein Kinase substrates were purchased from Promega Corp (Madison, WI), UBI, Inc. (Charlottesville, VA), EMD Biosciences, Inc. (San Diego, CA), and Sigma-Aldrich Corp. (St. Louis, MO). The library of pharmacologically active compounds (LOPAC) was purchased from Sigma-Aldrich Corp., St. Louis, MO.

8.5 KINASE-GLO REAGENT PREPARATION

Kinase-Glo Reagent is prepared as follows:

- Thaw the Kinase-Glo Buffer and equilibrate it to room temperature prior to use. For convenience the Kinase-Glo Buffer may be thawed and stored at room temperature for up to 48 h prior to use.
- Equilibrate the lyophilized Kinase-Glo Substrate to room temperature prior to use.
- Transfer the entire volume of Kinase-Glo Buffer into Kinase-Glo Substrate to reconstitute the lyophilized enzyme/substrate mixture. This forms the Kinase-Glo Reagent.
- Mix by gently vortexing, swirling, or inverting the contents to obtain a homogeneous solution. The Kinase Glo Substrate should go into solution easily, in less than 1 min.
- Kinase-Glo Reagent should be used immediately or dispensed into single-use aliquots and stored at −20°C.

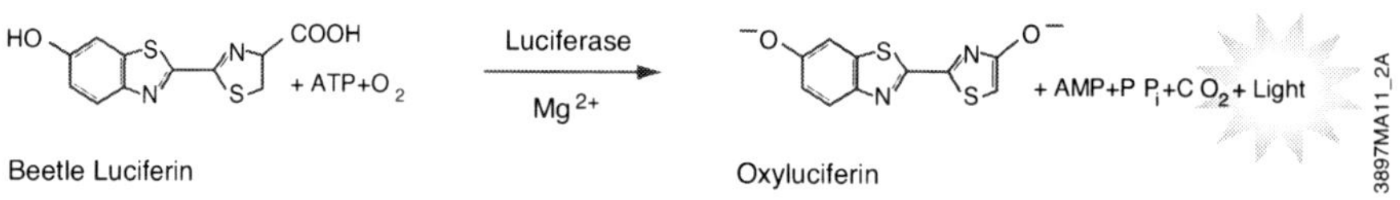

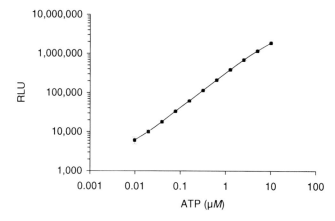

SCHEME 8.1 Luciferase reaction with luciferin, ATP, and oxygen will generate oxyluciferin and light. In the presence of excess luciferin and ambient oxygen, the amount of light generated is proportional to the concentration of ATP as shown in Figure 8.1.

FIGURE 8.1 Standard curve of relative luminescence as a function of ATP concentration. Standard curve was carried out using increasing concentrations of ATP at saturating luciferin concentration in the luciferase reaction. It is apparent that relative luminescence units (RLU) are proportional to ATP concentration; this relationship is linear up to 10 μM ATP.

8.6 PRINCIPLE OF THE ASSAY

The Kinase-Glo Luminescent Kinase Assay is a homogeneous, high-throughput screening (HTS) method for monitoring kinase activity by quantifying the amount of ATP remaining in solution following the kinase reaction. The principal reaction depicting the basis of the assay is shown in Scheme 8.1. The assay is performed in a single well of a 96- or 384- or 1536-well plate by adding an equal volume of Kinase-Glo Reagent to the completed kinase reaction in the well and measuring luminescence. The luminescent signal is correlated with the amount of ATP present, as shown in Figure 8.1. The luminescence readout is proportional to ATP concentration for at least four orders of magnitude range in ATP concentration up to 10 μM. If higher ATP concentrations are required, Kinase-Glo Plus can be used instead of Kinase-Glo, to extend the linearity of ATP up to 100 μM. Since the output luminescence is a measure of ATP remaining in the reaction, the more active the kinase the less luminescence signal is generated; hence, the activity of the kinase is reciprocally related to luminescence output.

The assay can be performed with virtually any kinase and substrate combination and does not require any radioactively labeled components or antibodies. The kinase substrate can be a peptide, protein, or lipid. The Kinase-Glo Reagent relies on the properties of a proprietary thermostable luciferase (UltraGlow™ Recombinant Luciferase) that is formulated to generate a stable "glow-type" luminescent signal and improve performance across a wide range of assay conditions. The signal produced by the luciferase reaction is stable, with a half-life greater than 4 h (Figure 8.2). This extended half-life eliminates the need for a luminometer with reagent injectors and provides for batch-mode processing of multiple plates. In addition, the unique combination of UltraGlow luciferase and proprietary buffer formulation results in luminescence that is much less susceptible to interference from library compounds than are other luciferase-based ATP detection reagents.

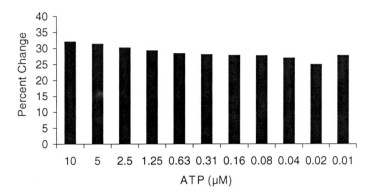

FIGURE 8.2 Signal stability was tested at various ATP concentrations to illustrate that signal output would not cause variability in interpreting the results. The signal is stable for at least 5 h, and minimal variation was observed at the different ATP concentrations tested.

The following are the recommended protocols we follow in developing kinase assays, testing for robustness of the assay (Z′), determining IC_{50} for kinase inhibitors, and screening combinatorial libraries for potential kinase modulators.

8.7 OPTIMIZING KINASE REACTION CONDITIONS

In order to get the maximum performance when using the Kinase-Glo Reagent, kinase reaction conditions need to be optimized with respect to the amount of ATP and kinase substrate. The ATP concentration that gives the maximum dynamic range is then used to establish the optimum substrate concentration. The optimal ATP and substrate concentrations can be used to establish the optimum enzyme concentration to be used for screening of potential kinase modulators. The following is an illustration of the optimization process for cAMP-dependent protein kinase (PKA); it is applicable to other enzymes of interest.

8.7.1 DETERMINING OPTIMAL ATP CONCENTRATION

Optimal ATP concentration is determined as follows:

- Twofold serial dilutions of ATP across the plate are prepared using as much PKA as practical and excess kinase substrate (Kemptide). As a control, make the same ATP titration without the kinase substrate or kinase present in the well. Allow the kinase reaction to consume as much ATP as possible. With less active kinases, this may take 1 to 2 h at room temperature. The reaction time can be shortened if the enzyme reaction is carried out at a higher temperature, such as 30 or 37°C.
- Add equal volume of Kinase-Glo Reagent to that of kinase reaction in each well. Mix the plate and incubate at room temperature for 10 min to stabilize luminescent signal. Because of the long half-life of the Kinase-Glo signal, plates may be left longer at room temperature before reading in a luminometer. Instrument settings depend on the manufacturer. An integration time of 0.25 to 1 sec per well should serve as a guideline.
- The optimal ATP concentration will give the largest change in luminescence when comparing the completed kinase reaction wells to wells that are missing the kinase or kinase substrate. Figure 8.3 shows that the dynamic range is highest at low ATP concentration, and it remains unchanged with increasing ATP concentrations until it reaches 10 μM; it decreases thereafter until it reaches its minimum at 100 μM. It is apparent that 10 μM of ATP is optimal for PKA since it satisfies the requirement for high dynamic range and it is near the Km_{ATP} for the enzyme [16].

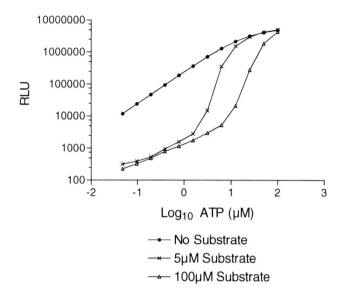

FIGURE 8.3 Optimization of ATP concentration in the kinase reaction. A sufficient amount of PKA and two concentrations of peptide substrate (Kemptide) were used to carry out the PKA assay at varying ATP concentrations to establish the concentration that gives the maximum dynamic range. The higher the ATP concentration, the higher the substrate concentration required for maximum dynamic range.

8.7.2 DETERMINING OPTIMAL SUBSTRATE CONCENTRATION

Optimal substrate concentration is determined as follows:

- Twofold serial dilutions of kinase substrate across the plate are prepared using as much kinase as practical and the optimal amount of ATP (determined above). As a control, do the same titration without kinase.
- Add a volume of Kinase-Glo Reagent equal to the volume of the kinase reaction in each well.
- Mix the plate and incubate at room temperature for 10 min to stabilize the luminescent signal. Because of the long half-life of the Kinase-Glo signal, the plates may be left longer at room temperature before reading in a luminometer.
- The optimal kinase substrate concentration will result in the largest change in luminescence when comparing kinase reaction wells to wells that do not contain the kinase.
- As shown in Figure 8.4, substrate concentration required for optimal kinase activity increases with increasing ATP concentration. We recommend that the ratio of concentrations of substrate: ATP in the reaction should be at least two (and preferably five) if the cost of substrate is not prohibitive.

8.7.3 DETERMINING THE OPTIMAL AMOUNT OF KINASE

The optimal amount of kinase is determined as follows:

- Twofold serial dilutions of kinase across the plate are prepared using the optimal amount of ATP and kinase substrate determined above.
- Add a volume of Kinase-Glo Reagent equal to the volume of the kinase reaction.
- Mix the plate and incubate at room temperature for 10 min to stabilize luminescence signal. Because of the long half-life of the Kinase-Glo signal, the plates may be left longer at room temperature before reading in a luminometer.

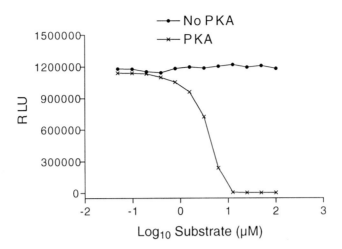

FIGURE 8.4 Optimizing substrate concentration in the kinase reaction. A sufficient amount of PKA and 5 μM ATP were used to establish the optimum substrate concentration. The figure shows that 10 μM of Kemptide gives maximum phosphorylation. This experiment should be repeated for each desired ATP concentration tested to establish the optimal substrate-to-ATP ratio.

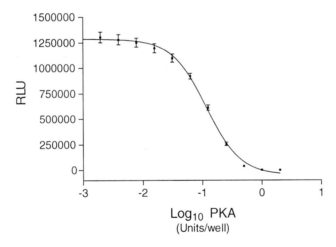

FIGURE 8.5 Enzyme titration for PKA was carried out using optimal ATP (5 μM) and peptide substrate concentrations (100 μM). One unit of PKA (pmol/min) gives maximum phosphorylation of the substrate at room temperature.

- The optimal amount of kinase to use in subsequent compound screens and IC_{50} determinations should be an amount that results in luminescence in the linear range of the kinase titration curve. As shown in Figure 8.5, the optimal enzyme required can be determined under optimal ATP and substrate concentrations.

8.7.4 DETERMINING Z' FACTOR

The Z' factor is determined as follows:

- Add 25 μL of 2× reaction mixture to each well of one half of the plate, and then add 25 μL of reaction mixture containing 2× the optimal amount of kinase and kinase substrate to each well of the other half of the plate.

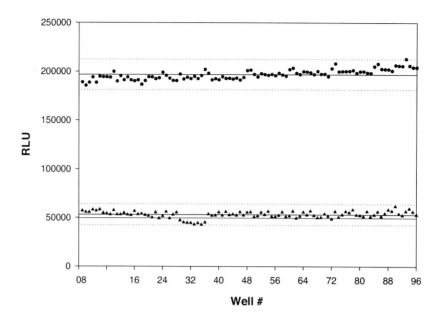

FIGURE 8.6 Z′ determination for PKA using Kinase-Glo. The kinase assay for PKA was determined in the presence (triangles) and absence (circles) of PKA in a 96-well plate. A Z′ value of 0.83 was obtained, indicating robustness of the assay in monitoring kinase activity.

- Add 25 μL of reaction mixture containing 2× the optimal amount of ATP to all wells.
- Mix the plate and incubate for the optimal reaction time and temperature.
- Add 50 μL of Kinase-Glo Reagent to all wells. Mix the plate, incubate for 10 min at room temperature, and read luminescence using a luminometer. Because of the long half-life of the Kinase-Glo signal, the plates may be left longer at room temperature before reading.

As shown in Figure 8.6, we often obtain a Z′ value of 0.7 and higher with kinases in the 96-well as well as 384-well plates. A Z′ value of 0.5 or higher is considered desirable for drug discovery programs and indicates robustness of the assay and tightness of experimental values [17].

8.8 SCREENING FOR KINASE INHIBITORS

The volumes provided in this protocol are intended for a 96-well plate. To perform the assay in a 384-well plate, reduce volumes fivefold:

- Add 5 μL of each compound to separate wells (controls contain 5 μL of solvent only).
- Add 20 μL of reaction mixture containing 2.5× the optimal concentration of kinase and kinase substrate.
- Add 25 μL of reaction mixture containing 2× the optimal concentration of ATP to all wells.
- Mix the plate and incubate at room temperature for the optimal reaction time.
- Add 50 μL of Kinase-Glo Reagent to all wells. Mix plate, incubate for 10 min at room temperature, and read luminescence using a luminemter. Because the signal is stable for hours, plates may be left longer at room temperature before reading. As shown in Figure 8.7, when no enzyme is included, a maximum luminescence signal is observed while the signal from a complete kinase reaction is at its minimum. If the library contains inhibitors of the kinase under study, then a luminescence signal will appear in between

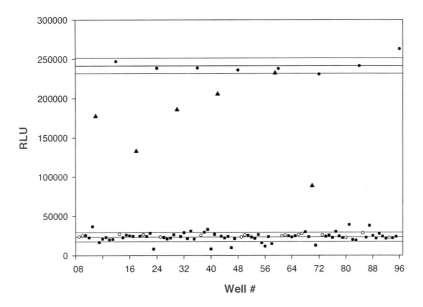

FIGURE 8.7 Screening of Library of Pharmacologically Active Compounds (LOPAC). Screening of LOPAC (Plate #6 of LOPAC) toward PKA was tested in a 96-well plate. Each well contained the LOPAC individual compounds at 10 μM final concentration. One group of control wells contained an equal volume of 2% vehicle solvent (DMSO) with kinase (open circles); a second group contained no enzyme (closed circles). Wells containing compounds without any effect on PKA are shown as rectangles, and wells containing compounds showing positive hits are shown as triangles.

the two extremes depending on the potency of the inhibitor; the more potent the inhibitor the higher the signal, while less potent inhibitors show low signal.

8.9 DETERMINING IC$_{50}$ VALUES

The following protocol is designed for a 96-well plate. To perform the assay in a 384-well plate, volumes should be reduced by fivefold:

- To each well add 20 μL of reaction mixture containing 2.5× the optimal concentration of kinase and kinase substrate and then add 5.0 μL of various inhibitor concentrations diluted in the above reaction mixture.
- Add 25 μL of reaction mixture containing 2× the optimal concentration of ATP. Mix the plate and incubate for the optimal reaction time.
- Add a 50 μL of Kinase-Glo Reagent to all wells. Mix and incubate at room temperature for 10 min to stabilize the luminescent signal. Luminescence is read using a luminescence reader. If desired, plates may be left longer at room temperature before reading because of the long half-life of the Kinase-Glo signal. Table 8.1 shows the titration profile for two known PKA inhibitors with apparent IC$_{50}$ for both similar to the values reported in the literature [16,18–20].

8.10 OTHER EXAMPLES OF KINASE-GLO THAT ILLUSTRATE THE DIVERSITY OF THE ASSAY

As mentioned earlier, the luminescent kinase assay offers several unique features such as versatility in choice of substrates (peptides, proteins, phosphopeptides, phospholipids, sugars, etc.) and speed

TABLE 8.1
IC$_{50}$ Values of LOPAC Hits[a]

Well #	Compound	Kinase-Glo (%)	ProFluor (%)
70	GW5074	29.1	4.2
18	H-7	49.6	5.5
6	HA-1004	70.2	17.6
30	H-8	74.1	20.6
42	H-9	83.3	38.8
59	U-73122	95.7	71.9

[a] IC$_{50}$ values of six hits identified from plate #6 of LOPAC using Kinase-Glo and IC$_{50}$ values for the same compounds using another kinase assay (Profluor PKA assay, Reference 15). The IC$_{50}$ values obtained by the two methods are very similar, confirming the validity of Kinase-Glo as a screening method.

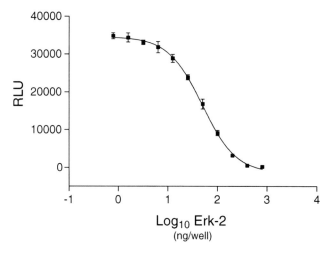

FIGURE 8.8 Kinase-Glo assay of MAPK (ERK). ERK enzyme activity was assayed using 60 μg of myelin basic protein (MBP) as substrate and 0.1 μ*M* ATP at room temperature for 2 h and the indicated amount of kinase.

by which optimization and screening for any kinase can be accomplished. To demonstrate the use of protein as substrate for a kinase, we carried out the enzyme assay of MAPK (ERK) using myelin basic protein (Figure 8.8). It is evident that the enzyme phosphorylates myelin basic protein (MBP) in a concentration-dependent manner indicating the feasibility of assaying the activity of this enzyme, not only with peptides as we have demonstrated (data not shown) and by other technologies but also with a protein substrate. This is only feasible by the Kinase-Glo and radioactive-based approaches. We also show that the activity of a kinase can be monitored with nonpeptidic, nonprotein substrates such as phospholipids for PI3 kinase and sugars for glucokinase. Figure 8.9 shows that the enzyme activity of PI3 kinase can be measured using phosphoinositide as substrate, and Figure 8.10 shows the activity of glucokinase using glucose as substrate. The results in Figure 8.9 and Figure 8.10 demonstrating the ability to carry out kinase assays for enzymes that use phospholipids and sugars illustrate the diversity of the substrates that can be used in the assay. Figure 8.11 shows the activity of GSK3 with a phosphopeptide as substrate that illustrates the capability of the assay to use phosphopeptides as substrate.

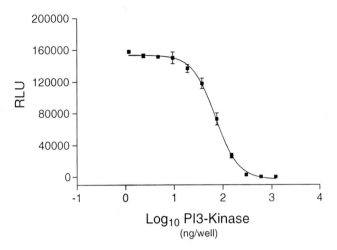

FIGURE 8.9 Kinase-Glo assay of PI3 kinase. PI3 kinase enzyme activity was assayed using 25 μg of phosphoinositide (PI) as substrate and 0.5 μM ATP and the indicated amount of kinase at room temperature for 90 min.

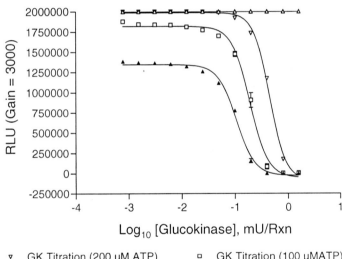

▽	GK Titration (200 μM ATP)	▫	GK Titration (100 μMATP)
▲	GK Titration with 20 mM Glucose (50 μM ATP)	▵	200 μM ATP alone

FIGURE 8.10 Kinase-Glo assay of glucokinase. Glucokinase enzyme activity was assayed using 50 (closed triangles), 100 (open rectangles), and 200 (open inverted triangles) μM ATP and 20 mM glucose for 30 min at room temperature. Control experiment was carried out in the absence of glucokinase (open triangles).

The ability to screen a library of peptides or proteins, rapidly and inexpensively, for the optimal substrate for an enzyme is another important feature of this assay. This is demonstrated with a very simple example of a src family protein tyrosine kinase with three different peptide substrates (Figure 8.12). These examples document the versatile nature of the assay and broaden the range of its applications to any kinase as long as its substrate is known and available.

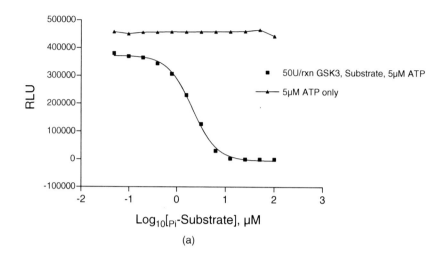

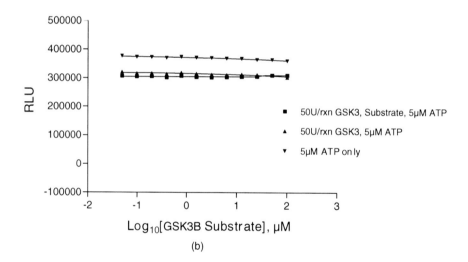

FIGURE 8.11 (a) Kinase-Glo assay of GSK3B with phosphopeptide substrate. GSK3B enzyme activity was assayed using 50 units/reaction of GSK3B (rectangles), and 5 μM ATP and varying concentrations of YRRAAVPPSPSLSRHSSPHQ(pS)EDEEE for 30 min at room temperature. A control without the kinase was also included (triangles). It is evident that the enzyme phosphorylates the phosphopeptide and that the phosphorylation is substrate dependent. (b) Kinase-Glo of GSK3B with peptide substrate. GSK3B was assayed using 50 units/reaction of GSK3B (rectangles), and 5 μM ATP and varying concentrations of GPHRSTPES-RAAV, for 30 min at room temperature. A control without the kinase was also included (triangles), and another control included 5 μM ATP only without substrate (inverted triangles). It is evident that the enzyme does not phosphorylate the peptide since it does not contain the *C*-terminal phosphate required for recognition by the enzyme. This shows an absolute stringent condition of substrate recognition by the enzyme.

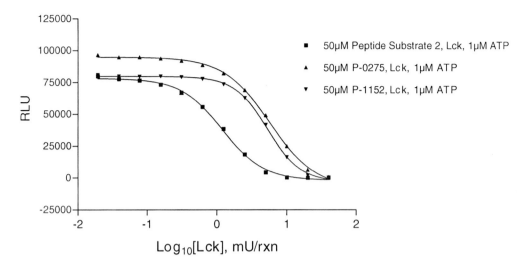

FIGURE 8.12 Screening for optimal substrate for Src tyrosine kinase. Screening was carried out using Kinase-Glo: Three substrates (E:Y, 4:1; A:E:K:Y, 6:2:5:1; and a proprietary peptide substrate [peptide substrate 2]) were assayed at 50 μM with 1 μM ATP and varying amounts of Src enzyme for 60 min at room temperature. It is apparent that the proprietary peptide shows an EC_{50} that is one fifth of those of the other two peptides and thus it is a better substrate. (EC_{50} = 1.195 mU/Rx for Substrate 2; EC_{50} = 5.685 for Poly (E-Y 4:1); and EC_{50} = 5.289 for Poly (A-E-K-Y 6:2:5:1).

8.11 GENERAL CONSIDERATIONS

Temperature. The intensity and rate of decay of the luminescent signal from the Kinase-Glo assay depends on the rate of the luciferase reaction. Environmental factors that affect the rate of the luciferase reaction will result in a change in the intensity of light output and the stability of the luminescent signal. Temperature is one factor that affects the rate of this enzymatic assay and thus the light output. For consistent results, equilibrate assay plates and reagents to room temperature prior to performing the assay. Insufficient equilibration may result in a temperature gradient effect between the wells in the center and on the edges of the plates.

Solvents. The chemical environment of the luciferase reaction will affect the enzymatic rate and thus luminescence intensity. Some solvents used to suspend the various chemical compounds tested may interfere with the luciferase reaction and thus the light output from the assay. Interference with the luciferase reaction can be determined by assaying a parallel set of control wells without kinase or kinase substrate. Dimethylsulfoxide (DMSO), commonly used as a vehicle to solubilize organic chemicals, has been tested at final concentrations of up to 2% in the assay and has a minimal effect on light output. Standard opaque-walled multiwell plates suitable for luminescence measurements are recommended for use. Consult the luminometer instrument manufacturer to find out which plates will work best for your particular instrument.

Inhibition of luciferase. Compounds that only inhibit the kinase will result in increased luminescence and are easily distinguishable from compounds that only inhibit luciferase activity, which decrease luminescence. Compounds that inhibit both the luciferase and kinase, however, might increase, decrease, or have no effect on luminescence depending on the level of inhibition directed toward the kinase and luciferase. To reduce this problem, the Kinase-Glo Reagent consists of a thermostable luciferase in a proprietary buffer that is more tolerant of compound interference.

In summary, to optimize the kinase assay for use with the luminescence reagent, run these two experiments:

- Using as much kinase as practical and excess kinase substrate, titrate ATP from 100 μM doing twofold dilutions across the plate. As a control, do the same ATP titration without the substrate or kinase present in the well. Allow the kinase enough time to consume as much ATP as possible. With poorly active kinases this may take several hours, so you might also try raising the reaction temperature and/or lower the ATP concentration.
- Repeat the above experiment titrating the amount of kinase substrate instead of ATP. The amount of ATP used in this second experiment should be the amount of ATP that provided the best fold decrease in light when wells with and without substrate or kinase are compared to each other.

REFERENCES

1. Noble MEM, Endicott JA, and Johnson LN (2004) Protein kinase inhibitors: insights into drug design from structure, *Science* 303:1800–1805.
2. Vlahos CJ, McDowell SA, and Clerk A (2003) Kinases as therapeutic targets for heart failure, *Nature Rev Drug Discovery* 2:99–113.
3. Manning G, Plowman GD, Hunter T, and Sudarsanam S (2002) Evolution of protein kinase signaling from yeast to man. *Trends Biochem Sci* 27:514–520.
4. Shears, SB (2004) How versatile are inositol phosphate kinases? *Biochem J* 377:265–280.
5. French KJ, Schrecengost RS, Lee BD, Zhuang Y, Smith SN, Eberly JL, Yun JK, and Smith CD (2003) Discovery and evaluation of inhibitors of human sphingosine kinase. *Cancer Res* 63:5962–5969.
6. Grimsby J, Sarabu R, Corbett WL et al. (2003) Allosteric activation of glucokinase: potential role in diabetes therapy. *Science* 301:370–373.
7. Force T, Kuida K, Namchuk M, Parang K, and Kyriakis JM (2004) Inhibitors of protein kinase signaling pathways. Emerging therapies for cardiovascular diseases. *Circulation* 109:1196–1205.
8. Goueli BS, Hsiao K, Tereba A, and Goueli SA (1995) A novel and simple method to assay the activity of individual protein kinases in a crude tissue extract. *Anal Biochem* 225:10–17.
9. Anderson SN, Cool BL, Kifle L, Chiou W, Egan DA, Barrett LW, Richardson PL, Frevert EU, Warrior U, Kofron JL, and Burns, DJ (2004) Microarrayed compound screening (μARCS) to identify activators and inhibitors of AMP-activated protein kinase. *J Biomolecular Screening* 9:112–121.
10. Cook, ND (1996) Scintillation proximity assay-A versatile high-throughput screening technology. *Drug Discovery Today* 1:287–294.
11. Turek TC, Small EC, Bryant RW, and Hill WAG (2001) Development and validation of a competitive AKT serine/threonine kinase fluorescence polarization assay using a product-specific antiphospho-serine antibody. *Anal Biochem* 299:45–53.
12. Sportsman JR, Gaudet EA, and Boge, A (2004) Immobilized metal ion affinity-based fluorescence polarization (MAP): advances in kinase screening. *Assay Drug Dev Tech* 2:205–214.
13. Morgan AG, McCauley TJ, Stanaitis ML, Mathrubutham M, and Millis SZ (2004) Development and validation of a fluorescence technology for both primary and secondary screenings of kinases that facilitates compound selectivity and site-specific inhibitor determination (2004) *Assay Drug Dev Tech* 2:171–181.
14. A practical guide to working with AlphaScreen™ (2003) PerkinElmer Life and Analytical Sciences, Boston, MA.
15. Kupcho K, Somberg R, Bulleit B, and Goueli SA (2003) A homogeneous, nonradioactive high-throughput fluorogenic protein kinase assay. *Anal Biochem* 317:210–217.
16. Taylor SS, Yang J, Wu J, Haste NM, Radzio-Andzelm E, Anand G (2004) PKA: a portrait of protein kinase dynamics. *Biochim Biophys Acta* 1697:259–269
17. Zhang J-H, Chung TDY, and Oldenburg KR (1999) A simple statistical parameter for use in evaluation and validation of high-throughput screening assays. *J Biomol Screen* 4:67–73.
18. Hidaka H, Watanabe M, and Kobayashi K (1991) Properties and use of H-series compounds as protein kinase inhibitors. *Methods Enzymol* 201:328–339.

19. Walsh DA and Glass DB (1991) Utilization of the inhibitor protein of adenosine cyclic monophosphate–dependent protein kinase, and peptides derived from it, as tools to study adenosine cyclic monophosphate–mediated cellular processes. *Methods Enzymol* 201:304–316.

20. Tamaoki T (1991) Use and specificity of staurosporine, UCN-01, and calphostin C as protein kinase inhibitors. *Methods Enzymol* 201:340–347.

9 Proteases as Drug Targets

Robin L. Thurmond and James P. Edwards

CONTENTS

9.1 INTRODUCTION

The study of proteases is one of the richest in the history of biochemistry, beginning in the late 1800s with the observations of tissue degradation *in vitro* (see [1] for a review). Proteases make up the largest class of enzymes, with over 1600 proteases identified from over 1700 organisms. Human proteases account for approximately 500 of that number and represent an attractive area for new drug discovery. One of the key developments in the study of the function of proteases occurred in the 1930s when the first synthetic substrates were developed [2–5]. This is important in two ways. First, small synthetic substrates are essential for the development of technically easy and inexpensive high-throughput assays. Second, the availability of well-defined substrates has enabled the detailed understanding of protease catalytic mechanisms. Indeed, more is probably known about the catalytic mechanism of proteases than any other class of enzymes. When using small-molecular-weight substrates many proteases follow straightforward kinetics. This is a great help in characterizing potential inhibitors. In addition, many proteases are small compact proteins that lend themselves to structural analysis, which can be useful in the design of novel inhibitors. With all of these advantages there has been great success in developing inhibitors for a number of different proteases; however, this success has not been easily translated into clinically useful drugs.

9.2 CLASSES

There are four main classes of proteases, which are organized by their catalytic mechanism. Aspartic proteases use two catalytic aspartic acid residues in the active site to coordinate the nucleophilic attack of the peptide bond by a water molecule. The classic aspartic protease is cathepsin D. With serine proteases the hydroxyl group on the active site serine acts as the nucleophile that attacks the peptide bond and forms a tetrahedral intermediate. Here a triad consisting of a serine, a histadine, and an aspartic acid residue characterizes the active site. The catalytic mechanism proceeds in such a way that a covalent acyl intermediate is formed between the substrate and the enzyme that subsequently is hydrolyzed to release the cleaved peptide substrate. Common serine proteases include trypsin, elastase, and chymotrypsin. In most cases threonine proteases are also included in the serine protease family since the catalytic mechanism is identical; the only difference is that a threonine is used instead of a serine. Threonine proteases are found as components of the proteasome. Cysteine proteases also have a catalytic triad and hydrolyze amide bonds in a manner similar to the serine proteases. In this case a thiolate ion on an active site cysteine residue is used to attack the peptide bond. Cathepsin B and the caspases are examples of cysteine proteases. The final class of proteases is metalloproteases. Here a metal atom (generally zinc) is used to coordinate the substrate and catalyze the nucleophilic attack of a water molecule on the peptide bond. Angiotensin-converting enzyme (ACE) is an example of a metalloprotease.

9.3 SUBSTRATE BINDING

Almost all proteases bind substrates in a broadly similar manner. The binding site consists of a series of subsites that, following the pioneering work of Schechter and Berger mapping the active site of papain, are called S_x sites and S'_x sites [6,7]. The S sites bind to peptide residues (P_x) on the N-terminal side of the scissile bond. The S' sites bind to peptide residues (P'_x) on the C-terminal side. The P1 or P1' residues are those residues located near the scissile bond. The subsites on the protease that complement the substrate residues are numbered … S3, S2, S1, S1', S2', S3' …. In general only a few of the substrate binding sites have strong preferences for a particular amino acid residue. For serine proteases and caspase-like cysteine proteases, the S1 site gives the most specificity, whereas the S2 site is the most important for members of the papain family of cysteine proteases. Many proteases have a somewhat broad selectivity for substrates and can cleave many sequences. Others, like some of the complement proteins, are much more selective. Most proteases readily cleave small peptides, which makes kinetic analysis and assay development easier. In addition, the identification of peptide substrates is often the starting point for the development of inhibitors. In fact, many of the protease inhibitors on the market are peptide based. On the other hand, the limited substrate specificity of some proteases can make it more difficult to develop selective drugs that target only a single protease. This may be one of the reasons why relatively few proteases are clinically validated targets.

9.4 REGULATION

All enzymes are regulated under normal conditions; however, proteases have the advantage of being regulated by endogenous competitive inhibitors [8,9]. Endogenous protease inhibitors such as serpins, cystatins, and tissue inhibitors of metalloproteases (TIMP) bind to the substrate-binding pocket on their respective protease, but in such a way that the binding is not conducive to hydrolysis. Therefore, the regulation of proteolytic activity by small-molecule inhibitors can mimic the natural regulation mechanisms. This, coupled with the fact that well-defined binding pockets exist for most proteases, has fostered the hope that proteases are amenable to inhibition by small-molecule drugs and that this inhibition can have desirable physiological effects.

9.5 SMALL-MOLECULE INHIBITORS

There are several different ways to categorize small molecules that inhibit the proteolytic activity of proteases (for reviews, see [10–12]), but for the purpose of this discussion we shall follow the convention of Otto and Schirmeister [11]. Thus, the major inhibitor classifications can be defined by where the inhibitor binds (active site or allosteric), the nature of the binding interaction (covalent or noncovalent), and the reversibility of the binding (reversible or irreversible). Due to the well-defined active sites of most proteases, by far the most successful strategies for inhibitor discovery have been targeted at active site–directed inhibitors. Covalent inhibitors are characterized by the formation of a covalent bond, which is generally highly energetic, between inhibitor and protease. Noncovalent inhibitors interact with the protease solely through weaker bonds (hydrogen bonds and van der Waals forces). Reversible inhibitors are characterized by the ability to dissociate (either rapidly or slowly) from the protease, allowing catalytic activity to be regained. Note that reversible inhibitors can be either covalent or noncovalent. On the other hand, exposure of a protease to an irreversible inhibitor results in covalent modification of the protein that leads to *permanent* inhibition of proteolytic activity. Since cysteine and serine proteases use an active site residue (rather than a water molecule) to attack the carbonyl of the scissile bond of their substrates, these two classes of proteases have proven to be good targets for the design of irreversible inhibitors. Inhibitors that form covalent bonds to the protease are found in both categories but, by this definition, noncovalent inhibitors are always categorized as reversible. Inhibitors with extremely slow off-rates ("tight-binding inhibitors") are therefore classified as reversible, even though they may be virtually indistinguishable from irreversible inhibitors by kinetic analysis (for a discussion, see [13]).

9.6 ACTIVE SITE INHIBITORS

Active site–directed inhibitors of proteases have classically been designed using an endogenous or synthetic substrate as a starting point and replacing the scissile bond with a moiety that avidly binds the active site, sometimes referred to as a "warhead" [14]; for an early reference see [15]. For aspartyl proteases and metalloproteases, the active-site recognition element generally does not form a covalent bond to the peptide, and thus inhibition is reversible; the vast majority of serine and cysteine protease inhibitors use an active-site recognition element that forms a covalent bond to the active site residue that can be either reversible or irreversible. The most common active-site recognition elements in aspartyl protease inhibitors are the hydroxyethylamine or 1,3-diamino-2-propanol motifs. Originally identified in the naturally occurring aspartyl protease inhibitor pepstatin, these isosteres have yielded potent inhibitors of numerous aspartyl proteases, most notably HIV protease (for a review, see [16]). Alternatively, inhibitors of metalloproteases take advantage of the active-site metal atom by incorporating a metal-chelating group such as a hydroxamic acid, carboxylic acid, or boronic acid. Irreversible inhibitors of proteases contain an electrophile that, upon reaction with the protease, forms a nonhydrolyzable adduct. As noted above, this approach works particularly well for serine and cysteine proteases [17]. Typical electrophiles that form nonhydrolyzable adducts include α-haloketones, ketohydrazides, diazoketones, and epoxide derivatives, examples of which are found in both cysteine and serine protease inhibitors. Since the utility of irreversible protease inhibitors as human therapeutics is questionable, much recent research has been directed toward the identification of inhibitors with hydrolyzable (reversible) electrophiles, such as nitriles, aldehydes, and ketoheterocycles.

9.7 SOURCE OF HITS

9.7.1 STRUCTURE-BASED DRUG DESIGN

Like all drug discovery programs, the development of protease inhibitors as therapeutic agents requires a starting point (or "lead") for the medicinal chemist. Often, the structure of substrates of the target protease can serve as this starting point [18–20]. The downside to this approach is that the resulting molecules themselves are very peptide-like and seldom have good pharmaceutical properties, although investigators continue to make progress in this area [21]. Natural products have also proved a fertile source of leads for protease inhibitors, such as pepstatin for aspartyl proteases (*vide infra*) or snake venom peptides for ACE inhibitors [22]. High-throughput screening has yielded leads for proteases as diverse as human cytomegalovirus (HCMV) protease [23] and cathepsin S [24]. Recently, the combination of improved computer algorithms with the wealth of structural information on proteases has led to the "in silico" design of new inhibitors [25].

Regardless of the mode of inhibition or the source of the seminal inhibitor, structure-based methods have proven invaluable to the design of potent and selective protease inhibitors [16]. Examples abound in the literature, from the very successful HIV protease programs (*vide supra*) to the extensive SmithKline Beecham (now GSK) program targeting inhibitors of the cysteine protease cathepsin K (c.f., [26–28]). As noted above, the compact nature and well-defined active site of many proteases make structure-based drug design particularly attractive. As described below, this strategy has yielded clinically important drugs.

9.7.2 COMBINATORIAL CHEMISTRY

At the other end of the spectrum, combinatorial chemistry also has a role in the discovery of protease inhibitors. The existence of active site-recognition elements specific for certain classes of proteases can be exploited in library design. Ellman and coworkers have described particularly elegant studies in this area, both for aspartyl proteases [29] and cysteine proteases [30]. For the aspartyl protease cathepsin D, a 204-member library of 1,3-diamino-2-propanols afforded nano molar inhibitors [29], while a library of α-amido ketones afforded potent and selective inhibitors of the cysteine protease cruzain [31].

9.8 PROTEASE INHIBITORS AS DRUGS

The development of small-molecule protease inhibitors has led to many successful therapeutics. In 2002, protease inhibitors made up almost 10% of the top 200 prescriptions in the U.S. While protease inhibitors have been quite successful on the market, the current drugs target a relatively small number of different proteases. The largest class of these inhibitors targets the angiotensin-converting enzyme (ACE). ACE is responsible for the production of angiotensin II, which is a potent vasoconstrictor. Therefore, in the clinic ACE inhibitors are used primarily to treat hypertension and acute myocardial infarction. There are a number of different approved ACE inhibitors on the market, but all of them employ a chelating group that binds to the active site zinc atom. The first ACE inhibitor on the market was captopril (Bristol-Myers Squibb), which was launched in 1980. The development of this class of compounds started with the identification of the inhibitory properties of snake venom peptides [32–34]. It was soon discovered that the *C*-terminal fragment of one of these peptides, consisting of Trp-Ala-Pro, was an inhibitor of ACE [22]. Cushman et al. [35] started with the Ala-Pro *C*-terminal dipeptide, reasoning that if ACE was a dipeptyl peptidase like carboxypeptidase A, then this should be sufficient to yield inhibition. A second key insight was the observation that benzylsuccinic acid was a weak inhibitor of carboxypeptidase A and that this inhibition was achieved by the interaction of the carboxyl group with the catalytically important zinc ion in the enzyme [36]. Replacement of the alanine amino acid residue of the dipeptide ACE inhibitor with a 3-mercapto-2-methylpropanoyl-proline zinc chelating group led to a very potent inhibitor both *in vitro* and *in vivo*, which became known as captopril [35,37]. The discovery of other inhibitors followed a similar path, with the next being enalapril, which was based on a tripeptide structure and replaced the mercapto chelating group with a weaker chelator yielding substituted *N*-carboxymethyl dipeptides [38]. To date there are 21 marketed ACE inhibitors [39–42]. However, there are still some class-specific side effects, such as cough and angioedema, that are of concern. It turns out that ACE is a two-domain enzyme with two active sites. These two active sites may have different biological functions and it may be possible to target them specifically and thus reduce some of the side effects [42].

HIV protease inhibitors are another large set of marketed protease inhibitors. The development of these inhibitors is a remarkable example of drug discovery and development both in terms of the diverse methods used to reach the same endpoints and the speed in which they reached the market [43]. The HIV protease was first discovered in 1985, soon after the sequencing of the viral genome was completed. It only took 10 years for the first inhibitors to reach the market. The first was saquinavir (Roche) in 1995, followed closely by indinavir (Merck), ritonavir (Abbott), and nelfinavir (Pfizer). To date there are ten HIV protease inhibitors used clinically. Saquinavir was developed using a mechanism-based strategy [44]. The investigators started with the fact that the protease had the unusual ability to cleave peptides with a proline residue in the P1' position. Combining hydroxyethylamine transition-state mimics with proline led to early leads. Finally, replacement of the proline with (S,S,S)-decahydroisoquinoline-3-carbonyl led to saquinavir [45,46]. This discovery combined the use of peptidomimetics with transition-state analogues to develop inhibitors. Indinavir (Merck) was also developed starting with a transition-state mimic, but in this case the lead molecule came out of a previous program targeting another aspartyl protease, renin [47].

Ritonavir (Abbott) owed much of its development to structural-based drug design [48]. The HIV protease was known to be a C_2-symmetric homodimer with the active site of both monomers combining to form the complete active site [49,50]. This became the impetus for the development of symmetric protease inhibitors, which, while yielding inhibitors with high affinity, lacked acceptable in vivo profiles due to low bioavailability or rapid clearance [51–53]. Crystal structures of some of the symmetrical compounds complexed with the enzyme showed that they could bind in an asymmetrical fashion [54,55]. This expanded the possibility of adding groups to increase bioavailability without sacrificing potency and led directly to the development of ritonavir [56–58]. Atazanavir (Bristol-Myers Squibb) was also developed using x-ray structural data of inhibitor-protease complexes and has the first HIV protease drug approved for a once-a-day dosing schedule [59].

High-throughput screening has also been used to discover HIV protease inhibitors. One example is tipranavir (Boehringer Ingelheim) [60]. The development of this drug started with the identification of a 4-hydroxycoumarin derivative (warfarin) from an HTS campaign [61]. Screening of related compounds helped develop an initial structure-activity relationship (SAR), which guided the identification of tipranavir [62].

The most recent protease inhibitor to gain regulatory approval is the proteasome inhibitor, bortezomib (Millennium). The proteasome is a multicomponent protease responsible for the degradation of cytoplasmic proteins. For many years it was viewed as a general housekeeping mechanism for removing damaged or misfolded proteins. However, it is now clear that the proteasome plays a major regulatory role in controlling cell-cycle progression, apoptosis, and the inflammatory response by targeted degradation of proteins [63]. This is a fairly central function, and complete inhibition of the proteasome leads to cell death. Proteasome subunits are upregulated in many cancer cells; these cells appear to be more sensitive to inhibition of the protease activity. This has been the basis for developing proteasome inhibitors to block cancer cell proliferation [64–67]. As mentioned previously, the catalytic activity of the proteasome is similar to that of the serine proteases except it contains a threonine residue in the active site instead of a serine. Early proteasome inhibitors were peptide aldehydes that form hemiacetal adducts with the active site threonine [68,69]. However, aldehydes are not ideal drug candidates since they also react with cysteine proteases and have poor metabolic stability. Adams et al. [70] replaced the aldehyde group with boronic acids and developed bortezomib, a dipeptidyl boronic acid. The boronic acid yields much more potent inhibitors with better selectivity and metabolic stability. Bortezomib is now on the market for the treatment of multiple myeloma.

A few other scattered protease inhibitors are on the market, some of which are only approved in a few countries. Heparin, which is an indirect inhibitor of the serine protease thrombin, has been used for many years as an anticoagulant. More recently, direct thrombin inhibitors have come into use [71]. Hirudin is a small protein derived from leeches, which irreversibly inhibits thrombin and is used as an anticoagulant [71]. The only small molecule inhibitor of thrombin that is approved is argotroban, which is a arginine-based peptidomimetic [72–74]. Some nonspecific serine protease inhibitors like gabexate mesilate, nafamostat mesilate, and camostat mesilate are used in Japan for the treatment of pancreatic disorders [75–77]. Tetracyclines, such as doxycycline, are some of the most commonly used antibiotics, but they have also been used for many years for the treatment of diseases like rheumatoid arthritis and periodontitis, where an infectious agent was thought to contribute. However, it has been discovered that tetracyclines are potent inhibitors of matrix metalloproteases, and it has been proposed that efficacy in these non–infection-related diseases may be related to this activity [78–80]. The first elastase inhibitor, silvelestat, was launched in Japan in 2002 for the treatment of respiratory distress syndrome. This small molecule compound is a pivaloyloxy benzene derivative that functions as an enzyme-acylating agent [81–84].

While there have been many successes bringing protease inhibitors to the market, they have been confined to only a few targets. However, work is reported from various pharmaceutical companies on around 70 different proteases, which represents about 10 to 15% of the total number of publicly acknowledged drug targets. Some of the old targets may still yield new breakthroughs. The development of drug resistance necessitates the introduction of new HIV protease inhibitors, and there are new approaches to developing ACE inhibitors with reduced side effects. There are also plenty of promising new areas. Inhibitors of neprilysin, a zinc metalloprotease, are in late-stage clinical trials for the treatment of heart failure and hypertension [85]. Inhibitors of endothelin converting enzyme and renin are also being explored for these indications [41,86,87]. In addition to HIV protease inhibitors, inhibitors are being developed against the rhinovirus 3C protease [88], fungal proteases [89], and a variety of parasitic proteases [90]. There are several protease inhibitors in development for the treatment of allergy and asthma including those targeting tryptase and chymase [91,92]. Both prolyl endopeptidase and BACE (β-site amyloid precursor protein cleaving enzyme) inhibitors are being developed for the treatment of cognition disorders and Alzheimer's disease [93,94]. Dipeptidyl peptidase IV inhibitors are in the clinic for the treatment of type 2 diabetes [95]. Osteoporosis has been targeted in the clinic with cathepsin K inhibitors [96–100]. Inhibitors of methionine aminopeptidase and cathepsin B are being explored for utility in treating cancer [101–103]. Cathepsin S and TNFα-converting enzyme (TACE) inhibitors are in development for the treatment of inflammation and autoimmune diseases [102,104,105].

9.9 CONCLUSION

It is clear that the development of protease inhibitors as novel therapeutic agents still has enormous potential. However, there have also been many disappointments, including the well-publicized failures to bring any matrix metalloprotease or caspase inhibitor to the market after over a decade of work in numerous laboratories. Thus, in spite of the characteristics that make it easy to discover highly potent protease inhibitors (or perhaps because of some of them), it is difficult to meet all of the requirements necessary to develop safe and effective drugs. Nevertheless, the tremendous success of ACE and HIV protease inhibitors demonstrates that the goal of developing clinically important drugs that target proteases can be accomplished; as more of the physiological function of proteases is uncovered, more successes are expected.

REFERENCES

1. Barrett, A.J. Introduction to the history and classification of tissue proteinases. *Research Monographs in Cell and Tissue Physiology* 1977; 2:1–55.
2. Hofmann, K. and Bergmann, M. Specificity of trypsin. II. *Journal of Biological Chemistry* 1939; 130:81–86.

3. Fruton, J.S., Bergmann, M., and Anslow, W.P., Jr. The specificity of pepsin. *Journal of Biological Chemistry* 1939; 127:627–641.

4. Bergmann, M. and Fruton, J.S. Proteolytic enzymes. XIII. Synthetic substrates for chymotrypsin. *Journal of Biological Chemistry* 1937; 118:405–415.

5. Bergmann, M., Zervas, L., and Fruton, J.S. Proteolytic enzymes. VI. The specificity of papain. *Journal of Biological Chemistry* 1935; 111:225–244.

6. Schechter, I. and Berger, A. On the size of the active site in proteases. I. Papain. *Biochemical and Biophysical Research Communications* 1967; 27:157–162.

7. Berger, A. and Schechter, I. Mapping the active site of papain with the aid of peptide substrates and inhibitors. *Philosophical Transactions of the Royal Society of London, Series B: Biological Sciences* 1970; 257(813):249–264.

8. Bode, W. and Huber, R. Natural protein proteinase inhibitors and their interaction with proteinases. *European Journal of Biochemistry/FEBS* 1992; 204:433–451.

9. Laskowski, M., Jr. and Kato, I. Protein inhibitors of proteinases. *Annual Review of Biochemistry* 1980; 49:593–626.

10. Muscate, A. and Kenyon, G.L., Approaches to the rational design of enzyme inhibitors. In: M.E. Wolff, Editor. *Burger's Medicinal Chemistry and Drug Discovery*, Fifth Edition, Volume 1: Principles and Practice. New York: John Wiley & Sons, 1995, 733–782.

11. Otto, H.-H. and Schirmeister, T. Cysteine proteases and their inhibitors. *Chemical Reviews* 1997; 97:133–171.

12. Veber, D.F. and Thompson, S.K. The therapeutic potential of advances in cysteine protease inhibitor design. *Current Opinion in Drug Discovery & Development* 2000; 3:362–369.

13. Silverman, R.B. *The Organic Chemistry of Enzyme-Catalyzed Reactions*. San Diego: Academic Press, 2000, 570–591.

14. Perni, R.B., Pitlik, J., Britt, S.D., Court, J.J., Courtney, L.F., Deininger, D.D., Farmer, L.J., Gates, C.A., Harbeson, S.L., Levin, R.B., Lin, C., Lin, K., Moon, Y.-C., Luong, Y.-P., O'Malley, E.T., Rao, B.G., Thomson, J.A., Tung, R.D., Van Drie, J.H., and Wei, Y. Inhibitors of hepatitis C virus NS3.4A protease 2. Warhead SAR and optimization. *Bioorganic & Medicinal Chemistry Letters* 2004; 14:1441–1446.

15. Ringrose, P.S. Warhead delivery and suicide substrates as concepts in antimicrobial drug design. *Symposium of the Society for General Microbiology* 1985; 38:219–266.

16. Babine, R.E. and Bender, S.L. Molecular recognition of protein–ligand complexes: applications to drug design. *Chemical Reviews* 1997; 97:1359–1472.

17. Powers, J.C., Asgian, J.L., Ekici, O.D., and James, K.E. Irreversible inhibitors of serine, cysteine, and threonine proteases. *Chemical Reviews* 2002; 102:4639–4750.

18. Ripka, A.S. and Rich, D.H. Peptidomimetic design. *Current Opinion in Chemical Biology* 1998; 2:441–452.

19. Kim, D.H. Design of protease inhibitors on the basis of substrate stereospecificity. *Biopolymers* 1999; 51:3–8.

20. Fairlie, D.P., Tyndall, J.D.A., Reid, R.C., Wong, A.K., Abbenante, G., Scanlon, M.J., March, D.R., Bergman, D.A., Chai, C.L.L., and Burkett, B.A. Conformational selection of inhibitors and substrates by proteolytic enzymes: implications for drug design and polypeptide processing. *Journal of Medicinal Chemistry* 2000; 43:1271–1281.

21. Tyndall, J.D.A. and Fairlie, D.P. Macrocycles mimic the extended peptide conformation recognized by aspartic, serine, cysteine, and metallo proteases. *Current Medicinal Chemistry* 2001; 8:893–907.

22. Cushman, D.W., Pluscec, J., Williams, N.J., Weaver, E.R., Sabo, E.F., Kocy, O., Cheung, H.S., and Ondetti, M.A. Inhibition of angiotensin-converting enzyme by analogs of peptides from *Bothrops jararaca* venom. *Experientia* 1973; 29:1032–1035.

23. Gopalsamy, A., Lim, K., Ellingboe, J.W., Mitsner, B., Nikitenko, A., Upeslacis, J., Mansour, T.S., Olson, M.W., Bebernitz, G.A., Grinberg, D., Feld, B., Moy, F.J., and O'Connell, J. Design and syntheses of 1,6-naphthalene derivatives as selective HCMV protease inhibitors. *Journal of Medicinal Chemistry* 2004; 47:1893–1899.

24. Thurmond, R.L., Sun, S., Sehon, C.A., Baker, S.M., Cai, H., Gu, Y., Jiang, W., Riley, J.P., Williams, K.N., Edwards, J.P., and Karlsson, L. Identification of a potent and selective noncovalent cathepsin S inhibitor. *Journal of Pharmacology and Experimental Therapeutics* 2004; 308:268–276.

25. Kallblad, P., Todorov Nikolay, P., Willems Henriette, M.G., and Alberts Ian, L. Receptor flexibility in the in silico screening of reagents in the S1′ pocket of human collagenase. *Journal of Medicinal Chemistry* 2004; 47:2761–2767.

26. Yamashita, D.S., Smith, W.W., Zhao, B., Janson, C.A., Tomaszek, T.A., Bossard, M.J., Levy, M.A., Oh, H.-J., Carr, T.J., Thompson, S.K., Ijames, C.F., Carr, S.A., McQueney, M., D'Alessio, K.J., Amegadzie, B.Y., Hanning, C.R., Abdel-Meguid, S., DesJarlais, R.L., Gleason, J.G., and Veber, D.F. Structure and design of potent and selective cathepsin K inhibitors. *Journal of the American Chemical Society* 1997; 119:11,351–11,352.

27. Thompson, S.K., Halbert, S.M., Bossard, M.J., Tomaszek, T.A., Levy, M.A., Zhao, B., Smith, W.W., Abdel-Meguid, S.S., Janson, C.A., D'Alessio, K.J., McQueney, M.S., Amegadzie, B.Y., Hanning, C.R., DesJarlais, R.L., Briand, J., Sarkar, S.K., Huddleston, M.J., Ijames, C.F., Carr, S.A., Garnes, K.T., Shu, A., Heys, J.R., Bradbeer, J., Zembryki, D., Lee-Rykaczewski, L., James, I.E., Lark, M.W., Drake, F.H., Gowen, M., Gleason, J.G., and Veber, D.F. Design of potent and selective human cathepsin K inhibitors that span the active site. *Proceedings of the National Academy of Sciences of the United States of America* 1997; 94:14,249–14,254.

28. Marquis, R.W., Ru, Y., Zeng, J., Trout, R.E.L., LoCastro, S.M., Gribble, A.D., Witherington, J., Fenwick, A.E., Garnier, B., Tomaszek, T., Tew, D., Hemling, M.E., Quinn, C.J., Smith, W.W., Zhao, B., McQueney, M.S., Janson, C.A., D'Alessio, K., and Veber, D.F. Cyclic ketone inhibitors of the cysteine protease cathepsin K. *Journal of Medicinal Chemistry* 2001; 44:725–736.

29. Lee, C.E., Kick, E.K., and Ellman, J.A. General solid-phase synthesis approach to prepare mechanism-based aspartyl protease inhibitor libraries. Identification of potent cathepsin D inhibitors. *Journal of the American Chemical Society* 1998; 120:9735–9747.

30. Lee, A., Huang, L., and Ellman, J.A. General solid-phase method for the preparation of mechanism-based cysteine protease inhibitors. *Journal of the American Chemical Society* 1999; 121:9907–9914.

31. Huang, L., Lee, A., and Ellman, J.A. Identification of potent and selective mechanism-based inhibitors of the cysteine protease cruzain using solid-phase parallel synthesis. *Journal of Medicinal Chemistry* 2002; 45:676–684.

32. Ferreira, S.H., Bartelt, D.C., and Greene, L.J. Isolation of bradykinin-potentiating peptides from *Bothrops jararaca* venom. *Biochemistry* 1970; 9:2583–2593.

33. Ferreira, S.H. A bradykinin-potentiating factor (BPF) present in the venom of *Bothrops jararaca*. *British Journal of Pharmacology and Chemotherapy* 1965; 24:163–169.

34. Bakhle, Y.S. Conversion of angiotensin I to angiotensin II by cell-free extracts of dog lung. *Nature* 1968; 220:919–921.

35. Cushman, D.W., Cheung, H.S., Sabo, E.F., and Ondetti, M.A. Design of potent competitive inhibitors of angiotensin-converting enzyme. Carboxyalkanoyl and mercaptoalkanoyl amino acids. *Biochemistry* 1977; 16:5484–5491.

36. Byers, L.D. and Wolfenden, R. Binding of the byproduct analog benzylsuccinic acid by carboxypeptidase A. *Biochemistry* 1973; 12:2070–2078.

37. Ondetti, M.A., Rubin, B., and Cushman, D.W. Design of specific inhibitors of angiotensin-converting enzyme: new class of orally active antihypertensive agents. *Science* 1977; 196:441–444.

38. Patchett, A.A., Harris, E., Tristram, E.W., Wyvratt, M.J., Wu, M.T., Taub, D., Peterson, E.R., Ikeler, T.J., ten Broeke, J., Payne, L.G., Ondeyka, D.L., Thorsett, E.D., Greenlee, W.J., Lohr, N.S., Hoffsommer, R.D., Joshua, H., Ruyle, W.V., Rothrock, J.W., Aster, S.D., Maycock, A.L., Robinson, F.M., Hirschmann, R., Sweet, C.S., Ulm, E.H., Gross, D.M., Vassil, T.C., and Stone, C.A. A new class of angiotensin-converting enzyme inhibitors. *Nature* 1980; 288:280–283.

39. Gante, J. Peptide mimetics — tailor-made enzyme inhibitors. *Angewandte Chemie* 1994; 106:1780–1802.

40. Leung, D., Abbenante, G., and Fairlie, D.P., Protease inhibitors: current status and future prospects. *Journal of Medicinal Chemistry* 2000:305–341.

41. Zaman, M.A., Oparil, S., and Calhoun, D.A. Drugs targeting the renin–angiotensin–aldosterone system. *Nature Reviews Drug Discovery* 2002; 1:621–636.

42. Acharya, K.R., Sturrock Edward, D., Riordan James, F., and Ehlers Mario, R.W. ACE revisited: a new target for structure-based drug design. *Nature Reviews Drug Discovery* 2003; 2:891–902.

43. Huff, J.R. and Kahn, J. Discovery and clinical development of HIV-1 protease inhibitors. *Advances in Protein Chemistry* 2001; 56:213–251.

44. Tomasselli, A.G. and Heinrikson, R.L. Targeting the HIV-protease in AIDS therapy: a current clinical perspective. *Biochimica et Biophysica Acta* 2000; 1477:189–214.
45. Tomasselli, A.G., Thaisrivongs, S., and Heinrikson, R.L. Discovery and design of HIV protease inhibitors as drugs for the treatment of AIDS. *Advances in Antiviral Drug Design* 1996; 2:173–228.
46. Redshaw, S., Roberts, N.A., and Thomas, G.J. The road to Fortovase. A history of saquinavir, the first human immunodeficiency virus protease inhibitor. *Handbook of Experimental Pharmacology* 2000; 140:3–21.
47. Lin, J.H., Ostovic, D., and Vacca, J.P. The integration of medicinal chemistry, drug metabolism, and pharmaceutical research and development in drug discovery and development. The story of Crixivan, an HIV protease inhibitor. *Pharmaceutical Biotechnology* 1998; 11:233–255.
48. Wlodawer, A. and Vondrasek, J. Inhibitors of HIV-1 protease: a major success of structure-assisted drug design. *Annual Review of Biophysics and Biomolecular Structure* 1998; 27:249–284.
49. Navia, M.A., Fitzgerald, P.M.D., McKeever, B.M., Leu, C.T., Heimbach, J.C., Herber, W.K., Sigal, I.S., Darke, P.L., and Springer, J.P. Three-dimensional structure of aspartyl protease from human immunodeficiency virus HIV-1. *Nature* 1989; 337:615–620.
50. Wlodawer, A., Miller, M., Jaskolski, M., Sathyanarayana, B.K., Baldwin, E., Weber, I.T., Selk, L.M., Clawson, L., Schneider, J., and Kent, S.B.H. Conserved folding in retroviral proteases: crystal structure of a synthetic HIV-1 protease. *Science* 1989; 245:616–621.
51. Erickson, J.W. Design and structure of symmetry-based inhibitors of HIV-1 protease. *Perspectives in Drug Discovery and Design* 1993; 1:109–128.
52. Erickson, J., Neidhart, D.J., VanDrie, J., Kempf, D.J., Wang, X.C., Norbeck, D.W., Plattner, J.J., Rittenhouse, J.W., Turon, M. et al. Design, activity, and 2.8 .ANG. crystal structure of a C2 symmetric inhibitor complexed to HIV-1 protease. *Science* 1990; 249:527–533.
53. Reedijk, M., Boucher, C.A.B., van Bommel, T., Ho, D.D., Tzeng, T.B., Sereni, D., Veyssier, P., Jurriaans, S., Granneman, R., et al. Safety, pharmacokinetics, and antiviral activity of A77003, a C2 symmetry-based human immunodeficiency virus protease inhibitor. *Antimicrobial Agents and Chemotherapy* 1995; 39:1559–1564.
54. Dreyer, G.B., Lambert, D.M., Meek, T.D., Carr, T.J., Tomaszek, T.A., Jr., Fernandez, A.V., Bartus, H., Cacciavillani, E., Hassell, A.M., et al. Hydroxyethylene isostere inhibitors of human immunodeficiency virus-1 protease: structure–activity analysis using enzyme kinetics, x-ray crystallography, and infected T-cell assays. *Biochemistry* 1992; 31:6646–6659.
55. Dreyer, G.B., Boehm, J.C., Chenera, B., DesJarlais, R.L., Hassell, A.M., Meek, T.D., Tomaszek, T.A., Jr., and Lewis, M. A symmetric inhibitor binds HIV-1 protease asymmetrically. *Biochemistry* 1993; 32:937–947.
56. Kempf, D.J., Marsh, K.C., Denissen, J.F., McDonald, E., Vasavanonda, S., Flentge, C.A., Green, B.E., Fino, L., Park, C.H., et al. ABT-538 is a potent inhibitor of human immunodeficiency virus protease and has high oral bioavailability in humans. *Proceedings of the National Academy of Sciences of the United States of America* 1995; 92:2484–2488.
57. Kempf, D.J., Sham, H.L., Marsh, K.C., Flentge, C.A., Betebenner, D., Green, B.E., McDonald, E., Vasavanonda, S., Saldivar, A., Wideburg, N.E., Kati, W.M., Ruiz, L., Zhao, C., Fino, L., Patterson, J., Molla, A., Plattner, J.J., and Norbeck, D.W. Discovery of ritonavir, a potent inhibitor of HIV protease with high oral bioavailability and clinical efficacy. *Journal of Medicinal Chemistry* 1998; 41:602–617.
58. Kempf, D.J. Discovery and early development of ritonavir and ABT-378. *Infectious Disease and Therapy* 2002; 25:49–64.
59. Bold, G., Faessler, A., Capraro, H.-G., Cozens, R., Klimkait, T., Lazdins, J., Mestan, J., Poncioni, B., Roesel, J., Stover, D., Tintelnot-Blomley, M., Acemoglu, F., Beck, W., Boss, E., Eschbach, M., Huerlimann, T., Masso, E., Roussel, S., Ucci-Stoll, K., Wyss, D., and Lang, M. New aza-dipeptide analogs as potent and orally absorbed HIV-1 protease inhibitors: candidates for clinical development. *Journal of Medicinal Chemistry* 1998; 41:3387–3401.
60. Wroblewski, T., Graul, A., and Castaner, J. PNU-140690. Antiviral for AIDS; HIV-1 protease inhibitor. *Drugs of the Future* 1998; 23:146–151.
61. Thaisrivongs, S., Tomich, P.K., Watenpaugh, K.D., Chong, K.-T., Howe, W.J., Yang, C.-P., Strohbach, J.W., Turner, S.R., McGrath, J.P., et al. Structure-based design of HIV protease inhibitors: 4-hydroxycoumarins and 4-hydroxy-2-pyrones as nonpeptidic inhibitors. *Journal of Medicinal Chemistry* 1994; 37:3200–3204.

62. Thaisrivongs, S., Skulnick, H.I., Turner, S.R., Strohbach, J.W., Tommasi, R.A., Johnson, P.D., Aristoff, P.A., Judge, T.M., Gammill, R.B., Morris, J.K., et al. Structure-based design of HIV protease inhibitors: sulfonamide-containing 5,6-dihydro-4-hydroxy-2-pyrones as nonpeptidic inhibitors. *Journal of Medicinal Chemistry* 1996; 39:4349–4353.

63. Hershko, A. and Ciechanover, A. The ubiquitin system. *Annual Review of Biochemistry* 1998; 67:425–479.

64. Murray, R.Z. and Norbury, C. Proteasome inhibitors as anticancer agents. *Anti-Cancer Drugs* 2000; 11:407–417.

65. Almond, J.B. and Cohen, G.M. The proteasome: a novel target for cancer chemotherapy. *Leukemia* 2002; 16:433–443.

66. Adams, J. Proteasome inhibitors as therapeutic agents. *Expert Opinion on Therapeutic Patents* 2003; 13:45–57.

67. Delcros, J.G., Floc'h, M.B., Prigent, C., and Arlot-Bonnemains, Y. Proteasome inhibitors as therapeutic agents: current and future strategies. *Current Medicinal Chemistry* 2003; 10:479–503.

68. Loewe, J., Stock, D., Jap, B., Zwickl, P., Baumeister, W., and Huber, R. Crystal structure of the 20S proteasome from the archaeon T. acidophilum at 3.4 .ANG. resolution. *Science* 1995; 268:533–539.

69. Groll, M., Ditzel, L., Loewe, J., Stock, D., Bochtler, M., Bartunik, H.D., and Huber, R. Structure of 20S proteasome from yeast at 2.4 .ANG. resolution. *Nature* 1997; 386:463–471.

70. Adams, J., Behnke, M., Chen, S., Cruickshank, A.A., Dick, L.R., Grenier, L., Klunder, J.M., Ma, Y.-T., Plamondon, L., and Stein, R.L. Potent and selective inhibitors of the proteasome: dipeptidyl boronic acids. *Bioorganic & Medicinal Chemistry Letters* 1998; 8:333–338.

71. Kaplan, K.L. Direct thrombin inhibitors. *Expert Opinion on Pharmacotherapy* 2003; 4:653–666.

72. Chen, J.L. Argatroban: a direct thrombin inhibitor for heparin-induced thrombocytopenia and other clinical applications. *Heart Disease* 2001; 3:189–198.

73. Moledina, M., Chakir, M., and Gandhi, P.J. A synopsis of the clinical uses of argatroban. *Journal of Thrombosis and Thrombolysis* 2001; 12:141–149.

74. Walenga, J.M. An overview of the direct thrombin inhibitor argatroban. *Pathophysiology of Haemostasis and Thrombosis* 2002; 32(Suppl. 3):9–14.

75. Okajima, K., Uchiba, M., and Murakami, K. Nafamostat mesilate. *Cardiovascular Drug Reviews* 1995; 13:51–65.

76. Sargen, K. and Kingsnorth, A.N. Acute pancreatitis: an overview of emerging pharmacotherapy. *BioDrugs* 1998; 10:359–371.

77. Komoriyama, H., Tanaka, I., Ikezawa, H., Kanasugi, K., Hagiwara, M., and Yamaguchi, S. Continuous intraarterial infusion of protease inhibitors in acute pancreatitis. *Drugs of Today* 2001; 37:151–158.

78. Ryan, M.E., Greenwald, R.A., and Golub, L.M. Potential of tetracyclines to modify cartilage breakdown in osteoarthritis. *Current Opinion in Rheumatology* 1996; 8:238–247.

79. Ryan, M.E., Ramamurthy, S., and Golub, L.M. Matrix metalloproteinases and their inhibition in periodontal treatment. *Current Opinion in Periodontology* 1996; 3:85–96.

80. Wollaston, S.J. and Kalunian, K.C. Matrix metalloproteinase inhibitors as therapies for rheumatoid arthritis. *Modern Therapeutics in Rheumatic Diseases* 2002:135–145.

81. Kawabata, K., Suzuki, M., Sugitani, M., Imaki, K., Toda, M., and Miyamoto, T. ONO-5046, a novel inhibitor of human neutrophil elastase. *Biochemical and Biophysical Research Communications* 1991; 177:814–820.

82. Pradella, L. ONO-5046, Ono Pharmaceutical. *IDrugs* 2000; 3:208–222.

83. Ohbayashi, H. Novel neutrophil elastase inhibitors as a treatment for neutrophil-predominant inflammatory lung diseases. *IDrugs* 2002; 5:910–923.

84. Zeiher Bernhardt, G., Matsuoka, S., Kawabata, K., and Repine John, E. Neutrophil elastase and acute lung injury: prospects for sivelestat and other neutrophil elastase inhibitors as therapeutics. *Critical Care Medicine* 2002; 30:S281–S287.

85. Corti, R., Burnett, J.C., Jr., Rouleau, J.L., Ruschitzka, F., and Luescher, T.F. Vasopeptidase inhibitors. A new therapeutic concept in cardiovascular disease? *Circulation* 2001; 104:1856–1862.

86. Doggrell, S.A. The therapeutic potential of endothelin-1 receptor antagonists and endothelin-converting enzyme inhibitors on the cardiovascular system. *Expert Opinion on Investigational Drugs* 2002; 11:1537–1552.

87. Stanton, A. Potential of renin inhibition in cardiovascular disease. *Journal of the Renin Angiotensin Aldosterone System* 2003; 4:6–10.

88. Dragovich, P.S. Recent advances in the development of human rhinovirus 3C protease inhibitors. *Expert Opinion on Therapeutic Patents* 2001; 11:177–184.

89. Stewart, K. and Abad-Zapatero, C. *Candida* proteases and their inhibition: prospects for antifungal therapy. *Current Medicinal Chemistry* 2001; 8:941–948.

90. McKerrow, J.H., Caffrey, C.R., and Salter, J.P. Parasite proteases as targets for therapy. *Handbook of Experimental Pharmacology* 2000; 140:189–204.

91. Burgess, L.E. Mast cell tryptase as a target for drug design. *Drug News & Perspectives* 2000; 13:147–157.

92. Akahoshi, F. Chymase inhibitors and their therapeutic potential. *Drugs of the Future* 2002; 27:765–770.

93. De Nanteuil, G., Portevin, B., and Lepagnol, J. Prolyl endopeptidase inhibitors: a new class of memory enhancing drugs. *Drugs of the Future* 1998; 23:167–179.

94. Roggo, S. Inhibition of BACE, a promising approach to Alzheimer's disease therapy. *Current Topics in Medicinal Chemistry* 2002; 2:359–370.

95. Drucker, D.J. Therapeutic potential of dipeptidyl peptidase IV inhibitors for the treatment of type 2 diabetes. *Expert Opinion on Investigational Drugs* 2003; 12:87–100.

96. Gowen, M. Inhibition of cathepsin K — a novel approach to antiresorptive therapy. *Expert Opinion on Investigational Drugs* 1997; 6:1199–1202.

97. Smith, W.W. and Abdel-Meguid, S.S. Cathepsin K as a target for the treatment of osteoporosis. *Expert Opinion on Therapeutic Patents* 1999; 9:683–694.

98. Yamashita, D.S. and Dodds, R.A. Cathepsin K and the design of inhibitors of cathepsin K. *Current Pharmaceutical Design* 2000; 6:1–24.

99. Rotella, D.P. Osteoporosis: challenges and new opportunities for therapy. *Current Opinion in Drug Discovery & Development* 2002; 5:477–486.

100. Doggrell, S.A. Present and future pharmacotherapy for osteoporosis. *Drugs of Today* 2003; 39:633–657.

101. Michaud, S. and Gour, B.J. Cathepsin B inhibitors as potential anti-metastatic agents. *Expert Opinion on Therapeutic Patents* 1998; 8:645–672.

102. Bromme, D. Cysteine proteases as therapeutic targets. *Drug News & Perspectives* 1999; 12:73–82.

103. Bradshaw, R.A. Methionine aminopeptidase 2 inhibition: antiangiogenesis and tumour therapy. *Expert Opinion on Therapeutic Patents* 2004; 14:1–6.

104. Nelson, F.C. and Zask, A. The therapeutic potential of small molecule TACE inhibitors. *Expert Opinion on Investigational Drugs* 1999; 8:383–392.

105. Skotnicki, J.S. and Levin, J.I. TNF-a converting enzyme (TACE) as a therapeutic target. *Annual Reports in Medicinal Chemistry* 2003; 38:153–162.

10 A Comparison of Homogeneous Bioluminescent and Fluorescent Methods for Protease Assays

Martha A. O'Brien

CONTENTS

10.1 INTRODUCTION

Rapid and sensitive assays of proteolytic activity are necessary for general characterization of proteases and high-throughput screening for protease inhibitors. Traditionally, fluorimetric or colorimetric methods have been used to assay for protease activity. For cysteine and serine proteases, simple peptide-conjugated fluorophores or chromophores have provided easy, single-step assays. For aspartyl and metalloproteases that require amino acids P′ to the cleavage site for enzyme recognition, fluorescence resonance energy transfer (FRET)-based assays have been the method of choice. The sensitivity of fluorescent assays can be limited for a variety of reasons, however. Peptide-conjugated fluorophores can have residual fluorescence or spectral overlap with their cleaved fluorescent products, thus increasing background and reducing sensitivity [1,2]. Cells can exhibit autofluorescence, and compounds in natural product and synthetic chemical libraries frequently exhibit fluorescence that can cause assay interference [3]. Bioluminescence has been extensively exploited and optimized for gene reporter assays, but applications for protease assays have been limited. In the few reported examples of bioluminescent protease assays, the assays were two-step assays with limited sensitivity that would not be ideal for high-throughput screening [4,5]. We wanted first to determine whether bioluminescent protease assays could be configured in a homogeneous format, and second, whether this bioluminescent format could improve sensitivity, while potentially avoiding the interference and background associated with fluorescence.

Bioluminescent substrates were made for several proteases, including caspases, dipeptidyl peptidase IV (DPPIV), and calpain, by coupling appropriate peptides to aminoluciferin using standard Fmoc chemistry. A caspase-3 assay was the initial model for demonstrating that the protease and luciferase could be combined in a single-step assay [6]. The caspase-3 substrate, Z-DEVD-aminoluciferin, was combined with a stabilized luciferase [7] and other necessary reaction components to make a single reagent for quantifying caspase activity. This coupled-enzyme format was subsequently applied to several other proteases by combining the appropriate peptide-conjugated aminoluciferin substrates and the stabilized luciferase in a single-step method. These one-step bioluminescent assays were compared directly to fluorescent assays and evaluated for sensitivity, dynamic range, and assay speed. The bioluminescent coupled-enzyme system demonstrated very low background, resulting in an increased dynamic range and improved detection sensitivity. These homogeneous assays reach maximum sensitivity when the protease and luciferase reach steady-state; this occurs quickly (≤ 1 h), and then the signal is quite stable for several hours. In contrast, the sensitivity of fluorescent assays is dependent upon accumulation of cleaved substrate. In all cases, the luminescent assays were found to be more rapid and sensitive than analogous fluorescent assays that use rhodamine 110 or coumarin-based substrates.

10.2 MATERIALS AND METHODS

10.2.1 REAGENTS

The peptide-conjugated aminoluciferin substrates were synthesized using standard Fmoc chemistry, as previously described for the caspase-3/7 substrate, Z-DEVD-aminoluciferin [6]. Z-LETD-aminoluciferin and Z-LEHD-aminoluciferin were synthesized for caspase-8 and caspase-9 assays, respectively. Gly-Pro-aminoluciferin was made as a dipeptidyl peptidase IV substrate and Suc-LLVY-aminoluciferin and Z-QEVY-aminoluciferin were synthesized for a calpain assay. All substrates were purified by preparative reverse-phase high performance liquid chromatography (HPLC) to >90% purity. The fluorescent substrates tested for caspase-3 were (Z-DEVD)$_2$-rhodamine 110 and Z-DEVD-AMC (Promega Corp., Madison, WI). Z-LETD-AFC and Z-LEHD-AFC (Sigma, St. Louis, MO) were used for caspase-8 and caspase-9, respectively; GP-AMC (Sigma) for DPPIV, and Suc-LLVY-AMC (Calbiochem, San Diego, CA) for calpain. Recombinant enzymes used were caspase-3 (Biomol Research Laboratories, Inc., Plymouth Meeting, PA or Upstate Biotechnologies, Inc., Lake Placid, NY), caspase-8 and caspase-9 (Biomol Research Laboratories, Inc.), and dipeptidyl peptidase IV (R&D Systems, Minneapolis, MN). Calpain I (Calbiochem) was purified from porcine erythrocytes. A recombinant luciferase that was molecularly evolved for stability (Ultra-Glo™, Promega) was used for all assays.

10.2.2 ASSAY CONDITIONS

For the bioluminescent caspase assays, the substrates, Z-DEVD-aminoluciferin, Z-LETD-aminoluciferin, or Z-LEHD-aminoluciferin, were combined with luciferase, dithiothreitol (DTT), and other components necessary for the luciferase reaction, then lyophilized to make the Caspase-Glo

Substrates (Promega). The lyophilized Caspase-Glo substrates were resuspended in Caspase-Glo buffers that were optimized for the combined caspase and luciferase activities as well as cell lysis to make the Caspase-Glo 3/7, Caspase-Glo 8, and Caspase-Glo 9 Assay Systems (Promega). For the other assays, the substrates were combined with the Luciferin Detection Reagent (Promega) in an appropriate buffer system. The Luciferin Detection Reagent contains luciferase and necessary components for the luciferase reaction. The stabilized luciferase was used at a final concentration of 50 µg/ml for the caspase-3/7, DPPIV, and calpain assays and at 200 µg/ml for the caspase-8 and -9 assays. The DPPIV substrate, GP-aminoluciferin (10 µM final), was combined with the Luciferin Detection Reagent in 100 mM Tricine (pH 8.4) and 50 mM $MgSO_4$. The calpain substrates, Suc-LLVY-aminoluciferin or Z-QEVY-aminoluciferin (10 µM final), were combined with the Luciferin Detection Reagent in 100 mM Hepes (pH 7.5), 50 mM $MgSO_4$, and 2 mM $CaCl_2$. For negative controls, the $CaCl_2$ was omitted. For assays using recombinant caspase enzymes, the enzymes were diluted in 10 mM HEPES (pH 7.5) and 0.1% Prionex® (Pentapharm, Centerchem, Inc., Norwalk, CT). Caspase-3 in U/mL refers to Biomol enzyme, and caspase in mU/mL refers to Upstate enzyme. For comparative purposes, units are translated into protein amounts for all caspase assays. The DPPIV enzyme was diluted in 25 mM TRIS (pH 8.0) as per the manufacturer's instructions, and 0.1% Prionex was added as a carrier. The calpain I enzyme was diluted in 10 mM Hepes (pH 7.5), 10 mM DTT, 1 mM ethylenediaminetetraacetic acid (EDTA), and 1 mM EGTA . An equal amount of the substrate/luciferase mixtures was added to recombinant or purified enzyme in white multiwell plates, and the plates were shaken for 30 sec on a plate shaker and incubated at room temperature (22°C) until being read on a luminometer. Luminescence was recorded on a Dynex MLX using Revelation software, a BMG Fluostar, or a Berthold ORION. Results were comparable on the various luminometers. Luminescence was recorded at various times after the addition of the substrate–luciferase reagents.

The fluorescent caspase-3 substrate, (Z-DEVD)$_2$-rhodamine 110, was used as part of the Apo-ONE™ Homogeneous Caspase 3/7 Assay (Promega) or the Homogeneous Caspase Assay (Roche Applied Science, Indianapolis, IN) in the buffers provided by the manufacturers. The Z-DEVD-AMC (Promega) was used at 50 µM in the same buffer as the (Z-DEVD)$_2$-rhodamine 110. The Z-LETD-AFC and Z-LEHD-AFC were used at 50 and 100 µM, respectively, with 10 mM DTT (AMRESCO, Solon, OH) in the Caspase-Glo 8 and 9 Buffers (Promega). The fluorescent DPPIV substrate, GP-AMC (10 µM final), was diluted in 25 mM TRIS (pH 8.0), 0.1% Prionex. The calpain substrate, Suc-LLVY-AMC (10 µM final), was diluted in 100 mM HEPES (pH 7.5), 0.1% Prionex, and 2 mM $CaCl_2$. Fluorescence assays were performed in multiwell plates similarly to the luminescent assay. After incubation at room temperature, fluorescence was recorded on a Labsystems Fluoroskan Ascent or a Cytofluor II.

For all the assays, the background was determined in the absence of the protease. For the calpain assay, background was also determined in the absence of calcium. Signal-to-noise ratios (S/N) were calculated as per Zhang et al. [8]. S/N = mean signal − mean background/standard deviation of background. The limits of detection for all the assays were determined as an S/N ratio of 3, i.e., the mean signal was at least three standard deviations above the mean background.

10.2.3 Cell Assay

Human T cell–derived Jurkat cells (ATCC, Manassas, VA) were grown in RPMI-1640 (Hyclone, Logan, UT, and Sigma) plus 10% fetal bovine serum (Hyclone). To induce apoptosis, the cells were treated with an IgM monoclonal antibody against the Fas receptor, anti-Fas mAb (clone CH-11, Medical and Biological Laboratories, Nagoya, Japan, or Pan Vera, Madison, WI), at 100 ng/ml for 3 to 4.5 h at 37°C. Cells were left untreated as a negative control, and culture medium only, without cells, was also included as a background negative control.

10.2.4 Z' FACTOR CALCULATIONS

Recombinant caspase-3 or a buffer negative control was added to wells in either 96- or 384-well plates at various concentrations in volumes of 100 and 25 μL, respectively. A half-plate (96-well) or a quarter-plate (384-well) was used for each caspase dose. An equal volume of the Z-DEVD-aminoluciferin/luciferase reagent or the (Z-DEVD)$_2$-rhodamine 110 reagent (Apo-ONE™ Homogeneous Caspase 3/7 Assay, Promega) was added to the wells (200 and 50 μL total in 96- and 384-well plates, respectively) and luminescence or fluorescence was recorded after 1 h. Fluorescence was recorded at additional later time points. Z' factor was calculated as per Zhang et al. [8]; $Z' = 1 - [(3\sigma_{c+} + 3\sigma_{c-})/(\mu_{c+} - \mu_{c-})]$, where σ_{c+} and σ_{c-} are the standard deviations of positive and negative controls, and μ_{c+} and μ_{c-} are the mean values of positive and negative controls.

10.2.5 LOPAC SCREEN

The 640 compounds of the RBI Library of Pharmacologically Active Compounds for High-Throughput Screening (LOPAC™) (Sigma) were diluted to 100 μM stocks in 10% dimethylsulfoxide (DMSO) and stored at –20°C in eight 96-well plates. Replicate plates were set-up with 10 μL of each compound. Control wells had 10 μL of 10% DMSO. Forty microliters of caspase-3 in 10 mM HEPES and 0.1% Prionex was added to each well for a final concentration of 1 mU/well or 10 mU/well (Upstate) with 20 μM compound in 2% DMSO in 50 μL. After addition of the caspase substrate, the concentration of the compounds was 10 μM in 1% DMSO. The first and last columns of each plate were control wells. The positive controls had caspase-3 in 2% DMSO without any test compound. The negative controls were either just vehicle (no caspase or compound) or vehicle + caspase-3 + known inhibitor. The caspase-3 inhibitors used as controls were Ac-DEVD-CHO and Z-VAD-FMK. They were both used at final concentrations of 10 μM. Two separate screens were performed. In the luminescent screen, 1 mU/well caspase-3 was used and an equal volume (50 μL) of the Z-DEVD-aminoluciferin/luciferase reagent was added to all wells. In the second fluorescent screen, 10 mU/well caspase-3 was used and an equal volume of the (Z-DEVD)$_2$-rhodamine 110 reagent (Apo-ONE Homogeneous Caspase 3/7 Assay, Promega) was added to all wells. Luminescence and fluorescence were recorded after 1 h at room temperature.

10.3 RESULTS

10.3.1 BIOLUMINESCENT, HOMOGENEOUS PROTEASE ASSAY CONCEPT

Caspases are cysteine proteases that are key mediators of apoptotic cell death. Caspase-3 is a key executioner caspase, and an increase in its activity is considered a sensitive monitor of apoptosis induction for general cytotoxicity screening. Caspase-3 is also a potential drug target for inhibitors that could minimize cell death in pathological conditions. Several high-throughput caspase-3 assays have been developed recently using a variety of formats [9–13]. For these reasons, caspase-3 was a useful model for testing a bioluminescent, homogeneous protease assay.

The bioluminescent protocol is outlined in Figure 10.1. The protease cleaves the peptide-conjugated aminoluciferin, whereby aminoluciferin becomes available as a substrate for luciferase and generates light in the presence of magnesium, ATP, and oxygen. Intuitively, one might assume that performing the assay in two steps, whereby the protease cleaves its substrate and allows aminoluciferin to accumulate prior to adding the luciferase, would provide the most sensitive assay. Contrary to intuition, this is not the case. Using a caspase-3/7 substrate as a model, we compared a one-step and a two-step assay. Figure 10.2 illustrates the advantages of the one-step assay.

In the two-step assay, when caspase 3/7 is incubated with its substrate for an hour prior to adding the luciferase reagent, as expected, the initial signal is high. However, in the absence of caspase, the signal is also quite high initially. In both cases (with or without caspase), the signal then drops rapidly. In the case of the no-caspase control, the initial high signal can be attributed

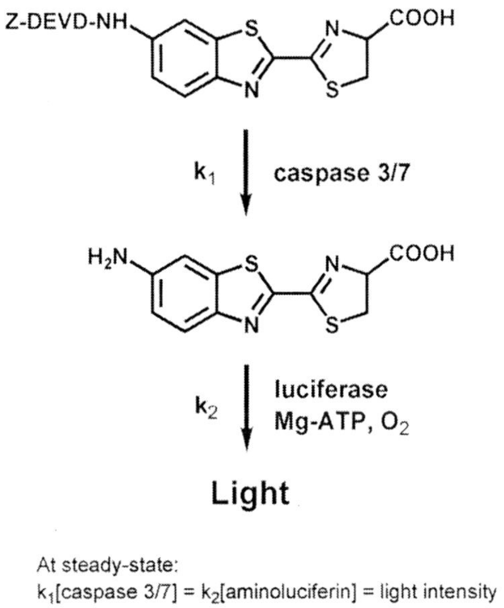

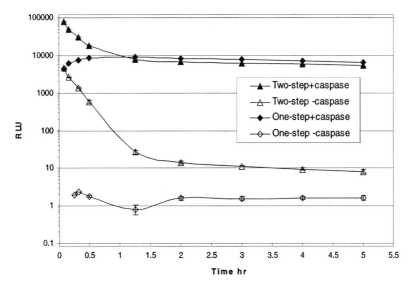

FIGURE 10.1 Scheme of the bioluminescent, coupled-enzyme, homogeneous format. An example is illustrated for a caspase-3/7 assay. Caspase-3 cleaves the DEVD from the aminoluciferin, and luciferase immediately consumes the aminoluciferin. The light emitted is directly proportional to the rate of caspase cleavage of the substrate. (From O'Brien MA et al., *J Biomed Screen* 2005;10:137–148. With permission.)

FIGURE 10.2 A homogeneous one-step, luminescent caspase assay compared to a two-step format. For the homogeneous format, the caspase 3/7 substrate, Z-DEVD-aminoluciferin, was combined with the luciferase in a 2× buffer before adding to wells containing either 10 U/ml (0.6 ng/ml) of caspase-3 or buffer only. After adding the substrate/luciferase mixture, RLUs generated with buffer only controls (open diamonds) and caspase test samples (closed diamonds) were monitored on a Dynex MLX luminometer at several time points for 5 h. For the two-step assay, the substrate was added to wells containing 10 U/ml of caspase-3 or buffer for 1 h prior to adding the luciferase in 2× buffer. RLUs generated with buffer-only controls (open triangles) and caspase test samples (closed triangles) were monitored as for the one-step format. The signal:background ratio is higher for the one-step assay at all time points. (From O'Brien MA et al., *J Biomed Screen* 2005;10:137–148. With permission.)

to contaminating free aminoluciferin in the preparation of peptide-conjugated aminoluciferin. What is striking is that by HPLC we could not detect any contaminating free aminoluciferin. The bioluminescence is so sensitive that significant signal can be generated from even trace amounts of aminoluciferin contamination. For the results with caspase, the initial high signal is a combination of accumulated free aminoluciferin due to caspase cleavage and contaminating free aminoluciferin in the original substrate.

In the one-step assay, the problem of contaminating free aminoluciferin is inherently eliminated. For this assay, the Z-DEVD-aminoluciferin substrate was incubated with luciferase overnight before running the caspase-3 assay. In the absence of caspase, the signal was very low because any contaminating, free aminoluciferin had been consumed by luciferase prior to beginning the assay. When caspase was added, the signal rapidly climbed until the steady-state between caspase cleavage of substrate and luciferase oxidation of aminoluciferin was reached (Figure 10.2). The signal generated with caspase in the two-step assay eventually reaches this plateau as well, because with the addition of luciferase, the accumulated aminoluciferin is consumed and a coupled-enzyme reaction is created. In the two-step assay, the signal in the absence of caspase eventually reaches the baseline as in the one-step assay, but this has not occurred even after 5 h, and thus the background is consistently lower in the one-step format (Figure 10.2). The reduction in background afforded by the one-step assay led to higher signal-to-background ratios at all time points and an improved overall sensitivity. This homogeneous, one-step assay also provided a very stable signal that can be attributed to both the steady-state nature of the coupled-enzyme assay and to the luciferase enzyme evolved for enhanced stability [7]. The limitations for utilizing this homogeneous, bioluminescent protease assay format are twofold:

- The protease assay buffer must be compatible with luciferase activity.
- The protease should not require P′ sites on the substrate for recognition/cleavage.

Serine and cysteine proteases generally should work with this method.

10.3.2 BIOLUMINESCENT ASSAY PERFORMANCE COMPARED TO FLUORESCENCE

10.3.2.1 Caspase 3/7 Assay

10.3.2.1.1 Recombinant Enzyme Assay

We compared the homogeneous bioluminescent caspase-3/7 assay to comparable fluorescent assays that utilize rhodamine 110 and aminomethyl coumarin (AMC) as the reporters. All of the substrates were based on the DEVD peptide sequence, the optimized recognition sequence for caspases 3 and 7 [14]. The assays were run in parallel, and signal-to-noise (S/N) ratios were calculated to enable meaningful comparisons between the luminescent relative light units (RLU) and the fluorescent relative light units (RFLU). A titration of recombinant caspase-3 from 0.0005 to 10 U/ml was added to 96-well plates, and after adding the appropriate substrate mixtures, readings were taken on either a luminometer or a fluorimeter at various times. At both 1 and 3 h after adding reagent, the S/N ratios for the luminescent assay were much greater than those for the fluorescent assays using either the (Z-DEVD)$_2$-rhodamine 110 substrate or the Z-DEVD-AMC substrate (Figure 10.3a). For the Z-DEVD-AMC substrate, the doses of caspase-3 tested here were below the limit of detection at 1 h and just barely above the limit of detection after 3 h. The limit of detection was defined as the caspase concentration giving a S/N ratio of at least three (dotted line, Figure 10.3a). As expected, between 1 and 3 h, the S/N ratio increased for the fluorescent assays, whereas it remained the same for the luminescent assay; even after 3 h, however, the fluorescent assays are still significantly less sensitive than the luminescent assay. The limit of detection for the luminescent assay was 0.002 U caspase-3 for both time points. This is about 75 times more sensitive than the (Z-DEVD)$_2$-

rhodamine 110 assay at 1 h, approximately 20 times more sensitive at 3 h, and ≥1000-fold more sensitive than the Z-DEVD-AMC assay at both time points (Figure 10.3a).

10.3.2.1.2 Cell Assay

To determine whether the luminescent caspase assay would demonstrate similar improvements over fluorescence in a cellular assay, a model apoptosis cell culture system was used to compare the two formats. This was an important consideration because caspase-3 assays are widely used as a sensitive monitor of apoptosis induction as an indicator of cytotoxicity in drug screening. We compared the bioluminescent assay to the more sensitive fluorescent assay that utilized the (Z-DEVD)$_2$-rhodamine 110 substrate. Jurkat cells, treated with anti-Fas to induce apoptosis or left untreated as controls, were serially diluted and added to 96-well plates. Both assay systems contain cell lysis reagents, so they can be used as single-step reagents for cell-based assays. The assays were run in parallel, and luminescence and fluorescence results were again plotted as S/N ratios. The S/N ratios at 1 h after adding the luminescent assay reagent were 20-fold higher than those for (Z-DEVD)$_2$-rhodamine 110 (Figure 10.3b). Likewise, the detection limit for apoptotic cells was 20-fold lower for the bioluminescent assay. At an S/N ratio of three, the luminescent assay could detect caspase activity in ~30 induced cells, whereas the fluorescent assay had a lower limit of detection of about 600 induced cells (Figure 10.3b). Improved sensitivity in a caspase assay used for cytotoxicity screening has important ramifications including:

- Being able to detect lower level cytotoxicity
- Being able to use fewer cells and less test compound in a drug screen

10.3.2.1.3 Z′ Factor Analysis

Assay suitability for HTS is often assessed using Z′ factors [8]. We calculated Z′ factors by testing recombinant caspase-3 at varying concentrations in 96-well plates for both the bioluminescent assay and the rhodamine-based fluorescent assay (Table 10.1). Z′ values ranged from 0.97 to 0.90 for the bioluminescent assay, when using caspase-3 at concentrations between 100 and 0.1 U/ml as the positive control and buffer without caspase as the negative control. As expected, the Z′-factor depended on the concentration of the caspase, with higher concentrations providing better Z′ values. An exception was for the highest concentration of caspase (100 U/ml), where the Z′ value decreased due to cross-talk between the negative control wells immediately adjacent to the very bright positive control wells. For the fluorescent assay, the Z′ factors ranged from 0.94 for 10 U/ml caspase to <0.5 at the 0.1 U/ml caspase concentration. Unexpectedly, the Z′ did not improve with longer incubation times using the Z-(DEVD)$_2$-rhodamine 110 substrate. Although the signal-to-background ratio increased with time for the fluorescence assay, the variability between wells also increased (Table 10.1). The Z′ values for bioluminescent assays in 384-well plates were similar to Z′ values for assays in 96-well plates, indicating that the assay can be readily miniaturized.

10.3.2.1.4 LOPAC Screen

A possible concern for using a coupled bioluminescent assay for screening inhibitors could be false positives due to inhibition of the luciferase itself. Using a stabilized luciferase in the assay formulation, which should be more resistant to potential structural perturbations caused by compounds from the libraries, minimizes this possibility. We examined the extent of this drawback in a simulated screen for inhibitors by testing the bioluminescent assay in a screen of the LOPAC collection and comparing it to a similar screen using the fluorescent assay [6]. The library was screened at a final concentration of 10 μM in 96-well plates. Recombinant caspase-3 was used at 0.1 ng/well for the bioluminescent assay and at 1 ng/well for the fluorescent assay. The tenfold greater concentration of caspase was needed for the fluorescent assay to achieve an appropriate signal-to-background ratio. The positive controls were caspase-3 in the compound vehicle (1% DMSO), and the negative controls were vehicle only.

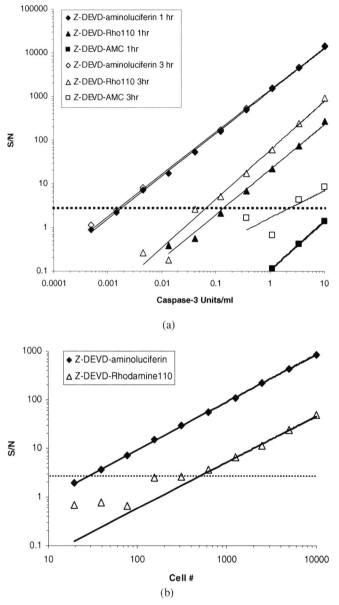

(a)

(b)

FIGURE 10.3 (a) A comparison between luminescent and fluorescent caspase assays using recombinant caspase-3. The assays were run simultaneously using the same titration of caspase-3 in 96-well plates. After addition of the appropriate substrate reagents, readings were taken at 1 and 3 h. Results were plotted as S/N. Background RLUs or RFLUs were determined from wells without caspase-3. The limit of detection for each assay is defined as the amount of caspase giving a S/N ratio ≥3 (dotted line). One unit of caspase-3 = 60 pg. (b) A comparison between luminescent and fluorescent caspase assays using apoptotic Jurkat cells. The assays were run simultaneously using the same anti-Fas treated Jurkat cells serially diluted in 96-well plates. After addition of the appropriate substrate reagents, readings were taken at 1 h. Results were plotted as S/N. Background RLUs or RFLUs were determined from wells containing culture medium without cells. The limit of detection is defined as the number of cells giving a S/N ratio ≥3 (dotted line). For the fluorescent results, only the points above the limit of detection were used to determine the linear trend since values below the limit of detection deviate significantly from the linear trend. (From O'Brien MA et al., *J Biomed Screen* 2005;10:137–148. With permission.)

TABLE 10.1

Z′ Factor Analysis Comparison for Luminescent and Fluorescent Caspase Assays[a]

Caspase-3	Z-DEVD-Aminoluciferin			Z-(DEVD)$_2$-Rhodanine 110	
	1 hr		1 hr	3 hr	18 hr
	384 Well	96 Well	96 Well		
100 U/ml (7 ng/ml)	0.88	0.9	ND	ND	ND
10 U/ml (0.7 ng/ml)	0.91	0.97	0.94	0.93	0.88
1 U/ml (0.07 ng/ml)	ND	0.96	0.73	0.78	0.7
0.1 U/ml (7 pg/ml)	ND	0.94	<0.5	<0.5	<0.5

Note: ND indicates not determined.

[a] The Z′ values were calculated as per Zhang JH, Chung TD, Oldenburg KR. *J Biomol Screen* 1999;4:67–73 for different concentrations of caspase-3 using both the luminescent and fluorescent caspase assays in both 96-well and 384-well plates. Wells without caspase constituted the negative controls.

Source: From O'Brien MA et al., *J Biomed Screen* 2005;10:137–148. With permission.

TABLE 10.2

Percentage False Hits in Screens for Caspase-3/7 Inhibitors Using the 640-Compound LOPAC[a]

	LOPAC Screen # Hits (2SD Below the Mean of All Samples)	(640 Compounds) # in Both Assays	False Hits	% False Hits
Z-DEVD-aminoluciferin	6	1	5	0.8
Z-DEVD$_2$-rhodamine 110	9	1	8	1.3
	# Hits (3SD Below the Mean of Positive Controls)			
Z-DEVD-aminoluciferin	18	10	8	1.3
Z-DEVD$_2$-rhodamine 110	55	10	45	7

[a] The library was screened with both the luminescent and fluorescent caspase assays. Hits were calculated in two different ways to determine a false hit rate. Hits were either two standard deviations below the mean of all compounds or three standard deviations below the mean of the positive (+caspase) controls. If a hit was detected in both the luminescent and fluorescent assays, it was not considered to be a false hit.

Source: From O'Brien MA et al., *J Biomed Screen* 2005;10:137–148. With permission.

The data were analyzed in two different ways (Table 10.2). First, a hit was defined as a sample compound that gave a value 2σ or more below the mean of all samples. Out of 640 compounds, there were six hits using the bioluminescent assay and nine hits for the fluorescent assay. One compound gave a hit in both assays; this compound was considered to be active (potentially a true inhibitor), while the others are considered false hits. The five false hits detected with the Z-DEVD-aminoluciferin were picked up in other luminescent assay screens of the LOPAC collection and have been determined to be inhibitors of the luciferase reaction (unpublished results). The eight false hits detected with the Z-DEVD-rhodamine 110 are presumed to be due to fluorescence interference. Thus we observed false hit rates of 0.8% in the luminescent screen and 1.3% in the fluorescent screen. Secondly, the results were analyzed using a definition of a hit as 3σ below the mean of the positive controls. In this case, there were 18 hits with the luminescent substrate, but 55 hits with the fluorescent substrate. Ten of these hits overlapped in both assays; therefore, this

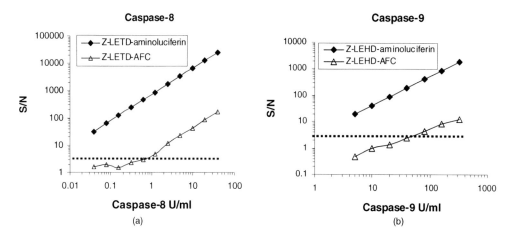

FIGURE 10.4 A comparison between (a) luminescent and fluorescent caspase assays using recombinant caspase-8 and (b) recombinant caspase-9. The luminescent and fluorescent assays were run simultaneously using the same titration of caspase-8 or -9 in 96-well plates. After addition of the appropriate substrate reagents, readings were taken at 1 h on a BMG Fluostar luminometer and a Cytofluor II fluorimeter. Results were plotted as S/N. Background RLUs or RFLUs were determined from wells without caspase-8 or -9. The limit of detection for each assay is defined as the amount of caspase giving a S/N ratio ≥ 3 (dotted lines). Ten U of caspase-8 = 2.5 ng. Ten U of caspase-9 = 140 ng.

gives a false hit rate of 1.3% for the bioluminescent assay and 7.0% for the fluorescent assay. The bioluminescent assay gave fewer false hits regardless of the criteria used, ranging from 0.8 to 1.3% compared to 1.3 to 7.0% for the fluorescent assay.

10.3.2.2   Caspase-8 and Caspase-9 Assays

Bioluminescent substrates were also synthesized for caspase-8 and caspase-9, Z-LETD-aminoluciferin and Z-LEHD-aminoluciferin, respectively [15]. Comparisons to fluorescent caspase-8 and -9 assays were carried out with the fluorescent substrates Z-LETD-AFC and Z-LEHD-AFC and using recombinant enzyme. The assays were run in parallel and signal-to-noise (S/N) ratios were calculated. A titration of recombinant caspase-8 from 0.04 to 40 U/ml was added to 96-well plates and after adding the appropriate substrate mixtures, readings were taken on either a luminometer or a fluorimeter at various times. For caspase-9, a titration of 5 to 320 U/ml caspase was used. At 1 h after adding reagent, the S/N ratios for both luminescent assays were much greater than those for the fluorescent assays. The limit of detection was again defined as the caspase concentration giving a S/N ratio of at least three (dotted line, Figure 10.4). For caspase-8, the limit of detection using the Z-LETD-AFC substrate was ~0.6 U/ml, whereas the limit of detection was well below the lowest 0.04 U/ml concentration of caspase-8 for the luminescent assay (Figure 10.4a). Likewise, for the caspase-9 assay, the limit of detection was not reached for the luminescent assay, so it is well below 5 U/ml, while the limit of detection for the fluorescent caspase-9 assay was ~100 U/ml (Figure 10.4b). The sensitivity improvement of the bioluminescent format was applicable not only to a caspase-3 assay but also to caspase-8 and caspase-9 assays.

10.3.2.3   DPPIV Assay

Dipeptidyl peptidase IV is a serine protease that is an important drug target for type 2 diabetes due to its role in inactivation of peptides of the glucagon family. DPPIV has specificity for cleavage of Xaa-Pro and to a lesser extent Xaa-Ala on the N-terminal end of peptides [16]. GlyPro-AMC is widely available as a DPPIV substrate. A GlyPro-aminoluciferin substrate was synthesized and used in a homogeneous assay in comparison to the GlyPro-AMC substrate. The bioluminescent format was run in one half of a 96-well plate, and the fluorescent format was run in the other half. Fluorescence and luminescence were recorded at various times and plotted as S/N ratios (Figure 10.5a). After 30 min, the S/N ratios are 100-fold greater for the bioluminescent assay than for the fluorescent assay. The S/N ratios of the fluorescent assay improve after 5 h as the cleaved fluorophore accumulates, but they are still tenfold lower than for the luminescent assay (Figure 10.5a). The half-life for the luminescent signal is >5 h (Figure 10.5b). The stability of the luminescent signal over time leads to relatively consistent S/N ratios over several hours (Figure 10.5). Concomitant with the improved S/N ratios in the bioluminescent assay is a decrease in the limit of detection compared to the fluorescent assay. Even after 5 h, the limit of detection for DPPIV is more than tenfold lower in the bioluminescent assay (Figure 10.5a). To check for linearity over the broad dynamic range, the titration results were plotted on a log_{10} scale, and a linear regression trendline was calculated. For the reading taken at 30 min, the R^2 value was 0.9999 and the slope was 0.95 (Figure 10.5a). These results indicate linearity over a broad range of DPPIV concentration. As was demonstrated for the caspases, a homogeneous bioluminescent assay for the serine protease, DPPIV, showed significant improvements in the sensitivity, stability of signal, and speed to maximum sensitivity.

10.3.2.4 Calpain Assay

The calpains, a family of calcium-activated cysteine proteases, are ubiquitously expressed, but their physiological roles are not well defined [17]. The most extensively studied calpains, μ-calpain (I) and m-calpain (II), are likely to have important roles in many pathological conditions where calcium homeostasis is involved, such as neurodegenerative diseases, neural ischemia, and spinal cord injury. Calpain assays have been problematic because, in general, labeled-peptide substrates for calpain have limited sensitivity and apparent K_m values that are orders of magnitude higher than those of the natural protein substrates [17,18]. The rapid autolysis and inactivation of calpains upon calcium activation have made it difficult to develop sensitive assays. The most sensitive assays utilize fluorescently labeled protein substrates, such as MAP2 or casein, in an autoquenching format [19,20]. The MAP2 assay, using a naturally occurring substrate, is the more sensitive with limits of detection reported to be ~1 nM, but the assay requires purified MAP2, which is not readily available. More recently, a FRET-based calpain assay was developed using a P3-P3' peptide sequence from the naturally occurring substrate, α-spectrin [21]. The FRET-based peptide assay is somewhat less sensitive, with a limit of detection of 20 nM, but it is reported to be more specific for calpain.

Two bioluminescent peptide substrates for calpain were made. The first, Suc-LLVY-aminoluciferin, was based on the standard peptide calpain substrate, Suc-LLVY-AMC [18]. The second, Z-QEVY-aminoluciferin, was based on the P4-P1 cleavage site sequence in α-spectrin [21]. In a direct comparison to the Suc-LLVY-AMC, the bioluminescent substrates gave dramatically better results (Figure 10.6). The fluorescent assay had a very limited linear dynamic range and poor sensitivity. In contrast, the bioluminescent assay gave a broad dynamic range, and the S/N ratios were ≥100-fold higher than those of the fluorescent assay [Figure 10.6a]. The limit of detection for the bioluminescent calpain assay was less than 1 nM; this was substantially more sensitive than the FRET peptide assay (limit of detection of 20 nM [21]) and even more sensitive than the assay using fluorescently-labeled MAP2, which is reported to be the most sensitive [19]. In contrast to the other protease assays, the signal is less stable for the calpain assay. After 1 h, the S/N ratios have dropped by ~50% (Figure 10.6a). This may be the result of calpain autolysis. Recent studies indicate that autolysis upon calcium activation leads to instability of the calpain heterodimers and subsequent loss of proteolytic activity [22]. In this case, where the protease of interest exhibits significant instability, the rapid time to maximal sensitivity in the homogeneous, bioluminescent assay provides an advantage over fluorescent assays.

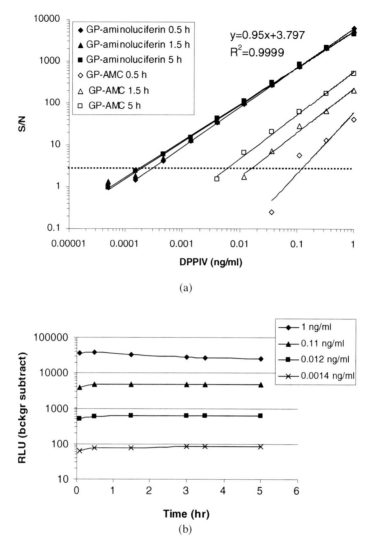

FIGURE 10.5 (a) A comparison between luminescent and fluorescent DPPIV assays using recombinant DPPIV. The assays were run simultaneously using the same titration of DPPIV in 96-well plates. After addition of the appropriate substrate reagents, readings were taken at 30 min, 1.5 h, and 5 h. Results were plotted as S/N. Background RLUs or RFLUs were determined from wells without DPPIV. The limit of detection for each assay is defined as the amount of caspase giving a S/N ratio ≥ 3 (dotted lines). The slope and R^2 value are shown for the 30 min results. (b) Signal stability in the homogeneous, bioluminescent DPPIV assay. DPPIV activity was measured at several concentrations and monitored over time after addition of the substrate/luciferase mixture. The background RLUs (no DPPIV) were subtracted from all values.

The usefulness of the homogeneous, bioluminescent calpain assay was also demonstrated with a calcium titration. The Z-QEVY-aminoluciferin substrate was compared to the Suc-LLVY-AMC over a broad range of calcium concentration with a fixed amount of calpain (Figure 10.6b). Note that the assays were run in the presence of EDTA and EGTA, so the absolute amount of available calcium is not known and the apparent EC_{50} is higher than that established for μ-calpain. Nevertheless, the homogeneous, bioluminescent assay generated a robust dose-response within 10 min whether or not the background was subtracted, whereas the fluorescent assay gave a weak dose–response that was only detected after background was subtracted and only after 2 h. The high background in the fluorescent assay obscured the calcium titration results.

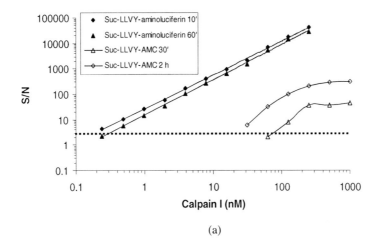

(a)

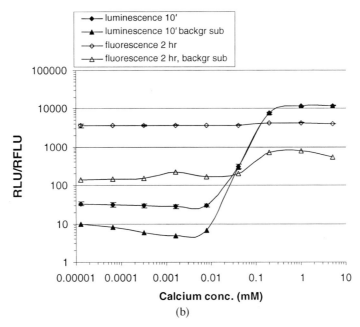

(b)

FIGURE 10.6 (a) A comparison between luminescent and fluorescent calpain assays using purified calpain I. The assays were run simultaneously using the same titration of calpain I in 96-well plates. After addition of the appropriate substrate reagents, readings were taken at several time points between 10 min and 2 h. Results were plotted as S/N. Background RLUs or RFLUs were determined from wells without calpain I. The calpain titration was also run in the absence of calcium. This gave background values for all calpain concentrations. The limit of detection for each assay is defined as the amount of calpain I giving a S/N ratio $\geq$ 3 (dotted lines). (b) A calcium titration comparison for the luminescent and fluorescent calpain assays. For the luminescent assay, the Z-QEVY-aminoluciferin substrate was used with 25 nM calpain I. For the fluorescent assay, the Suc-LLVY-AMC substrate was used with 75 nM calpain I. The same calcium titration was used for both assays run in parallel in 96-well plates. As a negative control, the calcium titration was also run in the absence of calpain I. The RLUs and RFLUs were monitored at several time points on a Dynex MLX luminometer and a Labsystems Fluoroscan Ascent fluorimeter. The signal for the luminescent assay reached a maximum in 10 min. A difference in the calcium titration was observed only after 2 h for the fluorescent assay.

10.4   LIMITATIONS

The basis of bioluminescent protease technology is that the luciferase will not recognize peptide-conjugated aminoluciferin as a substrate. This fact poses a limitation regarding proteases that require P′ sites for recognition. The cleavage needs to occur just adjacent to the aminoluciferin in order to generate a viable luciferase substrate, so only protease peptide substrates with sequences N-terminal to the cleavage site will be functional. Furthermore, to achieve the benefits of the homogeneous format, the protease buffer system needs to be compatible with the luciferase assay. The stabilized luciferase assay system tolerates a wide range of buffer components including reducing agents (e.g., DTT), DMSO, high $CaCl_2$, and lysis detergents. However, unusual protease assay conditions, such as very high or very low pH requirements, would have to be tested for compatibility.

10.5   CONCLUSIONS

We have developed single-step, bioluminescent protease assays and shown that they have several advantages over comparable fluorescent assays. For the bioluminescent caspase-3/7 assay, we demonstrated improved sensitivity, speed, stability, and Z′ values, and a lower false hit rate in a test inhibitor screen when compared to the comparable fluorescent assay. This assay was tested using recombinant enzyme and a model cell system, and in both situations the luminescent assay demonstrated improved performance over available fluorescent assays. For caspase-8, caspase-9, DPPIV, and calpain, the bioluminescent assays also resulted in improved sensitivity, speed, and stability when compared to fluorescent assays. These improvements are the combined result of the inherent bioluminescent chemistry, the homogeneous format, and the use of a stabilized luciferase. The noted advantages, which should not be unique to the proteases tested here, can directly impact HTS, further enabling drug discovery by reducing costs, adding convenience, and saving time. This bioluminescent format should be widely applicable to proteases that do not have strict requirements for P′ amino acids in their peptide substrates.

ACKNOWLEDGMENTS

The author wishes to acknowledge numerous people involved in the ideas and development of the bioluminescent protease assays. Keith Wood and Dieter Klaubert were instrumental in developing the homogeneous, bioluminescent protease assay concept. Bill Daily, Eric Hesselberth, and Mike Scurria synthesized and purified the aminoluciferin-conjugated substrates. Bob Bulleit, Terry Riss, Rich Moravec, Drew Niles, Laurent Bernad, Jean Osterman, Mary Sobol, Kay Rashka, Deborah Bishop, Tracy Worzella, Brad Larson, Brian McNamara, Marni Amburn, and Pam Guthmiller contributed significantly to the development and production of the Caspase-Glo Assays. Thanks to Bob Bulleit and Keith Wood for a careful reading of this manuscript.

REFERENCES

1. Leytus SP, Melhado LL, Mangel WF. Rhodamine-based compounds as fluorogenic substrates for serine proteinases. *Biochem J* 1983;209:299–307.
2. Liu J, Bhalgat M, Zhang C, Diwu Z, Hoyland B, Klaubert DH. Fluorescent molecular probes V: a sensitive caspase-3 substrate for fluorometric assays. *Bioorg Med Chem Lett* 1999;9:3231–3236.
3. Grant SK, Sklar JG, Cummings RT. Development of novel assays for proteolytic enzymes using rhodamine-based fluorogenic substrates. *J Biomol Screen* 2002;7:531–540.
4. Monsees T, Miska W, Geiger R. Synthesis and characterization of a bioluminogenic substrate for alpha-chymotrypsin. *Anal Biochem* 1994;221:329–334.
5. Monsees T, Geiger R, Miska W. A novel bioluminogenic assay for alpha-chymotrypsin. *J Biolumin Chemilumin* 1995;10:213–218.

6. O'Brien MA, Daily WJ, Hesselberth PE, Moravec RA, Scurria MA, Klaubert DH, Bulleit RF, Wood KV. Homogeneous, bioluminescent protease assays: caspase-3 as a model. *J Biomed Screen* 2005;10:137–148.

7. Hall MP, Gruber MG, Hannah RR, Jennens-Clough ML, Wood KV. Stabilization of firefly luciferase using directed evolution. In: Roda, A. et al., (eds.): *Bioluminescence and Chemiluminescence — Perspectives for the 21st Century*. Chichester, U.K.: John Wiley & Sons, 1998:392–395.

8. Zhang JH, Chung TD, Oldenburg KR. A simple statistical parameter for use in evaluation and validation of high-throughput screening assays. *J Biomol Screen* 1999;4:67–73.

9. Tawa P, Tam J, Cassady R, Nicholson DW, Xanthoudakis S. Quantitative analysis of fluorescent caspase substrate cleavage in intact cells and identification of novel inhibitors of apoptosis. *Cell Death Differ* 2001;8:30–37.

10. Waud JP, Bermudez Fajardo A, Sudhaharan T, Trimby AR, Jeffery J, Jones A, Campbell AK. Measurement of proteases using chemiluminescence-resonance-energy-transfer chimaeras between green fluorescent protein and aequorin. *Biochem J* 2001;357:687–697.

11. Preaudat M, Ouled-Diaf J, Alpha-Bazin B, Mathis G, Mitsugi T, Aono Y, Takahashi K, Takemoto H. A homogeneous caspase-3 activity assay using HTRF technology. *J Biomol Screen* 2002;7:267–274.

12. Karvinen J, Hurskainen P, Gopalakrishnan S, Burns D, Warrior U, Hemmila I. Homogeneous time-resolved fluorescence quenching assay (LANCE) for caspase-3. *J Biomol Screen* 2002;7:223–231.

13. Gopalakrishnan SM, Karvinen J, Kofron JL, Burns DJ, Warrior U. Application of micro arrayed compound screening (µARCS) to identify inhibitors of caspase-3. *J Biomol Screen* 2002;7:317–323.

14. Thornberry NA, Rano TA, Peterson EP, Rasper DM, Timkey T, Garcia-Calvo M, Houtzager VM, Nordstrom PA, Roy S, Vaillancourt JP, Chapman KT, Nicholson DW. A combinatorial approach defines specificities of members of the caspase family and granzyme B. *J Biol Chem* 1997;272:17,907–17,911.

15. Niles A, Moravec R, Scurria M, O'Brien M, Riss T. A novel homogeneous bioluminescent caspase-8 activity assay [Poster 274]. Paper presented at the Keystone Symposia: Molecular Mechanisms of Apoptosis, Banff, Alberta, February 2003.

16. Wiederman, PE, Trevillyan JM. Dipeptidyl peptidase IV inhibitors for the treatment of impaired glucose tolerance and type 2 diabetes. *Curr Opinion Invest Drugs* 2003;4:412–420.

17. Goll DE, Thompson VF, Li H, Wei W, Cong J. The calpain system. *Physiol Rev* 2003;83:731–801.

18. Sasaki T, Kikuchi T, Yumoto N, Yoshimura N, Murachi T. Comparative specificity and kinetic studies on porcine calpain I and calpain II with naturally occurring peptides and synthetic fluorogenic substrates. *J Biol Chem* 1984;259:12,489–12,494.

19. Tompa P, Schád E, Baki A, Alexa A, Batke J, Freidrich P. An ultrasensitive, continuous fluorometric assay for calpain activity. *Anal Biochem* 1995;228:287–293.

20. Thompson VF, Saldana S, Cong J, Goll DE. A BODIPY fluorescent microplate assay for measuring activity of calpains and other proteases. *Anal Biochem* 2000;279:170–178.

21. Mittoo S, Sundstrom LE, Bradley M. Synthesis and evaluation of fluorescent probes for the detection of calpain activity. *Anal Biochem* 2003;319:234–238.

22. Li H, Thompson VF, Goll DE. Effects of autolysis on properties of µ- and m-calpain. *Biochim Biophys Acta* 2004;1691:91–103.

11 Scintillation Proximity Assay (SPA) Receptor Binding Assays

Jeffrey R. Cook, Robert Graves, Molly J. Price-Jones, Jenny A. Berry, and Kelvin T. Hughes

CONTENTS

11.1 INTRODUCTION

Receptor binding assays (RBA) encompass a variety of techniques used to study ligand–receptor interactions. Activation of the receptor–ligand complex and its downstream cellular pathway(s) often leads to the changes in cellular function that may be related to a disease phenotype. Understanding this process can lead to the development of new therapeutic agents used to treat a variety of diseases. Scintillation proximity assay (SPA) bead-based technology has been successfully applied to study receptor–ligand binding interactions in a wide range of assays. These beads, which are size excluded to approximately 3 to 5 microns, can be coated with various capture molecules that facilitate the binding of specific biomolecules. The basic principle of SPA employs immobilizing a receptor by a specific capture molecule, typically wheat germ agglutinin (WGA)–coated

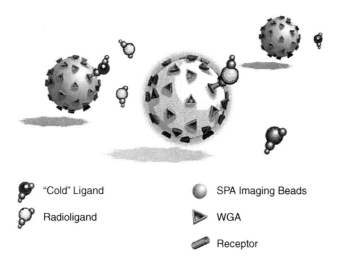

"Cold" Ligand   SPA Imaging Beads

Radioligand   WGA

Receptor

FIGURE 11.1 SPA receptor binding assay.

SPA beads. Isotopes commonly used with SPA include ^{33}P, ^{14}C, ^{35}S, ^{3}H, and ^{125}I; the latter two isotopes, tritium and iodine 125, are most commonly used for receptor binding assays. The general concept of a SPA receptor binding assay is depicted in Figure 11.1. Once a specific membrane receptor is bound to the bead, it can be challenged by an appropriate radiolabeled molecule or by a specific competitor. If the radiolabeled molecule or ligand binds to the bead via its receptor, energy from the decaying isotope is transferred to the scintillant within the bead, at which point light will be emitted from the bead, and a specific signal will be detected that is proportional to the amount of radiolabeled compound bound to the bead [1].

11.2 SPA BEAD TYPES

SPA beads can be divided into two core types depending on the type of instrumentation used to detect them. SPA beads counted with a conventional photo multiplier tube (PMT) are made of either polyvinyltoluene (PVT) or yttrium silicate (YSi). These are known as conventional SPA beads. The other SPA bead type is detected with a charge-coupled device or CCD. These beads are known as red-shifted or Imaging beads and are made of either polystyrene (PS) or yttrium oxide (YOx). PVT and PS beads are composed of a plastic medium while YSi and YOx are composed of a glass matrix. The main difference between conventional SPA and Imaging beads is the wavelength of light emitted from the beads. Figure 11.2 shows the light emission spectra for the various SPA beads. Standard photomultiplier tubes (PMTs) are more sensitive to light in the blue region of the emission spectrum; therefore, conventional SPA beads were developed to emit light in this region, at 400 to 450 nm.

Conventional liquid scintillation counters and microplate-based scintillation counters such as the TopCount™ and Microbeta® from PerkinElmer (Boston, MA) are well suited to read conventional SPA beads. However, various types of CCD chips or detectors can be more sensitive to light in the red region of the electromagnetic (EM) spectrum, rather than the blue. Therefore, technologies that emit light in the red region can take advantages of CCD-based cameras as the main mode of detection. These cameras or detectors offer several advantages over standard PMTs, two of which are higher throughput data acquisition and the ability to miniaturize high-throughput screening (HTS) assays to the 1536-well format. Since many compounds used during drug screening are

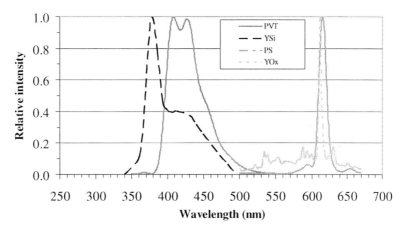

Emission Spectra for SPA and Imaging Beads

FIGURE 11.2 SPA bead emission spectrum. There are four core bead types, two based on use with a conventional PMT reader such as the Topcount™ or the Microbeta®. These beads emit light around the 360 to 400 nm range. Imaging beads are more red-shifted and emit light at around 615 nm.

yellow to orange in color and therefore absorb blue light, SPA imaging technology can decrease the number of false positive "hits" as compared to conventional SPA technology, due to the shift of its light emission into the red region of the electromagnetic spectrum [2]. Instruments that incorporate CCD technology are the LEADseeker™ Multi-Modality Imaging System from GE Healthcare (Piscataway, NJ) and the ViewLux™ ultraHTS Microplate Imager from PerkinElmer.

Since SPA is a solid phase bead-based proximity technology, it is important to optimize the amount of receptor with respect to the immobilizing surface. The surface area of the beads is finite, and the signal attainable therefore depends on the number of receptors (not cell membrane protein) that are specifically bound to the bead. Also, the affinity of the ligand for the receptor and the specific activity of the ligand play important roles in a successful SPA assay. The advantages of using this technology are that one can simply add all the components for a receptor binding assay, which include receptor complex, radiolabeled compound, inhibitors, and/or competitors, and appropriate SPA beads into a single well. The reaction is allowed to incubate for the desired time, after which it can be read by the appropriate detector. The assay is simple, fast, and homogeneous, with no need to separate unbound radiolabeled material from bound isotope.

11.3 RECEPTOR SOURCES

The section reviews the major sources of receptors used for SPA binding assays. In general, the more receptors available for use in an assay, the greater the probability that a successful receptor binding assay will be developed. Therefore, the aim is to obtain as many receptors as possible per unit measurement. Measurement is usually expressed in fmol or pmol of receptor per milligram of membrane protein. To help maximize the number of membrane receptors used for SPA, a number of techniques exist to concentrate receptors used for various technologies. These include affinity or diethylaminoethyl (DEAE) cellulose chromatography or differential centrifugation [3,4].

11.3.1 WHOLE TISSUE EXTRACT

Membrane receptors can be derived from whole tissue preparations and used for a SPA receptor binding assay [5]. Using this source, one can obtain information about the status of the receptor in its native environment. One critical factor in using whole tissue preparations as a source of receptors is that the tissue source must have a sufficiently high number of receptors expressed. Factors that affect signal intensity will be discussed later. In general, though, the choice of tissue type should correlate to the highest possible expression of an associated receptor in the tissue. If the numbers of receptors that may be available for a SPA assay are limiting, those samples should be purified by the appropriate means [3,4].

11.3.2 WHOLE CELLS

Cultured whole cells, either adherent or suspension culture, can be utilized for SPA receptor binding assays [6,7]. No fractionation of the cell is required in order to use this format. Similar to using tissue extracts, however, you may need to concentrate the receptor. Using a cell line that overexpresses the receptor of interest is a routine method, and using flow cytometry to segregate a cell population based on an expressed phenotype is another popular technique. It should be noted that once cells are removed from optimal growing conditions, which may not be normally used for screening, cells could enter the apoptotic pathway, thus decreasing their assay performance. This may be a factor in designing a suitable protocol for the drug screening process.

11.3.3 CELL MEMBRANES

Cell membrane preparations are most commonly used when performing SPA receptor binding assays [8,9]. One factor in developing a successful receptor binding assay is to obtain cell membranes that retain the receptor in its native conformation so as to not perturb ligand binding. One method that is commonly used to disrupt cellular architecture but yet retain receptors in their native state is to use nitrogen disruption [10,11]. For this method one can use the Parr Cell Disruption Bomb, model # 4639 (Parr Instrument Company).

11.4 NITROGEN DISRUPTION: EXAMPLE ASSAY PROTOCOL

Grow an appropriate quantity of cells for your experimental needs: for example ~50 T-175 flasks or 10 roller bottles (900 cm^2 surface area), each containing about 200 mL of culture medium. After cells reach the desired confluency, they are ready to be harvested. *Note*: It will be necessary to determine that receptor density is not adversely affected by the confluency of your cell culture environment. Before harvesting cells, wash each vessel twice with phosphate-buffered saline (PBS), pH 7.4. Scrape each vessel into cold PBS and pellet the cells by gentle centrifugation for 5 min, 400 g (1200 r/min in a Beckman J-6M). Resuspend the cells in 50 mM Tris-HCL/5 mM MgCl$_2$ pH 7.4 (50 mL per 9000 cm^2). Let the pellet stand on ice for 20 min and split it into four 12.5-mL portions. Process each portion in the nitrogen "bomb" cell disruptor for 5 min at 800 to 900 psi. Centrifuge the pooled homogenates at 4°C for 5 min at 1600 rpm in the Beckman J-6M centrifuge or another suitable centrifuge. Discard the pellet and dilute the supernatant to 50 mL with 50 mM Tris-HCl pH 7.4, 5 mM MgCl$_2$. Determine the protein concentration with a methodology appropriate for your lab (for example, Biorad protein determination assay). Dispense 5-mL aliquots into appropriate vials and freeze at –70°C. Membranes should be should be stable for several months under these conditions. It is advisable to check the activity of each new membrane preparation against an existing receptor binding assay to be sure the activities are comparable. On the day of use, thaw aliquots, vortex briefly, and use as appropriate.

11.5 BEAD BINDING CAPACITY

The capacity for WGA-coated SPA and Imaging beads to bind receptors is 10 to 30 μg of membrane protein per milligram of bead. Beads can bind membranes from a number of different cell lines including SF9 insect cells, Chinese Hamster Ovary (CHO) cells, and HEK293 cells. The binding capacity of polylysine SPA beads is about 10 μg of membrane protein per milligram of bead. The overall binding capacity is dependent on the level of the N-acetyl-β-D-glucosamine residues present on the cell surface. This will be dependent on the cell type used.

11.6 SOLUBLE RECEPTORS

SPA beads can be used to bind receptors from tissues and cell membranes that have been solubilized by a variety of detergents as long as the biologic relevance of the receptor is not altered or affected by that detergent and subsequent purification processes [12]. It is important that solubilized receptors are sufficiently enriched so that the SPA bead can capture a substantial quantity of receptor. This allows for a variety of possible capture mechanisms to be used. For example, purified receptors can be biotinylated for use with streptavidin-coated beads [13]. Also, receptors or receptor ligands can be expressed as fusion proteins, specifically using glutathione s-transferase (GST) and 6X His tag fusion protein expression systems. For these types of assays, glutathione and copper-coated SPA beads can be used, respectively. In addition, antibodies can be used to selectively capture receptors in a crude membrane preparation [12]. A variety of SPA beads and reagents available from GE Healthcare exist to accommodate these various capture techniques and assay formats.

11.7 ASSAY DEVELOPMENT

The development of SPA receptor binding assays requires much the same effort as more traditional methods, such as a filtration assay. To that end, it is advisable that the main sources of information for setting up a new SPA receptor binding assay be the literature and past assay development (a reference assay) employing the use of the specific receptor–ligand complex being investigated. In nearly all instances, the conditions that were employed in the traditional method can be carried over to the SPA format. Since SPA is a homogeneous technology it is important to consider the addition of protease inhibitors to stabilize components of the assay critical to the success of the assay. This may prevent the degradation of various components, especially the receptor and ligand complex. Also, as mentioned previously, every opportunity should be taken to maximize the amount of receptor used. This general principle will maximize the specific signal that one obtains when using a homogeneous technology with receptor binding assays. Below are various factors to consider when setting up a new SPA assay (whether conventional or imaging based) or troubleshooting an existing assay.

11.7.1 RADIOLABEL AND LIGAND SELECTION

The selection of the appropriate radiolabel or isotope to use with any type of SPA assay is dependent on several factors. Essentially, any isotope that has the mean path length of isotopic decay of approximately 100 microns in an aqueous environment can be used. Tritium and Iodine 125 are the two isotopes most commonly used for receptor binding assays. Initial considerations for which isotope to use should focus on whether the desired binding ligand exists in a radiolabeled format. If is does not, then it will have to be labeled by either chemical or *de novo* synthesis with the desired isotope, if possible. If the desired isotope cannot be used due to chemical or biological constraints, then the isotope with the closest specific activity should be chosen.

Once a suitable target has been identified and the ligand selected, knowing the receptor density and ligand affinity can greatly aid in the development of the assay. If ^{125}I (~2000 Ci/mmol) is used as the radiolabel, then typical expression levels in the region of 50,000 receptors per cell are required. This corresponds to approximately 200 fmol or receptor/mg of membrane protein. Higher expression levels, close to 500,000 receptors per cell, are required if ^{3}H-ligands (20 to 80 Ci/mmol) are used. This corresponds to densities greater than 2 pmol or receptor/mg of membrane protein and must correlate to the binding capacity of the SPA beads used. Below are general guidelines for choosing the best isotope based on receptor density and relative affinity of the ligand for the receptor. It is important to note that overall assay performance will dictate which isotope may be best for a particular assay:

- If ligand affinity is high in conjunction with a high receptor density, then either ^{125}I or ^3H radiolabels can be used.
- If ligand affinity is low in conjunction with a high receptor density, then ^3H ligands are preferred.
- If ligand affinity is high in conjunction with a low receptor density, then ^{125}I radiolabeling is preferred.
- If ligand affinity is low in conjunction with a low receptor density, then a low signal is likely, and another receptor source and/or ligand should be considered.

The distinction between a high-affinity ligand vs. a low-affinity ligand is somewhat undefined. In general, though, a ligand with an affinity in the low nanomolar range would be considered a high-affinity ligand. Conversely, a ligand that binds to its receptor with a low affinity would be approaching the micromolar range.

11.7.2 BEAD SELECTION

Choosing the right bead type is an important part of configuring a successful SPA assay. Before going on, it will be useful to highlight the various bead types available for conventional and imaging-based SPA assays. Table 11.1 and Table 11.2 give a list of conventional SPA (Table 11.1) and SPA/Imaging (Table 11.2) bead types that can be used for various types of receptor assays. If desired, SPA and Imaging beads can be manufactured with custom coatings to facilitate the unique capture of various biomolecules. The binding capacity of each bead type will ultimately depend on assay conditions; therefore, optimal binding capacity must be determined experimentally from matrix experiments.

TABLE 11.1
SPA Bead Type vs. Binding Capacity

SPA Bead Type	Relative Binding Capacity
Donkey antirabbit, YSi and PVT	0.4 μg antibody/mg bead
Sheep antimouse, YSi and PVT	0.4 μg antibody/mg bead
Donkey antisheep, YSi and PVT	0.4 μg antibody/mg bead
Protein A, YSi and PVT	0.4 μg antibody/mg bead
Anti–guinea pig, PVT	0.4 μg antibody/mg bead
Wheat germ agglutinin, PVT and YSi	10 to 30 μg membrane protein/mg bead
Streptavidin, PVT	100 pmol biotin/mg bead
Streptavidin, YSi	200 pmol biotin/mg bead
Gluthathione, PVT	0.7 pmol GST/mg bead
Gluthathione, PVT	14 to 21 pmol GST/mg bead
WGA PEI type A, PVT	10 to 30 μg membrane protein/mg bead
WGA PEI type B, PVT	10 to 30 μg membrane protein/mg bead
Copper his-tag, PVT and YSi	150 pmol his tag/mg bead
Polylysine, YSi	10 μg membrane protein/mg bead
Poly cationic PEI, YSi	Application dependent
Poly cationic PEI, PVT	Application dependent
Protein binding, YSi	Application dependent

TABLE 11.2
Imaging Bead vs. Binding Capacity

SPA Imaging Beads	Relative Binding Capacities
Wheat germ agglutinin, PS and YOx	10 to 30 µg membrane protein/mg bead
Streptavidin, PS	200 pmol biotin/mg bead
Streptavidin, YOx	300 pmol biotin/mg bead
Protein A, YOx	0.4 µg antibody/mg of bead
Copper his-tag, YOx and PS	150 pmol his tag/mg bead
Polyethyleneimine, PS	10 to 30 µg membrane protein/mg bead
Protein binding, YOx	Application dependent
WGA PEI type A, PS	10 to 30 µg membrane protein/mg bead
WGA PEI type B, PS	10 to 30 µg membrane protein/mg bead
Polylysine, PS	10 to 30 µg membrane protein/mg bead
Donkey antirabbit, PS and YOx	0.3 to 0.6 µg antibody/mg bead

Wheat germ agglutinin (WGA) and polylysine are the two most commonly used capture methods for membrane preparations. Polylysine SPA beads are positively charged and bind to negatively charged membrane preparations. The binding capacity of these beads for crude membrane protein is about 10 µg of membrane protein per milligram of bead. The nature of this interaction means that polylysine YSi beads are less selective in their binding preference than WGA beads for receptor–membrane complexes.

WGA is the most popular capture molecule system for receptor binding assays. WGA binds to the N-acetyl-β-D-glucosamine residues present on cell membranes or receptors and sometimes can be found in the ligand itself [5,14,15]. The binding capacity for WGA SPA beads (both conventional and imaging) is approximately 10 to 30 µg of membrane protein per milligram of bead. It is important to note that it may be difficult to use WGA beads with ligands or labeled proteins, which themselves are glycoproteins. Although not common, it is always possible that WGA could also act as an inhibitor to the receptor–ligand binding interaction. It is therefore important to ascertain whether the WGA–receptor interaction interferes with the receptor–ligand binding process.

If interfering factors do exist within an assay that result in either a high nonspecific binding (NSB) of the radiolabeled ligand to the bead or the inability of the cell membrane to bind to the WGA moiety, then one can try polyethyleneimine (PEI)-coated WGA SPA beads. PEI is a long-chain polycation polymer of ethylene imine that has been used extensively to decrease nonspecific binding with receptor assays. Its structure is given in Figure 11.3.

The designation "Type A WGA PEI SPA bead" refers to the PEI being added onto the bead after the covalent addition of WGA. Similarly, "Type B WGA PEI SPA bead" refers to PEI being added on the bead before the covalent addition of WGA. The order of addition of the PEI onto the bead can make a significant difference in the NSB profile for that specific bead.

From the possible bead types that can be used for a receptor binding assay, it is recommended that the assay developer choose at least three different types and test them in a compatibility assay that could be used to capture the receptor–ligand complex. The aim is to predict which of the

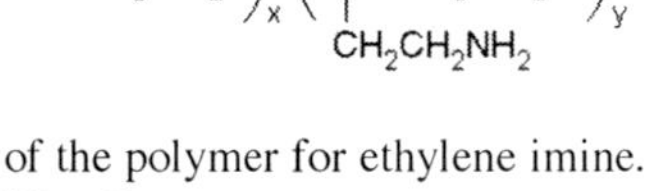

$$\left(-NH\ CH_2\ CH_2 - \right)_x \left(\begin{array}{c} N - CH_2\ CH_2 - \\ | \\ CH_2 CH_2 NH_2 \end{array} \right)_y$$

FIGURE 11.3 The chemical structure of the polymer for ethylene imine. Its total molecular weight can vary widely from 600,000 to well over 1 million Da.

selected bead types will be most appropriate for your radiolabeled ligand by examining the non-specific interactions of the radiolabeled ligand with the bead and its assay environment. These results represent only the nonspecific interaction of the ligand with the bead in the absence of any receptor and should be used only as a guide to the initial choice of bead to be used in the assay.

A typical compatibility testing protocol is given below; it should be modified to suit individual experimental parameters:

- HEPES buffer (50 mM), pH 7.4
- One milligram beads per well in a 96-well format
- One hundred thousand liquid scintillation cpm of each ligand per well
- Final volume of 200 μL
- Incubate for 60 min at room temperature without agitation
- Count on a microplate scintillation counter

Additives to standard 50 mM HEPES buffer can include:

- NaCl, 10 mM
- NaCl, 100 mM
- MgCl$_2$, 5 mM
- MgCl$_2$, 50 mM
- Bovine serum albumin (BSA), 0.1%
- PEI, 0.1%
- TRITON, 0.1%

The first step would be to profile your ligand against the bead alone to determine the lowest amount of NSB. This would include combining bead, buffer, and label. One should then use the bead type that gives the lowest NSB for continued assay development. If the NSB is less than 10 to 20% of total liquid scintillation counts added, it might not be a problem in the assay unless unknown factors in the membrane preparation contribute to or facilitate the binding of the labeled ligand to the bead. If the NSB is higher than 20% or becomes an issue during assay development, then additional steps can be taken to reduce it as outlined in Section 11.7.3.

To help find a compatible bead type for a particular RBA, it is suggested that one use the SPA select-a-bead kits available from GE Healthcare. These kits contain five to six different bead types commonly used for receptor binding assays.

11.7.3 Assay Buffer

The assay buffer is a central part of the homogeneous SPA assay. As a general rule, all assay components present in the reference assay will be amenable for use with SPA technology. The literature or current assay practices are the best sources for your buffer formulations. Some of the most common buffers used for receptor binding assays include TRIS, HEPES, PBS, MOPS, and PIPES. The use of protease inhibitors is highly recommended and will be highlighted later. In addition to protease inhibitors, any necessary cofactors or agents needed to enhance receptor binding while decreasing nonspecific binding of radiolabeled components to the bead are highly recommended.

Various methods can be used to help reduce the level of NSB in a receptor binding assay. NSB occurs by several mechanisms, the most common of which is when the radiolabel ligand binds directly to the bead or some component that is bound to the bead. Another possible cause of high NSB is due to unincorporated isotope or label binding to the bead. Lastly, radioligand degradation can sometimes free the isotope from the ligand. The freed isotope can then bind to the bead. This is most commonly seen when older preparations of a radiolabeled ligand are used.

To reduce NSB in a receptor-binding assay, several techniques can be used. The addition of "cold" ligand to the assay is the most direct and effective method but this has to be carefully considered, as it should not compete for active receptor binding sites. To this end, the order of addition of cold ligand to the assay has to be carefully considered. Another method that might prove useful is to use a "ligand mimetic." These are compounds that are structurally and chemically similar to the radiolabeled ligand but do not affect its direct binding to the receptor.

One of the most common methods to help reduce NSB is to add a nonspecific protein carrier such as BSA, peptone, or casein. The effective concentration must be determined experimentally but usually will start around 0.01% (w/v) in the final assay buffer. Small amounts of various detergents may also be used to help reduce NSB. They include PEI, CHAPS, SDS, Triton® X-100, Tween®-20, and others, again in similar concentrations that start at 0.001%. It is important to note that the detergent used not disrupt the receptor–ligand interaction or membrane structure. Other components that can be added to the assay buffer to help reduce NSB or increase your specific signal include DTT, dimethylsulfoxide (DMSO), and various salts such as KCl, NaCl, $MgCl_2$, and $MnCl_2$. Additionally, various cellular components may also reduce NSB. These can include RNA, DNA, and proteins similar to the ligand found in the assay. In addition to NSB, the volume of the assay should be the highest that it allows. This will keep the nonproximity effect (NPE) to a minimum when using ^{125}I or higher-energy isotopes.

11.7.4 RECEPTOR/MEMBRANE AMOUNT

Receptor number or density is a critical factor for establishing any successful RBA, regardless of whether it is an SPA receptor binding assay. As stated earlier, the specific activity of a receptor to be used in a binding assay is usually expressed in fmol or pmol of receptor per milligram of membrane protein. This figure will be used in conjunction with the binding capacity of SPA beads, which is 10 to 30 μg membrane protein per milligram of bead. Therefore, the concentration of the membrane preparation expressed in milligrams per milliliter or micrograms per milliliter must be aligned to the volume of membrane needed in the assay based on the concentration, assay format, 96, 384, etc., and to the amount of bead to be used as well as to the overall volume of the assay. For most experiments, the optimal amount of receptor/membrane protein complex will have to be determined experimentally by titration of assay components. As can be seen in Figure 11.4, using increasing concentrations of membrane protein will yield a greater signal as long as there is enough bead in the well to capture all membrane used. This will be discussed further under bead optimization.

Most often the scientific literature will contain information that can aid the SPA assay developer. Calculations from filter binding assays that have determined various kinetic parameters such as the B_{max} (receptor concentration), and ligand affinity (K_d), along with nonspecific assay background and amount of isotope used, can be beneficial.

11.7.5 SPA ASSAY FORMATS

There are three different types of SPA assay formats that give the assay developer the choice on how to configure the best assay possible. Either the receptor can be precoupled to the bead (precoupled format), the bead can be added to the well at the same time as other reagents (T = 0), or the bead can be added to a preequilibrated receptor–ligand complex (delayed addition). The choice is dependent on many factors related to assay design and preference as described below. The three formats are shown in Figure 11.5.

For the precoupled assay format, SPA beads are incubated with an appropriate level of membrane for a minimum of 4 h at room temperature or overnight at 4°C. During the incubation it may be beneficial to ensure that the bead–membrane solution is constantly being mixed to ensure adequate binding of all components to the bead. Afterwards, gently centrifuge the bead mixture for 3 to 5 min. A good rule of thumb is to treat the beads as if they were whole cells. This will

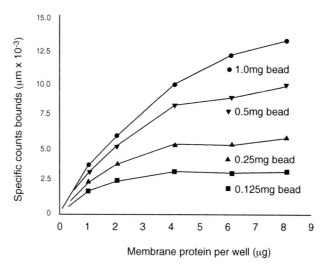

FIGURE 11.4 Matrix experiment for optimization of the bead-to-membrane ratio for the binding of (3-[^{125}I]iodotyrosyl^4) [sar^1, ile^8] angiotensin II to bovine adrenal cortex membranes. Assays were performed in a 96-well plate with 200 μL assay buffer (phosphate buffered saline, pH 7.3, containing 0.1% (w/v) BSA and 5 mM EDTA) containing 169 pM (3-[^{125}I]iodotyrosyl^4) [sar^1, ile^8] angiotensin II, varying concentrations of bovine adrenal cortex membranes, and different weights of WGA-PVT beads. NSB was defined using 50 μL of 2 μM [sar^1, ile^8] angiotensin II. Assays were counted using a PerkinElmer TopCount scintillation counter.

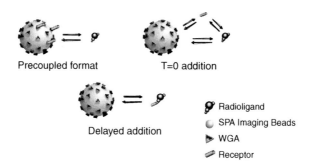

FIGURE 11.5 Different assay formats that can be used to configure SPA assays.

help ensure that the membrane–receptor complex will not dissociate from the bead. Once the bead–membrane mixture has been pelleted, remove the supernatant. Beads should be washed twice in a suitable buffer (screening buffer is usually used) to remove unbound membrane, and any component(s) that could interfere in assay performance. The advantages of the precoupled format may allow for a lower NSB, a better signal-to-noise ratio, and the possible use of lower amounts of radiolabeled ligand and bead. Also, this mixture can be either used immediately, stored frozen at –80°C, or lyophilized as appropriate. The stability of the receptor complex is the determining factor when choosing a storage method. *Note*: One will not be able to do classical Michaelis–Menton kinetics with this format, and protease inhibitors should be included in the assay buffer to help prevent component degradation.

The next format is the T = 0 format of assay configuration. This is the most popular format due to its ease of automation. All components of the assay are added at the same time with a slight excess of bead (~10%) to capture the entire membranereceptor–ligand complex. The T = 0 format

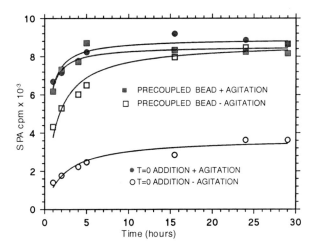

FIGURE 11.6 The binding of [^{125}I]ET-1 ligand to porcine lung membranes was evaluated over time in conjunction with the effect of agitation and bead addition format. For this example, polylysine-coated YSi beads were used.

includes SPA beads, membrane, hot ligand, NSB competitors if needed, buffer, test compound, and any necessary cofactors. Coupling of membrane receptor complexes to beads occurs simultaneously with the addition of ligand. This method needs to be optimized in order to ensure assay performance as all components are added at the same time. An excess of any one component or varying the order of addition of assay reagents may lead to decrease in assay performance. An important point to note is that if too much cold ligand is added before or at the same time as adding hot ligand, then the cold ligand will compete for available receptors, thus lowering the signal-to-noise ratio. The appropriate amount of cold label, if needed, should be determined in a matrix-based experiment. Figure 11.6 shows the difference in assay performance between two types of bead coupling formats. In this instance the precoupled format gave better assay results than the T = 0 format.

The last type of format to be discussed is the delayed addition format. With this format, the membrane and ligand are equilibrated in the fluid phase first before being added to beads. This format is similar to that of a more traditional assay format such as a filtration-binding assay. Using the delayed addition format may give lower NSBs and improve signal-to-noise ratios. With this format, bead amount must also be optimized in order to ensure that enough beads are present, usually in excess of ~10% to capture all available product. As this is the case, volume changes that can occur upon bead addition may change binding dynamics. Therefore, care must be taken that this change in volume does not disrupt the K_d of the final assay.

11.7.6 SPA BEAD OPTIMIZATION

The choice of the best SPA bead type is an important part of configuring a successful SPA assay. The other two parameters previously discussed were the amount of radiolabeled ligand to be used along with the appropriate quantity of membrane protein (Figure 11.4). The performance in a typical SPA assay may vary from bead type to bead type. This may be due to the amount of membrane protein bound to the bead and the type of coating applied to the bead such as PEI. Variations between cell types can be due to the presence of differing amounts of N-acetyl-β-D-glucosamine residues within the membranous structure of different cell types. Figure 11.7a and Figure 11.7b indicate the selection of the best Imaging bead type for the binding of [^{125}I-HIS^9] gherlin to the G-protein–coupled receptor (GPCR) growth hormone receptor [16].

For this assay, 1 μg of human ghrelin receptor membrane complex was incubated with 200 μg of the various Imaging beads along with human [^{125}I-HIS^9]-ghrelin (Amersham Biosciences, IM347)

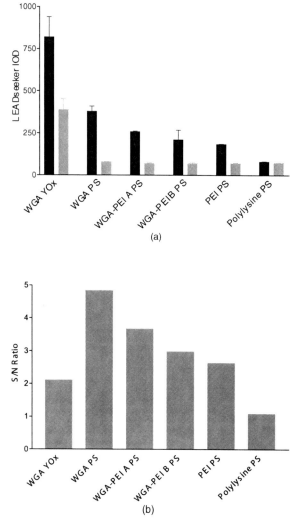

FIGURE 11.7 Evaluation of SPA imaging beads. Six different SPA imaging beads from the LEADseeker select-a-bead kit were evaluated, and the relative signal-to-background ratio was determined. As can be seen in (a), the WGA YOx beads give the higher reading of integrated optical intensity units or IODs. When combined with the NSB for this ligand, the overall signal-to-noise ratio is greater with the WGA PS beads, as seen in (b). This bead type was used for further assay development.

in 50 μL of assay buffer (50 mM Tris-HCl, pH 7.5, 5 mM MgCl$_2$, 1 mM ethylenediaminetetraacetic acid (EDTA), 0.1% BSA, and 1% protease inhibitor cocktail). Nonspecific binding (NSB) was determined in the presence of 1 μM human unlabelled ghrelin. The assays were performed using a solid white 384-well Costar® 3705 microplate. The assay was incubated for 1 h at room temperature and then imaged on the LEADseeker multimodality imaging system (MMIS) for 5 min (3 × 3 binning w/quasi-coincidence averaging).

Once the best bead type is determined for a set amount of membrane protein, then overall amount of bead can be determined in conjunction with the binding capacity for that bead type (Figure 11.4). Table 11.3 indicates the various ranges of beads and membrane protein levels that can be used as an initial starting point for assay optimization in a matrix-based experiment for both SPA and Imaging SPA assays. Other factors that can influence the overall bead and membrane amounts include K$_d$ or binding affinity of the ligand, isotope used, and the amount of NSB per bead type.

TABLE 11.3
Recommended Bead vs. Membrane Amount for the Conventional SPA Bead and the Imaging SPA Bead

Well Format	Conventional SPA		Imaging SPA	
	Bead Amount	Membrane Amount[a]	Bead Amount	Membrane Amount[a]
96	0.5 to 3.0 mg	5.0 to 90.0 μg	250.0 μg to 2.0 mg	2.5 to 60 μg
384	50.0 μg to 2.0 mg	0.5 to 60.0 μg	50 to 750 μg	0.5 to 22.5 μg
1536	NA[b]	NA[b]	10 to 300 μg	0.1 to 10.0 μg

Note: SPA technology offers a good deal of flexibility in reagent amount that can be used for any one particular assay. This table indicates the approximate starting amounts for SPA beads dependent on the type of modality and plate density used. Other factors that can influence the overall bead and membrane amounts include K_d or binding affinity of the ligand, the isotope used, and the amount of NSB per bead type.

[a] Membrane amounts were determined from the binding capacity of SPA beads used for receptor binding assays. This range is between 10 and 30 μg of membrane protein per milligram of bead.
[b] NA indicates that conventional SPA assays are not routinely performed with this format.

In general, lower amounts of bead and membrane are used in conjunction with higher-energy isotopes such as ^{125}I, whereas higher amounts of bead and membrane are used with ligands labeled with a low-energy isotope such at tritium. Please note that it is not advisable to go below 100 μg of beads per well in a 96-well format. This may lead to variations in counting efficiency, especially if high-energy isotopes are used. To determine the optimal amount of bead and membrane to be used in conjunction with other assay components, it is advised to configure a matrix experiment that will establish the best overall acceptable signal. This can be done manually or with the use of software platforms that can automate the process, such as Automated Assay Optimization or AAO software from Beckman Coulter Inc. (Fullerton, CA).

11.7.7 SPA Acceptable Signal

The goal of the assay development process is to configure an assay with an acceptable signal-to-noise (background) ratio. In spite of all best efforts if this is not seen, then one should reevaluate the amount of membrane and bead and the format used in the assay. If a signal window is present but is not as high as anticipated, or the background is higher than expected, then one can troubleshoot the existing assay. Below are a few areas that can be reviewed to ascertain the robustness of an assay:

- The specific activity of the ligand if a low signal is seen
- The ability of the bead to bind a captured molecule
- The ability of the captured molecule to interact with label
- The biological relevance of the labeled ligand to interact either with isotope or capture molecule; for example, biotin labeling

An important part of the assay development process that often gets overlooked is whether or not the labeling process alters the affinity of the ligand for the receptor. This needs to be checked if recent changes to the ligand have been made or if a new batch of labeled ligand has been received. If a high background signal is noticed (high NSB), then the ability of the label to bind directly to the bead itself may need to be examined. Also, the label may be binding directly to the microplate well, thus giving a high background. Lastly, although rarely seen, cross-talk associated with using high amounts of high-energy isotopes may be a factor. If this is seen then various correction factors often can be applied via the instrument software.

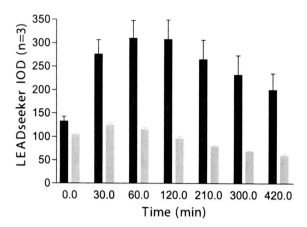

FIGURE 11.8 SPA imaging bead assay time course for binding of [^{125}I]β-endorphin to mu -opioid receptor. (■) Assay totals with 0.1 mg PS-WGA bead, 0.5 nM ligand, and 1 μg receptor; (■). NSB wells contained 10 μM Naltrexone. Maximum signals start to decrease after 2 h, but the signal:noise (S:N) ratio stays relatively the same. Data represent means ± standard error of the mean (SEM) (n = 3).

11.7.8 DETERMINATION OF STABLE COUNTING WINDOW

If an acceptable signal is seen, then the stability of the counting window should be determined. Such parameters include the time necessary for SPA beads to bind all available membrane. This usually occurs within 30 to 60 min but can take longer if either YSi or YOx beads are used. Also, temperature, agitation, and the K_d of the ligand must be considered.

Figure 11.8 highlights the time required to obtain maximal signal and binding of a β-endorphin to its mu-opioid receptor [17]. Experiments were performed to optimize the amounts of assay components and to identify the most suitable membrane-binding bead types using GE Healthcare select-a-bead kits. The final optimized assay used 1 μg of receptor incubated with 0.5 nM (50 nCi) [^{125}I]β-endorphin (Amersham, 2000 Ci/mmol) and 0.1 mg PS-WGA SPA imaging beads, in 50 μl of buffer [50 mM Tris-HCl, pH 7.4, 1 mM EDTA, 10 mM MgCl$_2$, and 0.1% BSA (w/v)], in a Matrix ScreenMate® 384-well solid white plate. The plates were shaken for 2 h at room temperature; the SPA imaging beads were then imaged using LEADseeker MMIS. Exposure time was 10 min per plate.

As seen in Figure 11.8, any assay and its associated signal window must be stable over a defined period of time to enable investigations for screening purposes. If a decrease in signal occurs over time, this may indicate the need for protease inhibitors or other components to stabilize the assay. Figure 11.9 shows a time course for the binding ^{125}I Bolton and Hunter–labeled Substance P ([^{125}I]BHSP) to the human neurokinin-1 (NK-1) receptor expressed in a human NK$_1$R CHO cell line [18]; the counts are stable for several hours when the membranes are prepared in the presence of a cocktail of protease inhibitors. In the absence of these inhibitors, the signal increases initially but is not stable and decreases steadily over a period of several hours.

In addition to determining parameters that stabilize the assay, the assay should be read under stable, equilibrium conditions. These conditions are usually when beads have settled to the bottom of the microplate. Plastic beads (PVT and PS) will settle to the bottom in 4 h whereas glass beads (YSi and YOx) will settle to the bottom in approximately 10 to 15 min. As an alternative to waiting for the plastic beads to settle, they can be counted while maintained in suspension. To increase the speed at which SPA assays can be read, the microplate can be centrifuged, which results in beads (all bead types) being packed at the bottom plate. If a stable signal is not seen when beads have reached equilibrium counting conditions, then one should reevaluate the ligand, receptor, and buffer conditions.

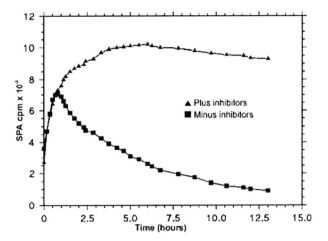

FIGURE 11.9 Time course of binding of [^{125}I]BHSP to hNK$_1$ R CHO membranes by SPA. Assays were performed in 200 μL assay buffer (50 mM Tris/HCl, pH 7.4, 0.1% (w/v) BSA, 1 mM EDTA, 2 mM MnCl$_2$) in the presence or absence of 400 μg/mL bacitracin, 20 μg/mL leupeptin, 40 μg/mL chymostatin, and 1 mM Pefabloc™ SC) containing 141 pM [^{125}I]BHSP, 6 μg hNK$_1$R CHO cell protein, and 0.5 mg WGA SPA beads. Nonspecific binding was determined in the presence of 50 μM substrate P. Assays were counted using a Wallac MicroBeta scintillation counter.

11.7.9 ASSAY VALIDATION

Once an acceptable signal is obtained, the assay must be validated. This can include using appropriate negative controls (cell lines that do not express the cloned or studied receptor) along with determining various kinetic and binding parameters for the ligand. This may include IC$_{50}$/K$_i$ values for known drugs or ligands and establishing competition binding curves. In addition, saturation binding experiments to establish or confirm K$_d$s and B$_{max}$s should be determined and related to known values in the literature. This can be achieved by measuring an assay sample that has radioisotope bound to the beads but no unbound radioactivity present, first as SPA counts and then in another counter standardized to give dpm values for that radioisotope. One parameter that is often used to measure the quality of an assay used for screening is called the Z-factor [19]. This statistical measurement is defined as Z′ = 1 − [(3)SD of signal + (3)SD of control]/(mean of signal − mean of control). The Z factor or score is a parameter for the quality of the assay itself without the intervention of test compounds. A Z of 1 represents the most upper limit for this score. If it falls to less than 0.5, it may indicate that the assay may be in need of further optimization before it is robust enough for the screening environment.

11.8 SCREENING

Once the assay has been validated and all parameters are fully optimized, the assay is ready to be screened or otherwise evaluated. One feature of high-volume screening is that the SPA beads must be kept mixed and in suspension for uniform delivery into 96-, 384-, and 1536-well formats. An ever-growing number of types of liquid handling instrumentation can accomplish this task. Some of these include Beckman (Fullerton, CA), V&P Scientific (San Diego, CA), and Genomic Solutions® (Ann Arbor, MI). It is important to note once again that changing the order of addition or combining steps not previously validated may impact the overall assay performance.

11.9 SUMMARY

The development of receptor binding assays using SPA technology should follow a logical flow. Each developer will most likely have a unique method of assay development. It is important that the development of the assay follow a defined method of experimental progression, keeping in mind the parameters mentioned in this document. Doing so can shorten the design of new SPA assays and help troubleshoot existing assays.

The results of the SPA development process should be an assay:

- With an acceptable level of signal
- With an adequate signal-to-noise ratio
- With the required sensitivity
- With adequate stability over time
- That is amenable for miniaturization and hence cost effective
- Comparable to other assay formats

SPA technology overcomes the disadvantages of traditional filtration techniques for receptor assays. No separation step is required, and thus radioactive handling and waste generated are minimized. SPA provides a method that is simple and rapid and suited to automated HTS. This chapter has highlighted the key issues for the development of a SPA receptor binding assay.

ACKNOWLEDGMENTS

The authors gratefully acknowledge many past and present members of the GE Healthcare Bio-Sciences Development groups in Cardiff and Piscataway who have contributed to the work in this chapter. The authors are particularly grateful to Alison Harris and John Ireson for their contributions. This chapter is dedicated to Lori, Seamus, and Keara Ahern.

REFERENCES

1. Nelson, N., A novel method for the detection of receptors and membrane proteins by scintillation proximity radioassay. *Anal. Biochem.* 165, 287–293, 1987.
2. Cook, L. et al., Benefits from using imaging based detection (LEADseeker) and red-shifted emission for radiometric assay screening. Poster presentation at SBS Annual Meeting, Baltimore, MD, 2001.
3. Venter, J. C., *Receptor Purification Techniques*, John Wiley & Sons, New York, 1984.
4. Selinsky, B. S., *Membrane Protein Protocols: Expression, Purification, and Characterization*, Methods in Molecular Biology Series, Vol. 248, 2003.
5. Berry, J. A. et al., Scintillation proximity assay: competitive binding studies with [^{125}I]-endothelin in human placenta and porcine lung. *J. Cardiovasc. Pharmacol.* 17 (suppl. 7), S143–S145, 1991.
6. Fuhlendorff, J. et al., [Leu31], Pro34 Neuropeptide Y — A specific Y1 receptor agonist. *Proc. Natl. Acad. Sci. USA* 87: 182–186, 1990.
7. McKinnon, M. et al., An interleukin 5 mutant distinguishes between two functional responses to human eosinophils. *J. Exp. Med.* 186, 121–129, 1997.
8. Koller, K. J. et al. A generic method for the production of cell lines expressing high levels of 7-transmembrane receptors. *Anal. Biochem.* 250, 51–60, 1997.
9. Giulumian, A. D. et al., Role of ET-1 receptor binding and [Ca^{2+}]$_i$ in contraction of coronary arteries from DOCA-salt hypertensive rats. *Am. J. Physiol. Heart Circ. Physiol.* 282, H1944–H1949, 2002.
10. Hunter, M. J. and Commerford, S. L., Pressure homogenization of mammalian tissues. *Biochim. Biophys. Acta* 47:580–586, 1961.
11. Dowben, R. M. et al., Isolation of liver muscle polyribosomes in high yield after cell disruption by nitrogen cavitation. *FEBS Lett.* 2, 1–3, 1968.

12. Guo, Q. et al., Biochemical and genetic characterization of a novel human immunodeficiency virus type 1 inhibitor that blocks gp 120-CD4 interaction. *J. Virology* 77, 10,528–10,536, 2003.

13. Jone, S. A. et al., The pregnane X receptor: a promiscuous xenobiotic receptor that has diverged during evolution. *Mol. Endocrinol.* 14, 27–39, 2000.

14. Espinosa, J. F. et al., NMR ivestigation of protein–carbohydrate interactions. *Eur. J. Biochem.* 267, 3965–3978, 2000.

15. Lui H. W. et al., GP-38 and GP-39, two glycoprotiens secreted by the human epididymis, are conjugated to spermatozoa during maturation. *Molec. Human Reprod.* 6, 422–428, 2000.

16. Lowitz K. et al., Comparison of human ghrelin receptor binding assays using SPA and the LEAD-seeker™ multimodality imaging system. Poster presentation at SBS 9th Annual Meeting, Portland, OR, September 21–25, 2003.

17. Cook, J. et al. A comparison of imaging and PMT-based detection for a radiometric and FP GPCR binding assay screened against the LOPAC library. Poster presentation at SBS 9th Annual Meeting, Portland, OR, September 21–25, 2003.

18. Yokota, Y. et al., Molecular characterization of functional cDNA for rat substance P receptor. *J. Biol. Chem.* 264, 17,649–17,652, 1989.

19. Zhang, J. et al., A simple statistical parameter for use in evaluation and validation of high-throughput screening assays. *J. Biomolec. Screening* 4, 67–73, 1999.

12 Radioligand Binding Filtration Assay: Full Automation

Stephen K.-F. Wong

CONTENTS

12.1 SUMMARY

This chapter describes the development of the first fully automated workstation for hands-free radioligand filtration binding assays. The workstation consists of three separate instruments linked together to perform the following operations: a TECAN Genesis to perform liquid handling, incubation, and scheduling operations; a TECAN Genmate 96 well pipettor for liquid handling using disposable pipet tips; and a Brandel automated harvester to perform rapid filtration. A custom-designed tip holder was built at Genmate for loading disposable pipet tips by the 96-well pipettor.

The dopamine D3 receptor was used to develop the automated assay. In this example, assays for 84 compounds with six concentrations that span six logs can be completed within 4 h. The automated workstation increases reproducibility and throughput, reduces hands-on time, and minimizes time of exposure to radiation and biological hazards. The development of this workstation in an ordinary laboratory setting enables rapid determination of the structure–activity relationship of ligands for many membrane-bound therapeutic targets.

12.2   INTRODUCTION

Radioligand binding assay using rapid filtration is the most widely used assay to characterize potency and selectivity of compounds for membrane-bound therapeutic targets such as cell surface receptors for neurotransmitters and hormones. The assay can also determine receptor density as well as the kinetics of binding to the receptor. The availability of radioligands for many membrane-bound receptors and the wealth of information to establish the principles of the filter binding assay [1] made the assay a method of choice for pharmacological characterization of compounds.

The filter binding assay involves incubating membrane preparations with radioligand in the presence of varying concentrations of compounds. Upon completion of incubation, radioligand bound to the membrane-bound receptors is separated from the free radioligand by rapid filtration through glass fiber filters. The amount of radioligand bound to the target receptor is determined by scintillation counting. The procedure involves liquid handling, incubation, and filtration, all of which are typically performed manually. The complexity of the operation makes it difficult to obtain reproducible results. Although variability of the assay can be minimized by the use of automated liquid handling workstations, filter plates, semiautomatic harvesters, and multichannel pipets, the variability of the data can be up to two- to threefold between experiments, even for an experienced operator.

In addition to the variability of the assay, another problem is that the constant hands-on operation of the assay will expose the operator to radiological and biological hazards. This is especially apparent in a large assay (e.g., 10 or more microtiter assay plates). Shielding to minimize exposure to radioactivity is inconvenient since it is necessary to transport the assay plates manually from one operation to the next.

Recent advances in combinatorial organic synthesis (see [2] for review) have significantly increased the number of compounds to be screened. Not only do these advances add more compounds to the sample bank for lead identification, the number of novel compounds synthesized from chemistry labs for lead optimization has also been increased. The increases in production of compounds and the speed at which compounds are made put tremendous challenges on pharmacology labs to rapidly produce biology data to guide the chemistry synthetic efforts. Enhancements in throughput and reproducibility of data are the goal for all pharmacology laboratories in the pharmaceutical industry.

Toward this goal, we have recently developed, in a normal laboratory setting, the first fully hands-free automated workstation to perform radioligand filter binding assays [3]. Compared to manual operations, the fully automated assay has increased assay throughput, improved reproducibility, and greatly decreased hands-on time. This automated operation is readily applicable to any filtration radioligand binding assay and can be adapted to a variety of other similar assays. Also, the automated system can decrease the time required to train a new operator. This chapter will focus on the technical aspects of development of the automated workstation.

12.3   COMPONENTS OF THE FULLY AUTOMATED WORKSTATION

The fully automated workstation was built around the Genesis liquid handler (TECAN, Durham, NC), which performs liquid handling, incubation, and scheduling operations. Two additional com-

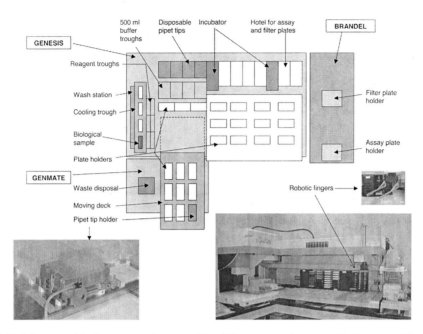

FIGURE 12.1 Layout of the instrumentation to enable a fully automated receptor binding assay. Two TECAN liquid handlers (Genmate and Genesis) and a Brandel automated harvester are linked as shown. The various workstations and racks are indicated in the figure. Liquid handling operations were performed by Genesis and Genmate, while Brandel was used to harvest the samples onto filter plates. The top figure is the diagrammatic presentation of the setup. The insets show an actual picture of the instrument and modified robotic fingers (right corner) and the custom-built holder for the disposable pipet tip cartridges (left corner).

ponents were added: a Genmate 96-well automated pipettor (TECAN) to perform liquid handling using disposable pipet tips, and a Brandel automated harvester (Gaithersburg, MD) to harvest the contents from a 96-well microtiter assay plate to a filter plate.

12.3.1 GENESIS

The Genesis liquid handler provides the flexibility and capacity to develop the automation. The 2-m system used in our laboratory can store up to 48 microtiter plates, 15 plate-holding positions for microtiter and deep-well plates, two shaking incubators, a cooling trough for storing the receptor preparation, four racks of pipet tips (36 cartridges of 96 pipet tips), and troughs for storage of reagents and buffers (see Figure 12.1). Water is used as the system fluid of the workstation. To ensure continuous operation, an automated water feeding system was linked to the reservoir to supply water to the system. All the liquid waste (radioactive and nonradioactive) is fed into a drain on the bench top certified for radioactivity and chemical disposal.

To transfer objects (microtiter plates, deep-well plates, and pipet tip cartridges) between the workstations, each of the pair of robotic fingers at the Genesis was extended by 7.5 cm (Figure 12.1). This enables the fingers to have access to all plate holding positions in Genmate and Brandel. To minimize bending of the fingers during gripping as a result of the extension, the thickness of the fingers was increased from 5 to 7 mm, and the fingers were also reinforced by a crossbar (Figure 12.1). The force, speed, and width of gripping of the fingers can all be adjusted by the Genesis operating software (Gemini 3.2).

Liquid handling in the Genesis is controlled by the liquid handling arm, which has independent Z- and variable Y-span capabilities to move the eight fixed pipet tips. This enables liquid handling to and from racks or troughs of any geometry. Volumes delivered by the fixed tips are controlled by the 1-mL syringes. The volume delivered in the automated assay is between 25 and 900 μL, allowing addition of small volumes of reagents into the microtiter plates, as well as transferring large volumes of buffer to deep-well plates for serial dilutions.

Among the various liquid handling tips (stainless steel, Teflon-coated) available from TECAN, ceramic-coated tips were selected because of the ease of cleaning. This is important as the fixed tips are being used to pipet biological samples and significant lipid and protein buildup develops in the tips upon usage. We use warm detergent to flush the tubing and the fixed tips once every 3 months.

The incubator on the deck of Genesis provides the temperature control as well as a shaking platform of the incubation. The incubator has the capability to control temperature from room temperature to 50°C. Most of the incubations in our radioligand binding studies are at 37°C. If room temperature can be used for the incubation without shaking, the binding assay reaction can be performed at the locations where the microtiter plates are stored.

A cooling trough was used in the setup to stabilize the biological activity of the sample. This is important since the assay sometimes takes 4 to 5 h to finish. The trough was linked to a cooling circulating water bath that continuously circulates water at 2°C to the cooling trough.

The Genesis liquid handler is a reliable instrument that needs little maintenance. Besides quarterly cleaning of tubings and fixed tips, the other regular maintenance includes annual replacements of the syringe for system liquid.

12.3.2 USE OF DISPOSABLE PIPET TIPS

In the initial developmental phase of the automated workstation, the fixed tips in Genesis were used to perform all the liquid handling functions (i.e., pipetting of buffers and reagents, serial dilutions of compounds) with the tips washed with dimethylsulfoxide (DMSO) and then with water. While this washing procedure is sufficient to pipet aqueous buffers, there is significant carryover of compounds during serial dilutions. This is especially apparent for compounds with low solubility in aqueous buffer, suggesting that the DMSO wash is not sufficient to clean compounds from the fixed tips. Moreover, a liquid handling operation with this washing step takes a long time to complete (up to 7 h for 84 compounds for liquid handling alone) and generates a large amount of DMSO liquid waste. This prompted the use of disposable pipet tips in the development of the automated workstation.

The main issue in using disposable pipet tips is the limited tip storage space in Genesis. This hurdle is overcome by the use of pipet tips available in stackable cartridges, a 96-well pipetting workstation that is compatible with the pipet tips, and a custom-built pipet tip holder to hold the cartridge in place for the pipettor head to pick up the pipet tips.

12.3.2.1 Choice of Pipet Tips

Pipet tips (GPS-250) from Rainin (Oakland, CA) were selected because this type of pipet tip is available in stackable cartridges (Spacesaver™ tip racks). Each rack contains nine cartridges of 96 pipet tips, and as many as four racks (36 cartridges) can be stored on the Genesis deck as shown in Figure 12.1. Each cartridge of tips can be transferred from the rack to the holder (see below) at Genmate by the robotic fingers in TECAN Genesis.

12.3.2.2 Genmate 96-Well Pipettor

The use of the Genmate 96-well pipettor, instead of an eight-channel pipettor, reduces time for liquid handling. The reasons for the choice of Genmate are compatibility to the Rainin pipet tips, the availability of software to integrate to the Genesis operations, and flexibility in pipetting. The instrument can be programmed to pick up 96 pipet tips, single or multiple rows, single or multiple columns, and even down to a single pipet tip. It can also be programmed to pipette 384-well plates in quadrants. The speed of pipetting and movements of the pipettor can all be adjusted. This flexibility enables serial dilution within a single plate or across multiple plates, using solvents with different viscosities. There is a waste slot at Genmate to discard the cartridge after usage. A hole was cut beneath the instrument for waste collection.

To maintain a firm seal between the pipettor head and pipet tips, it is necessary to slow the speed of the pipettor to gently load pipet tips. This is especially important when picking up 96 pipet tips. The speed of the pipettor for loading is adjusted to about 1.6 mm/sec. This slow speed will also minimize damage to the holder (see below) when picking up the pipet tips. To ensure reliability in pipetting, o-rings of the 96-well pipettor head need to be inspected and replaced on a regular basis (e.g., one to two times per year). The other factor that affects the accuracy and reproducibility of the pipettor is the volume for pipetting. In our experience, there is good accuracy and reproducibility (CV < 5%) for pipetting 20 to 25 μL with blow-out of the liquid into an empty well. For volumes less than 10 μL, the instrument maintains its reproducibility by pipetting the content into a well with buffer and having a mixing step afterwards.

12.3.2.3 Holder of Pipet Tip Cartridge

A custom-designed holder (Figure 12.1) was built to hold the pipet tip cartridge at the GENMATE. This holder has an air pressure–driven clamp that is controlled by the Genesis operating software. As there is as much as 900 lb of force generated from the 96-well pipettor to load the pipet tips, the pipet holder needs to be well supported. The holder is made of cast aluminum, and the plate in contact with the cartridge is supported by six steel rods at the interior of the holder.

12.3.3 BRANDEL AUTOMATED HARVESTER

To enable the robotic fingers of Genesis to access the area in the automated Brandel harvester for filtration and washing, the manufacturer was asked to make a slight modification. The plastic housing of the tubing on the left side was removed so the platform for harvesting can be of closer proximity to Genesis (see Figure 12.1). The other information required is the type of filter plate that will be used; this defines the specification of the Brandel harvester. The filter plate has to be compatible with the scintillation counter used, and the filter membrane used in the plate should be strong enough to withstand the pressure from washing and vacuum during the harvesting procedure. Packard (Meriden, CT) Unifilter plates with GF/B filters and Packard TOPCOUNT.NXT were used in our setup. Integration of Brandel to Genesis also requires linking the scheduling software of Genesis to Brandel through the RS 232 interface. The software linking both instruments was developed by Pfizer. The software has been made available to both TECAN and Brandel.

The key to obtaining reproducible data is to have a good seal between rubber gaskets in the harvester at the harvesting location and the filter plate. This can be accomplished by regular (quarterly) replacement of the gaskets.

12.3.4 CHOICE OF MICROTITER PLATES

Though the dimensions of most commercially available microtiter plates and deep-well plates are standardized, our experience is that there are minor variations among different vendors; the user should test out different plates before setting up the assays. The footprint of the microtiter plate and the deep-well plates should be compatible with the plate holders in Genesis, Genmate, and Brandel. For 96-well assay plates (i.e., plates for membrane incubation with radioligand and unlabelled compounds), the diameter of the individual well should be as wide as possible for easy access by Genesis, Genmate, and Brandel. Each well should be able to hold at least 300 μL, as the volume for incubation is 250 μL. The height of the plates should also not be more than 15 mm so the plates can be moved in the out of the shaking incubator easily. Round-bottom microtiter plates should be used to facilitate the mixing of reagents during the incubation and washing of wells during the harvesting steps. Polypropylene is the material of choice due to its flexibility and strength.

Corning (Corning, NY) round-bottom polypropylene microtiter plates (catalogue no. 3371) were used. For the deep-well plates used for compound dilution, v-bottom polypropylene deep-well plates from Waters (Franklin MA, catalogue no. WAT-58958) were used.

12.4 MEMBRANE PREPARATION

Due to the limited capacity of the filter membrane (less than 0.1 mg protein/well) to harvest membrane sample, the automated binding assay is mostly suitable for recombinant receptors that are over-expressed in culture cells ($\geq$ 0.5 pmol/mg). The procedure to prepare crude membrane fractions used for the automated filter binding assay is similar to that of the manual procedure. As the binding sites of most membrane-bound receptors are sensitive to proteolytic degradation, trypsin should not be used for the harvesting. Cells can be detached from the tissue culture plates by incubation with Mg^{2+}-free phosphate-buffered-saline (pH 7.4), or with cell dissociation buffer (Gibco, Grand Island, NY; Cat no. 13151-014). Alternatively, cells can be harvested by scraping from the plates, although this is not feasible for cells growing in triple flasks and cell factories with multiple layers of substrates.

After harvesting, the cells are homogenized with a Polytron in the presence of hypotonic buffer (e.g., 50 mM Tris, 120 mM NaCl, 5 mM KCl, 2 mM $CaCl_2$, 5 mM $MgCl_2$, pH 7.4). This is also the incubation buffer for the assay. After washing the cells with the buffer at least two times by centrifugation (e.g., 20,000 r/min, 4°C for 10 min), the crude membrane preparation is resuspended in assay buffer at a concentration $\geq$ 10 mg protein/mL and stored in frozen aliquots at –80°C. This preparation is stable for several months. On the day of assay, the frozen membrane preparation is diluted to the required volume by the incubation buffer.

12.5 ASSAY SETUP

12.5.1 PREPARATION

As the automated assay is designed to be a walk-away system, it is necessary to set up all the instruments and add all the reagents before the start of the assay. The checklist includes:

1. Cooling of cooling trough by the refrigerated reservoir
2. Pumping the ice-cold wash buffer into the Brandel harvester
3. Putting assay plates, filter plates, deep-well dilution plates in Genmate and Genesis
4. Adding the reagents (polyethyleneimine, incubation buffer, DMSO, dilution buffer (e.g., 10% DMSO) to the buffer troughs in Genesis (Figure 12.1)
5. Dilution of radioligand in incubation buffer and transfer into the designated reagent trough in Genesis (Figure 12.1)
6. Reconstitution and thawing of crude membrane fraction by addition of assay buffer, and addition into the cooled trough in Genesis
7. Machine initialization of Genesis, Genmate, and Brandel
8. Making stock solutions of test compounds (see below) and putting them onto the Genmate deck

12.5.2 PREPARATION OF STOCK SOLUTION

Before the assay, test compounds are first dissolved in 100% DMSO as 1 or 10 mM stock. To maximize the chances that compound can be completely dissolved, 1 mM stock concentration in 100% DMSO is preferred. At least 100 μL of the stock solution is needed to ensure that the compound can be pipetted by the pipettor. To allow easy access by the Genmate 96-well pipettor, Matrix (Hudson, NH) 0.75 mL Screenmates tubes (catalogue no. 4171) are used to store and dissolve the compounds. The highest possible concentration of DMSO should be used to increase solubility of compounds in the assay. The amount of DMSO used in the assay will depend on the tolerability of the receptor toward DMSO. The final DMSO concentration is 1% for most of our receptor binding assays.

12.5.3 SERIAL DILUTIONS

To minimize any potential artifacts due to DMSO on the pharmacology of any given target, the DMSO concentration (e.g., at 10% DMSO) is maintained constant during the incubation step. This is accomplished by first diluting the stock concentration (in 100% DMSO) tenfold in water. Subsequent serial dilutions are carried out in the dilution plates containing 10% DMSO. During the incubation with the biological sample, the compounds will be diluted another tenfold with buffers without any DMSO, resulting in a final DMSO concentration of 1%.

Serial dilution can be carried out either within the same deep-well dilution plate or across different deep-well plates. Depending on the number of compounds to be tested in the assay, the sequence of addition should be adjusted to minimize the time needed for the assay. For equal to or fewer than 14 compounds, it is more efficient to perform serial dilutions within a single deep-well plate, using a single column of pipet tips for each dilution. For 15 to 84 compounds, dilution across different deep-well plates is desirable, using all 96 pipet tips together. A protocol for more than 14 compounds is described below.

12.5.4 PLATE LAYOUT

The layout of compounds in the assay plates determines how the dilution and transfer of compounds should be carried out. The data analysis software will also be based on the layout. As the assays last several hours, the B_{max} of the biological sample can differ significantly between the first and last assay plates due to the denaturation of the receptor. It is therefore essential that each assay plate have its own controls for total and nonspecific bindings. Figure 12.2 shows the plate map of an assay plate. It contains six wells with vehicle for total binding and six wells containing a high concentration of a compound that can block all the binding to the target of interest to determine nonspecific binding.

To accomplish this plate format, 450 μL of water is pipetted to the first deep-well dilution plate. Row D is left empty. All the other dilution plates have a similar layout, except that vehicle with the appropriate DMSO concentration (e.g., 10% DMSO) is added to the wells as indicated in Figure 12.2. In the first dilution, 50 μL of compound is transferred from rows A to C of the stock plate to rows A to C of the first dilution plate and the well is mixed, followed by transferring rows D to G of the stock plate to rows E to H of the dilution plate. After each mixing, the pipet tips are discarded. Subsequent serial dilutions are carried out using all 96 pipet tips, with all of the compounds in each dilution plate transferred from plate to plate. In this format, the same seven compounds will be in the same column on every dilution plate, except that the concentrations at each plate will be different. Upon completion of serial dilution, 25 μL of the diluted compound (in a single column) will be transferred in duplicates from the dilution plates to one assay plate. The transfer will start from the plate containing the least concentrated compounds to plates with the higher concentrations. In this way, the same set of pipet tips can be used. Row D of the assay plate is empty as there is no liquid present in row D of the deep-well dilution plate. After the compounds are added to each plate, vehicle and reagent for nonspecific binding will be added to row D of the assay plate.

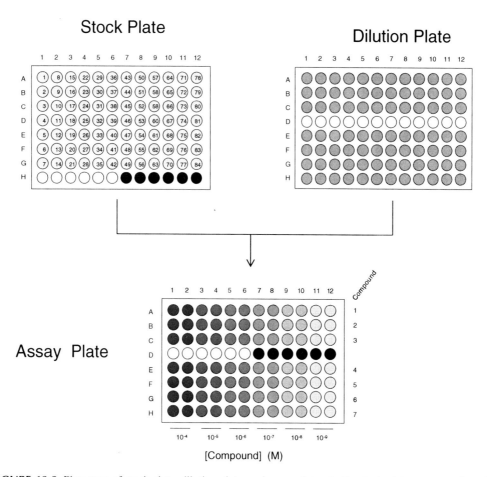

FIGURE 12.2 Plate map of stock plate, dilution plate, and assay plates. In the stock plate, compound stocks (with number denoted in the figure) are stored in the 0.75-mL tube rack. Vehicle (white well) and reagent for nonspecific binding (black well) are at row H. For the dilution plate (2 mL deep-well plate), water or vehicle is added to all the wells except row D. In the microtiter assay plate, increasing compound concentrations are indicated by increasing intensity of shading of the wells. The wells at row D for total and nonspecific binding are indicated by white and black wells.

12.6 ASSAY PROTOCOL

Figure 12.3 shows the flow chart of assay setup. The process can be divided into two components:

1. Liquid handling (dilution and transfer of compounds into microtiter plates) which is performed in serial fashion (i.e., one plate after another)
2. An incubation/harvesting step in which addition of radioligand, addition of membrane, incubation, and harvesting are carried out in a parallel manner (i.e., multitasking)

An example of how the automated process was carried out is shown below.

12.6.1 LIQUID HANDLING

Before dilution of compounds, 120 μL/well of 0.5% polyethyleneimine was added to the Unifilter GF/B plates using the fixed tips of Genesis. The function of the polyethyleneimine is to reduce nonspecific binding of the radioligand to the filter membrane. In addition, the presence of liquid in the filter plate would increase the seal between the filter plate and the rubber gasket of the Brandel harvester at the beginning of the harvesting cycle. Each fixed tip was then flushed with 5 mL of water (system liquid), followed by 1 mL of DMSO (stored in buffer trough), then with 30 mL of water. Water (450 μL) was added to the deep-well plate for the first dilution. Vehicle (450 μL, 10% DMSO) was then added to each well of five other 2-mL deep-well plates (waters) for subsequent dilutions of compounds. The deep-well plates were then transferred by the robotic fingers to the Genmate deck for dilutions.

Test compounds (1 mM stock concentration, 200 to 300 μL) were dissolved in 100% DMSO and stored in 0.75 mL Matrix deep-well plates. Disposable tips were used for all dilutions and transfer operations of compounds, using the custom-designed pipet tip holder as shown in Figure 12.1. Compounds were serially diluted (tenfold dilution) six times in 10% DMSO (50 into 450 μL). The diluted compounds were transferred (25 μL/well in duplicates) to each microtiter assay plate. After the transfer, each assay plate was transferred back to the microtiter plate hotel in Genesis. Upon the completion of compound transfer to all assay plates, six replicates of 25 μL of 100 μM butaclamol (reagent for nonspecific binding) in 10% DMSO and 25 μL of 10% DMSO (vehicle for total binding, six replicates) were added to each microtiter assay plate from the stock compound plate by Genmate (Figure 12.2).

12.6.2 INCUBATION AND HARVESTING

Twenty-five microliters of [^{3}H] 7-OH-DPAT (0.4 to 0.6 nM final concentration) was added to each well followed by 150 μL assay buffer. The binding reaction was initiated by the addition of 50 μL membrane preparation (3 to 5 μg/well), which was kept at 2°C in the cooling trough. Incubation was for 15 min at 37°C in the incubator, after which the reaction was terminated by rapid filtration through the polyethyleneimine-soaked Unifilter GF/B filter plates prepared earlier. The plates were washed six times (200 μL per wash per well) with ice-cold 50 mM Tris buffer at pH 7.4 in the Brandel harvester. Filter plates were transferred back to the microtiter plate hotel and air-dried overnight at room temperature. Afterwards, 40 μL of MicroScin 0 (Packard) was pipetted into each well of the Unifilter plate and counted in the Packard TOPCOUNT.NXT counter.

12.6.3 SCHEDULED OPERATION

Figure 12.3 shows a graphic presentation of the scheduled operation for 84 compounds. Dilution of 84 compounds and subsequent transfer to 12 microtiter plates were performed first. Addition of radioactivity and membrane preparations, incubation, and filtration operations were all scheduled and staggered such that there was an identical time of incubation (15 min) for each microtiter plate. The total time of the automated operation is about 4 h for 84 compounds.

12.7 DATA ANALYSIS

Scintillation counts from the TopCount were recorded in a computer and transferred online to another computer, which had the data analysis software. K_i values were determined by nonlinear regression (see [3] for reference). Analysis of the data and plotting of graphs for 84 compounds can be completed within 5 min.

12.8 ACCURACY

As published previously [3], the K_d and K_i generated by the fully automated workstation are comparable to published values of known compounds. Table 12.1 shows the K_i of several D3 ligands determined by the automated assay. The values are similar to the published K_i values for these compounds (see [3] for references) and thus validated the fully automated procedure. A decrease

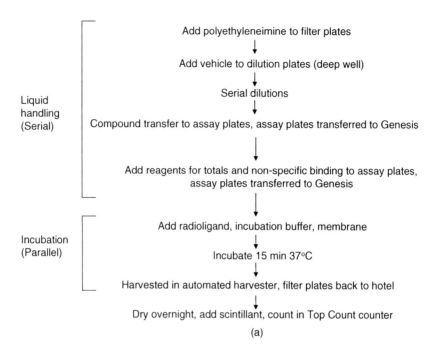

FIGURE 12.3 Flow chart and scheduling of the robotic operations. (a) A flow chart for the liquid handling and subsequent incubation and harvesting operations is shown. (b) The GANTT chart for the operations is shown. There is identical time of incubation (15 min) for each microtiter plate. The later staggered operations are shown with bars. (From Shrikhande, A., Courtney, C., Smith, D., Melch, M., McConkey, M., Bergeron, J., and Wong, S.K. 2002. *Biotechniques*. 33:932–937.)

of about 35% of the total binding was observed between the first and the 12th plate. This is likely due to denaturation of D3 upon storage in the assay buffer at 4°C. The decrease in receptor number did not affect the K_i values, as internal controls (total and nonspecific binding) were present in each plate.

12.9 REPRODUCIBILITY

The data generated from the automated workstation are highly reproducible. Figure 12.4 shows the binding isotherms when haloperidol and L-741,626 were tested multiple times across 12 assay plates. The C.V. of individual data points between individual duplicates is less than 10% (C.V. = standard error/mean * 100%). The apparent K_i values of haloperidol and L-741,626 are 5.0 ± 0.6 nM (N = 60) and 96.3 ± 15.3 nM (N = 24), respectively, and are similar to the published K_i values of these two compounds [4,5]. The C.V. of the K_i values is between 12 and 15% for both compounds. This is superior to data generated by manual operation (C.V. about 100%). The highly automated nature of the assay described here minimizes variability of data as a result of the manual liquid handling. This enables an inexperienced operator, within a short training time (1 to 2 weeks), to produce high-quality data with a throughput that is required for many drug discovery projects.

12.10 TIME OF OPERATION

Minimal hands-on time was required as compared to manual operation. The total time for hands-on operation is about 1 h (Table 12.2). This compares favorably to the manual operation, which requires 3 hours constant hands-on time to complete a similar task. The robotic operation thus saves more than 70% of hands-on time. More importantly, the time of exposure to the radioligand

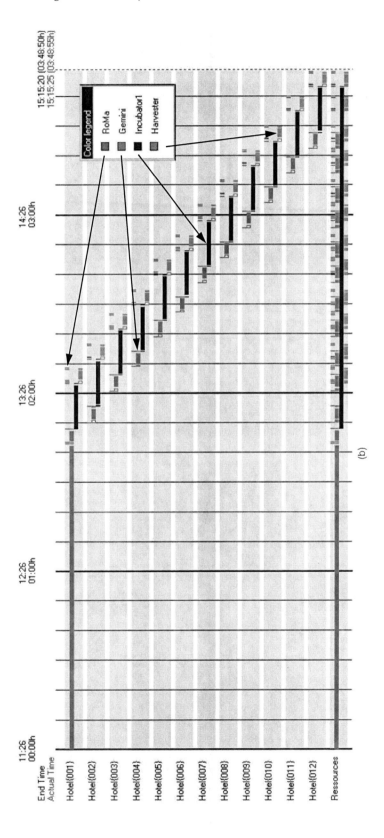

FIGURE 12.3 (CONTINUED)

Table 12.1

Comparison of K_i of D3 Ligands from Automated Assay and Published Literature[a]

Compound	K_i (nM)	
	Automated	Published
7-OH-DPAT	4.5 ± 2.9	2.2
Quinelorane	4.6 ± 0.1	6.1
Quinepirole	14.4 ± 1.4	43
PD-128907	2.2 ± 0.6	1.8
Raclopride	3.8 ± 0.4	1.4
Risperidone	10.7 ± 2.9	11
Clozapine	124.5 ± 22.9	88
Bromocriptine	10.6 ± 2.8	7.4

[a] The automated ligand binding assay was performed as described in the text. The K_i values of different D3 ligands shown are the mean ± standard deviation of three repeated determinations from the same experiment. The results shown are typical of two independent experiments. The K_i values determined are compared with published values. (See Shrikhande, A., Courtney, C., Smith, D., Melch, M., McConkey, M., Bergeron, J., and Wong, S.K. 2002. *Biotechniques.* 33:932–937 for references.)

and the biological sample is reduced by more than 95%. Although the total time for the automated operation (5 h) is longer than that for manual operations (3 h), the fully automated operation has the distinct advantage of enabling the operator to walk away once the assay is initiated.

12.11 THROUGHPUT

The throughput of the assay is only limited by the space available at Genesis to store the assay and filter plates. In its current format, the equipment enables K_i determination of 84 compounds per run. Two to three runs can be performed in a working day. The present format can be modified to single-point percent inhibition experiments, where 12 microtiter plates can be screened per run. If the assay is a homogeneous assay (such as the scintillation proximity assay, see below), the throughput of the assay can be doubled, as filter plates are not needed. The throughput can be further increased by the attachment of external plate stackers.

12.12 FUTURE DEVELOPMENT

The success in developing the automated assay can lead to development of many other automated assays. The universal application of the rapid filtration assay for membrane-bound receptors would enable parallel screening of many different targets. We have modified the cooling trough to store 96 different biological samples, so selectivity of a given compound across many targets can be profiled in a single run (one concentration per target, quadruplicate determinations). In addition, the scheduling operation of the assay allows identical incubation times between plates and enables automation of functional assays where termination of incubation is required at a fixed time interval.

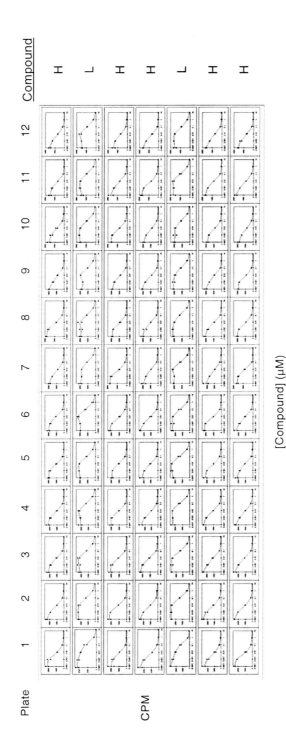

Plate

CPM

[Compound] (μM)

FIGURE 12.4 Competition binding of haloperidol and L-741,626 using the automated ligand binding assay. Haloperidol (H) and L-741,626 (L) at increasing concentrations (horizontal axis: 0.1 nM to 10 μM) were incubated with [^{3}H]-7-OH DPAT (0.4 nM) and membranes expressing D3 receptors. The two compounds were tested multiple times in 12 microtiter plates as indicated by the rows labeled (H) and (L). The amount of bound [^{3}H]-7-OH DPAT (CPM) is shown in the vertical axis of the binding isotherm. (Data published in Shrikhande, A., Courtney, C., Smith, D., Melch, M., McConkey, M., Bergeron, J., and Wong, S.K. 2002. *Biotechniques*. 33:932–937.)

TABLE 12.2
Minutes Required to Determine K_i of 84 Compounds by Manual
Procedures and by the Fully Automated Ligand Binding Assay

Operation	Manual Hands-on	TECAN Hands-on	TECAN Machine
Preparation of reagents	45	45	
Dilution/transfer/incubation/filtration	135	0	230
Addition of scintillation fluid	30	15	
Total (min)	190	290	

Source: Shrikhande, A., Courtney, C., Smith, D., Melch, M., McConkey, M., Bergeron, J., and Wong, S.K. 2002. *Biotechniques.* 33:932–937.

REFERENCES

1. Bylund, D.B. and Yamamura, H.I. 1990. In: *Methods in Neurotransmitter Receptor Analysis* (Yamamura, H.I. et al. Ed.), pp. 1–35. Raven Press, New York.
2. Ganesan, A. 2002. Recent developments in combinatorial organic synthesis. *Drug Discovery Today.* 7(1):47–55.
3. Shrikhande, A., Courtney, C., Smith, D., Melch, M., McConkey, M., Bergeron, J., and Wong, S.K. 2002. Fully automated radioligand binding filtration assay for membrane-bound receptors. *Biotechniques.* 33:932–937.
4. Bowery, B.J., Razzaque, Z., Emms, F., Patel, S., Freedman, S., Bristow, L., Kulagowski, J., and Seabrook, G.R. 1996. Antagonism of the effects of (+)-PD- 128907 on midbrain dopamine neurones in rat brain slices by a selective D2 receptor antagonist L-741,626. *Br. J. Pharmacol.* 119:1491–1497.
5. Sokoloff, P., Giros, B., Martres, M.P., Bouthenet, M.L., and Schwartz, J.-C. 1990. Molecular cloning and characterization of a novel dopamine receptor (D3) as a target for neuroleptics. *Nature.* 347:146–151.

13 Nuclear Receptors as Drug Targets

Patricia D. Pelton

CONTENTS

13.1 ABSTRACT

Nuclear receptors are ligand-activated transcription factors that are involved in a number of physiological functions and disease states such as inflammation, cancer, and metabolism. Nuclear receptors are the target of several currently marketed therapeutics, and many pharmaceutical and academic laboratories are actively screening for novel modulators of these receptors. They are considered to be desirable drug targets, in part, because the ligands that activate them are generally of low molecular weight. This review will give a general overview of the nuclear receptor classes — their structure, function, and potential as drug targets.

13.2 INTRODUCTION

Nuclear receptors are ligand-activated transcription factors that regulate the expression of genes involved in development, metabolism, and disease states. These receptors have allosteric interactions with ligand, DNA, and transcription cofactor proteins. A number of lipophilic hormone ligands, including the steroids (estrogen, progesterone, androgens, and glucocorticoids), thyroid hormone, and vitamins A and D, were discovered in the early part of the century based upon their effects on development, differentiation, and organ physiology. These hormones also played roles in disease and some, for example the glucocorticoids, were used as therapy before their molecular targets were identified. The first receptors to be cloned in the mid-1980s were the estrogen and glucocorticoid receptors, followed closely by the thyroid receptor [1–4]. Subsequently, the retinoic acid receptor was cloned, and a family of receptors emerged [5]. This review will provide an overview of the nuclear receptor superfamily and summarize the key aspects of the biology of these receptors that have made for development of unique screening tools. Some of these tools take advantage of their interactions with other proteins to develop tissue- and gene-selective modulators with the goal to improve both clinical efficacy and safety.

TABLE 13.1
Classes of Nuclear Receptors

Class	Name	Examples	Dimerization Type	DNA Binding (Repeat Type)
I	Steroid receptors	Glucocorticoid, estrogen, androgen progesterone, mineralocorticoid	Homodimers	Inverted
II	Retinoid X receptor (RXR) heterodimers	Thyroid, liver X receptor, peroxisome proliferator activated receptors, vitamin D, retinoic acid	Heterodimers	Direct (some inverted)
III	Dimeric	RXR, HNF-4	Homodimers	Direct
IV	Monomeric	NGFI-B, SF-1	Monomers	Extended

13.3 NUCLEAR RECEPTOR CLASSES

To date, 48 members of the nuclear receptor superfamily are known. Nuclear receptors can be grouped in four classes; classification is based to some extent on ligand binding, but DNA binding and dimerization properties are the primary bases for classification [6,7] (Table 13.1). The first class comprises the steroid receptors, which are homodimeric in the active state. The second class is the retinoid X receptor heterodimers. RXR has been termed a master regulator [6] and is an obligate partner with a number of other receptors, many of which have only recently been deorphanized, including the peroxisome proliferator activated receptors (PPARs), the liver X receptors (LXRs), the farnesoid X receptor (FXR), the retinoic acid receptors (RARs), the thyroid receptors (TRs), and the vitamin D receptor. A third group includes the nonsteroid homodimers such as HNF-4 and RXR (which are present as both hetero- and homodimers). A fourth class exists as monomers. Some nuclear receptors may not require a ligand to function, such as Nurr1 [8], or may have a nondisplaceable ligand present in the ligand-binding pocket, such as a fatty acid, as has been reported for HNF-4 [9].

13.4 GENERAL STRUCTURAL FEATURES OF NUCLEAR RECEPTORS

Nuclear receptors range in size from 50,000 to 100,000 Da, and each receptor is divided into different functional domains. There is an N-terminal activation domain, also called activation function-1 (AF-1), a DNA binding domain (DBD), a hinge region, and the ligand binding domain (LBD) [6]. At the carboxy terminus, which is still considered part of the LBD, there is a second activation function domain (AF-2). The DBD is the most highly conserved domain and consists of two zinc finger motifs typical of DNA binding proteins [10,11]. The zinc finger motifs bind to hexameric sequences (response elements) as repeats based upon the dimerization mode of the DBD. The binding affinity of the receptors for DNA is more efficient as dimers with the exception of a few receptors, which bind as monomers [12,13]. As mentioned above, the steroid receptors are homodimers and bind to the DNA as symmetrical (palindromic) repeats of 5'-AGAACA-3' (except for the estrogen receptor, which is 5'-AGGTCA-3') and a spacing of one nucleotide. RXR heterodimers bind to direct repeats (DRs) having a generic sequence 5'-AGGTCA-3' but with a differing number of spaces in between depending upon the receptor [6,7]. For example RXR/TR heterodimers bind to repeats with a spacing of four nucleotides (DR-4), while RXR/PPARs can bind to repeats with either one- or two-nucleotide spacing (DR-1 or DR-2). The DNA binding of monomers such as ROR, SF-1, HNFs, and NGFI-B is the same as for the RXR heterodimers, but the selectivity comes from flanking sequences at the 5' end.

The most relevant domain of the nuclear receptors for targeting drugs is the LBD. In recent years, crystal structures for several LBDs have been obtained both in the presence and absence of agonists, antagonists, and coregulatory proteins [14–20]. Even though the sequences of the LBDs are quite varied among the different nuclear receptors, they all share a similar structure with 12 alpha helices, 11 of which are folded into a tightly compacted tertiary structure. The major helices involved in ligand binding are H3, H4, H11, and H12. H12 contains the AF-2 domain, which is crucial for the protein–protein interactions that occur after ligand binding. When in an unliganded state, H12 is situated like a lid over the ligand binding pocket. Ligand binding results in conformational changes and alters the orientation of both H11 and H12 [for reviews see 21,22]. Repositioning of H12 results in dissociation of corepressor proteins and an exchange of the corepressor for coactivator protein(s). A "charge clamp" is formed between residues in the AF2 domain of H12 and LXXLL motifs (also known as NR boxes) on the coactivator proteins. Many antagonists have a bulky side chain that effectively blocks H12 from assuming the same orientation as in the agonist conformation. H12 instead occupies the binding surface of the LXXLL motif of the coactivator, thereby preventing the coactivators from binding. H11 can also play an important role in the determination of ligand activity. In addition to either agonist or antagonist conformations, ligands can also be labeled as partial agonists with varying potency but lower efficacy than full agonists.

The dimeric interactions of the receptors also play a key role in the response of the receptors to ligands, particularly in the case of the RXR heterodimers. Several of the RXR heterodimers are considered permissive in that activation of either heterodimer partner by a ligand can result in activation of the receptor dimer. Additive or synergistic effects can occur when ligands of both heterodimer partners bind. This has been shown, for example, both *in vivo* and *in vitro* with PPAR/RXR and LXR/RXR heterodimers [23; unpublished observations, P. Pelton]. On the other hand, for RAR/RXR heterodimers, the RXR partner is considered to be silent and the receptor can only be activated by RAR ligands, although this is not always the case [24,25]. It is thought that H11 interactions contribute to this phenomenon because H11 has contact with the ligand, the dimer interface, and H12.

13.5 FROM LIGAND BINDING TO TRANSCRIPTION

In the absence of ligands, nuclear receptors exist in different states. For example, the steroid hormone receptors are cytosolic and are bound to chaperone proteins like hsp90 [26]. Ligand binding results in dissociation of the steroid receptor from these proteins; the receptor is translocated to the nucleus, where it interacts with the appropriate response elements of the target genes. Other nuclear receptors such as the RXR heterodimers are considered to reside mainly in the nucleus bound to the DNA. With these receptors, in the absence of ligand, corepressor proteins dock to a hydrophobic groove found in H3 and H4 with a common sequence motif known as a corepressor–nuclear receptor box (CoRNR box) [27]. This binding allows for the recruitment of transcriptional repressor proteins such as histone deacetylases (HDACs), which promote a condensed chromatin structure around the promoter region of the target DNA and thereby repress gene expression.

As described above, ligand binding induces a conformational change that allows for docking of coactivator proteins (or even corepressors in some contexts). Binding of the coregulator proteins results in the further recruitment of other proteins that are required for effective transcriptional activity. There are at least 50 coactivator proteins that have been identified to date, adding to the complexity and multiplicity of responses that can occur with nuclear receptor activation. At least three potential modes of coactivator interaction have been described and are reviewed in greater detail by Glass and Rosenfeld [28]. The first mode could be sequential, where a coactivator complex binds and completes a function such as an acetyltransferase reaction to remodel the chromatin; this complex then leaves and subsequent complex(es) come with the required factors for completion of the transcriptional process. A second model envisions more than one complex interacting at the same time in a combinatorial manner, while a third interaction may be parallel in nature, where ligand binding may result in the recruitment of more than one coactivator complex on the same promoter sequence. Not only can a number of coactivators and potential complexes be formed, there are also several ways in which the coactivator proteins themselves can be modified or modulated. The coactivator proteins can be translocated from the cytoplasm to the nucleus; like the nuclear receptors, they can also be phosphorylated or acetylated or the levels can be regulated by proteolysis. There are also tissue- and cell type–specific levels of different coactivators that can influence the response of the nuclear receptors to ligands. The important role many nuclear receptors play in development and homeostasis requires highly specific and finely tuned responses, which is evident from the degree of complexity involved in nuclear receptor interactions.

13.6   NUCLEAR RECEPTORS LIGANDS AND "REVERSE ENDOCRINOLOGY"

Traditionally in endocrinology, a hormone was characterized based upon its physiological effect, and subsequently the target (receptor) was identified. This was the case for the steroid hormone receptors. In some cases, therapeutics existed before the molecular target was known. For example, the fibrate drugs for the treatment of hypertriglyceridemia were prescribed long before they were found to be agonists of PPARα [29]. In recent years, however, with the cloning of several novel nuclear receptors, the situation has reversed. The receptors are used to identify new hormones or physiological sensors and together are used to further probe the physiological response of modulation of these receptors. This is termed "reverse endocrinology" [30]. For the PPARs, it was a bit different in that synthetic ligands already existed but the target was unknown. The proposed endogenous ligands for the PPARs are fatty acids and fatty acid metabolites; unlike the classic hormone ligands, these are of weak affinity [31,32]. The evidence to date suggests that no high-affinity hormone ligand may exist for these receptors, and their role may be more as nutrient sensors regulating the expression of genes involved in glucose and lipid metabolism.

Examples of "reverse endocrinology" include the identification of LXR and FXR as a new generation of steroid receptors. The receptors were cloned, and subsequently the endogenous ligands were identified as oxysterols and bile acids, respectively [33–36]. The physiological role of these receptors has been elucidated using knockouts and both endogenous and synthetic ligands as tools. These receptors play important roles in cholesterol homeostasis, and synthetic ligands or modulators of these receptors could have beneficial effects in metabolic diseases.

Nearly half of the 48 nuclear receptors identified to date do not have known endogenous ligands. As stated earlier, for some there may be no particular binding pocket for either an endogenous or synthetic ligand, or an endogenous ligand may be bound and not displaced by synthetic ligands. These receptors may not be optimal targets in the traditional sense, but it may be possible to modulate them by altering their protein–protein or DNA interactions. The remaining orphans could potentially be drug targets once a ligand or their role is identified.

13.7   NUCLEAR RECEPTORS AS DRUG TARGETS: THE NEXT GENERATION

As shown in Table 13.2, nuclear receptor ligands are currently used to treat diseases including cancer, inflammation, dyslipidemia, and diabetes. However, in all cases these ligands result not only in beneficial effects but also in annoying or severe side effects that limit their utility. The next generation of ligands will hopefully overcome some of these issues, at least to some extent by having tissue/gene selectivity. Chemistry efforts on a number of targets have shown this is possible and, in addition to being classified as agonists, partial agonists, or antagonists, these compounds can also be called selective nuclear receptor modulators. A highly publicized example is the SERMS

TABLE 13.2
Current Nuclear Receptor Drug Targets

Receptor	Disease Indication
Steroids	
Glucocorticoid (GR)	Immunology/metabolic
Progesterone (PR)	Reproduction, breast cancer
Estrogen (ERα, β)	Menopause, breast cancer
Androgen (AR)	Prostate cancer
RXR Heterodimers	
Thyroid (TRα, β)	Atherosclerosis, obesity, diabetes, metabolic syndrome
Retinoid receptor (RARα, β, γ)	Cancer, skin disorders
Retinoid X receptor (RXRα, β, γ)	Cancer, atherosclerosis, diabetes, metabolic syndrome
Liver X receptor (LXRα, β)	Atherosclerosis, diabetes, Alzheimer's disease
Farnesoid X receptor (FXR)	Dyslipidemia, liver disease
Peroxisome proliferator activated receptor (PPAR α γ, δ)	Metabolic syndrome, dyslipidemia, atherosclerosis, diabetes
Vitamin D receptor (VDR)	Osteoporosis
Nurr1	Parkinson's disease
Others	
Constituitive androstane receptor (CAR)	Xenobiotic metabolism
Pregnane X receptor (PXR)	Xenobiotic metabolism
Germ cell nuclear factor (GCNF)	

Note: Receptors in bold type are targets of marketed products.

(selective estrogen receptor modulators) such as Nalvodex (tamoxifen) and Evista™ (raloxifene). These compounds are currently used for the treatment of estrogen-dependent breast cancer, and they act as antagonists in breast and uterine tissue but as agonists in bone. X-ray crystallography studies of receptor/ligand complexes showed how the conformation of the receptor differs in the presence of an agonist like estradiol compared with these antagonists [17,20]. To make matters more interesting and complicated, recently another estrogen receptor (ERβ) was cloned [37,38]. Studies comparing the activity of different compounds on the two receptor subtypes have revealed some very interesting data. Ligand effects on ERβ appear to be highly dependent upon the cell/promoter context, and it is "easier" to shift from an agonist to an antagonist [39]. The identification of specific agonists/antagonists of these receptors will not only yield important information on the physiological role of these receptors but may also result in the development of improved SERMS.

Selective nuclear receptor modulators of PPARγ and LXR have also been reported. FMOC-L-leucine has low potency but full efficacy at the PPARγ receptor and has insulin-sensitizing effects. However, it is very weak in differentiating adipocytes, a key role of PPARγ [40]. This compound illustrates that it may be possible to obtain insulin sensitivity without significant adipocyte differentiation, which could result in less weight gain in patients, a problem with the currently marketed compounds. For the LXR receptor, the first reported ligand was TO-901317; it was noted that this compound induced lipogenesis resulting in hypertriglyceridemia, which, if replicated in the clinic, would preclude the use of this compound for the treatment of metabolic diseases [41,42]. Another compound with unrelated structure was subsequently reported and called GW3965 [43]. This compound was slightly weaker in potency when compared in cotransfection assays but showed good activity *in vivo* in preventing atherosclerotic lesion development in a mouse model of atherosclerosis and in its antidiabetic effects in rodent models of diabetes. Interestingly, however, GW3965 does not seem to possess the lipogenic activity of TO-901317, and it was recently reported that it acts as selective LXR modulator, with less activity in the liver compared with TO-901317 [44]. These data suggest that it will be possible to design LXR modulators devoid of lipogenic activity.

While the receptors have diverse actions and ligands, the common structural and functional features of nuclear receptors have greatly aided in the development of screening tools. Common assay platforms include cell-based cotransfection assays and the biochemical-based fluorescent polarization and time-resolved fluorescence assays [45]. These assays allow not only for the identification of agonists and partial agonists but can also identify antagonists when screened in the presence of an agonist. The biochemical-based assays also give information on the interactions of the receptor/ligand complex with different coactivator proteins/peptides. For many of these assays, short peptide sequences consisting of an LXXLL sequence from an NR box of a coactivator is sufficient to show interaction. Many examples of this type of assay are reported in the literature [45–47].

In addition to the assays described above, it is also now possible to screen in a relatively high throughput format for selective gene markers using the Quantigene bDNA technology ([48]; see also http://www.genospectra.com [49]). Selective gene induction or repression can be compared to vehicle controls without the need for RNA purification. The development of multiplexing capability for this assay also allows for comparison of multiple genes from one well.

Tissue selectivity can also be assessed using different primary cells or cell lines as shown for the LXR agonists described above. For the identification of future nuclear receptor modulators, simple binding assays will not be sufficient. In fact, many companies screen for nuclear receptor ligands using a cell-based approach.

In summary, targeting nuclear receptors offers the potential to develop therapies to treat myriad diseases, especially those where there are current unmet needs. The challenge will be to develop compounds that target desired gene targets, have appropriate physicochemical properties, and do not alter the metabolism of either their own or other concurrent medications. As we learn more about the role of nuclear receptors in metabolism, reproduction, and development, we have also learned and need to gain more information on their role in xenobiotic metabolism. Many compounds have overlapping activities on not only their target receptors, which themselves may regulate genes involved in Phase I and II drug metabolism, but also on novel xenobiotic nuclear receptors such as constituitive androstane receptor (CAR) and the pregnane X receptor (PXR) [49,50]. An important characterization of a compound before further development may also include screening for activity on xenobiotic receptors. Another area of concern is the widespread use of herbal supplements, many of which also interact with nuclear receptors. A good example is St. John's wort, which contains a ligand for PXR, inducing the expression of the cyp3A gene, which is responsible for metabolizing many currently prescribed drugs [51].

REFERENCES

1. Green S; Walter P; Greene G; Krust A; Goffin C; Jensen E; Scrace G; Waterfield M; Chambon P Cloning of the human oestrogen receptor cDNA. *Journal of Steroid Biochemistry* 1986; 24, 77–83.
2. Greene G L; Gilna P; Waterfield M; Baker A; Hort Y; Shine J Sequence and expression of human estrogen receptor complementary DNA. *Science* 1986; 231, 1150–1154.
3. Weinberger C; Hollenberg S M; Ong E S; Harmon J M; Brower S T; Cidlowski J; Thompson E B; Rosenfeld M G; Evans R M Identification of human glucocorticoid receptor complementary DNA clones by epitope selection. *Science* 1985; 228, 740–742.
4. Sap J; Munoz A; Damm K; Goldberg Y; Ghysdael J; Leutz A; Beug H; Vennstrom B The c-erb-A protein is a high-affinity receptor for thyroid hormone. *Nature* 1986; 324, 635–640.
5. Petkovich M; Brand N J; Krust A; Chambon P A human retinoic acid receptor which belongs to the family of nuclear receptors. *Nature* 1987; 330, 444–450.
6. Mangelsdorf D J; Evans R M The RXR heterodimers and orphan receptors. *Cell* 1995; 83, 841–850.

7. Mangelsdorf D J; Thummel C; Beato M; Herrlich P; Schuetz G; Umesono K; Blumberg B; Kastner P; Mark M et al. The nuclear receptor superfamily: the second decade. *Cell* 1995; 83, 835–839.

8. Wang Z; Benoit G; Liu J; Prasad S; Aarnisalo P; Liu X; Xu H; Walker N P C; Perlmann T Structure and function of Nurr1 identifies a class of ligand-independent nuclear receptors. *Nature* 2003; 423, 555–560.

9. Wisely G B; Miller A B; Davis R G.; Thornquest A D; Johnson, R; Spitzer T; Sefler A; Shearer B; Moore J T; Miller A B; Willson T M; Williams S P Hepatocyte nuclear factor 4 is a transcription factor that constitutively binds fatty acids. *Structure* 2002; 10, 1225–1234.

10. Hard T; Kellenbach E; Boelens R; Maler B A; Dahlman K; Freedman L P; Carlstedt-Duke J; Yamamoto K R; Gustafsson J A; Kaptein R Solution structure of the glucocorticoid receptor DNA-binding domain. *Science* 1990; 249, 157–160.

11. Luisi B F; Xu W X; Otwinowski Z; Freedman L P; Yamamoto K R; Sigler P B Crystallographic analysis of the interaction of the glucocorticoid receptor with DNA. *Nature* 1991; 352, 497–505.

12. Danielsen M; Northrop J P; Jonklaas J; Ringold G M Domains of the glucocorticoid receptor involved in specific and nonspecific deoxyribonucleic acid binding, hormone activation, and transcriptional enhancement. *Molecular Endocrinology* 1987; 1, 816–822.

13. Kumar V; Chambon P The estrogen receptor binds tightly to its responsive element as a ligand-induced homodimer. *Cell* 1988; 55, 145–156.

14. Wagner R L; Apriletti J W; McGrath M E; West B L; Baxter J D.; Fletterick R J A structural role for hormone in the thyroid hormone receptor. *Nature* 1995; 378, 690–697.

15. Renaud J P; Rochel N; Ruff M; Vivat V; Chambon P; Gronemeyer H; Moras D Crystal structure of the RAR-gamma ligand-binding domain bound to all-*trans* retinoic acid. *Nature* 1995; 378, 681–689.

16. Bledsoe R K; Montana V G; Stanley T B; Delves C J; Apolito C J; McKee D D; Consler T G; Parks D J; Stewart E L; Willson T M; Lambert M H; Moore J T; Pearce K H; Xu H E Crystal structure of the glucocorticoid receptor ligand binding domain reveals a novel mode of receptor dimerization and coactivator recognition. *Cell* 2002; 110, 93–105.

17. Brzozowski A M; Pike A C; Dauter Z; Hubbard R E; Bonn T; Engstrom O; Ohman L; Greene G L; Gustafsson J A; Carlquist M Molecular basis of agonism and antagonism in the oestrogen receptor. *Nature* 1997; 389, 753–758.

18. Darimont B D; Wagner R L; Apriletti J W; Stallcup M R; Kushner P J; Baxter J D.; Fletterick R J; Yamamoto K R Structure and specificity of nuclear receptor-coactivator interactions. *Genes and Development* 1998, 12, 3343–3356.

19. Xu H E; Lambert M H; Montana V G; Parks D J; Blanchard S G; Brown P J; Sternbach D D; Lehmann J M; Wisely G B; Willson T M; Kliewer S A; Milburn M V Molecular recognition of fatty acids by peroxisome proliferator-activated receptors. *Molecular Cell* 1999; 3, 397–403.

20. Shiau A K; Barstad D; Loria P M; Cheng L; Kushner P J; Agard D A; Greene G L The structural basis of estrogen receptor/coactivator recognition and the antagonism of this interaction by tamoxifen. *Cell* 1998; 95, 927–937.

21. Moras D; Gronemeyer H The nuclear receptor ligand-binding domain: structure and function. *Current Opinion in Cell Biology* 1998; 10, 384–391.

22. Weatherman R V; Fletterick R J; Scanlan T S Nuclear-receptor ligands and ligand-binding domains. *Annual Review of Biochemistry* 1999; 68, 559–581.

23. Mukherjee R; Davies P J A; Crombie D L; Bischoff E D; Cesario R M; Jow L; Hamann L G; Boehm M F; Mondon C E; Nadzan A M; Paterniti J R Jr; Heyman R A Sensitization of diabetic and obese mice to insulin by retinoid X receptor agonists. *Nature* 1997, 386, 407–410.

24. Germain P; Iyer J; Zechel C; Gronemeyer H Coregulator recruitment and the mechanism of retinoic acid receptor synergy. *Nature* 2002, 415, 187–192.

25. Pogenberg V; Guichou J-F; Vivat-Hannah V; Kammerer S; Perez E; Germain P; de Lera A R; Gronemeyer, H; Royer C A; Bourguet W Characterization of the interaction between retinoic acid receptor/retinoid X receptor (RAR/RXR) heterodimers and transcriptional coactivators through structural and fluorescence anisotropy studies. *Journal of Biological Chemistry* 2005; 280, 1625–1633.

26. Pratt W B The role of heat shock proteins in regulating the function, folding, and trafficking of the glucocorticoid receptor. *Journal of Biological Chemistry* 1993; 268, 21,455–21,458.

27. Hu X; Lazar M A The CoRNR motif controls the recruitment of corepressors by nuclear hormone receptors. *Nature* 1999; 402, 93–96.

28. Glass, C K; Rosenfeld M G The coregulator exchange in transcriptional functions of nuclear receptors. *Genes and Development* 2000; 14, 121–141.

29. Issemann I; Prince R A; Tugwood J D; Green S The peroxisome proliferator-activated receptor:retinoid X receptor heterodimer is activated by fatty acids and fibrate hypolipidaemic drugs. *Journal of Molecular Endocrinology* 1993; 11, 37–47.

30. Kliewer S A; Lehmann J M; Willson T M Orphan nuclear receptors: shifting endocrinology into reverse. *Science* 1999; 284, 757–760.

31. Xu H E; Lambert M H; Montana V G; Parks D J; Blanchard S G; Brown P J; Sternbach D D; Lehmann J M; Wisely G B; Willson T M; Kliewer S A; Milburn M V Molecular recognition of fatty acids by peroxisome proliferator-activated receptors. *Molecular Cell* 1999; 3, 397–403.

32. Kliewer S A; Sundseth S S; Jones S A; Brown P J; Wisely G B; Koble C; Devchand P; Wahli W; Willson T M; Lenhard J M; Lehmann J M Fatty acids and eicosanoids regulate gene expression through direct interactions with peroxisome proliferator-activated receptors α and γ. *Proceedings of the National Academy of Sciences of the United States of America* 1997; 94, 4318–4323.

33. Janowski B A; Willy P J; Devi T R; Falck J R; Mangelsdorf D J An oxysterol signalling pathway mediated by the nuclear receptor LXR alpha. *Nature* 1996; 383, 728–731.

34. Lehmann J M; Kliewer S A; Moore L B; Smith-Oliver T A; Oliver B B; Su J L; Sundseth S S; Winegar D A; Blanchard D E; Spencer T A; Willson T M Activation of the nuclear receptor LXR by oxysterols defines a new hormone response pathway. *Journal of Biological Chemistry* 1997; 272, 3137–3140.

35. Parks D J; Blanchard S G; Bledsoe R K; Chandra G; Consler T G; Kliewer S A; Stimmel J B; Willson T M; Zavacki A M; Moore D D; Lehmann J M Bile acids: natural ligands for an orphan nuclear receptor. *Science* 1999; 284, 1365–1368.

36. Wang H; Chen J; Hollister K; Sowers L C; Forman B M Endogenous bile acids are ligands for the nuclear receptor FXR/BAR. *Molecular Cell* 1999; 3, 543–553.

37. Kuiper G G; Enmark E; Pelto-Huikko M; Nilsson S; Gustafsson J A Cloning of a novel receptor expressed in rat prostate and ovary. *Proceedings of the National Academy of Sciences of the United States of America* 1996; 93, 5925–5930.

38. Mosselman S; Polman J; Dijkema R ER beta: identification and characterization of a novel human estrogen receptor. *FEBS Letters* 1996; 392, 49–53.

39. Katzenellenbogen B S; Sun J; Harrington W R; Kraichely D M; Ganessunker D; Katzenellenbogen J A Structure–function relationships in estrogen receptors and the characterization of novel selective estrogen receptor modulators with unique pharmacological profiles. *Annals of the New York Academy of Sciences* 2001; 949, 6–15.

40. Rocchi S; Picard F; Vamecq J; Gelman L; Potier N; Zeyer D; Dubuquoy L; Bac P; Champy M-F; Plunket K D; Leesnitzer L M; Blanchard S G; Desreumaux P; Moras D; Renaud J-P; Auwerx J A unique PPAR.γ ligand with potent insulin-sensitizing yet weak adipogenic activity. *Molecular Cell* 2001;8, 737–747.

41. Repa J J; Turley S D; Lobaccaro J A; Medina J; Li L; Lustig K; Shan B; Heyman R A; Dietschy J M; Mangelsdorf D J Regulation of absorption and ABC1-mediated efflux of cholesterol by RXR heterodimers. *Science* 2000; 289, 1524–1529.

42. Schultz J R; Tu H; Luk A; Repa J J; Medina J C; Li L; Schwendner S; Wang S; Thoolen M; Mangelsdorf D J; Lustig K D; Shan B Role of LXRs in control of lipogenesis. *Genes and Development* 2000; 14, 2831–2838.

43. Collins J L; Fivush A M; Watson M A; Galardi C M; Lewis M C; Moore L B; Parks D J; Wilson J G; Tippin T K; Binz J G; Plunket K D; Morgan D G; Beaudet E J; Whitney K D; Kliewer S A; Willson T M Identification of a nonsteroidal liver X receptor agonist through parallel array synthesis of tertiary amines. *Journal of Medicinal Chemistry* 2002; 45, 1963–1966.

44. Miao B; Zondlo S; Gibbs S; Cromley D; Hosagrahara V P; Kirchgessner T G; Billheimer J; Mukherjee R Raising HDL cholesterol without inducing hepatic steatosis and hypertriglyceridemia by a selective LXR modulator. *Journal of Lipid Research* 2004; 45, 1410–1417.

45. Schulman I G; Heyman R A The flip side identifying small molecule regulators of nuclear receptors. *Chemistry and Biology* 2004; 11, 639–646.

46. Drake K A; Zhang J-H; Harrison R K; McGeehan G M Development of a homogeneous, fluorescence resonance energy transfer–based in vitro recruitment assay for peroxisome proliferator-activated receptor delta via selection of active LXXLL coactivator peptides. *Analytical Biochemistry* 2002; 304, 63–69.

47. Zhou G; Cummings R; Li Y; Mitra S; Wilkinson H A; Elbrecht A; Hermes J D; Schaeffer J M; Smith R G; Moller D E Nuclear receptors have distinct affinities for coactivators: characterization by fluorescence resonance energy transfer. *Molecular Endocrinology* 1998; 12, 1594–1604.

48. Burris T P; Pelton P D; Zhou L; Osborne M C; Cryan E; Demarest K T A novel method for analysis of nuclear receptor function at natural promoters: peroxisome proliferator-activated receptor γ agonist actions on aP2 gene expression detected using branched DNA messenger RNA quantitation. *Molecular Endocrinology* 1999; 13, 410–417.

49. Watkins R E; Wisely G B; Moore L B; Collins J L; Lambert M H; Williams S P; Willson T M; Kliewer S A; Redinbo M R The human nuclear xenobiotic receptor PXR: structural determinants of directed promiscuity. *Science* 2001; 292, 2329–2333.

50. Wei P; Zhang J; Egan-Hafley M; Liang S; Moore D D The nuclear receptor CAR mediates specific xenobiotic induction of drug metabolism. *Nature* 2000; 407, 920–923.

51. Moore L B; Goodwin B; Jones S A; Wisely G B; Serabjit-Singh C J; Willson T M; Collins J L; Kliewer S A St. John's wort induces hepatic drug metabolism through activation of the pregnane X receptor. *Proceedings of the National Academy of Sciences of the United States of America* 2000; 97, 7500–7502.

14 Nuclear Receptor Scintillation Proximity Assays

David Powell and Molly J. Price-Jones

CONTENTS

14.1 ABSTRACT

The nuclear receptors form a superfamily of evolutionarily related members, divided into six subfamilies and 26 groups of receptors [1]. The family members share a common domain structure that includes a DNA binding domain (DBD) and a ligand binding domain (LBD). Functioning as ligand-activated transcription factors, the nuclear receptors have critical roles in both development and adult physiology. Family members are involved, among other things, in defense against xenobiotics pregnane X receptor (PXR) [2] and regulation of lipid metabolism liver X receptor

(LXR) [3] peroxisome proliferator activated receptor (PPAR) [4]. This chapter discusses the application of scintillation proximity assay (SPA) methods to the study of ligand binding to these receptors.

14.2 INTRODUCTION

Investigations into receptor–steroid interactions, for those receptors that have now been classified as nuclear receptors, have been ongoing for over 30 years. Early workers used techniques such as filter binding assays, improving their assay quality and reproducibility by using different filter materials to reduce background effects caused by the sticky nature of the ligands [5]. The alternatives at the time, which were used well into the 1990s, included charcoal [6] and gel filtration [7] methods. All these methods share the requirement for washing steps to remove unbound ligand, with a resulting increase in buffer usage and waste volume. Additionally, with the need for washes, centrifugation, and multiple postassay manipulations, they were unsuited to the higher throughput requirements of screening laboratories that emerged during the 1990s. The advent of SPA, a homogeneous technique ideally suited to screening applications, in the early 1990s for the first time offered researchers not only a way to increase their throughput, but also offered significant savings in time and waste.

14.3 METHODS

14.3.1 GENERAL REQUIREMENTS

The general requirements for SPA receptor binding assays (RBAs) are discussed in detail in Chapter 11 by Cook and coauthors. Briefly, the key requirements, regardless of the receptor type, are:

- A relatively pure source of receptor protein (either membrane bound or free)
- A radiolabelled ligand ($[^3H]$ and $[^{125}I]$ are most commonly used) that has suitable affinity for the receptor
- A means of coupling the receptor–ligand complex to the surface of the SPA bead

14.3.2 GENERAL METHODOLOGY

The general methods used for SPA with nuclear receptors are broadly similar to those used for cell surface–expressed receptors. However, the lack of glycoproteins on the membranes to enable membrane capture, together with relatively low expression levels, means that wheat germ agglutinin (WGA)–coated SPA beads are not generally useful. In addition, nuclear receptor preparations are not commercially available. To overcome this, workers have opted to clone either the separate LBDs or the full-length receptor protein sequence instead. These cloned domains and proteins are purified using integrated tags such as the six histidine tag (His_6) or glutathione S-transferase (GST). In some instances the purified receptor or receptor fragment has been further tagged (with biotin) to facilitate capture with streptavidin-coated SPA beads [8].

Once a receptor has been chosen, a radiolabeled ligand is required. Nuclear receptor ligands, due to their hydrophobic nature, are generally recognized to be "sticky," a factor which has led to issues in interpreting results with traditional separation assays [9]. Although SPA does not require separation of bound from free ligand it is not immune to this issue. However, it can be addressed by the use of appropriate buffer formulations. Another method that can be used for overcoming these issues with nonspecific binding (NSB) is to precouple the receptor to the beads. This step is carried out prior to incubation with the labeled ligand and further requires removal of unbound receptor. The process of precoupling helps to block NSB sites on the bead surface; this has the further advantages of reducing steps in an automated assay and reducing variation in data points when compared with a "T = 0" assay format [9] (see Chapter 11).

Factors influencing the choice of SPA bead types are discussed fully in Chapter 11 by Cook et al., but to summarize, the surface coating of the bead will largely be determined by the specific capture tag available on the receptor protein or LBD. Thus a His_6 tag requires copper chelate beads, and a biotin tag requires streptavidin-coated beads. Bead types used in SPA nuclear receptor assays are:

- Streptavidin-coated polyvinyltoluene (PVT) beads — these require the receptor preparation to have been biotinylated
- Copper chelate PVT beads — these require the receptor preparation to possess a His_6 tag motif
- Protein A–coated YSi beads — used for antibody capture assays
- Polylysine-coated yttrium silicate (YSi) beads — positively charged bead surface enables protein capture

The choice of ligand is influenced by the receptor system. Universally, the literature reports [³H]-labeled ligands; examples are listed here by receptor type:

- RXR: 9-*cis*-retinoic acid [9], 24(S)-25-epoxycholesterol [9], 24(S)-hydroxycholesterol [3], TO314407 [11], TZD [11], F_3methylAA [12]
- PPARγ: rosiglitazone [13], BRL49653 [7,8], GW2433 [14], AD-5075 [6]
- PPARα: GW362433X [13], GW2331 [14,15]
- PPARδ: GW362433X [13]
- PXR/SXR: SR12813 [16–18]
- CAR/PXR: clotrimazole [19]

With regard to an assay buffer, there are nearly as many variations in buffer composition as there are publications in the field. The vast majority are Tris based, with Hepes and phosphate buffers also noted. Assays are generally run in a fairly narrow pH range, between 7.0 and 8.0. Minor components usually include KCl, dithiothreitol (DTT), and BSA (bovine serum albumin) and appear to be dependent on the source of the assay reference used by any particular author. KCl is included to reduce NSB (nonspecific binding), as is BSA, while DTT should help to stabilize the LBD/receptor preparation itself. Full buffer recipes can be found in Section 14.5.

14.4   A BRIEF HISTORY OF NUCLEAR RECEPTOR SPA

The first SPA RBA for a nuclear receptor was described in 1998 [8], following the availability of BRL49653, a high-affinity ligand for PPARγ [7]. The authors cloned and expressed the PPARγ LBD as a fusion protein with the His_6 tag, which was used to purify the protein. The receptor fusion was then further modified by biotinylation, to enable capture with streptavidin-coated SPA beads (GE Healthcare RPNQ0007). The biotinylation was made necessary by the fact that the copper chelate beads were not available at that time. This assay has proven to be the model most used by subsequent researchers in the field; however, there have been variants, all of which have proven to be equally viable. Two variants on this assay were reported in 1999, one using a charge-based capture system with polylysine SPA beads [9] while the other used antibody capture of a GST fusion [11]. Copper chelate beads were widely available by 2000 and removed the need for biotinylation of receptor preparations. This reduction in the number of steps required led to their inclusion in the nuclear receptors assay area [10].

14.5 EXAMPLE ASSAY PROTOCOLS

14.5.1 Nichols et al., 1998

14.5.1.1 Materials

- Biotinylated human PPARγ LBD
- [³H]-BRL-49653 (5-[[4-[2-(methyl-2-pyridinylamino) ethoxy]-phenylmethyl]-2,4-thiazolidinedione)
- Streptavidin-coated PVT SPA beads (GE Healthcare RPNQ0007)
- Assay buffer (50 mM Tris, pH 8.0, 50 mM KCl, 2 mM ethylenediaminetetraacetic acid (EDTA), 5 mM CHAPS, 0.1 mg/mL BSA, 10 mM DTT)

14.5.1.2 Precoupling the Receptor

- Resuspend 50 mg SPA beads in 100 mL assay buffer.
- Split the bead slurry into two equal parts and transfer to 50-mL centrifuge tubes.
- Add biotinylated receptor stock at ~100 nM.
- Incubate 15 min, with gentle agitation every 5 min.
- Centrifuge tubes for 10 min at 1010 g. Gently decant supernatant and discard.
- Resuspend bead pellets in 50 mL assay buffer.
- Incubate 15 min, with gentle agitation every 5 min.
- Centrifuge tubes for 10 min at 1010 g. Gently decant supernatant and discard.
- Resuspend bead pellets in 50 mL assay buffer. Use immediately or store at 2 to 8°C. Authors reported receptor stability for at least 4 days under these storage conditions.

14.5.1.3 Assay Protocol

- To each well, add 50 μL precoupled receptor–bead suspension.
- Add 25 μL 40 nM [³H]-BRL49653 in assay buffer.
- Add 25 μL test compound in dimethylsulfoxide (DMSO)/assay buffer.
- Seal plate and incubate for 1 h at room temperature. Final assay concentrations: 250 μg receptor-bead (5 nM receptor), 10 nM [³H]-BRL49653. DMSO held to maximum 3% (v/v).
- Count plate, 1 min per well, using the appropriate plate-counting instrument.

An example of this protocol is shown in Figure 14.1.

14.5.2 Elbrecht et al., 1999

14.5.2.1 Materials

- GST-human PPARγ
- [³H] TZD (5-[4-[2-(5-methyl-2-phenyl-4-oxazolyl)-2-hydroxyethoxy]benzyl]-2,4-thiazolidinedione)
- Protein A-coated yttrium silicate SPA beads (GE Healthcare RPN143)
- Goat anti-GST antibody (GE Healthcare 27-4577-01)
- Assay buffer [10 mM Tris, pH 7.2, 1 mM EDTA, 10% (w/v) glycerol, 10 mM sodium molybdate, 1 mM DTT, 0.5 mM phenylmethylsulfonyl fluoride, 2 μg/mL benzamidine, 0.1% (w/v) dry milk powder]

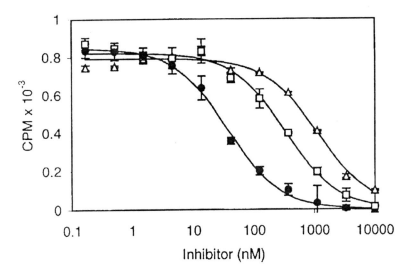

FIGURE 14.1 SPA competition binding data were generated using the protocol above. Three compounds — BRL49653 (closed circles), troglitazone (open squares), and arachidonic acid (open triangles) — were tested as inhibitors of radioligand binding. IC_{50} values: BRL49653, 36 ± 4 nM; troglitazone, 322 ± 42 nM; arachidonic acid, 1144 ± 220 nM. N = 2. (Reproduced with permission from Nichols JS, Parks DJ, Consler TG, Blanchard SG. *Anal Biochem* 1998; 257:112–119.)

14.5.2.2 Reagent Preparation

- Resuspend SPA beads in 25 mL assay buffer [minus dry milk powder but including sodium azide to a final concentration of 0.01% (w/v)]. This generates a 25 mg/mL suspension.
- Dissolve [^{3}H]TZD (stock was 21 Ci/mmol) in ethanol, at a radioactive concentration of 10.5 µCi/mL (500 nM).
- Dilute antibody stock 1:40 in assay buffer.
- Dilute receptor stock to 50 nM in assay buffer.

14.5.2.3 Assay Protocol

- To each well, add 10 µL of each of the following reagents: receptor, antibody, and [^{3}H]TZD.
- Make volume up to 70 µL with assay buffer.
- Add 5 µL test compound. Ensure DMSO concentration will not exceed 2% of final assay volume (not more than 2 µL DMSO per well).
  - For positive control add 5 µL assay buffer/DMSO.
  - For NSB add 5 µL 100 µM unlabelled TZD.
- Add 25 µL SPA bead suspension.
- Seal plate, incubate 24 h at 15°C, with shaking.
- Final assay concentrations: receptor, 5 nM; antibody, 400-fold final dilution; [^{3}H]TZD, 10 nM, bead, 0.5 mg/well.
- Count plate, 1 min per well, using appropriate plate-counting instrument.

For an example using this protocol, see Figure 14.2.

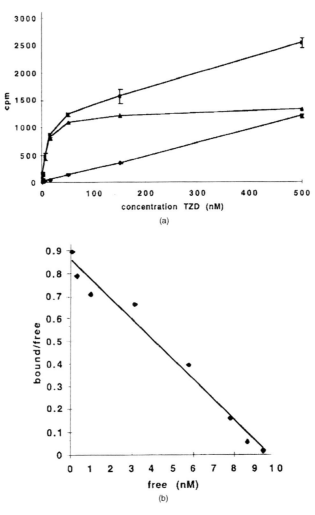

FIGURE 14.2 SPA saturation binding data for [³H]TZD: (a) shows the raw data (N = 3), nonspecific binding values were determined in the presence of a 100-fold excess of unlabelled TZD; and (b) shows the Scatchard analysis of these data (K_d = 11 nM). (Reproduced with permission from Elbrecht A, Chen Y, Adams A, Berger J, Griffin P, Klatt T, Zhang B, Menke J, Zhou G, Smith RG, Moller DE. *J Biol Chem* 1999; 274:7913–7922.)

14.5.3 JANOWSKI ET AL., 1999

14.5.3.1 Materials

- His₆-hLXRα-LBD or His₆-hLXRβ-LBD or His₆-hRXRα
- [³H]-24(*S*),25-epoxycholesterol or [³H]-9-*cis*-retinoic acid
- Polylysine coated yttrium silicate SPA beads (GE Healthcare RPNQ0010)
- Assay buffer [10 mM K₂HPO₄, 10 mM KH₂PO₄, 2 mM EDTA, 50 mM NaCl, 1 mM DTT, 2 mM CHAPS, 10% (v/v) glycerol, pH7.1]

14.5.3.2 Reagent Preparation

- Resuspend SPA beads in assay buffer (10 mg/mL stock)
- Dilute [^{3}H]-24(S),25-epoxycholesterol stock to 250 nM in assay buffer or dilute [^{3}H]-9-cis-retinoic acid to 50 nM in assay buffer
- Dilute receptor protein His$_6$-hLXRα-LBD (60 μg/mL) or His$_6$-hLXRβ-LBD (25 μg/mL) or His$_6$-hRXRα (25 μg/mL)

14.5.3.3 Assay Setup

- To each well, add 10 μL of each of the following reagents: receptor stock, appropriate [^{3}H] ligand.
- Make volume up to 75 μL with assay buffer.
- Add 5 μL test compound. Ensure DMSO concentration will not exceed 2% of final assay volume (not more than 2 μL DMSO per well).
    - For positive control, add 5 μL assay buffer/DMSO.
    - For NSB, add 5 μL 250 μM unlabeled 24(S),25-epoxycholesterol or 50 μM unlabeled 9-cis-retinoic acid.
- Add 20 μL SPA bead suspension.
- Seal plate, incubate 24 h at 15°C, with shaking.
- Final assay concentrations/masses: receptor (600 ng/well His$_6$-hLXRα-LBD, 250 ng/well His$_6$-hLXRβ-LBD and His$_6$-hRXRα), [^{3}H]-24(S),25-epoxycholesterol (25 nM) [^{3}H]-9-cis-retinoic acid (5 nM), bead (0.2 mg/well).
- Count plate, 1 min per well, using appropriate plate-counting instrument.

See Figure 14.3 for an example using this methodology.

14.6 TROUBLESHOOTING GUIDE

14.6.1 INTRODUCTION

With such a large range of both receptors and ligands, it is not possible to give advice on specific issues. The following is some general advice about how to handle commonly arising problems. It can be read in conjunction with the advice available in Chapter 11 by Cook et al.

14.6.2 POSSIBLE PROBLEMS

In very simple terms assays only suffer from two problems: low signal or high background or, in the worst-case scenario, both. The causes can be many and varied for each and will be heavily influenced by the exact reagents used in the assay, but the following is a brief overview of the issues to investigate:

- Low signal:
    - Insufficient radiolabeled ligand (Titrate ligand at constant receptor concentration.)
    - Ligand adsorbing to plate or pipetting equipment (Use treated plates; change plasticware.)
    - Degradation of receptor or ligand (Add protease inhibitors.)
    - Insufficient receptor (Titrate receptor at constant ligand concentration.)
    - Insufficient bead (Titrate bead at constant ligand and receptor concentrations.)
- High NSB:

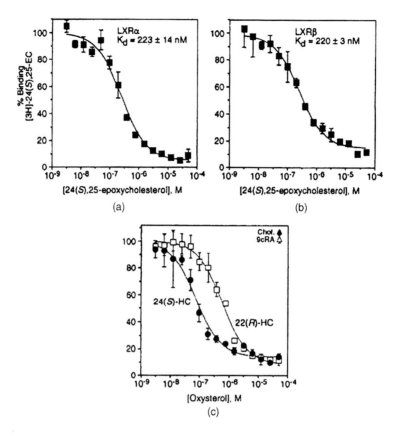

FIGURE 14.3 SPA competition binding data generated using the protocol in Figure 14.2. Plate (a) shows competition binding for 24(S),25-epoxycholesterol using the hLXRα-LBD. Plate (b) is the same data for the hLXRβ-LBD. Plate (c) shows competition binding for naturally occurring oxysterols to hLXRα. N = 3. (Reproduced with permission from Janowski BA, Grogan MJ, Jones SA, Wisely GB, Kliewer SA, Corey EJ, Mangelsdorf DJ. *Proc Natl Acad Sci* 1999; 96:266–271.)

- Radioligand binding directly to bead (Change bead core type or surface coating, use additives — salts at various concentrations, detergent to 0.05% (v/v) or BSA to 0.1% (w/v) — or precouple receptor if not already doing so. When precoupling it is vital not to use too great a spin speed during the centrifugation step and to ensure that the beads are properly resuspended before use. If possible, it may be better to use preoptimized ratios of bead and receptor.)
- Buffer formulation not optimized (Use additives; optimize buffer pH.)
- Too high a concentration of radiolabeled ligand (Titrate ligand at constant receptor and bead concentrations.)

14.7 MINIATURIZATION

14.7.1 INTRODUCTION

All of the example assays shown here have been optimized for 96-well plate formats. With the size of compound libraries only increasing, there is pressure on throughput. The best way to increase sample throughput is by miniaturization. The 384-well plate formats are fast becoming the industry standard, with 1536-well plates also gaining popularity. Typical assay volumes for 384-well plates are around 30 to 40 µL, though this can be pushed lower by the use of low-volume versions of 384-well plates. Assay volumes of 5 to 7µL are typically used in 1536-well plates.

SPA is still a good choice of assay platform for 384-well plates, but for 1536-well plates imaging systems are the only solution. (PMT-based systems do not exist for plates with this well density.) In fact, imaging is becoming the method of choice for 384-well plates, as it offers significant savings in data acquisition times. For imaging assays to work, specially designed imaging beads need to be used instead of standard SPA beads. This is because imaging systems have their optimal sensitivity at light wavelengths in the 600-nm region, rather than the 400-nm region used by PMT plate counters.

14.7.2 A Basic Guide to Nuclear Receptor Assay Miniaturization

There are two ways of moving to miniaturized formats; either an existing assay is miniaturized or a new assay is developed *de novo* in a high-density format.

Considering the first route, there are several different approaches, any of which could yield the desired result:

- Maintain reagent concentration at the same level as in the 96-well parent assay but reduce the volume of each reagent added such that the final volume desired can be reached.
- Maintain reagent mass per well but reduce the final assay volume. This results in using an increased reagent concentration and produces lesser savings.
- Use a mixed approach, with some reagents kept at the same concentrations, while others are increased. This requires full reoptimization of the assay, and care must be taken to avoid ligand depletion.

For an assay to be developed *de novo*, the simplest guidelines to follow are those that would normally be used, regardless of assay format. That is, each individual assay component should be tested in a series of matrix experiments, culminating in an optimized assay protocol.

Other factors to consider are the effects of altered surface-to-volume ratios in higher-density well formats (these may increase adsorptive loss of reagents to the well wall), different plastic surfaces (again may lead to adsorptive losses), the difficulty in mixing reagents in such a small volume (generally it is best to combine compatible reagents to reduce the number of additions but also to promote better reagent mixing). When moving from a PMT system to an imaging system, the core bead material is likely to be different and thus may exhibit different binding properties (potential for NSB to the bead). SPA Imaging beads largely mirror the same surface coatings that are available in SPA beads, increasing the likelihood of finding an appropriate bead surface.

14.8 CONCLUSIONS

This chapter details the successful development of assays for ligand binding to a number of different nuclear receptors. It shows that a variety of related approaches can be used. Once a receptor source has been selected and any subsequent manipulations effected (secondary tagging, e.g., biotinylation), and it has been paired with an appropriate ligand, assays can be developed in a variety of buffer backgrounds. The choice of SPA bead type is dictated by the properties of the receptor preparation. This information, together with that provided in the SPA receptor chapter and citations from the references included below, will enable the development of assays for other nuclear receptors not detailed here.

REFERENCES

1. Nuclear Receptors Committee. A Unified Nomenclature System for the Nuclear Receptor Subfamily. *Cell* 1999; 97:1–20.
2. Goodwin B, Redinbo MR, Kliewer SA. Regulation of CYP3A gene transcription by the Pregnane X receptor. *Annu Rev Pharmacol Toxicol* 2002; 42:1–23.
3. Schultz JR, Tu H, Luk A, Repa JJ, Medina JC, Li L, Schwender S, Wang S, Thoolen M, Mangelsdorf DJ, Lustig KD, Shan B. Role of LXRs in control of lipogenesis. *Genes Devel* 2000; 14:2831–2838.
4. Desvergne B, Wahli W. Peroxisome proliferator-activated receptors: Nuclear control of metabolism. *Endocr Rev* 1999; 20:649–688.
5. Santi DV, Sibley CH, Perriard ER, Tomkins GM, Baxter JD. A filter assay for steroid hormone receptors. *Biochemistry* 1973; 12:2412–2416.
6. Elbrecht A, Chen Y, Cullinan CA, Hayes N, Leibowitz MD, Moller DE, Berger J. Molecular cloning, expression, and characterization of human peroxisome proliferators activated receptors $\gamma1$ and $\gamma2$. *Biochem Biophys Res Comm* 1996; 224:431–437.
7. Lehmann JM, Moore LB, Smith-Oliver TA, Wilkison WO, Willson TM, Kliewer SA. An antidiabetic thiazolidinedione is a high-affinity ligand for peroxisome proliferators-activated receptor γ (PPARγ). *J Biol Chem* 1995; 270:12,953–12,956.
8. Nichols JS, Parks DJ, Consler TG, Blanchard SG. Development of a scintillation proximity assay for peroxisome proliferator-activated receptor γ ligand-binding domain. *Anal Biochem* 1998; 257:112–119.
9. Janowski BA, Grogan MJ, Jones SA, Wisely GB, Kliewer SA, Corey EJ, Mangelsdorf DJ. Structural requirements of ligands for the oxysterol liver X receptors LXRα and LXRβ. *Proc Natl Acad Sci* 1999; 96:266–271.
10. Urban F, Cavazos A, Dunbar J, Tan B, Escher P, Tafuri S, Wang M. The important role of residue F268 in ligand binding by LXRβ. *FEBS Lett* 2000; 484:159–163.
11. Elbrecht A, Chen Y, Adams A, Berger J, Griffin P, Klatt T, Zhang B, Menke J, Zhou G, Smith RG, Moller DE. L-764406 is a partial agonist of human peroxisome proliferators-activated receptor γ. *J Biol Chem* 1999; 274:7913–7922.
12. Menke JG, MacNaul KL, Hayes NS, Baffic J, Chao Y-S, Elbrecht A, Kelly LJ, Lam M-H, Schmidt A, Sahoo S, Wang J, Wright SD, Xin P, Zhou G, Moller DE, Sparrow CP. A novel liver X receptor agonist establishes species differences in the regulation of cholesterol 7α-hydroxylase (CYP7a). *Endocrinology* 2002; 143:2548–2558.
13. Thullier P, Brash AR, Kehrer JP, Stimmel JB, Leesnitzer LM, Yang P, Newman RA, Fischer SM. Inhibition of peroxisome proliferators-activated receptor (PPAR)-mediated keratinocyte differentiation by lipoxygenase inhibitors. *Biochem J* 2002; 366:901–910.
14. Xu HE, Lambert MH, Montana VG, Parks DJ, Blanchard SG, Brown PJ, Sternbach DD, Lehmann JM, Wisely GB, Willson TM, Kliewer SA, Milburn MV. Molecular recognition of fatty acids by peroxisome proliferators-activated receptors. *Mol Cell* 1999; 3:397–403.
15. Moya-Camarena SY, Vanden Heuvel JP, Blanchard SG, Leesnitzer LA, Belury MA. Conjugated linoleic acid is a potent naturally occurring ligand and activator of PPARα. *J Lipid Res* 1999; 40:1426–1433.
16. Jones SA, Moore LB, Shenk JL, Wisely GB, Hamilton GA, McKee DD, Tomkinson NCO, LeCluyse EL, Lambert MH, Willson TM, Kliewer SA, Moore JT. The Pregnane X receptor: a promiscuous xenobiotic receptor that has diverged during evolution. *Mol Endocrinol* 2000; 14:27–39.
17. Moore JT, Kliewer SA. Use of the nuclear receptor PXR to predict drug interactions. *Toxicology* 2000; 153:1–10
18. Staudinger JL, Goodwin B, Jones SA, Hawkins-Brown D, MacKenzie KI, LaTour A, Liu Y, Klassen CD, Brown KK, Reinhard J, Willson TM, Koller BH, Kliewer SA. The nuclear receptor PXR is a lithocholic acid sensor that protects against liver toxicity. *Proc Natl Acad Sci* 2001; 98:3369–3374.
19. Moore LB, Parks DJ, Jones SA, Bledsoe RK, Consler TG, Stimmel JB, Goodwin B, Liddle C, Blanchard SG, Willson TM, Collins JL, Kliewer SA. Orphan nuclear receptors Constitutive Androstane Receptor and Pregnane X Receptor share xenobiotic and steroid ligands. *J Biol Chem* 2000; 275:15,122–15,127.

15 Homogeneous Assay Development for Nuclear Receptor Using AlphaScreen™ Technology

Nathalie Rouleau and Roger Bossé

CONTENTS

The interaction between nuclear receptors (NRs) and their coactivators is a key step in transcription regulation. AlphaScreen™ (Amplified Luminescent Proximity Assay) is a nonradioactive homogeneous proximity assay that takes advantage of the receptor/coactivator interaction to identify and characterize compounds modulating NR activity. In these sections, estrogen receptor α (ERα) and retinoic acid receptors γ (RARγ) were chosen as models to demonstrate the different steps leading to the development of an AlphaScreen assay for NRs. The assays presented are performed with either a full-length receptor or a tagged receptor ligand binding domain (LBD) each interacting with a biotinylated peptide derived from a coactivator interacting domain sequence. The assays can differentiate between ligands that act as agonists vs. antagonists, since only agonists will allow recruitment of the coactivator sequence–derived peptide. The assay configuration offers flexibility to screen for antagonists targeting the ligand binding site in the ligand binding pocket of the LBD or the interaction interface between receptor and coactivator.

15.1 ALPHASCREEN GENERAL PRINCIPLE

AlphaScreen relies on the transfer of energy between an acceptor bead and a donor bead brought into proximity via a biological interaction. The donor beads are embedded with a photosensitizer (phtalocyanine), which converts ambient oxygen to an excited state singlet oxygen molecule upon illumination at 680 nm. Within its 4-μsec half-life, the excited oxygen diffuses approximately 200 nm in solution. If a biomolecular interaction drags an acceptor bead into close proximity of a donor bead, the excited singlet oxygen will transfer its energy to the acceptor bead. Three different fluorophores are present in the acceptor beads (thioxene, anthracene, and rubrene). Upon excitation by singlet oxygen, thioxene will emit light and induce a cascade of fluorescence resonance energy transfer inside the acceptor beads leading to the emission of light at 520 to 620 nm (Figure 15.1a). In the absence of any biological interaction, the singlet oxygen will return to ground state resulting in no signal production (Figure 15.1b).

Each donor bead contains a concentration of photosensitizer capable of generating up to 60,000 singlet oxygen particles per second resulting in substantial signal amplification. Because of this chemical energy transfer between acceptor and donor beads, the emission (500 to 600 nm) will be shorter than the excitation wavelength (680 nm), leading to reduced background. Moreover, the singlet oxygen produced has a half-life of around 0.3 sec, allowing measurements in a time-resolved mode.

15.2 ALPHASCREEN NUCLEAR RECEPTOR ASSAY DESIGN

Radioligand binding, reporter gene, coimmunoprecipitation, pull down, and gel shift are classical methods used to study and characterize NRs. Since these techniques are poorly automatable, the development of novel assays is required to speed up the identification of novel modulators for these important drug targets. AlphaScreen technology can be used to rapidly develop high-throuput screening (HTS) assays for NRs [1–3]. The principle of the assay involves two major steps: a ligand-activated biomolecular interaction between NR and its coactivator (Section 15.2.1), followed by the detection of that interaction using AlphaScreen beads (Section 15.2.2).

15.2.1 THE BIOMOLECULAR INTERACTION

It is well established that following agonist binding, allosteric changes in the LBD of NR will allow the interaction between the AF-2 domain in the LBD and the NR box present in the coactivator's structure [4–6]. In cells, this interaction will trigger a cascade of events leading to the transcriptional activation of NR target genes. A consensus sequence present on all NR coactivators, called the LXXLL motif, is sufficient for the interaction with the agonist-bound receptor LBD [7,8]. Other known regulators of NR signaling pathways are the corepressors that regulate receptor activity via a different molecular mechanism than that of a coactivator [9,10]. These different ligand-dependent interactions of receptor and regulatory proteins form the basis of the development of an AlphaScreen NR assay. Different assay components can be used depending on their cost and ease of production, availability, and stability. Here are some examples of such combinations:

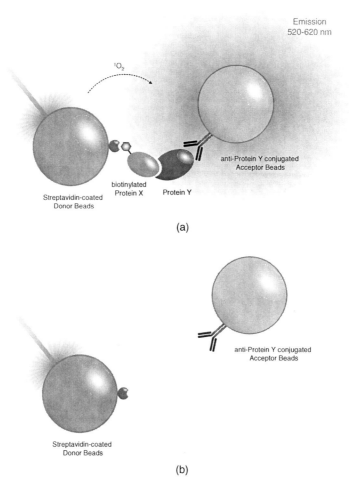

FIGURE 15.1 Scheme of AlphaScreen general principle: (a) Binding of biological partners will bring donor and acceptor beads into close proximity (200 nm) and generate fluorescent emission between 520 and 620 nm upon excitation at 680 nm and (b) in the absence of a biological interaction, the singlet oxygen will decay and no signal will be detected.

1. Interaction between agonist/antagonist bound receptors and coactivator/corepressor
2. Interaction between agonist-bound LBD and LXXLL motif-containing peptide

Depending on the detection reagents, the binding partners can be wild type, truncated, or tagged with, for example, biotin, FITC, GST, Flag, cmyc, HA, or $(His)_6$.

An assay based on the interaction between receptor and coactivator will generate a signal increase upon agonist binding and a signal decrease following antagonist binding. The interaction with corepressors could also be used to find ligands acting at a different molecular event of the receptor signaling cascade. Thus far, no assays using corepressors have been reported in the literature.

15.2.2 THE DETECTION REACTION

Following ligand binding, the interaction between the receptor and the coactivator can be detected using various strategies depending on the nature of the binding partners involved. Since off-the-shelf

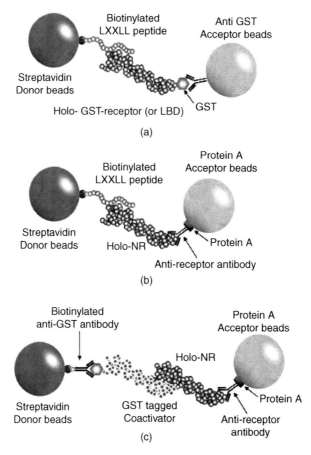

FIGURE 15.2 Schematic representation of the different detection formats available: (a) direct, (b) indirect, and (c) "sandwich" (see description in Section 15.2.2).

donor beads are available coated with streptavidin, one of the binding partners should be captured via a biotin moiety. Here are the different possible detection formats:

- Direct detection format
  - In this format, one of the binding partners is captured by antibodies directly coated on the acceptor beads, while the other is biotinylated and captured by the streptavidin donor beads (Figure 15.2a)
- Indirect detection format
  - In this format, one of the binding partners is captured by antibodies preadsorbed to protein A coating the acceptor beads, while the other is biotinylated and captured by the streptavidin donor beads (Figure 15.2b)
- "Sandwich" format
  - In this format, one of the binding partners is captured by antibodies preadsorbed to protein A coating the acceptor beads, while the other is captured by a biotinylated antibody captured by the streptavidin donor beads (Figure 15.2c)

15.3 ALPHASCREEN NUCLEAR RECEPTOR ASSAY DEVELOPMENT

15.3.1 GENERAL LABORATORY PRACTICE

15.3.1.1 Environmental Conditions

When choosing an environment for the readout, place the instrument in an area that is least prone to dramatic temperature fluctuations. Avoid manipulating reagents as well as reading the plates under bright light — for example, direct sun exposure. If possible, choose a location with dim light (around 100 lux) and/or apply green filters to the light fixtures.

15.3.1.2 Reagent Handling

The AlphaScreen beads (PerkinElmer, Boston) should be stored at 4°C unless specified otherwise and protected from exposure to light. (The plastic container in which the beads are provided is quite adequate for this purpose.) The beads are 200 nm in diameter and form a stable suspension. They are supplied in Proclin-300 (preservative) to prevent microbial growth. To avoid loss of material, it is recommended to centrifuge (pulse) down the beads. Keep the beads away from heat-generating sources. Although the beads themselves are not heat sensitive and the dyes and hydrogel coating them are very stable even up to 95°C [11], the proteins coating the beads may be sensitive to heat denaturation.

15.3.2 GENERAL CONCEPTS IN ALPHASCREEN ASSAY DEVELOPMENT

15.3.2.1 Instruments

Different instruments can be used to read an AlphaScreen reaction, such as the Fusion-Alpha Multilabel reader, the AlphaQuest HTS Instrument, and the new EnVision™ (PerkinElmer, Boston).

15.3.2.2 Antibodies in AlphaScreen Assay

Factors inherent to protein A antibody recognition should be considered when developing an assay using the indirect or "sandwich" formats. The protein A coating the acceptor beads does not recognize the Fc domain of antibodies from different species with the same efficiency. For example, rabbit Ig molecules and mouse IgG2a are recognized with higher affinity than other Igs. Of all antibodies, goat IgG and mouse IgG1 have the lowest affinity for protein A. For these last two classes of antibodies it is advisable to opt for protein G–coated or antispecies acceptor beads.

Monoclonal or affinity-purified polyclonal antibodies should be considered first during assay design. The presence of antibodies that do not recognize the antigen could act as a competitor in the detection reaction by saturating beads with nonspecific antibodies. However, antibodies taken directly from ascitic fluids have been reported to work well for some AlphaScreen applications and should not be completely disregarded.

15.3.2.3 Bead Binding Capacity and the "Hook" Effect

A phenomenon often referred to as the "hook" effect is common to most homogeneous proximity assays and is a direct consequence of the binding capacity of each acceptor and donor molecule. In the indirect detection assay using protein A capture, at a working concentration of acceptor beads (20 µg/mL), saturation by antibodies will occur at approximately 1 to 3 nM of antibody. On the other side, donor beads coated with streptavidin will become saturated at about 10 to 30 nM of biotinylated molecules. Upon AlphaScreen bead saturation, nonproductive interactions are generated resulting in loss of signal (Figure 15.3).

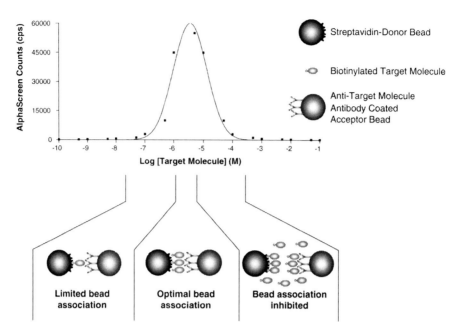

FIGURE 15.3 Graphic representation of the "hook" effect.

15.3.2.4 Interfering Compounds

Various agents should be avoided for use in an AlphaScreen assay. Detergents such as Triton X-100 and CHAPS should not exceed 0.1% (v/v). Preparations containing azide as a preservative agent are not recommended since azide is a potent scavenger of singlet oxygen. We strongly recommend eliminating the use of the following transition metal ions: Fe^{2+}, Fe^{3+}, Cu^{2+}, Ni^{2+}, and Zn^{2+}. These metals have been shown to be potent singlet oxygen quenchers in the millimolar and submillimolar ranges. For example, 100 μM of Fe^{2+} can be deleterious to an AlphaScreen assay.

15.3.3 ASSAY DEVELOPMENT: ERα AND RARγ MODELS

Two models will be used to illustrate the development of a NR AlphaScreen assay (these results can also be found in Reference 3):

- ERα assay using an indirect detection format
- RARγ assay using a direct detection format

The first model uses full length ERα captured by an antibody directed against the receptor, which interacts with a biotinylated LXXLL peptide (biotin-LXXLL) derived from the sequence of the coactivator SRC-1 (Figure 15.4a). The second model involves the interaction of a GST fusion of the RARγ LBD interacting with the same biotinylated peptide as in the ERα model (Figure 15.4b).

15.3.3.1 Order of Reagent Addition

All assays are performed in white opaque 384-well plates in a final volume of 25 μL using a buffer containing 25 mM Hepes (pH 7.4), 100 mM NaCl, 0.1% Tween-20, and 1 mM DTT. For both models, the different reagents are distributed in the plate in the following order:

To each well, 5 μL of receptor is added, followed by 5 μL of agonist. After agonist addition, 2.5 μL of buffer or competitor is added, followed by 5 μL of biotin-LXXLL peptide. Plates are sealed and incubated at room temperature for 1 h. After incubation, 7.5 μL of a mixture of

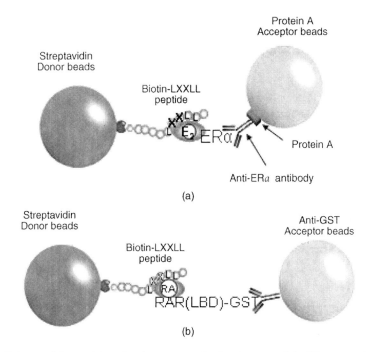

FIGURE 15.4 Scheme of the model used to demonstrate the development of the NR assay using AlphaScreen: (a) Indirect detection using full-length ERα and biotinylated LXXLL motif containing peptide and (b) direct detection using a GST fusion of the RARγ-LBD and biotinylated LXXLL sequence containing peptide.

streptavidin donor beads and protein A acceptor beads (preincubated with antireceptor antibody for 30 min) (ERα model) or 7.5 μL of streptavidin donor beads mixed with anti-GST acceptor beads (RARγ model) is added. Plates are sealed and incubated in the dark at room temperature for 1 h. The plates are read on an EnVision Alpha AlphaQuest HTS microtiter plate reader (or equivalent instrument).

In the process of assay optimization, different orders of addition should be tested. The order of addition that yields the highest signal window and dynamic range will depend on the nature of the interacting partners and the detection format. It has been demonstrated that NR assays will also perform well in volumes lower than 25 μL [1].

15.3.3.2 Binding Partner Titration (Biomolecular Interaction)

The first step in the assay development is the titration of each binding partner, at fixed concentrations of acceptor and donor beads (20 μg/mL each), and antireceptor antibody (ERα model only). Serial dilutions of the receptor are performed to obtain a final concentration ranging from approximately 100 pM to 30 nM. Serial dilutions of the biotin-LXXLL peptide are also performed in parallel to obtain a final concentration varying from approximately 1 to 30 nM. For the ERα model, the final concentration of antibody captured by the protein A acceptor beads is kept at fixed concentration (1 nM). Figure 15.5a and Figure 15.5b illustrate that for both models, 10 to 30 nM of receptor generated the highest signal with a signal-to-background ratio of 255 for ERα and 70 for RARγ. The sensitivity of the assay would allow one to use 10 times less material for a screening campaign. The signal reduction observed over 10 nM probably reflects the "hook" effect resulting from bead saturation.

15.3.3.3 Detection Reagent Titration (Beads and Antibody)

Even if most AlphaScreen assays use 20 μg/mL of acceptor and donor beads in a final volume of 25 μL, optimization of the acceptor:donor bead ratio can be performed. For the NR assays presented,

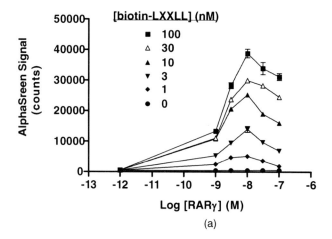

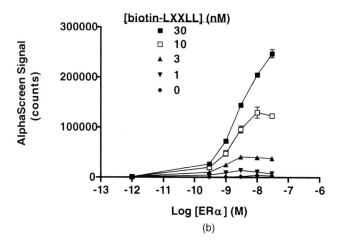

FIGURE 15.5 Binding partner titration. Increasing concentrations of (a) RARγ or (b) ERα (0 to 30 n*M*) were incubated with increasing concentrations of biotinylated SRC-1 LXXLL peptide (0 to 30 n*M*) in the presence of 10 μ*M* of each respective agonist. Generation of signal was determined after 1-h incubation at room temperature. Data are the means ± SEM of triplicate determinations from a single experiment, which is representative of three independent experiments.

no advantages were observed in changing this ratio. For the reasons mentioned in Section 15.3.2.2 and Section 15.3.2.3, titration of the antibody used in the capture must be performed going from 1 to 10 n*M*. In the ERα model, 1 n*M* of anti-ER antibody was found optimal.

15.3.3.4 Buffer Optimization

All reagents modifying pH and ionic strength are likely to influence the interaction between each of the binding partners. All NR assays developed using AlphaScreen were performed in 25 m*M* Hepes pH 7.4 supplemented with 100 m*M* NaCl. When high background is observed, detergent such as Tween-20 may be used at concentration under 0.1% (v/v). Bovine serum albumin (BSA) at concentration of 0.1% (w/v) is often sufficient to minimize nonspecific interactions. We have observed in several occasions that the presence of 1 m*M* DTT can improve the signal generated with GST fusion proteins.

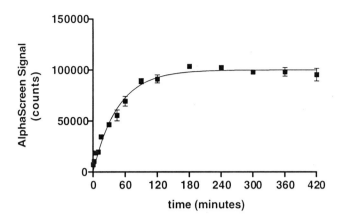

FIGURE 15.6 Time course of the NR-coactivator LXXLL peptide dimerization. Progress of the dimerization between agonist-bound NR and biotinylated peptide reaction was measured at different time points varying from 0 to 7 h. Each data point represents the mean of triplicates.

15.3.3.5 Time Course

During assay optimization the following order of addition can be used to look at the kinetics of binding partner interaction. To each well, 5 μL of receptor is added, followed by 10 μL of acceptor and donor beads, followed by 5 μL of agonist. The reaction is then started by the addition of 5 μL of biotinylated peptide and read at different time points. Figure 15.6 illustrates the dynamics of the binding of RARγ to the LXXLL-containing peptide and suggests that equilibrium is reached after 2 h.

15.3.3.6 Agonist Dose–Response Curve

Using the determined optimal concentration of binding partners and detection reagents (beads and/or antibody), an agonist titration curve can be performed. In the two assays presented here, a dose-dependent signal increase should be observed that reflects the affinity of the ligand for its receptor. For RARγ, *trans*-retinoic acid (RA) was titrated from 100 pM to 10 μM in the presence of 3 nM of receptor and 30 nM of biotinylated peptide (Figure 15.7a). As for ERα, titration of estradiol (E_2) and diethylstilbestrol (DES) from 10 pM to 1 μM was performed in the presence of ERα (3 nM) and biotinylated-SRC-1 LXXLL peptide (3 nM) (Figure 15.7b). Both direct (RARγ) and indirect (ERα) assays generated dose-response curves with EC_{50} in close proximity to those obtained using radioligand binding and florescence polarization (FP) assays (E_2 = 1.6 nM, DES = 1 nM, and RA = 5 nM). See Figure 15.7 [3].

15.3.3.7 Antagonist Dose–Response Curve

The execution of an antagonist competition curve should follow the same protocol as the agonist titration curve using saturating concentration of agonist. Different types of antagonists can be identified:

- Competitors for the ligand binding pocket
- Competitors interfering with the interaction between the agonist-bound receptor and its coactivator

Figure 15.8a illustrates the competition of two classical ERα antagonists. In this experiment, agonist (E_2 10 μM), receptor (3 nM), biotinylated peptide (3 nM), and beads (20 μg/mL) were kept at fixed concentration and antagonists titrated up to 10 μM. The AlphaScreen assay confirmed that

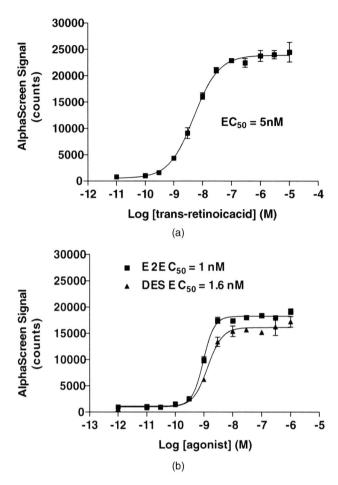

FIGURE 15.7 Agonist dose–response curve. For both the (a) ERα indirect assay and the (b) RARγ direct assay, agonist titration was performed using fixed concentrations of binding partners and detection reagents. EC_{50} values were estimated by nonlinear regression analysis using GraphPad Prism software (San Diego, CA). Data are the means ± SEM of triplicate determinations from a single experiment, which is representative of three independent experiments.

4-hydroxytamoxifen is a more potent antagonist than tamoxifen (Figure 15.8a). To simulate the mechanism of inhibition at the binding interface between the receptor and the coactivator, nonbiotinylated peptide derived from the LXXLL motif sequence was tested using the RARγ model (Figure 15.8b). RA (10 μM), receptor (3 nM), biotinylated peptide (30 nM), and beads (20 μg/mL) were kept at fixed concentration and antagonists titrated up to 10 μM. The peptide demonstrated an inhibition with an IC_{50} of 372 nM (Figure 15.8a), which would be comparable to the efficacy of tamoxifen on ERα.

15.3.3.8 Robustness and Assay Tolerance to Dimethylsulfoxide (DMSO)

Evaluation of the suitability of an assay developed for HTS can be assessed by z-factor studies. For both models using the same experimental design, comparison of signal generated in the presence of saturating concentration of agonist vs. antagonist was performed in 384-well plates. Intra- and interplate variability was tested as follows: for each plate, 48 wells were filled to measure agonist-bound heterodimerization (stimulated) and 48 different wells were filled to measure antagonist inhibition of agonist-induced heterodimerization (inhibited) using the optimized protocol. For both

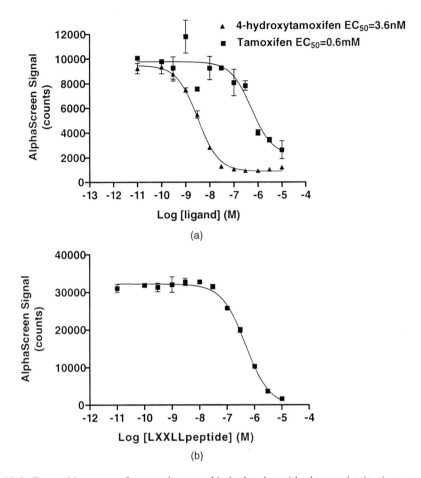

FIGURE 15.8 Competition curve of antagonist or nonbiotinylated peptide. Antagonist titration was performed for both (a) ERα indirect assay [antagonist tamoxifen (■), and 4-OH-tamoxifen (▲), and agonist DES (▼)] and the (b) RARγ direct assay. IC_{50} values were estimated for each antagonist by nonlinear regression analysis using Graph Pad Prism software (San Diego, CA). Data are the means ± SEM of triplicate determinations from a single experiment.

models, the intraplate variability generated CV 10% in the presence of agonist and 11% in the presence of antagonist, and Z' values >0.7 were generated (Figure 15.9) [3].

During assay development, since most compound libraries are present in DMSO, DMSO titration using concentrations ranging from 1 to 10% using the optimum assay conditions would be advisable.

15.4 SUMMARY

AlphaScreen is a homogeneous nonradioactive assay that can be used to develop sensitive assays to detect novel modulators of NRs. The NR assay takes advantage of an important step in receptor function, the agonist-dependent interaction between the receptor and its coactivator. Even though no data have been presented thus far, the use of corepressor and antagonist-bound receptor could be envisioned to find different modulators. The possibility to design multiple AlphaScreen assay formats (direct, indirect, and "sandwich") allows the development of assays in a short time period with a wide variety of reagents. In the design of a particular assay, a concept such as the "hook" effect and antibody affinity for protein A must be taken into consideration (Section 15.3.2.2 and

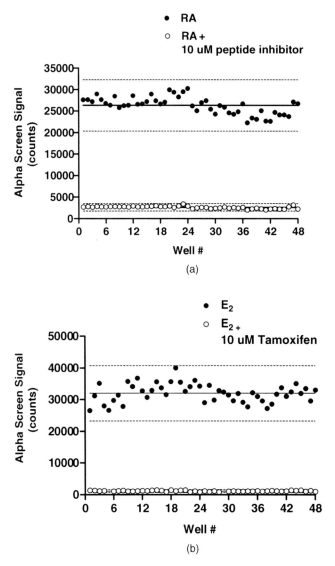

FIGURE 15.9 Assay robustness. Intraplate variation for the (a) RARγ assay in the absence or presence of 10 μ*M* of nonbiotinylated SRC-1 LXXLL peptide and the (b) ERα assay in the absence (●) or presence (○) of 10 μ*M* tamoxifen.

Section 15.3.2.3). When the decision has been made about assay format, the following development steps can be performed: binding partners titration (Section 15.3.3.2), detection reagents titration (Section 15.3.3.3), test of the order of addition (Section 15.3.3.1), buffer optimization (Section 15.3.3.4), time course (Section 15.3.3.5), agonist titration (Section 15.3.3.6), antagonist titration (Section 15.3.3.7), tolerance to DMSO, and variability assay (Section 15.3.3.8).

Pharmacologically, the example presented (ERα and RARγ), demonstrate the capacity of the technology to determine the relative affinity at which the ligands interact with their receptor. It was also possible to discriminate between different antagonist potencies, as shown with tamoxifen and 4-hydroxytamoxifen. Reproducibility of the assay, characterized by Z′ values over 0.5, indicates that the current format of RARγ and ERα assay should adapt well to integration into HTS protocols.

15.5   TROUBLESHOOTING GUIDE

The troubleshooting guide can be seen in Table 15.1.

TABLE 15.1
Troubleshooting Guide

Problem	Cause	Effect/Remedy
No signal	Reagents	Donor beads have been exposed to light/Use another lot of beads
		One of the binding partners is not biotinylated/Check biotinylation extent [HABA test (Pierce) or competition with the AlphaScreen TruHit Kit]
		Binding partners do not interact/Confirm their interaction using other techniques (e.g., Co-IP, GST pulldown, yeast two-hybrid assay)
	Assay	Change the order of addition
		Let binding partners incubate longer
	Buffer	Inhibitor or quenching component in assay buffer/Avoid reagents that quench singlet O_2 (see Section 15.3.2.4) and avoid components that absorb light between 520 and 680 nm
	Instrument/plates	Incompatible microtiter plates such as black plates/Use standard white solid opaque microtiter plates
		Plate reader error or failure/Consult instrument manual or call PerkinElmer customer service; if using Fusion-Alpha or EnVision make sure AlphaScreen mode is selected when setting up the instrument
Lower signal than expected	Reagents	Too low concentration of acceptor and/or donor beads/Start by using the recommended 20 µg/mL concentration
		Inappropriate concentration of binding partners/Titrate binding partners to determine optimal concentration.
		Degradation of bead conjugates/Store beads at 4°C in the dark
	Assay	Inappropriate assay buffer/Evaluate pH, buffering capacity, and salt concentration of the buffer, look at the concentration of reducing agents, detergents, ions, chelators
		Incubation time too short/Perform kinetics for binding partners interaction
	Instrument/plates	Incompatible microtiter plates such as black plates/Use standard white solid opaque microtiter plates
		Plate reader error or failure/Consult instrument manual or call PerkinElmer customer service; if using Fusion-Alpha or EnVision make sure AlphaScreen mode is selected when setting up the instrument
	Temperature	Abnormally low temperature prevailing in the room where the reader is located will result in decreased signal
Signal inconsistency	Plates/instrument	Warped or distorted plate/Avoid storage of microtiter plates under heavy objects or next to a source of heat
		Uneven plate molding/Try different plate manufacturer
		Light penetrates edges of plate/Be sure to cover plate during incubation and incubate plates in a dark environment
		Poorly fitted plate seals that are subject to evaporation/Use proper plate sealer
		Temperature drift in the reader (AlphaScreen counts increase with increasing temperature)
	Assay	Degradation of assay components/Prepare fresh buffer; store buffer without BSA at 4°C and use within 2 to 3 days

(continued)

TABLE 15.2 (CONTINUED)
Troubleshooting Guide

Problem	Cause	Effect/Remedy
High background	Assay	Nonspecific interaction between assay components/When using "sandwich" assay format make sure that protein A does not cross-react with the biotinylated antibody binding to the streptavidin donor beads. Preincubate the nonbiotinylated antibody before adding the biotinylated one, or presaturate protein A acceptor beads with nonspecific antibodies before adding the biotinylated antibody to the reaction
	Detection	Accidental exposure of acceptor beads to light just prior to reading (acceptor beads will autofluoresce for 2 to 3 min)/Readapt to dark at least 5 min prior to reading
	Temperature	Abnormally high temperature prevailing in the room where the reader is located will result in higher background
High degree of signal variability	Assay	Differences due to transfer from assay development to HTS laboratory/Ensure adequate training of the operators; consult PerkinElmer product support
		Evaporation/Use of adequate plate sealer.
		Mixing problems/For 96-well plates use a shaker during incubation; for higher-density plates, try to avoid addition steps lower than 5 μL
Unexpected gradient of signal across the plate	Plates	Uneven microtiter plates/Test different suppliers
		Plates are kept at too low temperature prior to reading/Plate chemistry designed to give best results at room temperature
	Robotic liquid dispensing	Inconsistent pipeting in wells, clogged liquid in head dispenser, uneven placing of plate on the dispenser platform, incorrect tip choice, automated dispenser program error/Call PerkinElmer technical support
	Reader	Improper temperature control inside the instrument/adjust the temperature control

REFERENCES

1. Glickman JF, Wu X, Mercuri R, Illy C, Bowen BR, He Y, Sills M. A comparison of ALPHAScreen, TR-FRET, and TRF as assay methods for FXR nuclear receptors. *J Biomol Screen* 2002; 7:3–10.
2. Wu X, Glickman JF, Bowen BR, Sills MA. Comparison of assay technologies for a nuclear receptor assay screen reveals differences in the sets of identified functional antagonists. *J Biomol Screen* 2003; 8:381–392.
3. Rouleau N, Turcotte S, Mondou M-H, Roby P, Bossé R. Development of a versatile platform for nuclear receptor screening using AlphaScreen™. *J Biomol Screen* 2003; 8:191–197.
4. Onate SA, Tsai SY, Tsa MJ, O'Malley BW. Sequence and characterization of a coactivator for the steroid hormone receptor superfamily. *Science* 1995; 270:1354–1357.
5. Wang JC, Stafford JM, Granner DK. SRC-1 and GRIP1 co-activate transcription with hepatocyte nuclear factor 4. *J Biol Chem* 1998; 273:30,847–30,850.
6. Zhu Y, Qi C, Calandra C, Rao MS, Reddy JK. Cloning and identification of mouse steroid receptor coactivator-1 (mSRC-1), as a coactivator of peroxisome proliferator-activated receptor. *Gene Expr* 1996; 6:185–195.
7. Heery DM, Kalkhoven E, Hoare S, Parker MG. A signature motif in transcriptional coactivators mediates binding to nuclear receptors. *Nature* 1997; 387:733–736.
8. Torchia J, Rose DW, Inostroza J, Kamei Y, Westin S, Glass CK, Rosenfeld MG. The transcriptional coactivator p/CIP binds CBP and mediates nuclear receptor function. *Nature* 1997; 387:677–684.
9. Horwitz KB, Jackson TA, Bain DL, Richer JK, Takimoto GS, Tung L. Nuclear receptor coactivators and corepressors. *Mol Endocrinol* 1996; 10:1167–1177.

10. Liu Z, Auboeuf D, Wong J, Chen JD, Tsai SY, Tsai MJ, O'Malley BW. Coactivator/corepressor ratios modulate PR-mediated transcription by the selective receptor modulator RU486. *Proc Natl Acad Sci USA* 2002; 99:7940–7944.

11. Beaudet L, Bedard J, Breton B, Mercuri RJ, Budarf ML. Homogeneous assays for single-nucleotide polymorphism typing using AlphaScreen. *Genome Res* 2001; 11:600–608.

16 Development of Nuclear Receptor Homogeneous Assay Using the LANCE™ Technology

Nathalie Rouleau, Pertti Hurskainen, Roger Bossé, and Ilkka Hemmilä

CONTENTS

16.1 LANCE GENERAL CONCEPT

16.1.1 FLUORESCENCE PRINCIPLE

16.1.1.1 Fluorophores

Fluorescence is the phenomenon in which absorption of light at a given wavelength by a fluorescent molecule is followed by the emission of light at longer wavelengths. Some of the advantages of fluorescence regarding screening assay development are its nonradioactive nature, its high sensitivity, and its safety. Safety refers to the fact that it is a noninvasive technique that does not affect

FIGURE 16.1 Jablonski diagram.

or destroy samples in the process. The fluorescence signal is proportional to the concentration of the substance being investigated, and techniques using fluorescence can accurately detect concentrations in the pico- and even femtomolar range.

The process responsible for fluorescence of fluorophores is well illustrated by a simple electronic-state diagram (Jablonski diagram) (Figure 16.1). When a light source, such as a laser, produces photons with certain energy (h_{EX}), fluorophores are transformed into an excited electronic singlet state (S_1'). During this excited state time (1 to 10 nsec), the fluorophores undergo conformational changes and the energy is partially dissipated, yielding a relaxed singlet excited state (S_1) with fluorescence emission (h_{EM}). The fluorescence quantum yield is defined by the ratio of the number of fluorescence photons that are emitted to the number of photons absorbed and is an indication of the efficiency at which this process occurs. The difference between the excitation and emission wavelengths of a fluorophore is called the Stokes' shift. When measuring fluorescence emission, an elevated Stokes' shift allows emission photons to be detected against a low background, i.e., isolated from excitation photons.

16.1.1.2 Fluorescence Energy Transfer (FRET)

FRET is a distance-dependent interaction between two dye molecules in which excitation is transferred from a donor molecule to an acceptor molecule. When the donor and acceptor dyes are different, FRET can be detected by:

- The appearance of sensitized fluorescence of the acceptor
- Quenching of donor fluorescence
- The ratio of sensitized fluorescence of the acceptor to quenching of the donor

The distance at which 50% of the energy is transferred is defined by the Forster radius (R_o), which is dependent on the spectral properties of the donor and acceptor molecules:

$$R_0 = (8.79 \cdot 10^{-5} \cdot J(\lambda) \cdot q_D \cdot n^{-4} \cdot \kappa^2)^{1/6}$$

where kappaE2 = dipole orientation factor, q_D = quantum yield of the donor, n = refractive index, and J(lambda) = spectral overlap integral (Figure 16.2).

R_0 values of 5 nm are typically observed when using conventional fluorophores, and R_0 values of 9 nm are observed for Eu cryptate/APC pairs [2].

During a FRET process, the transfer efficiency is defined as the probability of decay of the excited molecule due to transfer of energy compared to the total number of decay events (fluorescence radiation, energy transfer, various quenching processes, nonradiative processes, etc). Transfer efficiency is a function of the spectral characteristics of the donor and acceptor dyes, and it strongly depends upon the *R* separation distance of the donor and acceptor molecules:

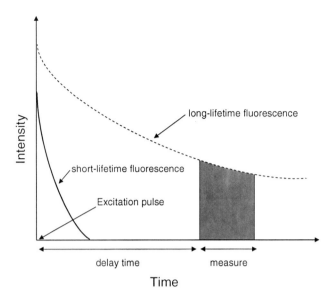

FIGURE 16.2 Graphical representation of the differences between short- and long-lifetime fluorescence.

$$E = \frac{R_0^6}{R^6 + R_0^6}$$

where R_0^6 and R^6 are the 6th powers of R_0 and R, respectively.

Since transfer efficiency depends on the 6th power of the separation distance, E quickly decreases with an increase of R. Thus, the maximum limiting distance depends on the numerical value of the R_0 characteristic distance. The maximum distance generally does not exceed 20 nm. The best performance is provided at 2 to 15 nm, which overlaps with the intramolecular distances and molecular "diameters" of macromolecules and separation distances of molecular aggregates and complexes.

FRET is a powerful technology to capture biological phenomena implicating biomolecular interactions. In summary, the following conditions must be in place for it to occur:

- Donor and acceptor molecules must be in close proximity (typically, 2 to 15 nm).
- The absorption spectrum of the acceptor must overlap the fluorescence emission spectrum of the donor (Figure 16.3).
- Donor and acceptor transition dipole orientations must be approximately parallel.

16.1.1.3 Time-Resolved FRET (TR-FRET)

Homogeneous assays particularly benefit from time-resolved fluorometry. However, conventional fluorometry employs fluorophores with short Stokes' shifts and decay times, making them sensitive to interference by autofluorescence, scattering and background of microtiter plates, and optics [1]. Temporal resolution in the micro- to millisecond time domains requires long lifetime fluorescence probes, short pulse excitation, and delayed detection. Lanthanides possess very specific magnetic and electronic properties playing an essential role in the temporal resolution in the millisecond range [2–4]. Europium chelates for LANCE assays have a typical narrow-banded emission with exceptionally long Stokes' shift and excited-state lifetime of hundreds of microseconds (Figure 16.2). Lanthanides' unique fluorescence properties, relatively small size, and hydrophilicity make them perfect labels for TR-FRET.

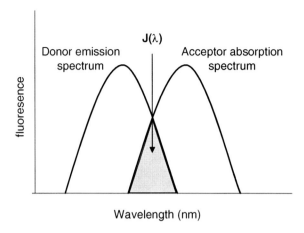

FIGURE 16.3 Graphical representation of the spectral overlap region [J(λ)] used in the calculation of FRET efficiency between two fluorophores.

16.1.1.4 Fluorescence Readers

Four elements are essential for detection of fluorescence:

- An excitation source
- A fluorophore
- Wavelength filters to isolate emission photons from excitation photons
- A detector that registers emission photons and produces a recordable output, usually as an electrical signal or a digital image

Compatibility of these four elements is essential for optimizing fluorescence detection. The excitation source intensity and fluorescence collection efficiency of the instrument are also important aspects. In solution, fluorescence intensity is linearly proportional to these parameters.

Different instruments have been developed to measure fluorescence in microtiter plates; they include the EnVision™, VICTOR™, Fusion™, and ViewLux™. The program provided with each instrument contains different default parameters designed for measuring TR-FRET with LANCE chelates.

16.1.2 Lance Assay Principle

In the Lance assay, a signal is generated when a donor molecule labeled with europium (Eu) chelate gets into proximity of an acceptor molecule labeled with allophycocyanin (APC). When a biological interaction brings the donor and the acceptor into close proximity, excitation of the Eu-chelate at 340 nm allows FRET to the acceptor APC molecule resulting in fluorescence emission at 665 nm. As mentioned earlier, long fluorescence decay by lanthanide chelates after excitation virtually eliminates background caused by emission of microtiter plate or buffer components. A long excited state also avoids interference from short-lived emission by APC directly excited without FRET. Moreover, a large Stokes shift minimizes cross-talk, resulting in a high signal-to-background ratio.

16.2 LANCE NUCLEAR RECEPTOR (NR) ASSAY DESIGN

Classical methods to screen for novel NR modulators included radioligand binding and reporter gene assay. Other techniques such as coimmunoprecipitation, pull down, and gel shift assay are also widely used to study the NR mechanism of action. However, the low throughput of these techniques has raised the need for novel assays that would speed up the identification of potential modulators of these important drug targets.

16.2.1 THE BIOMOLECULAR INTERACTION

It is well established that after agonist binding, allosteric changes in the ligand binding domain (LBD) of NR will allow interaction between the AF-2 domain in the LBD and the NR box present in the coactivator's structure [5,6]. In cells, this interaction will trigger a cascade of events leading to transcriptional activation of NR target genes. It has been demonstrated that a consensus sequence present on all NR coactivators, called the LXXLL motif, is sufficient for the interaction with the agonist bound receptor LBD [7]. Other known regulators of NR signaling pathways are the corepressors. Corepressors interact with antagonist-bound receptor resulting in an inhibition of transcription initiation [8]. These different ligand-dependent interactions form the basis of the development of LANCE NR assays. Different binding partners can be used depending on their cost, ease of production, availability, and stability. Here are some examples of such combinations:

- Interaction between agonist-bound receptor or receptor LBD and LXXLL motif-containing peptide
- Interaction between an apo or holo receptor and corepressor or corepressor interaction domain

Depending on the detection reagents, the binding partners can be wild type (WT), truncated, or tagged with, for example, biotin, c-myc, GST, FLAG, or $(His)_6$.

Different examples are presented in the literature to demonstrate the use of LANCE assays based on the interaction between receptor and coactivator-derived peptide [9–13]. These assays generate a signal increase upon agonist binding and a signal decrease following antagonist binding. The interaction with corepressors could also be used to find ligand acting at a different molecular event of the receptor-signaling cascade. Thus far, no high-throughput screening (HTS) LANCE assay using corepressors has been reported in the literature.

16.2.2 THE LANCE DETECTION REACTION

Most of the assays reported in the literature involve the interaction between biotinylated LXXLL peptides derived from a coactivator sequence and tagged receptor LBD. Following complex formation, the interaction is detected using Eu-labeled antibody and APC-labeled streptavidin. A large variety of LANCE Eu- and APC-labeled reagents is commercially available from PerkinElmer; they allow the capture of different tagged versions of the WT receptor, receptor LBD, coactivator, or corepressor domains, and LXXLL motif-containing peptides. Moreover, if both binding partners are biotinylated, the LANCE assay can be performed using commercially available Eu-streptavidin and streptavidin-APC [13].

16.3 LANCE NUCLEAR RECEPTOR ASSAY DEVELOPMENT

16.3.1 GENERAL LABORATORY PRACTICE

LANCE assays are performed in normal laboratory conditions. APC-labeled reagents should be stored at 4°C protected from light. LANCE Eu-labeled reagents are stored according to the manufacturer's instructions either at 4 or –20°C. LANCE reagents should never be diluted for storage. Eu-labeled reagents are supplied in Tris-HCl-based buffer containing sodium azide to prevent microbial growth.

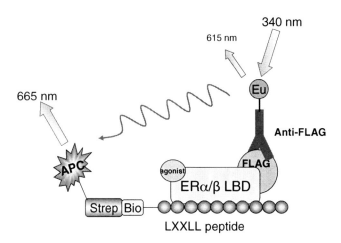

FIGURE 16.4 Scheme illustrating the LANCE NR assay principle.

16.3.2 EXAMPLE OF ASSAY DEVELOPMENT

To achieve desired optimal signal, sensitivity, reproducibility, and dynamic range, different parameters can be optimized, as presented in the following sections. The depicted model involves the interaction of a biotinylated LXXLL sequence containing peptide and a FLAG-tagged version of the ERα or ERβ LBD. Following agonist binding, the interaction between ligand-bound receptor and biotinylated peptide is detected using LANCE Eu-labeled anti-FLAG antibody and streptavidin-APC (Figure 16.4). The results presented have been taken from Liu et al., 2003 with kind permission [12].

16.3.2.1 Signal Detection

When using Eu as the donor and APC as the acceptor, TR-FRET measurement is performed at 665 nm and Eu fluorescence at 615 nm. Factory-set counting protocols for LANCE labels are available in VICTOR, EnVision, Fusion, and ViewLux instruments. In many published LANCE NR-coactivator recruitment assays, authors have also measured Eu fluorescence at 615 nm and expressed the data as the ratio of the emission intensity at 665 nm to that at 615 nm [12].

16.3.2.2 Buffer Conditions

Eu-W1024 chelates are stable at pH higher than 6, and Eu-W8044 at pH over 4. Most of the LANCE NR assays reported are developed using the Eu-W1024 chelate series at pH values between 7 and 8. Tris-HCl and HEPES buffer usually performs best in a LANCE assay. However, NR LANCE assays have been demonstrated to work well in phosphate-based buffers.

Chelating agents, such as ethylenediaminetetraacetic acid (EDTA), are often used as stopping agents in enzymatic reactions and as ion quenchers in screening assays. A NR LANCE assay often includes 1 to 2 mM EDTA in the assay buffer. However, it is recommended to not exceed 20 mM of EDTA when developing assays using W1024 chelates and 250 mM with W8044 at neutral pH. The assay buffer used by Liu and coworkers was composed of 20 mM HEPES pH 7.2 containing 1 mM EDTA, 50 mM KCl, 0.05% Nonidet P-40, 1 mM dithiothreitol, and 1 mg/mL bovine serum albumin (BSA).

Physiological concentrations of ions like K^+, Na^+, Mg^{2+}, and Ca^{2+} have had few effects on LANCE's chelate fluorescence when tested up to 50 mM. However, other ions such as heavy metal cations (Mn^{2+}, Cr^+, Co^{2+}, $Fe^{2+/3+}$, and Cu^{2+}) will quench negatively charged chelates by an ion-pair mechanism. This quenching can be compensated by the addition of 1.5 times more EDTA than the concentration of heavy metal ion just prior to measurement.

16.3.2.3 Reagent Order of Addition

Different orders of addition have been reported in the literature; the order of addition should be set on a case-to-case basis. The reactions can be carried out in volumes ranging from 200 down to 8 μL. The order of addition is described as follows:

- A reaction mix is prepared containing receptor, coactivator (or coactivator peptide), and the detection reagents (for example, streptavidin-APC and antitag-Eu antibody). The mix is loaded in the well containing ligand [12]. Plates are incubated from 1 h at room temperature (RT) to overnight at 4°C and read.
- Tagged receptor and antitag-Eu antibody are first incubated for 1 h at RT. At the same time, biotinylated coactivator peptide is incubated with streptavidin-APC. Following this incubation, the two reactions are combined in the presence of ligand and incubated for an additional 1 h at RT [14].

16.3.2.4 Reagent Titration

The presented guidelines should be followed for optimal assay performance. The concentration of Eu-labeled antibody should be kept fairly low (1 to 3 nM). Concentration of tagged receptor or tagged LBD would ideally be from 5 to 20 nM, resulting in efficient binding of antibody. Since the streptavidin-APC molecule (1:1 molar ratio) is able to bind around three biotinylated peptide molecules, it is essential to capture all the peptide to produce a performant assay.

As an optimization example, Liu et al. titrated the biotin-peptide and the streptavidin-APC to a fixed final concentration of:

- Eu-labeled anti-FLAG antibody (1 nM)
- Tagged ERβ LBD (5 nM)
- Saturating concentration of E2 (100 nM)

The biotinylated coactivator peptide was titrated from 5 to 175 nM against 15 and 100 nM of streptavidin-APC. The titration results established the optimal conditions at 100 nM of coactivator peptide and 30 nM of streptavidin-APC (Figure 16.5). Agonist (E2) titration was then performed against varying concentrations of ERβ LBD (5, 10, and 20 nM) using 1 nM of Eu-anti-FLAG, 30 nM of streptavidin-APC, and 100 nM of biotinylated coactivator peptide [12]. The results presented show that FRET increased when the concentration of ERβ LBD was increased. However, the lowest concentration of LBD (5 nM) already gave a satisfactory assay; this concentration was used in subsequent assays (Figure 16.6). It should be noted that Glickman et al., 2002 reported different reagent optimization procedures and also generated an adequate Lance assay [15].

16.3.2.5 Time Course

Incubation time can be optimized by looking at the kinetics of the biological interaction in the presence of saturating concentration of agonist. Reported incubation times in LANCE coactivator recruitment assays vary from 1 h at RT to overnight at 4°C. In the ER LBD model developed by Liu et al., the maximal signal was generated after overnight incubation at 4°C since E2 binding to ER LBD reaches equilibrium after 9 h [16].

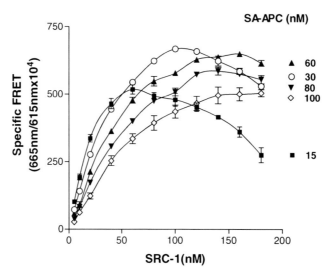

FIGURE 16.5 Optimization of relative concentration of biotin-LXXLL peptide (SRC-1) and streptavidin APC (SA-APC) using fixed concentrations of ERβ LBD (5 n*M*) and E$_2$ (100 n*M*). (Taken from Liu J, Knappenberger KS, Kack H, Andersson G, Nilsson E, Dartsch C, Scott CW. *Mol Endocrinol* 2003; 17:346–355.)

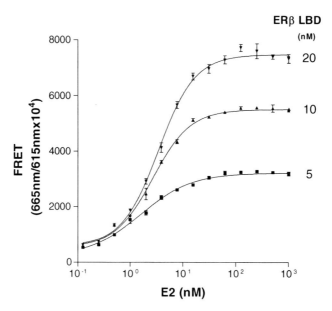

FIGURE 16.6 Agonist titration at different concentrations of ERβ LBD using the Lance Nuclear Receptor assay. The detected E$_2$ potency is not affected by variations in ER concentration. (Taken from Liu J, Knappenberger KS, Kack H, Andersson G, Nilsson E, Dartsch C, Scott CW. *Mol Endocrinol* 2003; 17:346–355.)

16.3.2.6 Agonist and Antagonist Titration

Once order of addition, time, detection reagents, and binding partner conditions have been optimized, agonist and antagonist titration should be performed. The EC$_{50}$ measured for a known agonist should reflect the affinity reported in the literature with other technologies. The EC$_{50}$ measured will reflect a combination of the affinity of the ligand for its receptor added to the affinity of the receptor

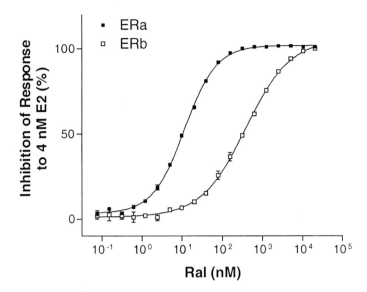

FIGURE 16.7 Effect of antagonist on agonist induced receptor–coactivator dimerization. Various concentrations of raloxifene were incubated with ERα and ERβ LBD, LXXLL peptide, and E_2. Raloxifene completely blocked E2-induced recruitment of the SRC-1derived peptide, with 30-fold selectivity for ERα LBD. (Taken from Liu J, Knappenberger KS, Kack H, Andersson G, Nilsson E, Dartsch C, Scott CW. *Mol Endocrinol* 2003; 17:346–355.)

for its binding partner. The results presented by Liu et al. demonstrate that LANCE is able to measure an affinity (EC_{50} 3.4 n*M*) for E2 that is in good agreement with other *in vitro* assays. In the LANCE ER assay raloxifene blocked E2-induced recruitment, yielding IC_{50} values of 10 and 300 n*M* for ERα and ERβ, respectively (Figure 16.7).

16.3.2.7   Robustness and Tolerance to Dimethylsulfoxide (DMSO)

Two crucial parameters in any HTS protocol are the robustness and tolerance to DMSO of the assay. Most of the compounds library formulations are provided in DMSO, which should not hamper the assay performance. Liu et al. reported a tolerance up to 5% DMSO for the LANCE NR assay developed with ER LBD.

A z′ factor over 0.5 is well established as a tool to predict the performance of an assay in HTS protocols. Typically, z′ factors in LANCE NR recruitment assays have been well above 0.5, indicating an assay with excellent precision in 384-well as well as in 1536-well plate formats. For the model we depicted [12], z′ values of 0.7 and 0.9 were reported even for low agonist concentrations generating a weak, twofold signal-to-background ratio (20 μL volume in a 384-well plate).

Stability of signal is an important factor when developing assays for HTS. In the LANCE ER coactivator recruitment assay, the signal was stable for more than 48 h, allowing batch processing in HTS.

16.4   TROUBLESHOOTING

Information on troubleshooting is provided in Table 16.1.

TABLE 16.1
Troubleshooting Guide

Problem	Cause	Effect/Remedy
No or lower signal than expected	Reagents	All the biotinylated peptide should be bound to the streptavidin-APC for optimal signal generation; perform a simultaneous titration of biotinylated peptide and streptavidin-APC to find optimal conditions; concentration of streptavidin-APC should be 0.3 to 0.5 times higher than the concentration of biotinylated peptide or protein
	Reagents	Both binding partners should be present in the correct concentrations; perform a simultaneous titration of ligand-bound receptor and coactivator using fixed concentrations of LANCE reagents
	Reagents	Presence of free biotin will interfere with signal generation; perform careful purification of biotin after protein or peptide labeling
	Reagents	Poorly biotinylated binding partner; check the extent of biotinylation of binding partners by, e.g., HABA test (Pierce)
	Buffer	Reagents are coated to the well during incubation times; Tween-20 or 40, CHAPS, nonidet P-40, Brij 35, or Triton X-100 at concentration ranging from 0.01 to 0.1% should be added to reaction buffer; BSA at concentration of 0.01 to 0.2% can be added to reaction buffer
	Buffer	Quenching component in the buffer; avoid use of heavy metal ions such as Mn^{2+}, Cr^+, Co^{2+}, $Fe^{2+/3+}$, and Cu^{2+}; if present use 1.5 times more EDTA than the concentration of heavy metal ion just prior to measurement
	Buffer	Inappropriate buffer composition; check for correct pH (around 7.4), buffering capacity, and salt concentration, as well as requirement for reducing agents, detergent, chelators
	Assay	Incubation time too short; perform a time course of the interaction of the two binding partners (receptor/coactivator).
	Instruments	Plate reader error or failure; consult instrument manual or call provider
High background signal	Reagents	Excessively high concentration of LANCE Eu-labeled component; do not exceed 5 nM final concentration of labeled antibodies
	Reagents	Excessively high concentration of biotin-coactivator or peptide (higher than 0.5 μM) producing artefactual interaction with the receptor in the absence of ligand; titrate the concentration of the coactivator

REFERENCES

1. Soini E, Hemmila I. Fluoroimmunoassay: present status and key problems. *Clin Chem* 1979; 25:353–361.
2. Mathis G. Rare earth cryptates and homogeneous fluoroimmunoassays with human sera. *Clin Chem* 1993; 39:1953–1959.
3. Selvin PR, Jancarik J, Li M, Hung L-W. Chrystal structure and spectroscopic characterization of a luminescent europium chelate. *Inorg Chem* 1996; 35:700–705.
4. Hemmila I, Mukkala VM. Time-resolution in fluorometry technologies, labels, and applications in bioanalytical assays. *Crit Rev Clin Lab Sci* 2001; 38:441–519.
5. Onate SA, Tsai SY, Tsai MJ, O'Malley BW. Sequence and characterization of a coactivator for the steroid hormone receptor superfamily. *Science* 1995; 270:1354–1357.
6. Wang JC, Stafford JM, Granner DK. SRC-1 and GRIP1 coactivate transcription with hepatocyte nuclear factor 4. *J Biol Chem* 1998; 273:30,847–30,850.
7. Heery DM, Kalkhoven E, Hoare S, Parker MG. A signature motif in transcriptional coactivators mediates binding to nuclear receptors. *Nature* 1997; 387:733–736.

8. Horwitz KB, Jackson TA, Bain DL, Richer JK, Takimoto GS, Tung L. Nuclear receptor coactivators and corepressors. *Mol Endocrinol* 1996; 10:1167–1177.

9. Tremblay GB, Kunath T, Bergeron D, Lapointe L, Champigny C, Bader JA, Rossant J, Giguere V. Diethylstilbestrol regulates trophoblast stem cell differentiation as a ligand of orphan nuclear receptor ERR beta. *Genes Dev* 2001; 15:833–838.

10. Coward P, Lee D, Hull MV, Lehmann JM. 4-Hydroxytamoxifen binds to and deactivates the estrogen-related receptor gamma. *Proc Natl Acad Sci USA* 2001; 98:8880–8884.

11. Drake KA, Zhang JH, Harrison, RK, McGeehan GM. Development of homogeneous, fluorescence resonance energy transfer–based *in vitro* recruitment assay for peroxisome proliferator-activated receptor δ via selection of active LXXLL coactivator peptides. *Anal Biochem* 2002; 304:63–69.

12. Liu J, Knappenberger KS, Kack H, Andersson G, Nilsson E, Dartsch C, Scott CW. A homogeneous *in vitro* functional assay for estrogen receptors: coactivator recruitment. *Mol Endocrinol* 2003; 17:346–355.

13. Jones SA, Parks DJ, Kliewer SA. Cell-free ligand binding assays for nuclear receptors. *Methods Enzymol* 2003; 364:53–71.

14. Lee G, Elwood F, McNally J, Weiszmann J, Lindstrom M, Amaral K, Nakamura M, Miao S, Cao P, Learned RM, Chen JL, Li Y. T0070907, a selective ligand for peroxisome proliferator-activated receptor, functions as an antagonist of biochemical and cellular activities. *J Biol Chem* 2002; 277:19,649–19,657.

15. Glickman JF, Wu X, Mercuri R, Illy C, Bowen BR, He Y, Sills M. A comparison of ALPHAScreen, TR-FRET, and TRF as assay methods for FXR nuclear receptors. *J Biomol Screen* 2002; 7:3–10.

16. Grill HJ, Manz B, Belovsky O, Krawielitzki B, Pollow K. Comparison of [^{3}H]oestradiol and [^{125}I]oestradiol as ligands for oestrogen receptor determination. *J Clin Chem Clin Biochem* 1983; 21:175–179.

17 The Emerging Role of Cell-Based Assays in Drug Discovery

Ralph J. Garippa

CONTENTS

17.1 WHY TURN TO CELL-BASED ASSAYS?

Cell-based assays have long been viewed as "three-in-one" assays [1]. That is, they yield the relevance of a functional readout in a self-contained living entity, they give an early indication of cytotoxicity (if the exposure is long enough and the compound concentration is high enough), and, if the target is an intracellular one, they give an approximation of the degree of cell penetration of the compound. Utilization of the cell is indeed a reductionism-based approach to drug discovery, where one depends upon the whole cell to be an impartial reporter of compound activity — a "sentinel" of a compound's impact on integrated steady-state biochemical events. In a sense, cells are the first "independent living subjects" to come in contact with chemical compounds that are destined to be drug candidates. With well-designed assays, investigators may first conduct a "proof of mechanism" study that links a drug modality with a target molecule. Once proven, such work generally leads to a "proof of concept" study in the whole animal, usually one with a set of symptoms and sequelae indicative of the particular disease state being addressed. Unlike their costlier animal counterparts, cell-based assays may be miniaturized and automated to a great degree. Of course there are always certain caveats and choices to be dealt with (suspension cells vs. adherent cells, effect of serum and media composition, passage number, etc.) and liabilities to consider (need for CO_2 exposure, humidity and temperature controls).

One may well ask, "Where else to turn"? The pros and cons of animal work have been debated elsewhere [2], and chip-based technologies [3,4] are still in their nascent stages, although the early results are impressive but isolated. Biochemical or noncell-based assays are extremely useful but cannot be expected to bear the entire burden of hit-to-lead and lead optimization strategies. In fact, cell-based assays and noncell-based assays are most useful when they are run in parallel. Cell-based assays are not a threat to replace biochemical assays but rather are supplemental and supportive when applied correctly. Typically, the two different types of screens will correlate well

[0.6 to 0.8] but not to unity. If the correlation were shown to be perfect, then there would be a costly redundancy in place. Therefore, one may use one type of assay to "cull" or on the other hand, to "advance" certain drug candidates from the rest of the pack. Accordingly, it is not surprising that the percentage of high-throughput screens (HTS) as cell-based assays has steadily risen in the past ~5 to 6 years, but it is not anticipated that the split with biochemical HTS will go beyond 50:50. Biochemical noncell-based assays are still, in general terms, cheaper, faster, simpler, more amenable to miniaturization, and ultimately easier to interpret.

17.2   WHEN IS IT APPROPRIATE TO INSERT A CELL-BASED ASSAY INTO ONE'S DRUG DISCOVERY STRATEGY?

Cell-based assays, which have long enjoyed a niche late in the drug discovery process (somewhere between lead identification and clinical candidate selection), have now infiltrated virtually every stage of drug development. As early as the target selection stage, one can see a positive effect of combining a cell-based functional screen as a next step following siRNA knockdown experiments [5]. A cell-based phenotypic change with the introduction of a specific siRNA adds credence to genotypic changes that can be measured using Taqman or single transcript expression profiling (STEP). Today, cell-based assays continue to make impressive inroads into the HTS arena [6]. Almost all of the major target classes (G protein coupled receptors [GPCRs], ion channels, transporters, kinases, phosphatases, and nuclear transcription factors) can be adapted to cell-based screening methodologies. In the 1970s and 1980s, cells were used as factories to generate cell lysates containing correctly folded GPCRs in a lipid environment for use in radiometric binding experiments. The filter-based or centrifugation-based separation steps were rate limiting, however, thereby preventing these assays from evolving at a pace with burgeoning compound libraries in big pharmaceutical houses. In the 1990s, the fluorescent imaging plate reader (FLIPR, Molecular Devices, Sunnyvale, CA) revolutionized the cell-based HTS field by integrating a charge coupled device (CCD) camera with on-board pipetting that could record (now in 96, 384, and 1536 microtiter plate well densities) kinetic readouts of calcium transients every 0.8 sec [7]. Daily outputs in FLIPR HTS were typically in the range of 30,000 to 40,000 data points per day per individual machine — good enough to screen most pharmaceutical compound libraries in a matter of weeks. In pushing the envelope even further, companies such as Norak Biosciences, Research Triangle Park, NC (RTP) and BioImage (U.K.) are now utilizing translocation and target-specific activation events read with high-speed confocal imagers (the InCell 3000, GE Biosciences U.K., and the Evotec Opera, Hamburg, Germany) to run screening campaigns with throughputs that match or exceed the FLIPR metrics [8,9]. Cell-based assays have also made their way onto ultra high throughput screening (uHTS) platforms such as the Carl Zeiss Multimode system and the Zymark Allegro system, utilizing whole-well readouts of fluorescence intensity or luminescence for throughputs exceeding 100,000 data points per day.

Perhaps in no other subdiscipline of cell biology are advancements so clearly seen as in the field of high-content screening (HCS). Simply, HCS is a battery of image-based cell assays that are facilitated by automation and *in silico* analysis tools. In an instance of science imitating art, HCS allows researchers to develop an assay directly from a photomicrograph or published microscopic image. In some cases, the image is one that originally drew the investigator's attention to a particular drug target. The credo, "if it can be seen by microscopy and localized using a fluorescent label, then it can be detected, measured, and quantified by HCS" has been proven true in recent published successes [10,11]. The spectrum of HCS-enabled cell-based assays continues to broaden the horizon, from nuclear translocation assays, plasma membrane internalization assays, angiogenesis/tube formation assays, neurite outgrowth, apoptosis/nuclear fragmentation, and quantification of mitotic figures. More sophisticated and multiplexed assays such as cell health (as an early indicator of general cytotoxicity or organ toxicity) and micronuclei detection (for relatively rare clastogenic and aneugenic events as an early index of genotoxicity) are bridging the gap between lead initiation and lead optimization strategies [12,13]. Even so, there will always be individual choices for the cell biologist and the project leader to make concerning the selection of cell, time of exposure to compound, and concentrations (doses) to be tested.

17.3 "FAIL FAST, FAIL CHEAP"

These were the rather shocking words of Dr. Lee Babiss of Roche [14] when as vice president of research and development in Nutley, NJ, he drew attention to the fact that it was desirable to terminate compound development (and often an entire project) when early evidence points to a high probability of failure later in the pipeline. A given clinical study costs approximately 16-fold more than its preclinical counterpart experiment. What was being advocated was not drug discovery failure, but rather the implementation of a cost-effective screening strategy that would eliminate nontractable chemistries as early as possible. Less than 3% of all Food and Drug Administration (FDA)–approved New Medical Entities (NMEs) in the last decade were withdrawn due to unexpected toxic (in some cases lethal) side effects [15] but that number alone underestimates the amount of lost monies in bringing unproductive compounds forward to Phase 0 to Phase 3 clinical efforts and late preclinical stages. So if the bottleneck of drug discovery has shifted out of the HTS area (with plating densities in the 1536 to 3456 range, who can argue this point?) and if the advent of the sequencing of the human and rodent genomes has opened access to targets once unobtainable, where did the bottleneck go? By all accounts, it is squarely lodged in the lead optimization and clinical candidate areas of drug discovery.

A trend we see to alleviate this downstream bottleneck is the use of cell-based assays as a prelude to clinical candidate selection. An example is the discovery that drugs that elicited a block of the inward rectifying potassium channel human ether-a-go-go related gene product (hERG) can lead to prolonged QT intervals in humans and dangerous arrhythmic episodes [16,17]. The gold standard for measuring these currents is pulled glass microelectrode impalement (top-down) of the cell via micromanipulation under microscope. Once again, a seemingly impossible task to automate has been automated to a large extent by several companies (MDC, Nanion, FlyIon, and Sophion) where "bottom-up" patch clamping on planar chips allows researchers to screen hundreds of compounds in situations where only dozens were possible by previous methodologies [18]. This development presents the possibility of avoiding a series of chemical hits post-HTS which possess an innate hERG liability within their core structure from ever being pursued at the expense of not developing other target-specific compound chemistries.

17.4 BETTER CELL-BASED ASSAYS THROUGH ADVANCES IN MOLECULAR BIOLOGY, CHEMISTRY, AND AUTOMATION

Advancements in related disciplines have greatly facilitated the adoption of cell-based assays in drug discovery. Many of the molecular biology tools used routinely today have influenced on assay design. For example, there are a host of cell lines available within commercial catalogs (ATCC) and individually developed in specific laboratories. These cells can be queried using STEP to determine whether the chosen cell line shows endogenous expression of a target transcript. Certain other cell lines such as HEK293, COS, and CHO have enjoyed their popularity as host cells for the expression of exogenous proteins when the search for an endogenous protein host proves fruitless. There is a wealth of vectors and promoters available for the cloning and expression of proteins of interest in cells. Of note are bicistronic vectors [19] that provide an estimate of target protein expression in parallel with an epitope-tagged or fluorescent protein–tagged reporter molecule. Inducible systems such as "tet-on, tet-off" methodologies (Invitrogen) offer the opportunity to express potentially cytotoxic proteins within cells during an assay window of screening. The methods of protein expression have advanced beyond simple Western blotting, particularly where flow-based technologies have been implemented. Fluorescence-activated cell sorting (FACS, Becton-Dickinson, Franklin Lakes, NJ) and the Personal Cell Analyzer (Guava, CA) allow researchers to quickly determine whether or not a protein of interest has been expressed in a cell and also the relative amount of a protein per cell. This in turn, enables a particular cell clone (a high vs. a low expresser) to be isolated for a particular assay type (signal transduction vs. receptor binding) that is relevant to the degree of expression. Transfection technologies now stretch beyond lipid-based reagents, electroporation, and protein precipitation. Most notably, viral transfection schemes (mammalianized baculovirus or BacMam, Semliki forest virus, and adenovirus) have found utility in transfecting a large percentage of numerous cell populations, primary and immortalized alike [20,21]. The use of stem cells and immortalized primary cells (Xenotech) creates a situation where researchers have access to a cell most representative of human physiology, expressing all of the relevant native proteins, ideally in amounts indicative of the steady state and under homeostatic control *in vivo*. Even primary cells such as hepatocytes, adipocytes, and muscle cells are enjoying a renaissance in drug discovery, moving into screening paradigms alongside well-established primary cell tools such as human umbilical vein endothelial cells (HUVEC).

The dyes, antibodies, fluorescent proteins, and particles that provide visualization and detection of cellular events are worth noting. It is a bit ironic that advances in chemistry allowed cell biologists to better address the needs of medicinal and combinatorial chemists within drug development programs. Specifically, the demonstration of Fluo-3 and Fluo-4 dyes for the measurement of GPCR signaling enabled the development of the FLIPR for high-throughput cell-based screening. Now, so-called no-wash dyes are available that make the assays even more amenable for HTS [22]. Similarly, the demonstration of Fura-based FRET dyes opened avenues of ion channel exploration in HTS via voltage ion probe reader (VIPR) instrumentation [23]. The isolation and cloning of nature's fluorescent proteins (luciferase from the firefly and aequorin from the jellyfish) and the fusion of these fluorescent proteins to targets of interest presented investigators with opportunities in live cell imaging previously unobtainable [24]. So-called "reporter gene assays" or RGAs have proven their worth in cell-based assays ranging from nuclear transcription factor activation to protein–protein interaction [25]. While antibody-based localization of a target is hardly new, there are new fluor-antibody pairs being developed in several vendor companies to keep pace with the number of newly discovered kinases, phosphatases, GPCRs, enzymes, ion channels, and transporters (the major classes of drug discovery targets). Beyond protein and dye-based detection schemes, there are fluorescent particle-based technologies such a QDots (Quantum Dot Corporation), which employ coated selenium-based microspheres that are fade resistant.

Perhaps it is the seemingly facile adaptation of fragile cells (long perceived as being exquisitely delicate to nuances of manual manipulation and shear stress) to manmade machines composed of metal, plastic, and glass that is the most surprising development in the last 15 years. For example, the advent of automated cell culture systems such as the SelecT from The Automation Partnership (U.K.) and the AcCellerator from RTS Thurnal (U.K.) have freed the hands of researchers from much of the rote work of cell passaging and plating as traditionally done by a technician seated squarely in front of a laminar flow hood. While these robotic instruments reflect the classical approach to cell passage and culture, recent developments have shown that cells can often be treated as simple "biochemical reagents" in complicated assays. Of note is the demonstration that division-arrested cells (CMT, Philipsburg, NJ) show full functionality in a variety of assays, from calcium flux experiments to reporter gene assays. Furthermore, Cowan and colleagues have now shown that frozen cell stocks that are plated and assayed 24 h post-thaw have nearly identical performance when compared side-by-side with passaged nonfrozen counterparts 24 h posttrypsinization [26].

17.5 WHERE DO WE GO FROM HERE?

The easy answer would be that systems biology, toxicogenomics, stem cell research, xenobiotic approaches, and label-free detection technologies would be among the most profound cell-based advancements over the next two decades. Systems biology will surely rely on a variety of cell-based assays to confirm, deny, or validate a number of pathway interrelationships that have been postulated using computational biology tools. The field of toxicogenomics, using animal-based gene expression profiles following toxicant exposure, will look to cell-based assays to validate results that often take weeks or months to manifest in a chronic *in vivo* setting. Stem cells, providing current governmental restrictions are lifted, offer great promise for treatment of illnesses such as Alzheimer's disease and other degenerative diseases, which currently have a largely unmet need. The field of xenobiotics holds the potential for using cells as agents in transplantation (for example, dopaminergic neurons in Parkinson's disease and insulin-producing pancreatic beta cells in diabetes), where a key deficit exists in production of an endogenous factor. Much as the Cytosensor microphysiometer offered a probe-free methodology to assess cell physiology [27], today's label-free detection technologies (MDX-Sciex) offer researchers the tantalizing ability to query cells and their associated native proteins without the necessity of intervention by dyes, proteins, or DNA manipulation.

In conclusion, the time is ripe to be extensively involved in cell-based assays for drug discovery. The tools of molecular biology once promised to researchers seemingly endless possibilities involving DNA and RNA manipulation two decades ago. Now, the burgeoning cell-based assay toolset available to cell biologists and screeners alike offers to deliver an unprecedented influx of value-adding data to usher in drug discovery in this new millennia.

REFERENCES

1. Coates J, Causey A. Taking drug discovery into clinical development. Drug Information Association seminar, Melbourne, Australia, Dec 18, 1998.
2. Arch JRS. Lessons in obesity from transgenic animals. *J Endocrinol Invest* 2002; 25:867–875.
3. Figeys D. Adapting arrays and lab-on-a-chip technology for proteomics. *Proteomics* 2002; 2:373–382.
4. Heller MJ. DNA microarray technology: devices, systems, and applications. *Annu Rev Biomed Eng* 2002; 4:129–153.
5. Semizarov D, Kroeger P, Fesik S. siRNA-mediated gene silencing: a global genome view. *Nucl Acids Res* 2004; 32:3836–3845.
6. Johnston PA, Johnston PA. Cellular platforms for HTS: three case studies. *Drug Disc Today* 7:353–363.
7. Schroeder K, Neagle BD. FLIPR: a new instrument for accurate high-throughput optical screening. *J Biol Screening* 1996; 1:75–80.
8. Cooke E-L, Ainscow E, Hargreaves A, Sullivan E, Alcock P, Ellston J, Peters S, Major J, Wannop J, Allen H, Plant D, Coudhry S, Hicks R, McCall E, Shaw J, Ronco L, Loomis C. G-protein-coupled receptor high throughput screen using Norak Transfluor technology and the In Cell Analyzer 3000. 9th Annual Society for Biomolecular Screening Conference, Portland, OR, Sept 21–25, 2003.
9. Garippa RJ. Use of the Evotec Opera confocal microscope imaging system to develop and complete an orphan GPCR high-throughput screening campaign using Norak Transfluor technology. 2nd Annual Cambridge Healthtech Institute High Content Analysis Conference, San Francisco, CA, Jan 25–28, 2005.
10. Guiliano K, Haskins J, Taylor DL. Advances in high content screening for drug discovery. *Assay Drug Dev Technol* 2003; 1:565–577.
11. Starkuviene V, Liebel U, Simpson JC, Erfle H, Poustka A, Weimann S, Pepperkok R. High-content screening microscopy identifies novel proteins with a putative role in secretory membrane trafficking. *Genome Res* 2004; 14:1948–1956.

12. Koppal T. Toxicogenomics warns of drug dangers. *Drug Disc Devel* 2004; 7:30–34.

13. Tencza S, Sipe MA. Detection and classification of threat agents via high-content assays of mammalian cells. *J Appl Toxicol* 2004; 24:371–377.

14. Boguslavsky J. Minimizing risks in hits to leads. *Drug Disc Devel* 2001; 4:26–30.

15. Schmid EF, Smith DA. Is pharmaceutical R&D just a game of chance or can strategy make a difference? *Drug Disc Today* 2004; 9:18-26

16. Gonzalez JE, Oades K, Leychkis Y, Harootunian A, Negulescu, PA. Cell-based assays and instrumentation for screening ion-channel targets. *Drug Disc Today* 1999; 4:431–439.

17. Finlayson K, Witchel HJ, McCulloch J, Sharkey J. Acquired QT interval prolongation and hERG: implications for drug discovery and development. *Eur J Pharmacol* 2004; 500:129–142.

18. Brueggemann A, George M, Klau M, Beckler M, Steindl J, Behrends JC, Fertig N. Ion channel drug discovery and research: the automated nano-patch-clamp technology. *Curr Drug Dis Technol* 2004; 1:1570–1638.

19. Royer Y, Menu C, Liu X, Constantinescu S. High-throughput gateway bicistronic retroviral vectors for stable expression in mammalian cells: exploring the biological effects of STAT5 overexpression. *DNA Cell Biol* 2004; 23:355–365.

20. Clay WC, Condreay JP, Moore LB, Weaver SL, Watson MA, Kost TA, Lorenz JJ. Recombinant baculoviruses used to study estrogen receptor function in human osteosarcoma cell. *Assay Drug Dev Technol* 2003; 1:801–810.

21. Sato Y, Shiraishi Y, Furuichi T. Cell specificity and efficiency of the Semliki forest virus vector-and adenovirus vector-mediated gene expression in mouse cerebellum. *J Neurosci Meth* 2004; 137:111–121.

22. Zhang Y, Kowal D, Kramer A, Dunlop J. Evaluation of FLIPR calcium 3 assay kit: a new no-wash fluorescence calcium indicator reagent. *J Biomol Screen* 2003; 8:571–577.

23. Gonzalez JE, Maher MP. Cellular fluorescent indicators and voltage/ion probe reader (VIPR™): tools for ion channel and receptor drug discovery. *Recept Chan* 2002; 8:283–295.

24. Wouters FS, Verveer PJ, Bastiaens PIH. Imaging biochemistry inside cells. *Trends Cell Biol* 2001: 11:203–211.

25. Schenborn E, Groskreutz D. Reporter gene vectors and assays. *Mol Biotechnol* 1999; 13:29–44.

26. Cowan C, Ouelette M, Payne R, Hudson C, Eckhardt A, Oakley R. Process improvements in cell delivery for GPCR high throughput screening. Tenth Annual Society for Biomolecular Screening Conference, Orlando, FL, USA, Sept 11–15, 2004.

27. McConnell HM, Owicki JC, Parce JW, Miller DL, Baxter GT, Wada HG, Pitchford S. The cytosensor microphysiometer: biological applications of silicon technology. *Science* 1992; 257:1906–1912.

18 The Preparation of Cells for High-Content Screening

Ann F. Hoffman

CONTENTS

18.1 INTRODUCTION

With the emergence of high-content screening (HCS), i.e., the automated imaging of cells to detect changes in cellular physiology, comes the need of the cell biologist to have and be able to accurately select the most appropriate cells and cellular markers on a large scale. This new era has opened the door for the large-scale automated analysis of cellular events such as the trafficking and translocation of proteins, cellular morphology, and colocalization of proteins within organelles, to name a few [1–3]. The new addition to this technology is that the individual scientist is now enabled to train his or her suite of technological robotic partners for the automation of the cell culture, the automation of the cellular assays on finely integrated instrument platforms, or for the automated analysis of the vast number of cellular images collected by HCS platforms, all to carry out detailed experimental protocols with high fidelity. This scenario, the change from the small laboratory bench to high-content/high-throughput screening (HCS/HTS) scales the throughput from a handful of microtiter plates to hundreds and thousands of microtiter plates full of compounds and requires new equipment, new methods of standardization and, equally important, new thinking to troubleshoot the issues that arise in scaling up cell biology to confidently quantify those cellular functions.

To adequately set up HCS within laboratories for primary or secondary screening in drug discovery, a few additional pieces of equipment are suggested that will lead to successful and efficient universal processing of HCS. During the implementation of HCS, the emphasis on process is critical as it enables all laboratory members with various types of expertise to participate, avoiding the trap of narrow specialization where one may hear the common phrase "We only want Sally to plate our cells," or "We only want Joe to analyze our results." The advantage becomes threefold:

- The conditions to which the cells are exposed, no matter what choice of cell type, are maximally optimized, resulting in reproducible growth and handling advantages.
- The assay is consistently executed.
- The emphasis on how the image analysis is to be interpreted can be straightforwardly deconvoluted with respect to the biology.

Emphasis on large-scale cellular techniques and integrating routine standardized processes will ultimately impact time and cost savings. It also provides an orderly, stepwise segmentation of the process such that troubleshooting of problems can be quickly effected.

18.2 BASIC EQUIPMENT NEEDS FOR THE HCS LABORATORY

The basic equipment needs for the HCS laboratory are:

- Cell freezers (–20°C and liquid nitrogen)
- Tissue culture facility, cell factories (Nunc), automated cell handling equipment
- Guava personal cell analyzer (PCA)
- Molecular biology reagents, laboratory, and tools
- Licenses for vectors, promoters, cell lines
- Transfection reagents and peripheral equipment (cloning cylinders, filter paper lifts)
- Fluorescent microscope 20, 40× objectives of highest numerical aperture (NA) quality
- Basic filter sets to include excitation filters to cover widely used fluorescent probes — typically, emission channels for blue, green, and red
- Liquid handling instrumentation, Multidrop or Multidrop/Titan Stacker (Titertek)
- High content screening platform(s) for the automated handling of stage movements, automated image collection, and image analysis software
- Integrated data management systems — ID Business Solutions, Inc. (IDBS), Activity-Base, and database (Oracle)
- Statistical software with standardized assessment tools to determine quality and reproducibility of assays
- Visualization tools — Spotfire, ISISDraw

18.3 APPROPRIATE CELL TYPES TO ESTABLISH HCS ASSAYS

A variety of commonly used cells, CHO, CHO-K1, HEK293, U-2 OS, and HeLa, are useful for HCS. Typically, many of these cell types display a phenotype of "flat extended nonclustered morphologies" as the cells proliferate. These cells are best suited for imaging platforms that are nonconfocal, as are other cell types with a similar gross morphology. Some of the clonal variants of HEK293 cells are less suitable for HCS, as their poor adhesion to microtiter plates (with or without the addition of extra-cellular matrix or poly-D lysine coating) tends to result in cells that easily detach during steps where media changes and washing are required. In all circumstances, it is important to establish for each cell type that, during the staining and fixing procedures, most if not all of the cells remain attached to the microtiter plate. Another consideration is that the cells remain displayed in an optimal form, that is, cells partially detached may appear with sickle-like morphologies. These phenotypes negatively impact upon the analysis and results, so further gentle optimization strategies must be put in place for these cell types before experimentation begins. In some cases, robotic automation may resolve the issues by decreasing the force of liquid addition to the cells; in other cases, cells must be grown on one of many types of extracellular matrix. Another remedy is to increase the fixation times to ensure proper cellular morphology. There are several commonly used fixation reagents including acetone, ice-cold methanol, paraformaldehyde, and combinations of 0.1% glutaraldehyde in 3% formaldehyde solution [4]. The single most commonly used fixative for our HCS applications is to quench the cellular reactions with a 10-min incubation of prewarmed, 37°C, 3% formaldehyde in phosphate-buffered saline containing calcium and magnesium (PBS). After the 10-min room temperature incubation, the microtiter plates are washed three times with PBS and sealed, and they may be stored for subsequent staining procedures.

As all HCS is dependent upon the clarity of the images collected over the various emission channels employed, significant care should be taken in correctly focusing and resolving the objects of interest for image analysis. Many of the HCS platforms employ an "autofocus" feature based upon fluorescent image contrast, particularly in the nuclear channel using Hoechst 33258, DAPI, or DRAQ-5 dyes. In this manner, for nonconfocal systems, the three-dimensionality of the cell is compressed in the z direction to amplify the signal(s). Under optimal conditions, where the cells are fairly flat, the quantification of the fluorescent probes identifying cytoplasm, plasma membrane, and nuclear space is rigorously collected. For the cell types mentioned previously that are described as "egg-like," defining a pattern where the nucleus appears as a raised yolk surrounded by a thinner cytoplasmic area (which is also spread over and under the nucleus), the nonconfocal systems are quite useful. Unfortunately for other applications of HCS performed on nonconfocal systems that typically require high resolution throughout the z-axis, the quantification of the fluorescent probe is often underestimated and occasionally inaccurately quantified.

For confocal systems, the emphasis on image autofocus is typically based upon an independent far-red laser beam reflecting off the microtiter plate's bottom surface. The z slice to be acquired is defined by the user at a specific distance from that surface. Thus, a virtual slice of the cell is focused upon, which would consist of approximately 2 microns for a 40× image to 4 microns for a 20× objective. The resultant images collected over the multiple fluorescent wavelengths, e.g., multiplexing, become the basis for the image analysis processing that defines the readout parameters. Advantages of the confocal systems and capturing "optical slices of cells" are that it allows the cell biologist access to a virtually unlimited listing of cell types and cellular phenotypes that are amenable to HCS/HTS. Even clustered cells, typical of human cancer cells such as RKO and HCT 116 and multilayered cell types, can then be addressed with confocal systems.

The superiority of confocal systems is further evidenced in their ability to inquire into subcellular events in "physiologically relevant cell types" — cancer cells such as SW-480, HT-29, skeletal muscle cells, hepatocytes, adipocytes, T and B cells, etc. [5]. Each series of optical sections that query the subcellular events within the cell are, in actuality, defined horizontal sections that are quantified without the interference of background out-of-focus light. The resulting advantage is one of higher resolution, while the definition of intracellular organelles, protein–protein interactions within these organelles, and assays recording translocation processes are done with a higher degree of precision/accuracy vs. the underestimate in certain cases of nonconfocal systems. The degree of flexibility that the confocal systems offer allow high-content experimentalists the ability to rapidly query from cell type to cell type and enhance their knowledge around pathways and compound effects. This allows questions to be addressed and answered in recombinant cell lines, multiple immortalized cell lines, and primary cell cultures.

Optical grating systems by Carl Zeiss are suited to strike a balance between confocal and nonconfocal systems [6]. The Zeiss Apotome insert offers this compromise as an addition for an easy-to-use bench-top microscope that allows "composite" z slices to be acquired in short time with the advantage of high objective resolution, 63×, and a high NA along with an automated stage. The use and impact of this type of system for HCS is still early, and as yet it has not been proven significantly advantageous over confocal HCS platforms. One reason for this is that during the image acquisition and reconstruction, greater than 50% of the transmitted fluorescent light is lost. Hence, cellular applications where the quantified fluorophores are weak emitters or the subpopulation of cells of interest displays a dim fluorescent signal may give poor results.

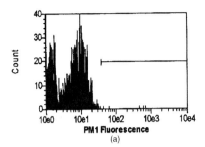 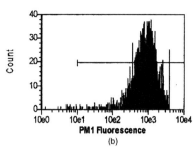

FIGURE 18.1 Experimental data illustrating the data output of HA-tagged protein expression using Guava's Express software on the PCA Instrument. Adherent cells were trypsinized and incubated with an anti-HA high-affinity rat monoclonal primary antibody (Roche), washed, and subsequently incubated with goat antirat R-phycoerythrin (R-PE, Southern Biotechnology Associates). (a) Wild type U 2-OS cells and (b) U 2-OS cells stably transfected with an orphan GPCR (oGPCR) containing an HA tag engineered into the *N*-terminal domain of the receptor.

18.4 CELL HANDLING PROTOCOLS

For basic cellular handling, evidence from multiple labs has sanctioned scaleup in T flasks, roller bottles, or cell factories to be equivalent and sufficient for HCS large-scale processes. Other fully automated systems such as the SelecT offered by The Automation Partnership (TAP) can robotically handle cellular counting, plating, and passaging cells while maintaining quality controls on cellular viability. For practical purposes, this chapter will adhere to more traditional means of cellular handling. Successful HCS assays generally rely on the premise that cells not exceed 90% confluence in the proliferating flask or 90% confluence in the microtiter plates. Furthermore, the cells are required to retain full expression of any protein labeled, whether fluorescent or introduced, for translocation or protein activation experiments. In our laboratory, we monitor recombinant receptor expression levels by using the Guava PCA instrument (Guava Technologies). Essentially, this microcapillary flow instrument allows the rapid determination of receptor expression via antibody-directed means as well as rapidly enables cellular counting and viability measurements. All of these measurements then serve as references for the quality of the cells being used in the HCS assay. Thus with daily evaluation of cell counting and cell viability, the Guava PCA cell analysis system has become a vital part of a standard operating procedure that is operator independent, especially when using the automated software Express™. Standard easy-to-implement protocols are available from the Guava website that result in a simple means of assessing and tracking cell health and protein expression over multiple cell lines and HCS/HTS campaigns (Figure 18.1 and Figure 18.2). This effort then accounts for standardization in cell number and ensures accurate plating efficiencies to maintain daily and weekly reproducibility in cell plates. Image analysis algorithms in use by the various HCS systems can be quite influenced by cell number, passage number, and cell density. All of these factors ultimately affect the cellular phenotype(s). These factors may also influence the cellular response to stimulation, particularly in areas of receptor activation and internalization. The under- or overestimate when quantifying a particular cellular function such as protein translocation can be detrimental to the identification of lead compound candidates, especially those with weaker functional activities. In some cases, the weaker compound activities may not be detected.

Nunc cell factories (Nunc, Thousand Oaks, CA) have been used in our laboratories to scale up the processing and production for cell-based screens in 384-well microtiter plates. Factory instruction guidelines are supplied for stepwise description of the use of these systems for maintaining and plating various cell types. In our format, we initiate one cell factory consisting of one monolayer of growth area, 632 cm^2 with 5×10^6 U-2 OS cells for propagation over a 3-day period in which a common harvest of 3×10^7 cells is achieved. This quantity of cells is sufficient to plate approx-

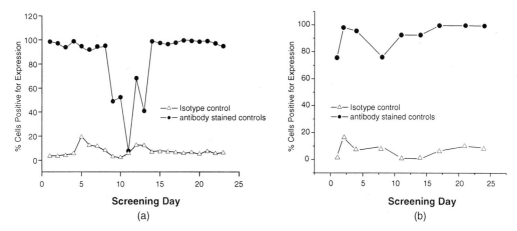

FIGURE 18.2 GPCR protein expression on the cell surface is monitored by quantifying the HA tag throughout two Norak Transfluor HCS/HTS screening campaigns. In (a) a dramatic decrease in the relative receptor expression was seen on days 9 to 13, indicating a problem with the cell passage, which was corrected on days 14 to 23. In (b), the receptor expression remained high on all screening days.

imately 20 384-well microtiter plates at a density of 4000 cells per well. One advantage of this system is that it is easily scaleable based upon the ratio of cell number seeded to the square centimeter area of the growth surface. Optimized growth for various cell types as well as the duration of propagation vary with cell type [7]. By using one single monolayer factory in combination with additional multilayered factories, the single factory can be used as a representative to microscopically monitor the extent of confluence. Thus, the conditions of growth on a per cell basis are effectively indistinguishable. Furthermore, the amount of effort devoted to harvesting cells decreases and a time savings is realized by quickly moving the cells into the microtiter plates. In this manner, we are able to confidently query cells in assays that measure translocation, cytotoxicity, and receptor internalization under conditions where the cells in all wells, in all microtiter plates, have been treated identically. The ability to perform cell-based screens in sets of 60 to 120 plates per day is successful when each plate is virtually identical with respect to cell number per well combined with the capacity to achieve a uniform biological cell response.

Typical plating of cells using instruments such as the Multidrop from Titertek demands routine calibration and care, as do all of the liquid handling instruments involved in HCS. Issues such as drifting of cell number per plate and clogging of tips may result in variations across the plate which impact upon the ability to scale up cell screens. Additionally, these issues impact data processing and can be identified subsequently as one reviews the plate-to-plate results in a number of visualization tools such as Spotfire.

18.5   A STEPWISE APPROACH TO EXECUTING HCS ASSAYS

- Define the cell type needed.
  - Identify which cell types contain the following items of interest: the protein being queried, the relevant signaling pathways, or the specific feature parameters that address the questions being asked of a particular HCS assay. Subsequently, one should evaluate a variety of these identified cell types to optimize the range of responses and choose accordingly.
  - Define the structure(s) or organelle(s) needed to evaluate the images for results.
  - Degree of translocation, physiological conditions, or is a reporter system needed?
- Determine time of compound preincubation and time of functional response.

- Set compound concentrations for cell exposure.
- Determine whether cell density influences the biological activation.
- Determine whether cell density influences the image analysis results.
- Define the HCS platform.
  - Determine which HCS platform meets the performance criteria needed for the HCS application being used, including image acquisition and processing times. As each platform has distinct specifications, define the size of the field of view, the objective and magnification, the numerical aperture, the degree of resolution and clarity required, the need for a confocal or nonconfocal platform, and appropriate filter sets, light source(s), laser(s), and lamp.
- Identify appropriate microtiter plate type.
  - Define whether the cell type of choice requires a particular extracellular matrix or adheres sufficiently to typical tissue culture treatments. Each HCS platform supports specific microtiter plate types based upon the differences in the physical instrumentation. Suppliers of these platforms will recommend compatible plate types.
- Establish which potential fluorophores are to be used.
  - Use minimally nonoverlapping excitation/emission fluorophores and optimize exposure times.
- Determine the detection system.
  - Antibodies, primary and secondary, cross-reactivity of antibodies, dyes, fluorophores, staining influences, staining of live cells or fixed cells.
- Establish parameters for statistical relevance.
  - Define the number of standards/controls, relative response per cell number per time and on a day-to-day basis, duplicates, triplicates, or more reference compounds.
  - Define the interassay and intraassay variability of the controls.
- Define which algorithm fit appropriately quantifies the assay.
  - Define the requirements of the HCS platform, exactness of the quantification, relative or abstract nature of the parameters within the cellular context, how nuclear borders and areas are defined, and whether it is by staining with Hoechst 33258 or other methods.
- Ascertain what quantitative results are necessary.
  - Define the functional results relative to the number of cells analyzed, the number of image fields analyzed, or the number of replicate wells analyzed. If replicates are to be used define that the output parameter(s) is an average of the number of wells evaluated.
    - Define whether there is classification of cells within the image field(s) to separate individual cell data into responding vs. nonresponding subpopulations.
    - Define the cellular responses that may be due to toxicity of the treatment as opposed to the cellular functional effect queried.
    - Utilize visualization software programs for data analysis and interpretation such as Spotfire DecisionSite (Spotfire, Inc.).
- Present data to colleagues.
  - Preempt questions involving the relevance of the multiparameter readouts by explaining the relevance of each individual measurement with images as visualization tools.
  - Define the underlying checks and assessments of aggregate well data outputs (i.e., for translocation assays, the cell number exceeds 300 cells per well, otherwise data are not reported).
  - Summarize the data in light of known reference compounds and literature results.
  - Along with presentation of numerical outputs, provide examples of the images with and without algorithm masking.
  - Define the known interferences of the assays and the accounting of possible variability.

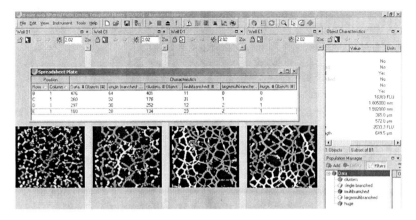

FIGURE 18.3 Screen shot of the Acumen Explorer HTS software when analyzing the endothelial tube formation assay. The data from four wells is shown, where HUVEC cells have been exposed to compound in descending concentrations (left to right). On the graphic user interface the visualizations of the pseudocolor images, the user defined object characteristics, the population settings, and the resultant spreadsheet data can be seen in a single screen.

18.6 EXAMPLES OF HCS ASSAYS

18.6.1 TUBE FORMATION ASSAY (TFA) USING THE ACUMEN EXPLORER HTS SYSTEM

Tube formation assays using human umbilical vein endothelial cells (HUVEC) have been used to assess the effect of compounds/agents on the angiogenic process. Typically, HUVEC cells are grown on a collagen-coated matrix in which cells reconstruct their morphology into lattice-like networks appearing as interconnected tube-like structures [8]. In contrast, HUVEC cells grown in the absence of the collagen extracellular matrix appear as a typical fibroblast-like confluent monolayer. The degree to which a compound may influence this process is an indication of its antiangiogenic potential. The quantification of this inhibition is determined by the number of cells forming networks, their interconnectedness, and the number of connection points, axis connections, and cell thickness of the tube structures. All or any of these parameters have been used to traditionally measure the extent of compound effect. The standard measurement of these parameters has been done on a field-by-field basis through manual curation, which is labor intensive. HCS platforms offer some if not all of these measurements as possible output results using automated and user-defined image analysis tools. In our laboratory, we have set up a rapid HCS method on the Acumen Explorer laser line scanning instrument using HUVEC cells by comparing untreated and compound-treated cells and measuring the degree of cellular interconnectedness (Figure 18.3). This HCS platform allows us the flexibility to use 24- and 48-well microtiter plates. Other HCS platforms are enabled for 6-, 24-, and 48-well plates offering additional assay development flexibility. To measure the extent of connectedness, the wells are treated with 1 μM calcein to fluorescently detect cells for identification. Use of precoated collagen plates from various suppliers has assisted in optimizing the process, and HUVEC cells are commercially available from multiple sources.

- Materials
  - BD Matrigel Matrix, phenol red–free (BD Biosciences)
  - Growth Factor Reduced (GFR) BD Matrigel Matrix, phenol red–free (BD Biosciences)
  - HUVECs
  - EGM bullet kit (Clonetics)

- MEM199 + 5% Fetal Bovine Serum (FBS)
- Cell culture plate (24- or 48-well)
- Needle and syringe (18G;1 mL)
- Calcein AM (Molecular Probes)
- Procedure
  - Thaw GFR Matrigel at 4°C overnight.
  - Using a 1-mL syringe with 18G needle, add 150 μL of GFR Matrigel/well onto a 48-well tissue culture plate (or 300 μL to a 24-well plate) and allow to solidify for 1 h at 37°C (incubator). *Note*: Start next step ~15 min before hour is up.
  - Trypsinize and resuspend HUVECs in MEM199 + 5% FBS. *Note*: The cells are grown in full EGM growth medium on Matrigel Matrix.
  - Centrifuge, resuspend (MEM199 + 5% FBS) and plate cells (250 μL at 2×10^5 cells/mL so that each well gets 5×10^4 cells/well) onto the solidified GFR Matrigel. *Note*: Double this for a 24-well plate (i.e., 500 μL or 1×10^5 cells/well) and plate no later than passage six.
  - If drugs are to be added to the cells, this is the point when they need to be added.
  - Mix gently so that the cells will be evenly distributed, and place in 37°C incubator.
  - Label the cells with a 1 μM calcien AM 30 min prior to imaging the cells.
  - Observe and take images at 24 or 48 h.

18.6.2 NUCLEAR TRANSLOCATION ASSAY

The stimulation of cells results in the initiation of numerous cascading downstream signaling responses that may include the activation of kinases, the redistribution of proteins to and from various organelles, and the translocation of transcription factors. Likewise, the activation of myristylation, farnesylation, and palmitylation serves to direct cytosolic proteins to other particular scaffolding proteins and membranes, plasma membranes, and nuclear membranes. Posttranscriptional modifications afford proteins new functions at specific intracellular sites. These cellular responses are typically the result of multiple protein and receptor interactions, which are completed over minutes to hours resulting in cellular physiological responses. We measure these parameters as a follow-up to many of our G-protein–coupled receptor and kinase drug targets as indications of efficacy. Thus we describe these translocation assays as platform assays, where they are used to evaluate compounds from multiple projects.

In setting up translocation assays, objectives should be identified to find the correct time course of the cellular events. The extent of translocation for any given protein in any given cell type needs to be empirically determined and may vary substantially (Figure 18.4). The example described below is a typical extracellular signal regulated kinase translocation assay, commonly referred to as an ERK translocation assay [9].

- Materials
  - HeLa cells
  - Growth media: Dulbecco's modified eagle media (DMEM), 10% heat-inactivated FBS, penicillin, streptomycin, L-glutamine
  - Fatty acid–free bovine serum albumin (BSA)
  - Hanks balanced salt solution (HBSS) containing 20 mM HEPES and 0.05% BSA
  - Phorbol myristic acetate (PMA)
  - Formaldehyde in phosphate-buffered saline (3%)
  - ERK translocation Hit-Kit assay (Cellomics)
- Procedure
  - Plate HeLa cells 5000 cells/well in growth medium.
  - Incubate cells overnight at 37°C.

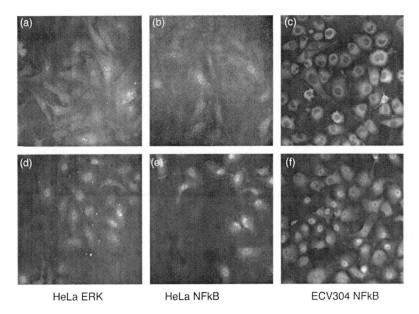

HeLa ERK HeLa NFkB ECV304 NFkB

FIGURE 18.4 Representative images of cytoplasm to nuclear translocations of ERK and NFkB collected on the nonconfocal ArrayScan II using the 3.1 Nuclear Translocation BioApplication (Cellomics, Inc.). Paired images are shown of the basal (a through c) and activated (d through f) states of the cells maximally stimulated with either TNFα (30 ng/mL) or PMA (100 ng/mL). The antibody-labeled redistribution of ERK in HeLa cells (a and d), NFkB in HeLa cells (b and e), and NFkB in ECV304 cells (c and f) is shown.

- Remove the medium and add 60 μL of 0.05% BSA in HBSS/HEPES.
- Add 20 μL of an appropriate concentration of each compound to the cell plate.
- Incubate cells with compounds for 30 min at 37°C.
- Add 20 μL of PMA at a final concentration of 100 ng/mL in the well or alternative stimulant to the plate and incubate for 30 min at 37°C.
- Remove the solution, and fix the cells using 3% formaldehyde in phosphate-buffered saline.
- Perform the ERK translocation Hit-Kit assay (Cellomics). (A full description of the materials and methods accompanies each kit for ease of use. The HitKit relies on antibody detection; thus, a permeabilization step is required for penetration of the antibodies within the cell as well as blocking reagents to minimize nonspecific binding events. The assay kit consists of an unlabeled ERK primary antibody and an Alexa Fluor® 488 conjugated secondary antibody.)

18.6.3 GPCR INTERNALIZATION ASSAY

G-protein–coupled receptors (GPCRs) have been and remain one of the most attractive drug targets. Many hormones and neurotransmitters exert their effects on cells via this family, also referred to as seven transmembrane-spanning proteins (7TM), through interaction at the cell surface. The development of various drugs that modulate, enhance, and activate these processes has proven value as therapeutics for a multiplicity of disease states. Once a primary interaction occurs between a GPCR and ligand (potentially "drug") a cascading sequence of events is triggered, which results in the activation of various signaling events and intracellular end products. These end points include changes in enzymatic activities, biochemical as well as cellular physiological events. The action of the ligand binding to a 7-transmembrane receptor causes a conformational change in the receptor and subsequent conversion of the guanine triphosphate (GTP) to guanine diphosphate (GDP) on

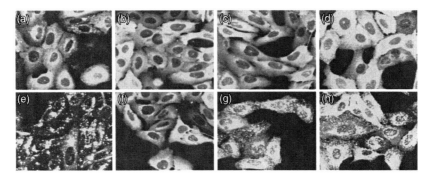

FIGURE 18.5 Confocal InCell 3000 images (GE Healthcare) of the resultant phenotypes of four different Norak Transfluor GPCR cell lines in the absence (a through d) and presence (e through h) of baculovirus administered constitutively active GRK, i.e., the LITe assay. Paired images of control and GRK-stimulated cells for oGPCR #1 (a and e), oGPCR #2 (b and f), oGPCR #3 (c and g), or oGPCR #4 (d and h) are shown after the overnight transient transfection of GRK via the BacMam system

its associated heterotrimeric G protein, whether that be Gi, Go, Gs, or Gq. As a result, the action of G protein–regulated kinases (GRK) occurs at the plasma membrane site, catalyzing the phosphorylation of particular amino acids on the C terminus of the 7TM receptor. A conformational change in the receptor facilitates the recruitment of beta-arrestin molecules from the cytoplasm to associate with the receptor:G protein complex, thus the process of receptor endocytosis begins. Endocytosis of the bound activated receptor into clathrin-coated pits and fully internalized vesicles proceeds, with its exact properties depending upon the degree of receptor phosphorylation and the affinity of the beta-arrestin for the phosphorylated sites on the receptor. The receptor can then recycle. In all cases this process, also known as "desensitization," returns the receptor to its initial preactivated or prestimulated equilibrium state over the course of hours [10].

Norak Biosciences (North Carolina) provides a system to screen 7TM receptors. The advantage of screening a 7TM receptor in the Norak Transfluor cell line is that the Transfluor U 2-OS cell line is stably transfected with a β-arrestin (arrestin2) conjugated to a green fluorescent protein (GFP). Receptor:agonist binding results in the β-arrestin migrating to the receptor of interest, and the redistribution of the fluorescently tagged complex can be fully assessed and quantified by the imaging technologies of HCS. The universal use of this technology for measuring receptor internalization has been demonstrated by Norak Biosciences for known and orphan GPCRs activated by various ligands (small molecules, proteins, peptides, or lipids [3,11]). We and others have corroborated this technology to be highly reproducible with a high signal-to-noise ratio and low background noise over a number of GPCRs. Norak has also developed a proprietary technique that we have found to be most helpful to standardize our efforts for defining potential orphan GPCR ligands when assessing receptor internalization using the Transfluor cells. This technique, called the LITe™ Assay, is an agonist-independent assay used to verify the transduction of β-arrestin-GFP to orphan GPCRs [11]. By transfecting cells with the ligand-independent translocation effector, a constitutively active GRK, it affords the experimentalist the ability to reproducibly initiate the orphan GPCR internalization process in the Transfluor-enabled cell line. This process then proceeds without the necessity of having a known ligand and may be used to define the phenotype that is associated with the ligand-activated internalization of the orphan receptor, whether it is endocytic pits, vesicles, or both (Figure 18.5).

- Materials
  - Norak Transfluor cell line containing a stably over-expressed orphan or known GPCR
  - Replacement buffer: modified eagle media (MEM) (no phenol red) + 10 mL HEPES per 500 mL (20 mM HEPES)
  - HBSS/HEPES buffer: 20 mL HEPES per liter (20 mM HEPES)

- – HBSS/HEPES/DMSO control buffer: 20 mL HEPES and 40 mL DMSO per liter
- – Fix: 500 mL HBSS/HEPES + 29 mL formaldehyde
- • Procedure
  - – Prepare 20 384-well microtiter plates containing 4000 Transfluor cells per well one day prior to assay.
  - – On the day of assay warm the replacement buffer and fix solutions to 37°C.
  - – Defrost compound plates (single-use plates of 2 μL of compound in DMSO).
  - – Add 50 μL HBSS/HEPES buffer to columns 5 to 24 (containing compound stocks).
  - – Add 50 μL HBSS/HEPES/DMSO control buffer to the empty columns 1 to 4.
  - – Initialize the Quadra (Tomtec Incorporated) and prepare water and waste reservoirs.
  - – By hand, flick off growth medium from the 20 microtiter plates containing cells, and add 25 μL replacement buffer to each well using a stackable Multidrop (this needs to be done quickly to avoid drying of the wells).
  - – Stack 20 compound plates (position 1) and 20 cell plates (position 3) on the Quadra. Fill DMSO reservoir, position 4 (for washing tips). *Note:* Stack with plate #s going high to low from bottom to top (i.e., plate #1 at bottom of stack).
  - – Run a 20-loop program that transfers 10 μL of control buffer or prediluted compound from one 384-well compound plate into one 384-well cell plate, rinse the Quadra tips with DMSO, rewash the tips three times with water rinses and sonication, and loop to continuously repeat the process for 20 compound and matched cell plates.
  - – When transfer of first 20 plates is complete (~40 min), remove cell and compound plates from stackers.
  - – Re-sort first 20 cell plates (plate #1 at bottom) in preparation for fixation.
  - – Stack 20 cell plates into position 3 (plate #1 at bottom). Switch reservoirs in position 4. Fill reservoir with warmed fix solution.
  - – Run "Add fix to cell plate" program on 20 cell plates.

18.7 QUALITY CONTROLS AND STANDARDS FOR HCS

The standardization to determine reproducibility of valid data for HCS can be addressed in relative or absolute terms. Typically for HCS/HTS, "relative terms" indicates a comparison to known controls or basal conditions that are sufficient to evaluate consistency of cellular responses on a daily basis. The quest to quantify data in absolute terms involves physical/chemical measurements including laser output efficiencies, photons per fluorophore, and accounting for spectral drift, as well as the daily fluctuation due to electrical conditions, charge-coupled device (CCD) cooling efficiencies, and quantum efficiencies within the cellular system. Nevertheless, the general HCS community has not yet called on these rigorous calibration metrics to define reference conditions other than to adhere to quantifying known relative cellular functional responses in cells. Future screens will likely determine the value of these calibrations and physical measurements.

As is standard at present, the relative signal fluorescence from a cell or the combination of all cells within a well that are untreated is typically compared to the signal fluorescence from a single cell or combined whole well response that has been activated or stimulated to undergo some relevant change. In many cases, commercially available compounds considered literature standards with respect to the known biology that is elicited are used to define the stimulated or activated condition. In an example of activating cells for a nuclear translocation event to measure ERK, there are many widely used positive controls including phorbol 12-myristate 13-acetate (PMA) and tumor necrosis factor-alpha (TNF-α). In other translocation biologies, certain compounds that block internalization are also used as surrogates and compound references (phenylarsine oxide, nonhydrolyzable GTP analogues, etc.). One key issue with respect to standardizing the output of HCS is the degree of variability within a well over a given number of cells. By evaluating the standard deviation of each of the feature parameters one can make an assessment of the degree of variability and the overlap

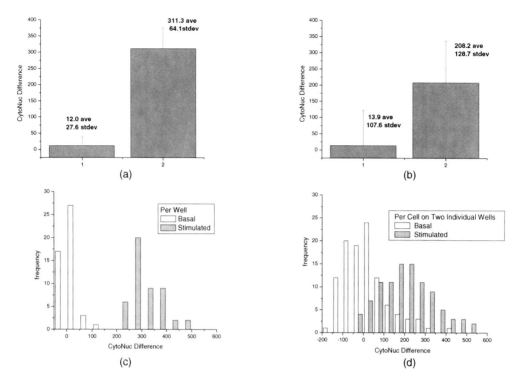

FIGURE 18.6 Graphs of typical cellular responses to translocation events are shown on a well basis and a single cell basis: (a) analysis of the average cytonuclear difference (translocation) of phosphoryated ERK from eight control wells and eight stimulated wells (average control 12.0 ± 27.6 SD, average stimulation 311.3 ± 64.1 SD); (b) frequency histogram analysis of the average cytonuclear differences of the well data described above; (c) analysis of the cytonuclear difference (translocation) of phosphorylated ERK from individual cells from a basal well and a stimulated well (average basal well of 106 cells 13.9 ± 107.6 SD, average stimuated well of 96 cells 208.2 ± 128.7 SD); and (d) single cell frequency histogram analysis of all of the cells from the two wells described above indicating the clear overlap of cellular responses.

between, for example, ERK translocating and ERK nontranslocating cells (Figure 18.6). Another useful method is to plot frequency histograms of the feature parameters of interest and establish the number of overlapping cells that display an intermediate response (Figure 18.7). In this manner, the experimentalist can define the degree of response on a per cell basis or the degree of response on a per well basis that is "necessary" to confirm a change from the basal or unstimulated population. In all cases, it is helpful to utilize methods such as signal-to-background ratio and z' factor [12] to evaluate the consistency of the HCS experiments (Figure 18.8).

An important quality control issue for the cell biologist is to identify and define the measurable biological features and parameters that the individual HCS platforms quantify. This is of particular importance for those entering the area of HCS. Since these numerical results are the outcome of the experiment, it is essential that they correlate well with the known biology to successfully integrate HCS into the workflow. Each HCS platform may identify "nuclear and cytoplasmic areas" for the determination of nuclear translocation events; however, in understanding the algorithms and the definitions one will find that for the different platforms each may be defined differently. For that reason, not all of the HCS platforms display equivalence in results, due to the variation in image analysis processing even though it may be the same biology (or for that matter, the same sample, Figure 18.9). Therefore, it is imperative that an understanding of the relevance of the numerical output of HCS and the biology be understood by the high-content experimentalist to resolve concerns and misunderstandings.

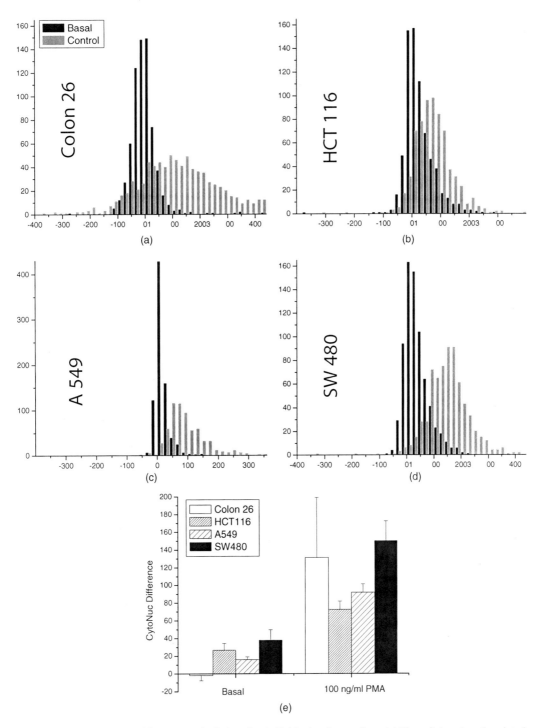

FIGURE 18.7 Frequency histograms depicting the individual cell-to-cell variability of the phosphorylated ERK translocation in basal and PMA-stimulated wells over a range of cell types: (a through d) Colon 26, HCT116, A 549, and SW480 cells were quantified for single-cell cytonuclear differences on the ArrayScanII using the Nuclear Translocation Bioapplication after a 30 min 100 ng/mL PMA stimulation and (e) the well averages of the cytonuclear difference for each treatment are plotted (n = 8 per treatment).

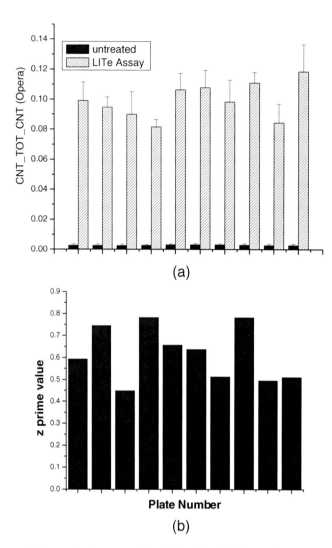

FIGURE 18.8 Bar graphs illustrating the comparison of signal and background responses and z′ factor analysis for 10 screening plates containing basal or background response wells (n = 60) and LITe assay or stimulated wells (n = 4) as evaluated on the EvoTec Opera confocal HCS instrument: (a) Norak Transfluor cells with a stably integrated GPCR are monitored for pit/vesicle signal (LITe assay) and background response (buffer addition) using the EvoTec GPCR Algorithm quantifying the fraction of the GFP signal in the detected spots (referred to as counts from centers per total counts) and (b) the same 10 plates are quantified for z′ for comparison.

18.8 SUMMARY

In HCS, as is in all types of assay development, it is important to perform multiple evaluations using repeats to evaluate the results from the range of quantitative markers recorded for any given biology measured. HCS by definition is not a single parameter analysis but is multiparameter by nature as it collects numerous parameters as output measurements. There are two methods of viewing the multi-parameter measurements:

- Data that determine a pass/fail result
- Data that classify the results based upon measured thresholds or values

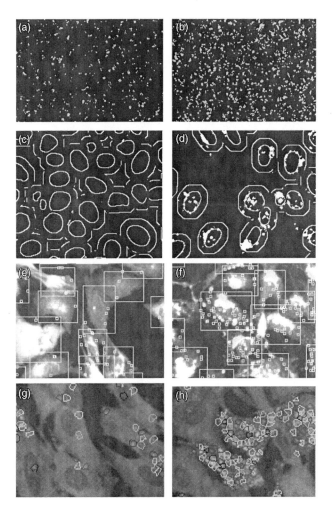

FIGURE 18.9 (See color insert following page 334.) Comparison of image analysis processing of Transfluor GPCR receptor internalization defining the masking of basal and LITe activated cell controls: (a and b) objects defined on the laser line scanner, the Acumen Explorer HTS: (c and d) spots and phase detection defined on the nonconfocal ArrayScanII instrument using the GPCR Bioapplication; (e and f) F-grains defined on the line-scanning confocal IN Cell 3000 from GE Healthcare using the Granularity module: and (g and h) counts from centers per total counts defined on the integrated Nipkow disc-based confocal EvoTec Opera system using the GPCR algorithm.

For the latter method, one effective method to classify or consolidate HCS data is to perform weighted or sequentially weighted analyses of the individual features and parameter results to obtain a single assessment of activity or a single output. A weighted analysis can take into account key influencing factors such as cell number, interfering fluorescence due to compound treatment, and the measurement of nuclear area, which decreases significantly when cells are exposed to cytotoxic compounds. This weighted method of evaluation then "qualifies" the final outcome of treatment as "appropriately active," "interfering," or "negative" by using a single output result. It also enables the classification of results, similar to "fingerprinting" the cellular response, as is performed when interpreting current microarray data.

Ultimately, it is the interpretation of the aggregate analysis of both the individual cell and well data that offers HCS the ability to evaluate many biological outcomes. The power to define responding and nonresponding populations by one or more parameters and the definition of classifying cells, evaluating signal magnitude changes, and assessing the time of biological response is achievable within standard operation of High Content Screening.

The true effectiveness of HCS lies in its positioning within the screening/discovery process. The high-fidelity information it supplies in relative abundance in multiple projects throughout discovery — target identification, screening, lead optimization, and specialized assays, cannot be supplied by other means. As a follow-up to a biochemical screening campaign, it adds functional cellular data, an estimate of compound permeability, and preliminary cellular toxicity data. When using HCS techniques for an HTS campaign, the advantages include the ability to follow up redundant signal transduction pathways within discrete cellular systems and further deconvolute the actions of compounds. Examples include determining compound effects on calcium pathways, cAMP production, Rac, Rho, f-actin polymerization, and cytoskeletal changes, including effects on activation and mobilization of transcription factors. Further downstream in drug discovery the profiling of multiple cell types with respect to function or cellular toxicity can now be automated and defined.

REFERENCES

1. Abraham VC, Taylor DL, Haskins JR. High content screening applied to large-scale cell biology. *Trends in Biotechnology* 2004; 22(1): 15–22.

2. Milligan, G. High-content assays for ligand regulation of G-protein-coupled receptors. *Drug Discovery Today* 2003; 8(13): 579–585.

3. Oakley RH, Hudson CC, Cruickshank RD, Meyers DM, Payne RE Jr, Rhem SM, Loomis CR. The cellular distribution of fluorescently labeled arrestins provides a robust, sensitive, and universal assay for screening G protein-coupled receptors. *Assay and Drug Development Technologies* 2002; 1(1–1): 21–30.

4. Allan V. *Protein Localization by Fluorescent Microscopy: A Practical Approach*, Oxford: Oxford University Press, 2000.

5. Pawley JB. *Handbook of Biological Confocal Microscopy*, 2d ed. New York: Plenum Publishing, 1995.

6. Lanni F, Wilson T. Grating image systems for optical sectioning fluorescent microscopy of cell tissues and small organisms. In: Yuste R, Lanni F, Konnerth A, eds. *Imaging Neurons: A Laboratory Manual*. New York: Cold Spring Harbor Laboratory Press, 2000:8.1–8.11.

7. Berger TG, Feuerstein B, Strasser E, Hirsch U, Schreiner D, Schuler G, Schuler-Thurner B. Large-scale generation of mature monocyte-derived dendritic cells for clinical application in cell factories. *Journal of Immunological Methods* 2002; 268(2): 131–140.

8. Yang S, Xin X, Zlot C, Ingle G, Fuh G, Li B, Moffat B, de Vos AM, Gerritsen ME. Vascular endothelial cell growth factor-driven endothelial tube formation is mediated by vascular endothelial cell growth factor receptor-2, a kinase insert domain-containing receptor. *Arteriosclerosis, Thrombosis, and Vascular Biology* 2001; 21(12): 1934–1940.

9. Horgan AM, Stork PJS. Examining the mechanism of Erk nuclear translocation using green fluorescent protein. *Experimental Cell Research* 2003; 285(2): 208–220.

10. Zhang J, Ferguson SSG, Barak LS, Aber MJ, Giros B, Lefkowitz RJ, Caron MG. Molecular mechanisms of G protein-coupled receptor signaling: role of G protein-coupled receptor kinases and arrestins in receptor desensitization and resensitization. *Receptors and Channels* 1997; 5(3–4):193–199.

11. Oakley RH, Cowan CL, Hudson CC, Loomis CR.: Transfluor® provides a universal cell-based assay for screening G protein-coupled receptors. *Handbook of Assay Development in Drug Discovery*, ed. L. Minor, in press, Taylor & Francis, 2005.

12. Zhang J, Chung TDY, Oldenburg KR. A simple statistical parameter for use in evaluation and validation of high throughput screening assays, *Journal of Biomolecular Screening* 1999; 4(2):67–73.

19 The Evolution of cAMP Assays

Patricia Kasila and Harry Harney

CONTENTS

19.1 INTRODUCTION

In 1957, adenosine $3',5'$ cyclic monophosphate (cAMP) was discovered by Earl Sutherland, who was awarded a Nobel Prize for his discovery (1971). This discovery of cAMP and adenylyl cyclase, the hormone-sensitive enzyme that synthesizes the cyclic nucleotide from ATP, led to the concepts of transmembrane signaling and of hormone-regulated synthesis of intracellular second messengers.

cAMP plays a critical role in the transmission of signals by functioning as a "second messenger" [1,2]. The binding of a hormone to its receptor can either enhance or inhibit the rate at which cAMP is produced. This is accomplished by altering the enzymatic activity of adenylyl cyclase [3]. By this mechanism, intracellular levels of cAMP are altered in response to hormonal stimulation. In turn, the intracellular level of cAMP regulates the enzymatic activity of a protein kinase, which phosphorylates other substances, setting off a cascade of cellular events, which leads to the expression of the hormones [4].

The purpose of this chapter is to discuss the evolution of cAMP assays — what technologies have been used since discovery and how the assays have evolved into homogeneous assays useful for high-throughput screening.

19.2 IN THE BEGINNING

In the 1960s, cAMP assays were performed by prelabeling the adenine nucleotides with a radioactive tag ($[^3H]$-adenine), incubating with hormones, and performing a separation on Dowex columns. The procedure was very tedious and time consuming. As time and technology progressed, radio-immunoassay and reporter gene assays became popular. Reporter gene assays are easily detected using conventional luminescence or fluorescence. However, these types of assays have the disadvantage of measuring downstream events requiring increased stimulation time, resulting in a higher rate of false positives.

For many years radioimmunoassays were used as the method of choice for measuring cAMP. Radioimmunoassay directly detects cAMP by competing a radiolabeled cAMP with cAMP from serum, urine, or tissue extracts for a limited amount of cAMP antibody. Although radioimmunoassays were a significant advantage over previous assays, there arose a need to develop functional cAMP assays to detect G-protein–coupled receptor (GPCR) stimulation in whole cells or membranes. Along with the need for functional cAMP assays, advances in cell culture techniques enabled researchers to develop stable cell lines expressing recombinant, highly overexpressed GPCRs.

19.3 TODAY

Most cAMP functional assays are designed to directly measure levels of cAMP produced upon modulation of adenylyl cyclase activity by GPCRs. The assays are based on the competition between cAMP produced by cells and exogenously added labeled-cAMP. The capture of cAMP is achieved using specific anti-cAMP antibodies that are coated, conjugated, or labeled. Most assay formats allow the measurement of both agonist and antagonist activity on $G\alpha_s$- and $G\alpha_i$-coupled receptors in whole cells and $G\alpha_s$-coupled receptors in membrane-based assays.

In the early 1990s radioimmunoassays were converted to scintillation proximity assays for higher throughput. Cellular assays were performed by seeding cells into tissue culture plates, incubating overnight, stimulating, lysing, and transferring the sample to the detection plate. In 1998, a patent was issued for the first fully homogeneous cAMP assay, which was developed for high throughput screening. Today, many assays, both radioactive and nonradioactive, exist that are highly suited for screening.

19.4 ASSAY CONSIDERATIONS

Depending on the receptor of interest and whether it is $G\alpha_s$ or $G\alpha_i$ coupled, there are several different ways to stimulate or inhibit cAMP production:

- *In the presence of forskolin.* Forskolin acts directly on the adenylyl cyclase to produce cAMP. As a result, more intracellular cAMP is present to compete with the exogenously labeled cAMP.
- *In the presence of an agonist ($G\alpha_s$-coupled receptor).* An agonist will stimulate the production of cAMP. As a result, more intracellular cAMP is present to compete with the exogenously added labeled cAMP.
- *In the presence of an antagonist ($G\alpha_s$-coupled receptor).* Cells or membranes are stimulated to produce intracellular cAMP with an agonist. An antagonist will inhibit agonist induced cAMP production. As a result, less intracellular cAMP is present to compete with the exogenously added labeled cAMP.
- *In the presence of an agonist ($G\alpha_i$-coupled receptor).* Cells are stimulated to produce intracellular cAMP with forskolin. An agonist will inhibit forskolin-induced cAMP production. As a result, less intracellular cAMP is present to compete with the exogenously added labeled cAMP.
- *In the presence of an antagonist ($G\alpha_i$-coupled receptor).* Cells are stimulated to produce intracellular cAMP with forskolin. The antagonist will block the effect of an agonist acting on the receptor observed by a decrease in the forskolin-induced cAMP production. As a result, the forskolin-induced cAMP production is restored.

Another consideration is whether to use whole cells or membranes. Whole cells can be difficult or easy to work with depending on the source. The cell source can be primary or established native or transected, transient or stable, and a human or nonhuman cell line. There are many suppliers of cell lines and membranes. The supplier will typically recommend the best growth conditions, including split ratios and references of where the cells may have been used in similar applications. Points to consider are ease of maintenance, growth rate, easy scaleup, clumping issues, suspension or adherence, and the number of cells needed/assay.

Instead of maintaining cell lines on a routine basis, membrane receptor preparations may be used. The use of cell membrane fragments in a cAMP assay is more technically challenging relative to whole-cell assays because the cell machinery for cAMP production must be reproduced in a membrane preparation. This requires the use of a regeneration buffer. Regeneration buffer containing various components is used to mimic the cell cytosol and allow for the production of cAMP from an agonist-induced activation of GPCRs expressed in cell membrane fragments. Some assay platforms may not be suited for membrane use.

19.5 ASSAY PLATFORM SELECTION

Many types of assay platforms are now available. When choosing the right type of assay platform, the following considerations are necessary:

- Reader availability
- Whole cells or membranes
- Attached cells or suspension cells
- Expression level of receptor
- Sensitivity of the assay
- Cost
- Radioactive or nonradioactive
- Heterogeneous or homogeneous
- Dimethylsulfoxide (DMSO) effects
- Miniaturizable
- Performance and reproducibility

19.6 ASSAY PROTOCOLS AND RESULTS FOR WHOLE CELL ASSAYS

Since most assays now need to be of higher throughput, this section will concentrate on homogeneous assays. Several different assays will be described with direct comparisons and manual comparisons. The assays that will be described are the following:

- PerkinElmer — Adenylyl Cyclase Activation FlashPlate Assay
- PerkinElmer — [FP]^2 cAMPfire Fluorescent Polarization Assay
- PerkinElmer — AlphaScreen cAMP Assay
- PerkinElmer — LANCE cAMP Assay
- CisBio — HTRF® cAMP Dynamic Assay
- CisBio — HTRF cAMP Femtomolar Assay
- Amersham — cAMP Fluorescence Polarization Biotrak Immunoassay System
- DiscoveRx — HitHunter™ cAMP Assay

For each assay, each cell line, and each receptor, assay optimization needs to be done. A cell or membrane titration with stimulation and incubation times should be performed. This was done for each of the assays described here. The studies were done using whole cells with overexpressed β2 adrenergic receptors.

The description and protocol for each assay platform were obtained directly from the product insert. This section is intended as a brief description of each protocol; it should not be used to perform the assays. Refer to the product inserts for complete instructions.

19.6.1 PerkinElmer — Adenylyl Cyclase Activation FlashPlate Assay

The Adenylyl Cyclase Activation FlashPlate Assay allows direct measurement of receptor-mediated adenylyl cyclase activation/inhibition in GPCRs that measure a true second messenger response (cAMP). The protocol is fully homogeneous and can be used with both whole cells and membrane preparations. These relevant functional data permit faster qualification of lead compounds in high-throughput screening than is possible with previously established conventional primary screens. This was the first truly homogeneous cAMP assay (Patent 5,739,001 issued April 14, 1998).

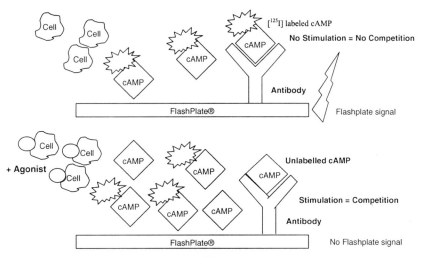

Agonist stimulated cAMP displaces [^{125}I] labeled cAMP from antibody.

FIGURE 19.1 Basic principle of the FlashPlate cAMP assay.

The assay is performed by adding live cells or membranes to the FlashPlate, stimulating (or stimulating and inhibiting), lysing, and detecting without wash steps. Cyclic AMP can be measured in the range of 10 to 1000 n*M*. The assay is accurate over a wide range of values and has a high degree of specificity. The basic principle is detailed in Figure 19.1.

19.6.1.1 Kit Components

The Adenylyl Cyclase Activation FlashPlate Assay (PerkinElmer Catalog #SMP701) kit includes the following components:

- cAMP Standard
- Stimulation Buffer
- [^{125}I]cAMP Tracer
- Detection Buffer
- Adenylyl Cyclase FlashPlates

19.6.1.2 Protocol

The protocol for this assay is as follows:

- cAMP standards or cells diluted in Stimulation Buffer (10 μL).
- Phosphate-buffered saline, or forskolin or agonist/antagonist (cells) (15 μL).
- Incubate for 30 to 60 min at room temperature.
- Detection Mix containing Detection Buffer and [^{125}I]cAMP (25 μL).
- Incubate for 2 h at room temperature.
- Read on a MicroBeta® or TopCount NXT® Microplate Scintillation Counter.

19.6.1.3 Results

See Figure 19.2.

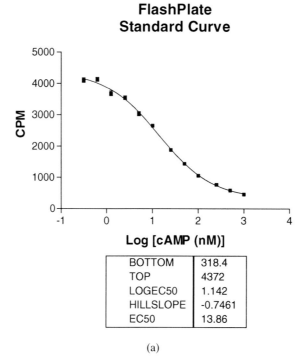

(a)

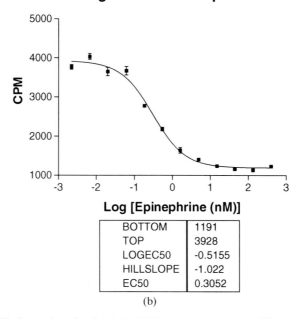

(b)

FIGURE 19.2 (a) A dilution series of unlabeled cAMP competing against [125I] cAMP. (b) Dose–response study with the β2 adrenergic receptor agonist, epinephrine.

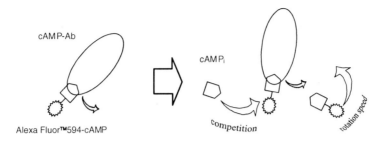

Low level of cAMP$_i$	High level of cAMP$_i$
cAMP-Ab is bound to Alexa Fluor™594-cAMP	Unbound Alexa Fluor™594-cAMP to cAMP-Ab
⇒ Low rotation speed, high polarization signal	⇒ Rapid rotation speed, low polarization signal

FIGURE 19.3 Basic principle of the fluorescence polarization cAMP assay.

19.6.2 PERKINELMER — [FP]² CAMPFIRE FLUORESCENT POLARIZATION ASSAY

This assay kit is designed for use in primary or secondary screening for $G\alpha_s$ and $G\alpha_i$ protein–coupled receptors. The assay quantifies cAMP generated in whole cells and cell membrane fragments following stimulation ($G\alpha_s$) or inhibition ($G\alpha_i$) of adenylate cyclase (AC) activity by a GPCR binding event. Production of cellular cAMP ($G\alpha_s$) or inhibition of chemically stimulated cAMP (using forskolin for $G\alpha_i$) can be quantified by its competition with Alexa Fluor™ 594-cAMP tracer for a limited number of cAMP antibody (cAMP-Ab) binding sites.

Fluorescence polarization is an empirical fluorescence detection technique that measures the parallel and perpendicular components of fluorescence emission generated by plane-polarized excitation light. Polarization values (measured in mP units) for any fluorophore-labeled complex are inversely related to the angular speed of molecular rotation of that complex. Since molecular rotation is, in turn, inversely related to molecular volume, a fluorescent tracer possesses a higher polarization value when it interacts with any molecule large enough to slow its rate of molecular rotation (e.g., an antibody). The magnitude of the polarization signal is thus used to quantitatively determine the extent of fluorescent tracer binding without the need for any filtration, wash, or separation step.

In this kit, quantification is based on the rotational properties of Alexa Fluor 594-cAMP tracer in solution. In the absence of cAMP, a large mole fraction of Alexa Fluor 594-cAMP is bound to the cAMP-Ab. This significantly slows its rotational properties, leading to a large polarization signal. Conversely, in the presence of cAMP, an effective competition can be realized, such that the mole fraction of Alexa Fluor 594-cAMP bound to cAMP-Ab is reduced, lowering the polarization signal. The basic principle is shown in Figure 19.3.

19.6.2.1 Kit Components

The [FP]² cAMPfire Fluorescent Polarization Assay (PerkinElmer Catalog #FPA203) kit includes the following components:

- Stimulation Buffer
- Lyophilized cAMP antibody
- Alexa Fluor 594-cAMP
- Detection Buffer
- cAMP Standard
- Assay Plates RM#1115

19.6.2.2 Protocol

The protocol for this assay is as follows:

- cAMP standards or forskolin or agonist/antagonist (10 μL).
- Cells diluted in cAMP antibody solution or cAMP antibody solution without cells (10 μL) added to the wells containing standards.
- Incubate for 30 to 60 min at room temperature.
- Alexa Fluor 594-cAMP working solution (20 μL).
- Incubate for 60 min at room temperature.
- Read on an appropriate fluorescence polarization reader (VICTOR^{2}V™, EnVision™, ViewLux™, or Fusion™).

19.6.2.3 Results

See Figure 19.4.

19.6.3 PERKINELMER — ALPHASCREEN cAMP ASSAY

The AlphaScreen cAMP assay has been designed to directly measure levels of cAMP produced upon modulation of adenylate cyclase activity by GPCRs. The assay is based on the competition between endogenous cAMP and exogenously added biotinylated cAMP. The capture of cAMP is achieved by using a specific antibody conjugated to acceptor beads. The assay is efficient at measuring both agonist and antagonist activities on Gα_i- and Gα_s-coupled GPCRs. Gα_s and Gα_i subunits act through the cAMP pathway by respectively activating or inhibiting adenylate cyclase, an enzyme catalyzing the conversion of ATP to cAMP. Cells are stimulated to either increase or decrease intracellular cAMP levels, followed by a combined cell lysis/detection step. For Gα_i-coupled receptors, an elevation in intracellular cAMP is stimulated using forskolin, resulting in a decrease in AlphaScreen signal due to an inhibition of association between the beads. However, when forskolin stimulation occurs concurrently with the agonist stimulation of a Gα_i-coupled receptor, the resultant decreased signal evoked by forskolin alone is inhibited, i.e., there is a signal increase relative to the forskolin-alone treatment. The basic principle is seen in Figure 19.5.

19.6.3.1 Kit Components

The PerkinElmer AlphaScreen cAMP Assay (PerkinElmer Catalog # 6760625D) kit includes the following components:

- Biotinylated cAMP
- Donor Streptavidin Beads
- Anti-cAMP Acceptor Beads
- cAMP Standard
- Control Buffer (10×)
- Tween-20 Solution (3%)

19.6.3.2 Protocol

The protocol for this assay is as follows:

- Cells/anti-cAMP Acceptor Beads or anti-cAMP Acceptor Beads only (standards) (5 μL).
- Forskolin or agonist/antagonist or cAMP standards (5 μL).
- Incubate for 30 to 60 min at room temperature.

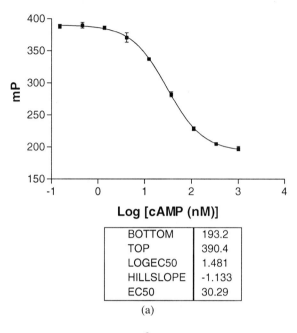

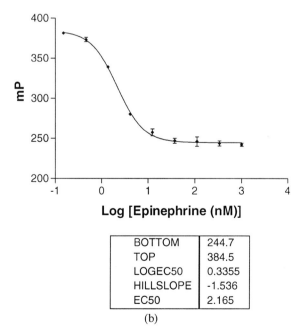

FIGURE 19.4 (a) A dilution series of unlabeled cAMP competing against fluorescent-labeled cAMP. (b) Dose–response study with the β2 adrenergic receptor agonist, epinephrine.

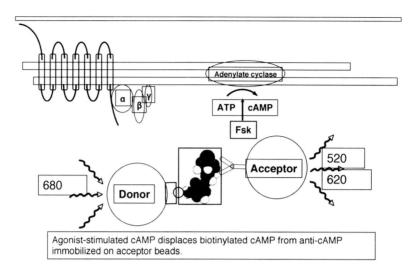

FIGURE 19.5 Basic principle of the AlphaScreen cAMP assay.

- Biotinylated-cAMP/Streptavidin Donor Beads Detection Mix (15 μL).
- Incubate for at least 60 min at room temperature.
- Read on a Fusion-α, AlphaQuest-HTS, or Alpha Envision microplate analyzer.

19.6.3.3 Results

See Figure 19.6.

19.6.4 PERKINELMER — LANCE cAMP ASSAY

Time-resolved fluorescence (TRF) assays exhibit low background and high signal-to-noise ratios, two attributes critical for robust HTS assays. LANCE™ refers to homogeneous TRF applications using techniques such as time-resolved-fluorescence resonance energy transfer (TR-FRET), where energy is transferred from europium to an acceptor, such as Alexa Fluor 647. A competitive binding assay to measure cell-derived cAMP using the LANCE platform has been developed.

The principle of the assay involves the loss of energy transfer as the quantity of cell- or receptor membrane–derived cAMP increases, thereby competing and therefore decreasing the amount of biotinylated-cAMP available to bind to the anti-cAMP antibody. Thus, this assay may be used as an alternative to other currently accepted platforms that measure cAMP and this assay may benefit from the advantages that TRF-based assays offer. The basic protocol is depicted in Figure 19.7.

19.6.4.1 Kit Components

The PerkinElmer LANCE cAMP Assay kit includes the following components:

- Alexa Fluor 647 Labeled Antibody
- Europium-Labeled Streptavidin
- cAMP Standard
- Biotinylated cAMP
- Detection Buffer

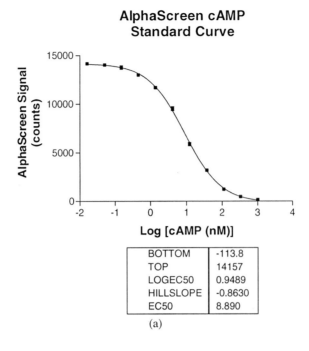

FIGURE 19.6 (a) A dilution series of unlabeled cAMP competing against biotin-labeled cAMP. (b) Dose–response study with the β2 adrenergic receptor agonist, epinephrine.

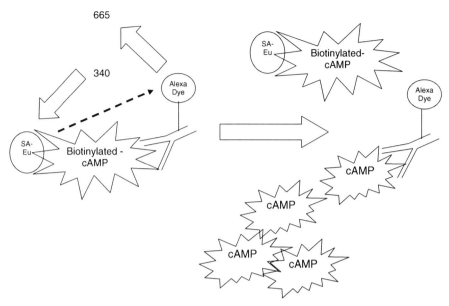

Agonist stimulated cAMP displaces Biotinylated-cAMP from Alexa Fluor labeled antibody

FIGURE 19.7 Basic principle of the LANCE cAMP assay.

19.6.4.2 Protocol

The protocol for this assay is as follows:

- cAMP standards or forskolin or agonist/antagonist (6 μL).
- Cells diluted in Stimulation Buffer containing cAMP antibody or Stimulation Buffer containing cAMP antibody without cells added to the wells containing standards (6 μL).
- Incubate for 30 to 60 min at room temperature.
- Detection Buffer containing biotinylated-cAMP/europium-streptavidin (12 μL).
- Incubate for 60 min at room temperature.
- Read on an appropriate reader.

19.6.4.3 Results

See Figure 19.8.

19.6.5 CisBio — HTRF cAMP Dynamic Assay

This kit is intended for the direct quantitative determination of cyclic AMP. The assay conditions have been optimized in order to reach a high signal-to-noise ratio and to enable cAMP assessment either on suspended or on adherent cells. The specific signal (i.e., energy transfer) is inversely proportional to the concentration of cAMP in the calibrator or sample. As with all other HTRF assays, the calculation of the fluorescence ratio (665 nm/620 nm) eliminates possible photophysical interferences and allows the assay to be unaffected by the usual medium conditions (e.g., culture medium, serum, biotin, colored compounds). The cAMP dynamic kit allows the measurement of agonist and antagonist effects on $G\alpha_s$- and $G\alpha_i$-coupled receptors in different cell lines. The basic assay is shown in Figure 19.9.

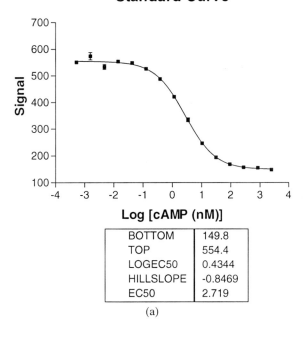

LANCE cAMP
Standard Curve

BOTTOM	149.8
TOP	554.4
LOGEC50	0.4344
HILLSLOPE	-0.8469
EC50	2.719

(a)

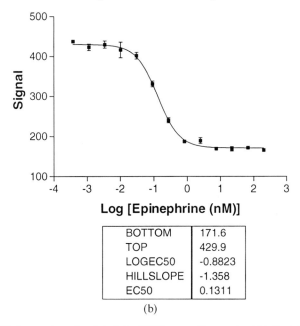

LANCE cAMP
Agonist Dose Response

BOTTOM	171.6
TOP	429.9
LOGEC50	-0.8823
HILLSLOPE	-1.358
EC50	0.1311

(b)

FIGURE 19.8 (a) A dilution series of unlabeled cAMP competing against biotinylated labeled cAMP. (b) Dose–response study with the $\beta 2$ adrenergic receptor agonist, epinephrine.

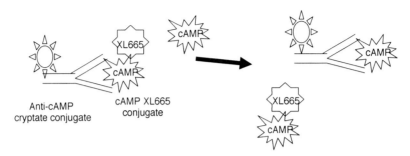

FIGURE 19.9 Basic principle of the CisBio cAMP assay.

19.6.5.1 Kit Components

The CisBio HTRF cAMP dynamic Assay (CisBio Catalog #62AM2PEB) kit includes the following components:

- Anti-cAMP Cryptate
- cAMP-XL665
- cAMP Calibrator
- cAMP Control
- Conjugate and Lysis Buffer
- Diluent

19.6.5.2 Protocol

The protocol for this assay is as follows:

- cAMP calibrators or cells (25 μL).
- Forskolin or agonist/antagonist or diluent to the standards (25 μL).
- Incubate for 30 to 60 min at room temperature.
- cAMP-XL665 (25 μL).
- Anti-cAMP cryptate (25 μL).
- Incubate for 60 min at room temperature.
- Read on an appropriate reader.

19.6.5.3 Results

See Figure 19.10.

19.6.6 CisBio — HTRF cAMP Femtomolar Assay

This kit is intended for the direct quantitative determination of cyclic AMP. The assay conditions have been optimized in order to reach maximum sensitivity (i.e., as low as 80 pM) and to enable cAMP assessment either on suspended or on adherent cells. A number of screens showed that the assay performances could allow the detection of very few cells per well and that the level of sensitivity was particularly adapted when cell production was limiting the process (e.g., transitory transfected cell lines). Its principle is the same as the cAMP dynamic assay (Figure 19.9).

19.6.6.1 Kit Components

The CisBio HTRF cAMP femtomolar Assay (CisBio Catalog #62AM1PEB) includes the following components:

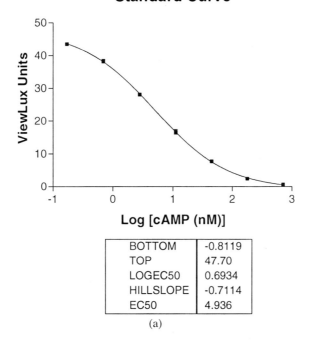

cAMP dynamic Assay
Standard Curve

BOTTOM	-0.8119
TOP	47.70
LOGEC50	0.6934
HILLSLOPE	-0.7114
EC50	4.936

(a)

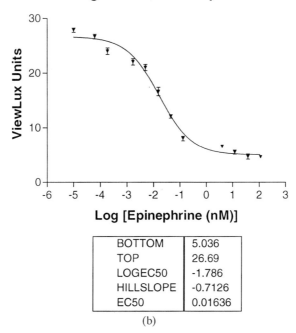

cAMP dynamic Assay
Agonist Dose Response

BOTTOM	5.036
TOP	26.69
LOGEC50	-1.786
HILLSLOPE	-0.7126
EC50	0.01636

(b)

FIGURE 19.10 (a) A dilution series of unlabeled cAMP competing against the XL665 cAMP conjugate. (b) Dose–response study with the β2 adrenergic receptor agonist, epinephrine.

- Anti-cAMP Cryptate
- cAMP-XL665
- cAMP Calibrator
- cAMP Control
- Conjugate and Lysis Buffer
- Diluent

19.6.6.2 Protocol

The protocol for this assay is as follows:

- cAMP calibrators or cells (25 μL).
- Forskolin or agonist/antagonist or diluent to the standards (25 μL).
- Incubate for 30 to 60 min at room temperature.
- cAMP-XL665 (25 μL).
- Anti-cAMP cryptate.
- Incubate for 60 min at room temperature.
- Read on an appropriate reader.

19.6.6.3 Results

See Figure 19.11.

19.6.7 AMERSHAM cAMP FLUORESCENCE POLARIZATION BIOTRAK IMMUNOASSAY SYSTEM

The cAMP Fluorescence Polarization (FP) Biotrak Immunoassay System is a homogeneous method of detecting and quantitating cAMP in a wide variety of samples. The kit is available in a 96- or 384-well format and uses the bright polarization dye, CyTM3B, for fluorescence detection. Cy3B is well suited for fluorescence polarization measurements due to the stability of its signal in a variety of solvent conditions, low nonspecific binding, and large polarization range.

Traditionally, cAMP requires extraction from cell culture prior to assay. However, the cAMP FP Biotrak Immunoassay System includes proprietary lysis reagents that release cAMP from the cell for direct measurement, reducing hands-on time and offering a simple one-stage protocol.

Quantitation of cAMP is measured as competitive displacement of Cy3B-cAMP from anti-cAMP antibody, whereby signal decrease is proportional to cAMP concentration. Known concentrations of cAMP are provided to generate a standard binding curve, and the fluorescence signal is detected as fluorescence polarization. These are:

- Homogeneous: no separation or washing steps.
- Proprietary lysis buffers: Offer one-stage protocol.
- Robust: Cy3B reduces interference from colored compounds, decreasing numbers of false positives.
- Reliable: Good Z' factors.
- Bright fluorescent dye: Cy3B has a high quantum yield and large signal window.
- Choice of protocols: Different sample types can be handled.

The assay is shown in Figure 19.12.

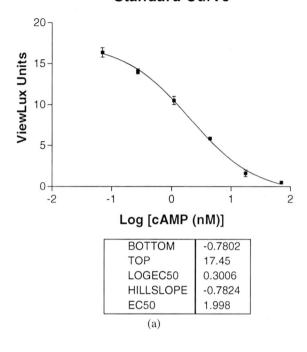

BOTTOM	-0.7802
TOP	17.45
LOGEC50	0.3006
HILLSLOPE	-0.7824
EC50	1.998

(a)

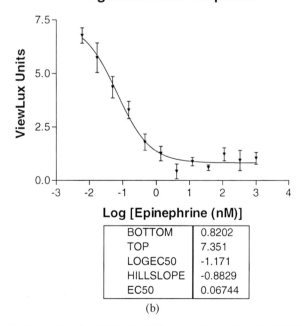

BOTTOM	0.8202
TOP	7.351
LOGEC50	-1.171
HILLSLOPE	-0.8829
EC50	0.06744

(b)

FIGURE 19.11 (a) A dilution series of unlabeled cAMP competing against the XL665 cAMP conjugate. (b) Dose–response study with the $\beta 2$ adrenergic receptor agonist, epinephrine.

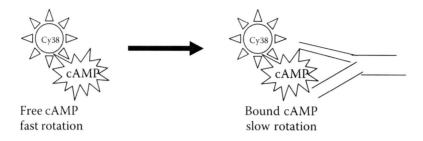

FIGURE 19.12 Basic principle of the Amersham fluorescence polarization cAMP assay.

19.6.7.1 Kit Components

The Amersham cAMP Fluorescence Polarization Biotrak Immunoassay System (Amersham Catalog #RPN3596) kit includes the following components:

- Assay Buffer Concentrate
- cAMP Standard
- Antiserum
- Cy3B Conjugate
- Lysis Reagent 1
- Lysis Reagent 2

19.6.7.2 Protocol

The protocol for this assay is as follows:

- Cells (20 μL).
- Forskolin or agonist/antagonist or diluent (10 μL).
- Incubate for 30 to 60 min at room temperature.
- Lysis reagent 1 (10 μL) and agitate.
- Standard (25 μL).
- Antiserum (10 μL).
- Cy3B Conjugate (10 μL).
- Incubate for 4 h at room temperature.
- Read on an appropriate reader.

19.6.7.3 Results

See Figure 19.13.

19.6.8 DISCOVERX — HITHUNTER CAMP ASSAY

The HitHunter cAMP assay is a homogeneous immunoassay where cAMP from cell lysates competes with ED-cAMP for binding to the anti-cAMP antibody. Upon antibody binding, the ED portion is incapable of association with EA, with the lack of complementation resulting in a low signal. The amount of ED-cAMP available for complementation to EA in the assay is proportional to the concentration of cAMP available to compete for antibody binding. The assay is shown in Figure 19.14.

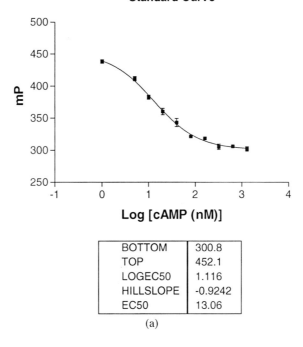

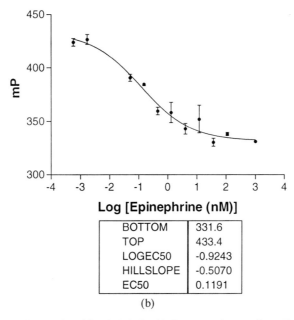

FIGURE 19.13 (a) A dilution series of unlabeled cAMP competing against Cy38-labeled cAMP. (b) Dose–response study with the β2 adrenergic receptor agonist, epinephrine.

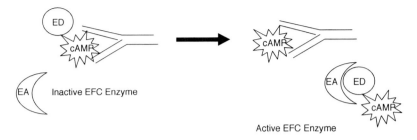

FIGURE 19.14 Basic principle of the DiscoveRx cAMP assay.

19.6.8.1 Kit Components

The DiscoveRx HitHunter cAMP Assay (DiscoveRx Catalog #90-0002) kit includes the following components:

- Lysis Buffer SUS/Ab
- cAMP Standard
- FL Substrate
- BKG Substrate
- ED Reagent
- ED/FL Substrate
- EA Reagent
- EFC Stopping Solution

19.6.8.2 Protocol

The protocol for this assay is as follows:

- Standards (15 μL).
- Cells (10 μL).
- Forskolin or agonist/antagonist to cells (5 μL).
- Incubate at 37°C for 30 min.
- Lysis buffer/antibody (10 μL).
- Incubate for 60 min at room temperature.
- ED/FL substrate (10 μL).
- EA reagent (10 μL).
- EFC stopping solution (10 μL) (optional).
- Read on an appropriate reader.

19.6.8.3 Results

See Figure 19.15.

19.7 ASSAY PROTOCOLS AND RESULTS FOR RECEPTOR MEMBRANE ASSAYS

To address receptor membrane assays, an example is described with protocol and results. The assay platform as described previously is the same with the exception of the protocol change to incorporate the ATP regeneration buffer. For each assay and each receptor membrane, assay optimization needs to be done. A membrane titration with stimulation and incubation times should be performed.

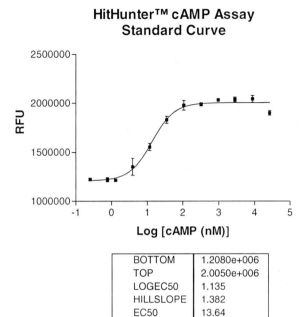

BOTTOM	1.2080e+006
TOP	2.0050e+006
LOGEC50	1.135
HILLSLOPE	1.382
EC50	13.64

(a)

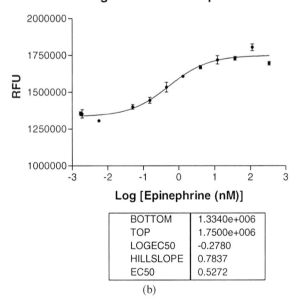

BOTTOM	1.3340e+006
TOP	1.7500e+006
LOGEC50	-0.2780
HILLSLOPE	0.7837
EC50	0.5272

(b)

FIGURE 19.15 (a) A dilution series of unlabeled cAMP competing against ED-labeled cAMP. b) Dose–response study with the β2 adrenergic receptor agonist, epinephrine.

This was done for the assay described here. The studies described here were done using receptor membranes with overexpressed β2 adrenergic receptors (HEK PEAK cell membrane preparation expressing recombinant β2 adrenergic receptors from PerkinElmer Life and Analytical Sciences) (RB-HBE2M, lot 1769).

19.7.1 PERKINELMER — ADENYLYL CYCLASE ACTIVATION FLASHPLATE ASSAY – RECEPTOR MEMBRANES

19.7.1.1 Protocol

Additional buffers needed for receptor membrane assays:

- Basic Buffer: 20 mM Hepes, pH 7.4.
- ATP Regeneration Buffer: 20 mM Hepes, pH 7.4, 5 mM MgCl$_2$, 10 mM phosphocreatine, 10 units/mL creatine phosphokinase, 10 μM GTP, 200 μM ATP, 150 μM IBMX.
- Basic Buffer (10 μL).
- Forskolin, agonist/antagonist in ATP Regeneration Buffer (5 μL).
- Membranes or standards in Basic Buffer (5 μL).
- Incubate 60 min at room temperature.
- Detection Buffer (20 μL).
- Incubate 2 h at room temperature.
- Read on a MicroBeta or TopCount NXT Microplate Scintillation Counter.

19.7.1.2 Results

See Figure 19.16.

19.8 PRECAUTIONS

The first step in a successful assay is to read and follow the product insert. Many factors will influence assay performance. Each assay is different and has its own set of precautions. Pay special attention to the precaution section of each product insert.

Always refer to the product insert and reagent label for proper storage conditions. Some reagents are supplied lyophilized; the recommended storage temperature is 2 to 8°C. After reconstitution, the reagent may need to be aliquoted and stored frozen. The same is true for ready-to-use reagents. Shipping conditions may be ambient or refrigerated, but for long-term storage, the reagent may be required to be stored at frozen temperature.

Cell culture conditions are an important factor in a successful assay. Many vendors of cells will supply growth conditions that will help. The number of passages may be very important for optimal receptor expression. It is also very important to not destroy the receptors when removing them from the flask. One of the most common mistakes of researchers first starting to do cell-based assays is the assumed need to keep cells on ice prior to the assay. Cells do not like ice, and the response will not be optimal if any response occurs at all.

19.9 MANUAL CLAIMS

The specifics given for each assay provided by the manufacturer are shown in Table 19.1. Please note that at this time there are no manual claims for the LANCE cAMP assay since it is currently a custom product soon to be released as a catalog product.

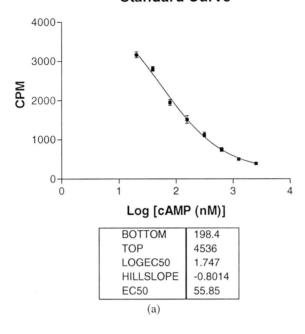

(a)

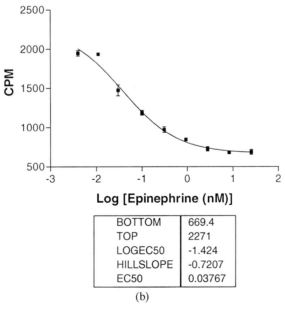

(b)

FIGURE 19.16 (a) A dilution series of unlabeled cAMP competing against [^{125}I] cAMP. (b) Dose–response study with the β2 adrenergic receptor agonist, epinephrine.

TABLE 19.1
Vendor Claims for cAMP Assays

Assay	PerkinElmer Life and Analytical Sciences Adenylyl Cyclase Activation FlashPlate Assay SMP701	PerkinElmer Life and Analytical Sciences [FP]2 cAMPfire FPA203	PerkinElmer Life and Analytical Sciences AlphaScreen cAMP 6760625D	CisBio cAMP Femtomolar 62AM1PEB	CisBio cAMP Dynamic 62AM2PEB	DiscoveRx HitHumter cAMP 90-0002	Amersham cAMP Fluorescence Polarization Biotrak Immunoassay System RPN3596
Format	FlashPlate	Fluorescence polarization	Coated beads	HTRF	HTRF	EFC fluorescent detection	Fluorescence polarization
Additional customer supplied equipment	PBS	PBS	Microplates, stimulation buffer, detection buffer	Microplates	Microplates	Microplates	Microplates
Primary antibody	Bound to FlashPlate	Lyophilized	Bound to beads	Lyophilized; labeled with cryptate	Lyophilized; labeled with cryptate	Concentrate	Lyophilized
Detection range (nM)	50 to 5000	0.017 to 1000	0.01 to 1000	0.07 to 70	0.17 to 712	0.13 to 2300	1 to 1280
Sensitivity	0.2 pmol/well	No claim	No claim	80 pM	No claim	No claim	0.1 pmol/well
IC$_{50}$	No claim	No claim	5 nM	3.46 nM	5.6 nM	No claim	No claim

Stability	Indicated on label	1 year at −20°C	No claim	Indicated on label	Indicated on label	Indicated on label	At least 4 weeks from date of dispatch
Incubation volume (mL)	0.05	0.04	0.025	0.05	0.05	0.048	0.04
# Reagents provided	5	6	4	6	6	8	6
	1 lyophilized; 1 concentrate; 3 ready to use	3 lyophilized; 3 ready to use	4 concentrates	4 lyophilized; 2 ready to use	4 lyophilized; 2 ready to use	5 concentrates; 3 ready to use	3 lyophilized; 1 concentrate; 2 solids
# Reagents requiring preparation	2	3	4	4	4	5	6
# of additions	3	3	3	4	4	5	4
Storage	2 to 8°C	−20 and 2 to 8°C	2 to 8°C	−20 and 2 to 8°C	−20 and 2 to 8°C	−20 and 2 to 8°C	2 to 8°C
Incubation temperature	Room temperature	Room temperature	Room temperature	Room temperature	Room temperature	37°C and room temperature	Room temperature
Detection time (h)	2 to 24	1	1	1	1	1 to 3	At least 4
Warnings	None	None	Beads are light sensitive	None	None	Keep reagents on ice; protect reagents from light	Do not agitate plate during incubation; incubate in the dark
On-site IC_{50} (nM)	13.9	30.3	6.0	2.0	4.9	16.6	13.1

REFERENCES

1. Sutherland, E. W., Robinson, G. A., and Butcher, R. W. Some aspects of the biological role of adenosine 3′, 5′-monophosphate (cAMP). *Circulation,* 37 (1968), 279–306.
2. Jost, J. P. and Rickenburg, H. A. cAMP. *Annual Review of Biochemistry,* 40 (1971), 741–774.
3. Perkins, J. P. (1973): Adenyl cyclase. In: Greengaard, P. and Robinson, G. A. (Eds.). *Advances in Cyclic Nucleotide Research,* Vol. 3:1–64. Raven Press, New York.
4. Langan, T. A. (1973): Protein kinase and protein kinase substrates. In: Greengaard, P. and Robinson, G. A. (Eds.). *Advances in Cyclic Nucleotide Research,* Vol. 3:99–153, Raven Press, New York.

VENDORS

- PerkinElmer Life and Analytical Sciences
  549 Albany Street
  Boston, MA 02118
- Amersham Biosciences U.K. Limited
  Amersham Place
  Little Chalfont
  Buckinghamshire HP7 9NA
  England
- DiscoveRx Corporation
  42501 Albrae Street
  Fremont, CA 94538
- CIS U.S., Inc.
  10 DeAngelo Drive
  Bedford, MA 01730

20 A Homogeneous, Fluorescent Polarization Assay for Inositol 1,4,5-Trisphosphate (Ins P$_3$)

Peter Fung, Rajendra Singh, Lindy Kauffman, Richard Eglen, and Tabassum Naqvi

CONTENTS

20.1 INTRODUCTION

G-protein–coupled receptors (GPCRs) are one of the largest classes of drug discovery targets [1,2]. GPCR ligands regulate cellular and physiological pathways by signaling through several second messengers, including cyclic AMP, inositol phospholipids, and calcium [3]. Quantitation of second messengers is frequently used as a means to screen and pharmacologically characterize GPCR ligands [4]. The GPCR signaling process occurs by two major pathways. GPCRs coupling to Gα_s and Gα_i proteins activate or inhibit, respectively, adenylate cyclase and subsequently change intracellular cAMP levels. GPCRs coupling to Gα_q or Gα_o proteins activate phosphoinositol phospholipase Cβ, which hydrolyzes phosphatidylinositol 4,5-bisphosphate (PIP$_2$) forming *sn* 1,2-diacylglycerol and inositol 1,4,5-trisphosphate (Ins P$_3$) [5]. Ins P$_3$ binds and opens an endoplasmic Ins P$_3$ gated calcium channel, causing release of bound calcium into the cytosol [6]. Several metabolic products of Ins P$_3$ also modulate cellular function, including inositol 1,3,4,5-P$_4$ (Ins P$_4$), which acts to facilitate Ins P$_3$-mediated calcium release synergistically [7].

There are several HTS assay systems to measure intracellular cyclic AMP as a marker of G$_s$- and G$_i$-coupled GPCRs [8]. In contrast, there are few assays available to selectively measure Ins P$_3$ to monitor Gq-coupled GPCR activation, particularly those suitable for automated HTS. Consequently, many HTS laboratories measure changes in intracellular calcium to assay G$_q$-coupled GPCRs using a fluorescent calcium-sensitive dye, loaded into intact cells as a cell-permeable ester. Real-time changes in the GPCR-induced signal are then determined in a microtiter plate using imaging instruments, such as a fluorescent imaging plate reader system (FLIPR, Molecular Devices Corp) [9].

Screening library compounds, however, may modulate intracellular calcium levels by other means than binding to the receptor, such as nonspecific blockade of calcium channels or exacerbated intracellular calcium release. Moreover, compounds that autofluoresce or quench fluorescence result in ambiguous changes in the assay signal and may manifest as false-positive or -negative hits. Consequently, several assays have been developed to measure GPCR-induced inositide phospholipid hydrolysis [10–12]. The majority of these assays involve radioactive measurements, many of which are suboptimal for high-volume screening.

20.2 MEASURING INOSITOL PHOSPHOLIPID HYDROLYSIS TO MONITOR GPCR ACTIVATION

A proportion of GPCRs that couple to $G\alpha_q$ proteins activate phospholipase C and mobilize Ins P_3 [5–7]. Measurement of GPCR-induced changes in phosphoinositide phospholipase C activity is frequently undertaken by measuring inositol phosphate production. Here, tritiated inositol is incorporated into the inositol phospholipids of the cell. Activation of the receptor results in release of radiolabeled Ins P_3. The experiments are conducted in the presence of lithium, which inhibits inositol monophosphate phosphatase, thereby blocking the cycle and increasing accumulation of the tritiated isotope at the monophosphate form. This radiometric approach is used in conjunction with scintillation proximity assay (SPA) technology (GE Healthcare) to provide a homogeneous platform more suitable for automation [13].

20.3 MEASURING INOSITOL PHOSPHATE LEVELS TO MONITOR GPCR RESPONSES

The measurement of the second messenger, Ins P_3 *specifically*, is undertaken differently and has traditionally been done using mass assays with gas liquid chromatography (GLC), anion exchange chromatography, or high performance liquid chromatography (HPLC) [14]. These techniques, while very sensitive, are not adaptable to assays requiring high throughput. The recognition that Ins P_3 binds to a specific intracellular receptor provides the basis for a radiometric competition-binding assay [5]. Here, tritium-labeled Ins P_3 is displaced from a crude preparation of the Ins P_3 receptor using a competition radioligand binding protocol [15,16]. A commercial version of this radioreceptor assay is available from GE Healthcare using bovine adrenal gland Ins P_3 receptor preparations. This format, again when used with SPA, is high throughput [6]. However, the economics of isotopic waste disposal emanating from high-volume screens remains a significant issue.

A nonisotopic assay for Ins P_3 is now available based on the AlphaScreen technology (PerkinElmer). This technique is an amplified luminescence assay that employs donor and acceptor beads. When the donor bead is excited with light at 680 nm, a photosensitizer converts O_2 to singlet oxygen. When two beads are in close proximity, the singlet oxygen produces a chemiluminescent signal in the acceptor bead, activating bead fluorophores and amplifying the signal. In an Ins P_3 assay, the two beads are held in close proximity by a biotinylated Ins P_3 molecule, as the donor bead is coated with streptavidin and the acceptor bead is coated with an Ins P_3-binding protein. In the absence of cell stimulation, a signal is seen. In the presence of free Ins P_3 from the cell, the donor and acceptor beads dissociate, and the signal proportionally decreases [17]. Echelon Biosciences have utilized the AlphaScreen assay format using a binding protein that binds a range of inositol phosphates, including IP$_2$ and IP$_4$. These cellular metabolites compete with a biotinylated inositol phosphate analog as described above [18]. This assay has an advantage in that it detects several phosphoinositols, although an extensive evaluation in HTS screens has not been reported to date. Despite the advantage of the AlphaScreen approach as a nonisotopic homogeneous assay technology, the signal is sensitive to compound quenching, and ambient fluctuations in light and temperature need to be carefully controlled [19]. The AlphaScreen Ins P_3 assay is also limited by

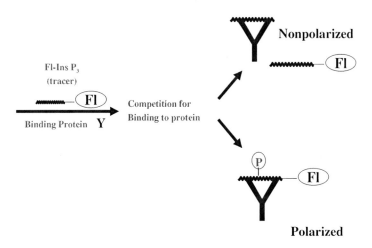

FIGURE 20.1 Schematic representation of the Ins P$_3$ FP assay principle.

in the number of cells per well, as matrix interferences from cell lysates reduce the signal. This issue, plus the instability of the Ins P$_3$ binding protein preparation, may cause variability in the assay performance and sensitivity.

20.4 HITHUNTER FLUORESCENCE POLARIZATION (FP) ASSAY FOR INOSITOL 1,4,5-TRISPHOSPHATE (INS P$_3$)

The HitHunter FP Ins P$_3$ assay from DiscoveRx is a competitive binding assay, in which cellular Ins P$_3$ displaces a fluorescent derivative of Ins P$_3$ from a specific binding protein. The assay measures changes in fluorescence polarization (FP), a single-wavelength ratiometric technique, in which a fluorescent derivative of Ins P$_3$ is used as a tracer. FP is determined as a ratio of fluorescence emissions in the vertical and horizontal planes. When fluorescent molecules are excited with polarized light, the degree to which the emitted light retains polarization reflects the rotation that the molecule underwent between excitation and emission. Small molecules rotate rapidly, and emitted light is random with respect to the plane of emission. When bound to a large protein (such as receptor or antibody), the molecule rotates much more slowly and the emitted light retains more of its polarization. This is measured as an increase in the FP signal.

When excited with polarized light, the emission from a fluorescent derivative of Ins P$_3$ (tracer) is depolarized compared to the exciting light, due to the rapid rotation of the molecule between excitation and emission. When the Ins P$_3$ derivative binds to a binding protein, the rotation time is reduced and a high polarization value is seen. In the assay unlabeled Ins P$_3$, either a standard Ins P$_3$ solution or derived from the cell lysate, displaces the tracer from the binding protein, and the rotation time increases and low FP signal is measured (Figure 20.1). By this means a calibration is generated to the standard Ins P$_3$ dilutions, and the molar concentration of Ins P$_3$ in the cell lysate determined by interpolation (Figure 20.2).

The critical components of the DiscoveRx assay are thus the fluorescent Ins P$_3$ tracer and the Ins P$_3$ binding protein, as shown in the protocol in Figure 20.3. In the case of the tracer, three dye conjugates have been developed including a green (fluorescein) derivative of Ins P$_3$ (Figure 20.4). As low concentrations of fluorescent tracers are used in the assay, the technique is sensitive to optical interference from screening library compounds. The ratiometric processing of the data corrects to some extent for fluorescent compounds. Artifacts or interferences can also be identified by measuring compound fluorescence in the absence of the Ins P$_3$ tracer. For this reason, the Ins P$_3$ assay has also been developed for a series of "red" tracers that are less prone to compound

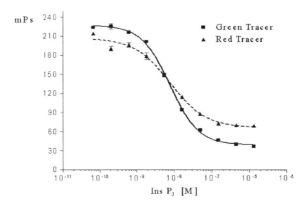

FIGURE 20.2 Ins P_3 standard curve. A standard curve was generated to measure levels of exogenously added Ins P_3. A high concentration of Ins P_3 at 7 μM was serially diluted 1:3 in Ins P_3 standard dilution buffer. Different concentrations of Ins P_3 were incubated with PCA, followed by the addition of the tracer and then the Ins P_3 binding protein. The reaction was read on a multiwell fluorescence polarization plate reader such as the Beckman-Coulter CRI Affinity or LJL Analyst. The majority of the experimental data for this publication was collected on a Beckman-Coulter CRI Affinity, unless noted. An IC_{50} of ~7 to 9 nM was observed when using either the green or red Ins P_3 fluorescent tracers.

TABLE 20.1
Assay Precision of Ins P_3 FP Assay Using Different Tracers

	Green Tracer	Red Tracer
mPs Low standard	238	196
mPs High standard	37	70
S/B ratio/μmPs	6/201	3/126
EC_{50}, nM Ins P_3	9	7
Average % CV of replicates	2	2
Z' Factor	0.97	0.92

Note: n = 4 replicates.

interference (Table 20.1). In all cases the sensitivity of the assay is similar, although changes in FP (denoted as the delta mP) vary according to the dye in question (Figure 20.5 and Table 20.2).

The FP Ins P_3 assay is performed in crude cell lysates, thereby avoiding laborious separation and filtration steps. It is therefore important that the Ins P_3 binding protein exhibit high affinity and selectivity for the D-*myo*-1,4,5-inositol-Ins P_3 isomer over other inositol polyphosphates. The buffer (20 mM HEPES, 150 mM NaCl, 1 mM DTT, 0.1% BGG, and 0.02% Tween 20, pH 7.5) used in the Ins P_3 assay is optimized to ensure high-affinity binding, and competition binding studies with several substituted inositol phosphates demonstrate that the Ins P_3 binding protein is specific for the D-*myo* 1,4,5 inositol Ins P_3 isomer (Table 20.3). In terms of stability, the performance did not change for 8 h at room temperature (Figure 20.6a) and can withstand multiple freeze/thaw cycles when stored at −80°C. The binding protein is also a stable reagent for more than 2 months at −80°C (Figure 20.6b).

In a similar fashion to many FP-based assays, the DiscoveRx Ins P_3 FP assay is amenable to assay automation systems. A representative standard curve dispensed by a BioMek 2000 liquid handler instrument is shown in Figure 20.7. Here, standard concentrations of Ins P_3 were run in replicates of 10. A coefficient of variance of 2% and a Z' factor of 0.92 to 0.97 are generally seen. Similar assay performances have been observed using either an Analyst FP reader or a CRI Infinity reader (Figure 20.8).

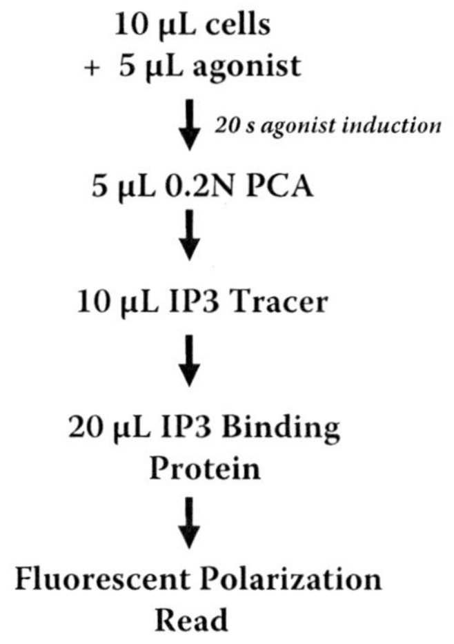

10 µL cells
+ 5 µL agonist

↓ *20 s agonist induction*

5 µL 0.2N PCA

↓

10 µL IP3 Tracer

↓

20 µL IP3 Binding
Protein

↓

Fluorescent Polarization
Read

FIGURE 20.3 HitHunter Ins P$_3$FP assay protocol. Schematic representation of the steps and additions made to measure levels of Ins P$_3$.

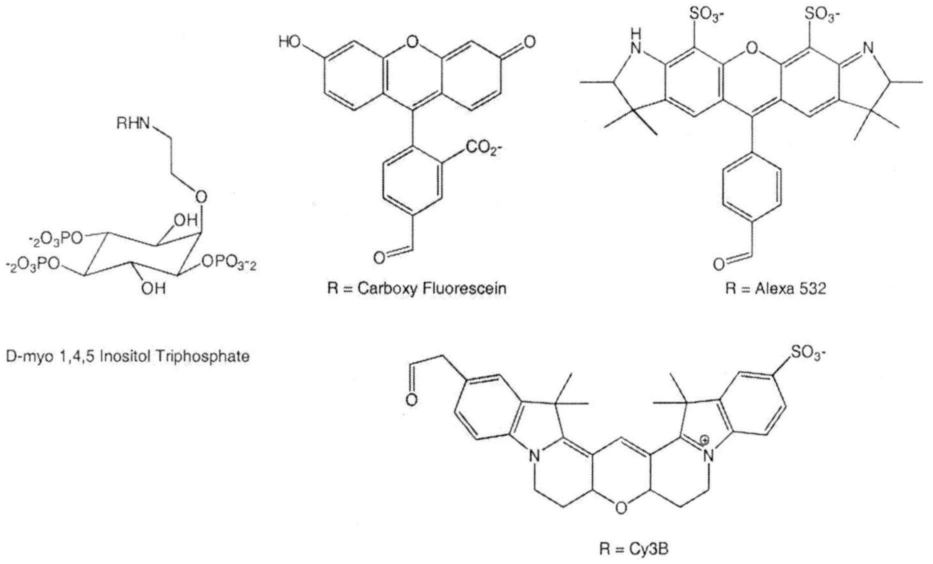

D-myo 1,4,5 Inositol Triphosphate

R = Carboxy Fluorescein

R = Alexa 532

R = Cy3B

FIGURE 20.4 Chemical structure of the Ins P$_3$ FP tracer. Amine derivatized D-*myo*-1,4,5-inositoltriphosphoric acid was reacted to each of the hydroxysuccinimide activated carboxy fluorescein, AlexaFluor, and Cy3B dyes separately in dry dimethyl formamide. Each of the Ins P$_3$ tracers was purified to 99.9% homogeneity by reverse phase HPLC on C18 column and triethyl ammonium acetate: acetonitrile gradient. The molecular weight of all the conjugates was corroborated by electrospray mass spectroscopy.

20.5 MEASURING GPCR AGONISM AND ANTAGONISM

In a similar fashion to other second messengers such as adenylate cyclase, basal and stimulated levels of Ins P$_3$ are highly dependent on cell number. To correlate cell number with Ins P$_3$ basal levels, three different CHO-M1 cell lines were studied using the green Ins P$_3$ tracer. As the cell number per well was increased from 5000 to 50,000, the basal levels of Ins P$_3$ increased in proportion (Figure 20.9A). These data indicate that the assay is applicable to a range of different cell densities.

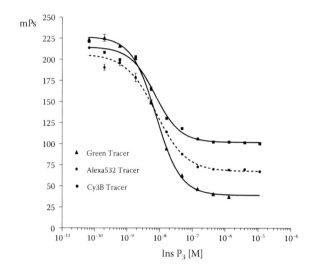

FIGURE 20.5 Alternative red-shifted dye tracers used for the Ins P_3 FP assay. To address issues of compound library interferences by autofluorescence or quenching that may occur using a fluorescein-based Ins P_3 tracer, two different red-shifted Ins P_3 tracers were synthesized and tested. Standard curves using the three different red-shifted dye tracers and Ins P_3 are shown. The assay was run following the above procedure, substituting the two different red-shifted dye tracers for the green Ins P_3 tracer. All subsequent data described herein were performed using the red dye tracer, Alexa 532. (From Eglen, RM. *Combin Chem & HTS*, 2005; 8:311–318. With permission.)

TABLE 20.2
Summary of Activity of Two Red-Shifted Dyes and the Green Ins P_3 FP Tracers

Dye	Ins P_3 (IC_{50}), nM	High mPs	Low mPs	Mean % CV	Z' Factor
Green tracer	8	227	39	2	0.91
Cy3B tracer	7	215	102	1	0.90
Alexa 532 tracer	8	206	67	2	0.95

TABLE 20.3
Selectivity of the Binding Protein to D-*myo*-Inositol 1,4,5-Trisphosphate (Ins P_3)

Compound	IC_{50} (nM)	Percent Cross-reactivity[a]
D-*myo*-Inositol 1,4,5-trisphosphate (Ins P_3)	10	100
D-*myo*-Inositol 1,3,4,5-tetrakisphosphate (Ins P_4)	>2000	0.23
D-*myo*-Inositol 1,4,6-trisphosphate (Ins P_3)	235	4.77
D-*myo*-Inositol 3,4,5-trisphosphate (Ins P_3)	> 2000	0.28
D-*myo*-Inositol 1,3-bisphosphate (Ins P_2)	> 2000	0.01
D-*myo*-Inositol 1,3,4-trisphosphate (Ins P_3)	> 2000	0.02
Inositol (IP)	> 2000	0.02

Note: n = 4 replicates.

[a] A competition binding assay was done in the presence of different inositol phosphates. The cross-reactivity for the Ins P_3 binding protein was determined by dividing the IC_{50} value of Ins P_3 by the IC_{50} value of other IPs and multiplying by 100.

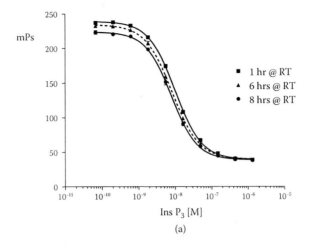

(a)

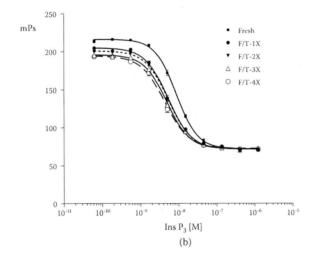

(b)

FIGURE 20.6 Stability of the Ins P$_3$ binding protein. The reagents were equilibrated at room temperature for 1, 6, and 8 h. After each time period, a standard curve titrating Ins P$_3$ was run. As shown in panel (a), the binding protein was stable over 8 h at room temperature. The sensitivity of the standard curve over 1, 6, and 8 h was 10, 8, and 9 nM, respectively. In panel (b), the Ins P$_3$ binding protein reagent was subjected to four freeze/thaw cycles ($-80°C$ to room temperature) and a standard curve was run. Freshly prepared Ins P$_3$ binding protein (closed square) IC_{50} = 9 nM, (closed circle) one freeze/thaw IC_{50} = 5 nM, (closed inverted triangle) two freeze/thaws IC_{50} = 5 nM, (open triangle) three freeze/thaws IC_{50} = 5 nM, (open circle) four freeze/thaws IC_{50} = 4 nM.

Experience has also shown that several different types of cell (CHO-K1, HEK 293 cells, and so on) can be used in the assay (Figure 20.9B).

The goal of a competitive Ins P$_3$ assay is to measure changes in cellular Ins P$_3$ concentration induced by GPCR agonist activation. It is well known that the cellular metabolism of Ins P$_3$ is extremely rapid; after an initial spike, the levels decline to a plateau, the height of which depends upon the cell type and perhaps cytosolic calcium concentration. In some cells, Ins P$_3$ peak levels oscillate in a frequency that directly correlates to the calcium oscillation frequency [20]. In an "end-point" assay for Ins P$_3$, such as those described in this chapter, it is important that the peak levels of Ins P$_3$ are reproducibly measured using assay conditions in which Ins P$_3$ metabolism is

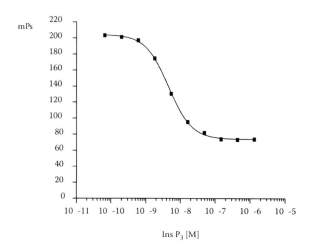

FIGURE 20.7 Automation of the HitHunter Ins P_3 FP assay. Dispensation of the reaction was done on a BioMek 2000; ten replicates for each standard concentration were run. The $IC_{50} = 5$ nM, the mean %CV was 2 and the Z′ factor was 0.90.

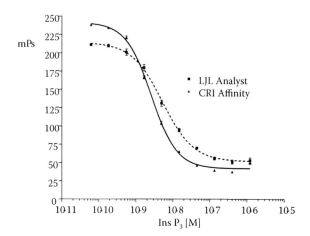

FIGURE 20.8 Comparison of two FP readers measuring the Ins P_3 FP assay. The Ins P_3 FP green tracer standard curve was run on an LJL Analyst GT and a Beckman-Coulter CRI Affinity. The LJL Analyst was set as follows: integration time = 100,000 μsec, G Factor 1.0. The filter set used in the CRI Affinity was excitation filter — fluorescein 485 nm, Emission filter — fluorescein 530 nm, and dichroic — fluorescein 505 nm. The exposure was set at 15 to 30 msec, and the focus was set at 2700 to 3200. In this particular experiment, the LJL Analyst run had an $IC_{50} = 5$ nM, with 5% mean CV and a Z′ factor = 0.87. For the CRI Affinity, the $IC_{50} = 3$ nM, with 3% CV and Z′ factor = 0.97.

arrested. To achieve this, the cell samples are rapidly deproteinized after agonist addition, using perchlortic acid (PCA; 0.2 N), which displaces Ins P_3 from the salts by acting as a chaotropic agent and terminates metabolic activity.

An important feature of using the assay in high-throughput robotic fluid dispensing systems is that the PCA needs to be added 20 to 30 sec after addition of the agonists, in order to measure the peak formation of Ins P_3. CHO-M1 cells induced with carbachol exhibited maximal Ins P_3 induction within 30 sec, followed by a rapid decline over the following 5 min (Figure 20.10). Similar findings are seen in histamine H_1 receptor cells (Figure 20.10).

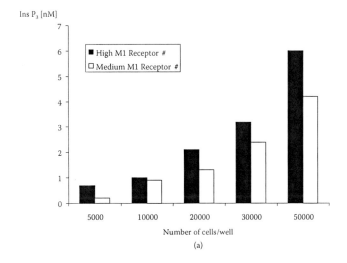

(a)

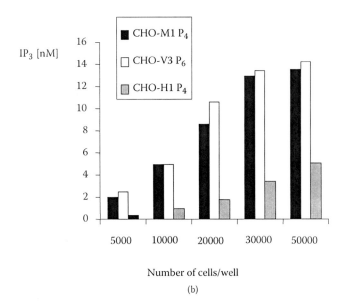

(b)

FIGURE 20.9 (a) Basal cellular Ins P_3 levels increase with cell number. Between 5000 and 50,000 CHO-K1 muscarinic M1 receptor cells (expressing either 1.5 or 8.3 pmol/mg protein of receptor) were assayed in triplicate to determine the basal amounts of Ins P_3 in the cell. No agonist was added to the cells in this experiment. Samples were assayed following the protocol shown in Figure 20.2. The amount of Ins P_3 was calculated from the standard curve run in parallel with the test conditions (data not shown). (b) Basal Ins P_3 levels in CHO-K1 cells expressing different G_q-coupled receptors. Between 5000 and 50,000 cells were assayed to measure the levels of Ins P_3 expressed by the cell lines in the absence of agonist addition. Samples were assayed in triplicate. The passage number of each cell line was noted (P_4 or P_6), as the age of the cell line can affect the expression levels of Ins P_3. The levels of detected Ins P_3 were calculated off a standard curve run in parallel with the experiment.

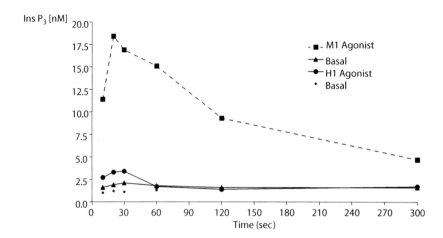

FIGURE 20.10 Monitoring agonist stimulated Ins P_3 levels in CHO-M1 cells and CHO-H1 cells over time. Twenty thousand CHO-M1 cells were treated with 1000 μM carbachol, and CHO-H1 cells were treated with 100 μM histamine. At the end of each noted time point, 0.2 N PCA was added to quench the reaction and the Ins P_3 FP assay was carried out as described above. The amount of Ins P_3 detected in the cells after the defined agonist stimulation period was calculated off an Ins P_3 standard curve run in parallel to the test samples. Samples were assayed in triplicate.

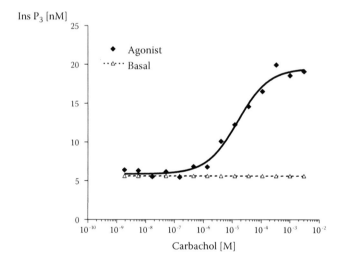

FIGURE 20.11 Agonist stimulation of CHO-M1 cells. Twenty thousand stably expressing CHO-M1 cells were treated with an increasing concentration of carbachol for 20 sec. PCA was immediately added after the agonist incubation period added to quench the reaction. The levels of Ins P_3 were extrapolated from a standard curve that was run in parallel (data not shown). The samples were assayed in triplicate. The IC_{50} of carbachol was determined to be 15 μM.

Agonist concentration response curves can be established using this assay with high precision. A prototypical receptor that induces formation of Ins P_3 is the muscarinic M_1 receptor. The agonist carbachol increased Ins P_3 levels approximately fourfold (Figure 20.11) with a potency (EC_{50}) of 7 μM [13]. The induction of Ins P_3 was antagonized by the muscarinic antagonist atropine with potencies in a range consistent with the literature (0.1 to 10 μM) (Figure 20.12) [21], using isotopic assay systems to detect Ins P_3. Similar data can be seen using a more potent agonist in a different receptor system (Figure 20.13), in which both full and partial agonists can be detected. Again, the responses were antagonized by compounds in a concentration-dependent fashion (Figure 20.14).

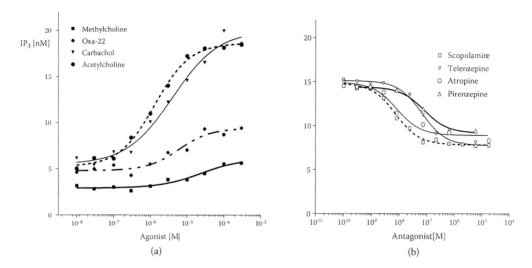

FIGURE 20.12 Agonism and antagonism of CHO-M1 receptor: (a) twenty thousand CHO- M1 cells were treated with increasing concentrations of the following known M1-specific agonists: methacholine {closed square, solid line}, OX-22 (cis-2-methyl-5-trimethylammoniummethyl-1,3-oxathiolane iodide) {closed diamond, dash line}, carbachol {inverted closed triangle, solid line}, and acetylcholine {closed circle}. (b) twenty thousand CHO-M1 cells were pretreated with increasing concentrations of the following muscarinic receptor antagonists: scopolamine hydrobromide {open square, dash line}, telenzepine {inverted open triangle, solid line}, atropine {open circle, solid line}, and pirenzepine {open triangle, solid line} for 30 min. The antagonists were washed from the cells, and then 300 μM carbachol (previously determined EC_{80} concentration) was added to the cells in fresh medium. The carbachol induction lasted for 20 sec, and the reaction was quenched by the addition of 0.2 N PCA. The levels of Ins P₃ detected by agonist and antagonist treatment were extrapolated from a standard curve run in parallel with the experiment. Both analyses were performed using the Ins P₃ green tracer. (From Eglen, RM. *Combin Chem & HTS*, 2005; 8:311–318. With permission.)

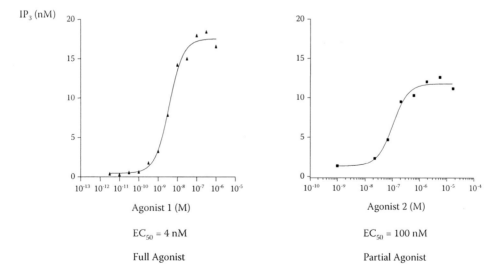

FIGURE 20.13 Concentration effect curves for Ins P₃ measurements of two antagonists at the same G-protein–coupled receptor. The agonist used to elicit the response was the full agonist shown in Figure 20.14.

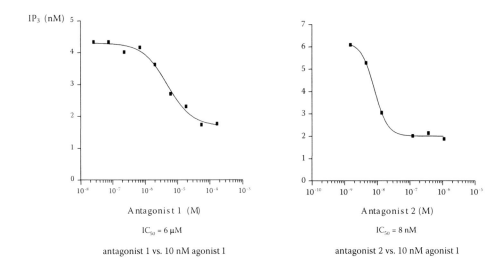

FIGURE 20.14 Concentration effect curve for Ins P_3 measurements at a Type II G-protein–coupled receptor. The agonist under investigation in the left panel is a full agonist with a high potency. The agonist under investigation in the right panel is a partial agonist.

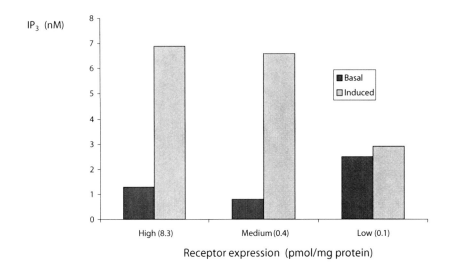

FIGURE 20.15 Correlation of receptor expression levels and detection of Ins P_3 levels in three different CHO-K1 cell lines expressing a muscarinic M_1 receptor. Three different stably transfected CHO-K1 cell lines were tested as they expressed different muscarinic M1 receptor levels [high = 8.3 pmol/mg-protein, medium = 0.4 pmol/mg-protein, and low = 0.1 pmol/mg-protein]. Twenty thousand cells per well were plated in two sets of triplicate wells. The samples were treated with either buffer (Basal) or 300 μM carbachol (induced). Shown are the results with the Ins P_3 green tracer. The concentration of Ins P_3 detected in the assay was determined from extrapolation from a standard curve run in parallel to the experimental conditions.

The low efficacy of some agonists at inducing Ins P_3 is due to the low receptor reserve associated with the response. Indeed, the response is much less well coupled to receptor activation than calcium is (see below). Consequently, it is anticipated that the maximal level of induction would be sensitive to the receptor expression levels in the cell line. This is indeed the case with CHO-M1 cells, as shown in Figure 20.15, where receptor expression levels of 0.4 pmol per mg protein and above are required.

20.6 COMPARISON OF AGONIST INDUCTION OF INS P$_3$ IN COMPARISON TO INTRACELLULAR CALCIUM

G_q-coupled GPCR stimulation ultimately causes the liberation of calcium from bound intracellular stores. When measuring a calcium response that is significantly downstream from the receptor, the GPCR response is highly amplified, resulting in potent agonist responses. As described above, the fluorescence imaging plate reader (FLIPR) is frequently used to measure calcium changes in living cells by means of calcium-specific fluorescent dyes (Figure 20.16). Comparison of several muscarinic agonists in assays measuring either Ins P$_3$ or calcium changes shows clearly marked differences in compound potencies (Figure 20.17). However, when equiactive agonist concentrations (such as the EC$_{80}$ concentration) are used to determine antagonist potency, similar values can be found (Figure 20.16). Thus, the values for a series of muscarinic potencies (IC$_{50}$) determined in an Ins P$_3$ assay compare well with values from a FLIPR experiment. A final point is that the rapid kinetics of either calcium release or changes in Ins P$_3$ does not allow sufficient time for the agonist to reach equilibrium with a preincubated antagonist, resulting in a state of hemiequilibrium in which the receptors are effectively bound irreversibly during the assay period. This is most noticeable using compounds of high affinity; therefore, depression in the agonist concentration response curve maxima will be observed in either assay. FLIPR analysis can be prone to compound interferences that modulate calcium levels resulting in false negatives or positives. It is anticipated that interferences of this nature would be much less with an Ins P$_3$ assay. Studies have confirmed that several calcium channel blockers interfere in the FLIPR assay, including verapamil, nifedipine, nimodipine, and nitrendipine. However, they did not influence the Ins P$_3$ stimulation, and were not therefore false negatives in this assay.

20.7 CONCLUSIONS

Measuring GPCR activation upon ligand addition via monitoring second messenger response is a commonly used technique in screening. In screening for ligands at G_q-coupled receptors, several methods have been developed to detect agonist induced changes in Ins P$_3$, PI, PIP$_2$, PLC, and calcium. These methodologies include both homogeneous and heterogeneous formats. The HitHunter FP Ins P$_3$ assay is a homogeneous assay that is a sensitive, nonisotopic high-throughput assay to measure Ins P$_3$. This assay is highly automatable and can be used with several cell lines expressing differing levels of GPCRs. The flexibility in the assay format provides for optimizing the sensitivity of the analysis for automation and miniaturization. The variety of tracers available for the assay may also reduce library compound interference.

ACKNOWLEDGMENTS

The authors wish to acknowledge their colleagues at DiscoveRx Corp. in the development of FP-Ins P$_3$ assay including Betty Bosano, Hyna Dotimas, Pyare Khanna, Vashti Lacaille, Sherrylyn de La Llera, Riaz Rouhani, and Inna Vainshtein.

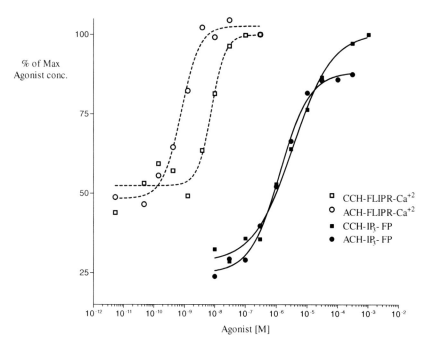

FIGURE 20.16 Agonist stimulation measured by HitHunter Ins P_3 FP and FLIPR analysis. For the FLIPR analysis (dashed line), CHO-M1 cells were plated at a density of 50,000 cells per well, while 20,000 cells per well were used in the Ins P_3 FP assay (solid line). The agonists carbachol (open or closed square) or acetylcholine (open or closed circle) were added to the cells for 20 sec, after which they cells were processed according to described protocols to measure changes in either calcium or Ins P_3 levels. For both analyses, all samples were assayed in triplicate. For FLIPR analysis, the average was taken for the peak fluorescent reading at each treatment while for Ins P_3 FP analysis, the mean FP values were extrapolated from a standard curve to determine the amount of Ins P_3. The results for both assays were normalized to the maximal agonist response.

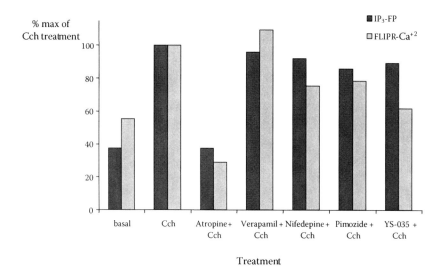

FIGURE 20.17 Effect of calcium channel blockers on the Ins P_3 FP assay. Four calcium channel blockers were examined in the HitHunter Ins P_3 FP assay. The number of cells used were 50,000 CHO-M1 cells (FLIPR analysis) or 20,000 CHO-M1 cells. As a control, carbachol was added to the cells at a concentration of 300 μM, and atropine was used a control antagonist at a concentration of 1 μM. Samples are assayed in triplicate. The different treatments are plotted against the percentage of the carbachol-alone treatment.

REFERENCES

1. Hopkins AL, Groom CR. The druggable genome. *Nat Rev Drug Discov.* 2002; 1:727–730.
2. Bleicher KH, Bohm HJ, Muller K, Alanine AI. Hit and lead generation: beyond high-throughput screening. *Nat Rev Drug Discov.* 2003; 2:369–378.
3. Strader CD, Fong TM, Tota MR, Underwood D, Dixon RA. Structure and function of G-protein–coupled receptors. *Annu Rev Biochem.* 1994; 63:101–132.
4. Dove A. Drug screening — beyond the bottleneck. *Nat Biotechnol.* 1999; 17:859–863.
5. Berridge MJ, Dawson RMC, Downer CP, Heslop JP, Irvine RF. Changes in the levels of inositol phosphates after agonist-dependent hydrolysis of membrane phosphoinositide. *Biochem J.* 1983; 212:473–482.
6. Berridge MJ. Inositol trisphosphate and calcium signaling. *Nature.* 1993; 361:315–325.
7. Michell RH. Inositol lipids in cellular signalling mechanisms. *Trends Biochem Sci.* 1992; 17:274–276.
8. Williams C. cAMP detection methods in HTS: selecting the best from the rest. *Nat Rev Drug Discov.* 2004; 3:125–135.
9. Kassack MU, Hofgen B, Lehmann J, Eckstein N, Quillan JM, Sadee W. Functional screening of G-protein–coupled receptors by measuring intracellular calcium with a fluorescence microplate reader. *J Biomol Screen.* 2002; 7:233–246.
10. Mullinax TR, Henrich G, Kasila P, Ahern DG, Wenske EA, Hou C, Argentieri D, Bembenek ME. Monitoring inositol-specific phospholipase C activity using a phospholipid FlashPlate(R). *J Biomol Screen.* 1999; 4:151–155
11. Liu JJ, Hartman DS, Bostwick JR. An immobilized metal ion affinity adsorption and scintillation proximity assay for receptor-stimulated phosphoinositide hydrolysis. *Anal Biochem.* 2003; 318:91–99.
12. Bembenek ME, Jain S, Prack A, Li P,Chee L, Cao W, Spurling H, Roy R, Fish S, Rokas M, Parsons, T, Meyers R. Development of a high-throughput assay for two inositol specific phospholipase Cs using scintillation proximity format. *Assay Drug Dev. Technol.* 2003; 1:435–443.
13. Brandish PE, Hill LA, Zheng W, Scolnick EM. Scintillation proximity assay of inositol phosphates in cell extracts: high-throughput measurement of G-protein–coupled receptor activation. *Anal Biochem.* 2003; 313:311–318.
14. Kuksis, A. *Laboratory Techniques in Biochemistry and Molecular Biology,* Vol. 30. Elsevier, Amsterdam, 2003.
15. Challiss RA, Batty IH, Nahorski SR. Mass measurements of inositol (1,4,5) trisphophate in rat cerebral cortex slices using a radioreceptor assay; affects neurotransmitters and depolarization. *Biochem Biophys Res Commun.* 1988; 15:684–691.
16. Williams D, Price Jones M, Hughes K. An homogeneous assay for the measurement of inositol-1,4,5-trisphosphate using scintillation proximity assay technology. Poster # P08012. 9th SBS Annual Conference, Portland, OR, Sept. 21–25, 2003.
17. Chelsky D, Bosse R, and Illy C. Alpha Screen HTS assay for IP3 (Abstract 10058).
18. 7th SBS Annual Conference and Exhibition, Baltimore, MD, Sept. 10–13, 2001.
19. Neilsen PO, Assis EF, Branch AM, and Dress BE. High-throughput nonradioactive inositol phosphate assay for GPCR inhibitor screening. Poster # P08064. 9th SBS Annual Conference, Portland, OR, Sept. 21–25, 2003.
20. Packard Biosciences. Analysis of potential compound interference of ALPHAscreen Signal (Application note, ASC-012). Packard Bioscience Company, Inc. Meriden, Connecticut, 2001.
21. Mishra J, Bhalla U.S. Simulations of inositol phosphate metabolism and its interaction with InsP(3)-mediated calcium release. *Biophys J.* 2002; 83:1298–1316.
22. Eglen RM. Functional G-protein–coupled receptor assays for primary and secondary screening. *Combin Chem & HTS.* 2005; 18:311–318.

21 Scintillation Proximity Assay of Inositol Phosphates

Wei Zheng and Philip E. Brandish

CONTENTS

21.1 INTRODUCTION

The second messenger molecules inositol (1,4,5)-trisphosphate (IP_3), diacylglycerol (DAG), and adenosine $3',5'$-cyclic monophosphate (cAMP) represent two major signal transduction pathways for G-protein–coupled receptors (GPCR). GPCRs fall into two broad categories: some activate the IP_3/DAG pathway through the G_q subtype of G protein which in turn activates the β-subtype of phospholipase C (PLC), while others stimulate cAMP production via the G_s subtype of G protein or inhibit the cAMP level via the G_i subtype of G protein (Figure 21.1a and Figure 21.1b). Measurements of inositol phosphate accumulation or increased intracellular calcium (mediated by the action of IP_3 on endoplasmic reticulum IP_3 receptors) are the standard readouts of receptor function for GPCRs coupling to G_q. Despite the relative difficulty (particularly in automated formats), measurement of IP_3 production stimulated by the activation of GPCRs is still very attractive because it is more proximal to the receptor activation than are measurements of downstream events in the signal transduction pathway.

In addition to GPCRs, protein tyrosine kinase–linked receptors such as the epidermal growth factor (EGF), platelet-derived growth factor (PDGF), nerve growth factor (NGF), and tropomyosin-related kinase (Trk) receptors signal through the PLC/IP_3/Ca^{2+} pathway by phosphorylation-mediated activation of PLC-γ and phosphatidylinositol-$3'$ kinase (PI3K) [1]. G-proteins are not linked to this tyrosine kinase signaling pathway (Figure 21.1c).

21.2 PHOSPHATIDYLINOSITOL (PI) TURNOVER

Phosphatidylinositols (PI), a minor class of phospholipids in cell membrane, are critically involved in the signaling of hormones and neurotransmitters. This class of phospholipids undergoes a continuous cycle of metabolism in cells through a series of kinases and phosphatases [2,3]. Briefly, phosphatidylinositol (4,5)-bisphosphate (PIP_2) is hydrolyzed to IP_3 and DAG by membrane-bound PLC. IP_3 is rapidly metabolized to inositol through a series of phosphatase reactions or by phosphorylation to inositol (1,3,4,5) tetrakisphosphate (IP_4) and subsequent dephosphorylation (Figure 21.2). Inositol, either from this pathway, or from *de novo* synthesis inside cells, is then incorporated into phosphatidylinositol (PI), which is phosphorylated to PIP_2 to complete the PI turnover cycle

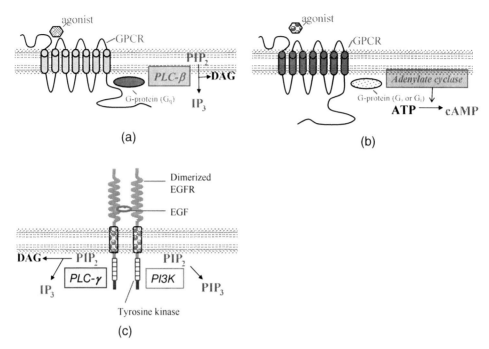

FIGURE 21.1 Schematic illustration of signal transduction pathways related to GPCRs and growth factor receptors. (a) G_q-coupled GPCR signaling pathway. Agonist binding to its receptor triggers the conformational change of a GPCR that leads to binding to G_q subtype of G proteins. This results in binding of GTP by the G_q subunit and activates the G protein subunit via dissociation of the β and γ subunits. The activated G_q subunit in turn stimulates the PLC-β, which hydrolyzes PIP_2 to IP_3 and DAG. Both molecules are second messengers: they act to increase intracellular calcium and activate protein kinase C, respectively. (b) G_s- or G_i-coupled GPCR signaling pathway. For a G_s-coupled GPCR, agonist binding results in the activation of adenylate cyclase (AC) to synthesize cAMP. For a G_i-coupled GPCR, agonist binding results in inhibition of AC activity and hence decreases cAMP levels. (c) EGFR signaling pathway. EGF (agonist) binding to EGFR induces the dimerization of EGFRs, resulting in activation of its tyrosine kinase domain located on the intracellular side of EGFR. The activated tyrosine kinase domain in turn activates PI 3-kinase and PLC-γ, another PLC family member that hydrolyzes PIP_2 to IP_3 and DAG.

(Figure 21.2). The short half-life (seconds) of IP_3 and the cyclic nature of this pathway mean that quantitation requires blockade of degradation and recycling. Lithium inhibits inositol monophosphatase and inositol polyphosphate-1-phosphatase [4], such that inclusion of lithium in assays inositol causes accumulation of phosphates (IPs) inside cells (Figure 21.2).

21.3 OVERVIEW OF IP ASSAYS

Receptor activation by its agonist can be determined by the measurement of PLC activity by assaying IP_3. In practice, total IPs are usually quantitated based on the assumption that the total mass of soluble IPs in the cell is derived from metabolism of IP_3. This is accomplished by including excess lithium in the assay medium (typically 10 mM). This signal reflects the PLC activity as the measurement of IP_3 mass produced. Generally, ^3H-*myo*-inositol is incubated with cells in culture resulting in incorporation of radiolabel into phosphatidylinositol lipids (including PIP_2) in the cell membrane. Two conventional methods have been used extensively to measure IP production in the last two decades. One is the high performance liquid chromatography (HPLC)-based separation of ^3H-IP_3 and other inositol phosphates from original ^3H-inositol in cell extracts [5–7]. This approach allows resolution and quantitation of individual inositol phosphates and as such represents a high

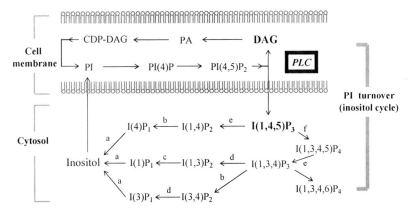

FIGURE 21.2 Schematic illustration of phosphatidylinositol turnover and lithium-sensitive enzymes. (a) Inositol monophosphatase (IMPase). (b) Inositol polyphosphate 1-phosphatase (IPPase). (c) Inositol polyphosphate 3-phosphatase. (d) Inositol polyphosphate 4-phosphatase. (e) Inositol polyphosphate 5-phosphatase. (f) Inositol (1,4,5) trisphosphate 3-kinase. IMPase and inositol polyphosphate 1-phosphatase are lithium-sensitive enzymes; the other enzymes are not sensitive to lithium. Abbreviations: I(1)P₁ etc., D-*myo*-inositol-1-phosphate. DAG, diacylglycerol. PA, phosphatidic acid. PI, phosphatidylinositol. CDP-DAG, cytidine 5′-diphospho-diacylglycerol. PLC, phospholipase C.

information content, low-throughput option. It has been used primarily for mechanistic studies of enzyme reactions and compound action in the PI turnover pathway. Other methods involve stepwise or total elution of inositol phosphates from anion exchange resins [6], which have high sample recovery and reproducibility with modest throughput, although it still involves a complicated assay procedure.

Although 96-well plate format assays have been introduced by two groups [8,9], these are not suitable for high-throughput screening (HTS) due to an excessive number of assay steps including cell washes, cell lysis, and separation of ³H-IPs from the original ³H-*myo*-inositol label and other lipids. A competitive binding assay exists that measures IP₃ based on the specificity of a fragment of the IP₃ receptor and biotinylated IP₃ (see the AlphaScreen section in the PerkinElmer Web site, www.las.perkinelmer.com). However, this assay is complex to run and expensive, and it was not consistent or reproducible in our hands. This may be related to the short half-life and capricious nature of IP₃ in cell extracts.

Recently, two methods for the assay of cellular inositol phosphates (a.k.a., PI turnover assay) with relatively high throughput and much-simplified assay formats have been reported [10,11]. Both methods use scintillation proximity assay (SPA) beads to capture and detect ³H-IPs selectively over ³H-inositol, which converts the labor-intensive separation step to one homogeneous detection step. The fact that essentially the same method was developed independently by two industry-based groups illustrated the burgeoning need to remove the barrier to a high-throughput and automatable PI turnover assays. The EC₅₀ values of agonists and antagonists, as well as antagonist pA₂ values evaluated in both studies, correlated well with values reported in preceding literature. The throughput of these SPAs is significantly better than that of previous quantitation methods. Ironically, we originally developed the SPA method as an approach to high-throughput screening for inhibitors of inositol phosphatase in a cell-based format. There was, and still is, no other way to do this. We have now reported the results of this robotic screen for inositol phosphatase inhibitors, which was run in a 384-well format with a throughput of 120 plates per day [7]. These recent advances in assay technologies have greatly improved the robustness of assay formats, improved screening throughput, and broadened the applicability of the PI turnover assay. Since the focus of this text is assay development, and since the great utility of the IP-SPA method is in HTS, the remainder of this text will focus on the technical details of the SPA method punctuated with example data from Merck labs collected in the 384-well semiautomated format.

TABLE 21.1
Yttrium Silicate Scintillation Proximity Assay Beads Efficiently Detect ^{3}H-*myo*-Inositol 1-Phosphate, but Not ^{3}H-*myo*-Inositol

	Radioactivity (cpm)		
	Microscint-20	YSi SPA beads	Efficiency (%)
^{3}H-*myo*-Inositol	9539 ± 151	260 ± 9.0	2.7
^{3}H-*myo*-Inositol 1-phosphate	7712 ± 604	4633 ± 331	60
Background	25 ± 6.1	2.7 ± 0.6	n.r.[a]

Note: Tests were carried out in white 96-well plates (Picoplate-96, Packard). For radioactivity measurements using yttrium silicate (YSi) scintillation proximity assay (SPA) beads (Amersham Biosciences), each well contained, in a final volume of 100 μL, 1 mg of YSi SPA beads suspended in 90 μL of water added to either 10 μL of 10 μM ^{3}H-*myo*-inositol (^{3}H-Ins, specific activity ~ 0.1 Ci/mmol), 1 mM ammonium phosphate, pH 8.0, or 10 μL of 10 μM ^{3}H-*myo*-inositol 1-phosphate (^{3}H-Ins-1-P, specific activity ~ 0.1 Ci/mmol), 1 mM ammonium phosphate, pH 8.0, or 10 μL of 1 mM ammonium phosphate, pH 8.0 (background). For comparison, the radioactivity of the ^{3}H-Ins and ^{3}H-Ins-1-P samples used was measured by adding 90 μL of scintillation fluid (Microscint-20, Packard) to 10 μL of each of the above solutions. Plates were sealed with adhesive, clear plastic cover sheets (Topseal-A, Packard). The contents were mixed by vigorous shaking on a microtiter plate shaker for 1 h, and radioactivity was measured using a Topcount NXT (Packard) liquid scintillation counter. YSi SPA beads were allowed to settle for 2 h before counting. Efficiency is the radioactivity measured using YSi SPA beads relative to the radioactivity measured using Microscint-20 scintillation fluid for the indicated sample. The values reported are means ± s.d. (n = 3).

[a] n.r., not relevant.

21.4 SPA-BASED INOSITOL PHOSPHATE ASSAY (IP-SPA)

The barrier to high-throughput for conventional, chromatography-based PI turnover assays is the requirement for separation of IPs from inositol. Functionally, that means washing an ion exchange matrix to which IPs are bound to remove the inositol. The premise for development of our IP-SPA method then was that a SPA bead could function as an ion exchange matrix, i.e., bind inositol phosphates but not inositol. Nonderivatized yttrium silicate SPA beads (Amersham Biosciences, catalog number RPNQ0013) are marketed for the capture and quantitation of RNA. We found that these beads, which are impregnated with cerium and carry a net positive charge, preferentially capture the negatively charged IPs compared to noncharged inositol (Table 21.1). The experimental data showed that the YSi beads detected ^{3}H-*myo*-inositol-1-phosphate (^{3}H-Ins-1-P) approximately twentyfold more efficiently than ^{3}H-Ins. Similarly, YSi beads could be used to detect ^{3}H-Ins-1,4-P$_2$, or ^{3}H-Ins-1,4,5-P$_3$, with a comparable signal magnitude being measured from each InsP at a given concentration and specific activity. Thus, YSi SPA beads can be used to detect radiolabeled inositol phosphates in the presence of radiolabeled inositol [10].

Data shown here and in published papers demonstrate the general applicability of this method across cell and receptor types [10,11]. In fact, the general method is the same as for the traditional PI turnover (see Figure 21.3), albeit with modifications to make it suitable for miniaturization and automation. Thus, the considerations regarding cell line selection in the case of recombinantly expressed receptors are the same as for any typical HTS assay. The examples to date have been with adherent cell lines. After drug treatment/receptor activation, the cells are lysed with a dilute (hypotonic) solution of formic acid (typically 10 to 100 mM) to extract the soluble inositol phosphates and inactivate inositol phosphatases. The cell membranes will contain radiolabeled inositol phospholipids, which if released as particulates could interfere with detection of the soluble inositol phosphates. This is not an issue in the conventional chromatography PI turnover method because soluble inositol phosphates are eluted from the resin whereas insoluble particulates are

IP₃ assay design

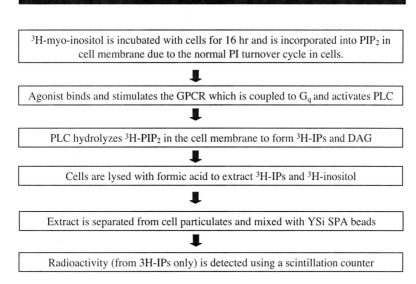

FIGURE 21.3 Procedure for measurement of GPCR activity using IP-SPA.

not. We found that after lysis with formic acid, cell bodies would remain on the bottom of the plate for cells that were adherent under typical culture conditions. In fact, in establishing a screen for inositol phosphatase inhibitors, we selected the T24 parental cell line, among other reasons, because it is strongly adherent. This property was a selection criterion in preparation of the M1-T24 cell line used for the inositol phosphatase screen [7] and in the example data shown in Figure 21.4. This feature of the assay method necessitated the aspiration of the supernatant lysate and mixing with the YSi SPA beads in a separate plate. We have not investigated ways to permit addition of the SPA beads directly to the well containing lysate and cells. However, Liu et al. reported successful incorporation of this into their assay method using polyvinyltoluene (PVT) SPA beads [11].

We originally developed the IP-SPA method in a 96-well format. In the process of adapting the assay to the 384-well format, we systematically tested several variables in ways expected to lead to increased signal and decreased liquid handling steps. These were the amount of ³H-Ins per well used to label the cells (1 to 10 μCi), the volume of medium in which the ³H-Ins was delivered to the cells (80, 120, or 200 μL), and the volume of formic acid used to lyse the cells (50, 75, 100, 150 or 200 μL). In each case, the change in the measured signal was proportional to the change in the given variable over the range tested. This means that there is a lot of room in this method for increasing the signal magnitude, and this is important for the assay of endogenous receptors, or poorly expressed/coupled recombinant receptors. Increasing the amount of radiolabel used was a very easy way to boost the signal, but in HTS the cost of doing this can be prohibitive. We have used as much as 10 μCi per well to label cells when the sample extract was destined for HPLC chromatography. We tested the time required to label the cells to equilibrium, but found that this took >7 h. Therefore, for the sake of convenience, we labeled cells overnight. Lowering the concentration of formic acid used to lyse the cells from 100 to 10 mM did not change the final signal measured (data not shown). This was important because the final concentration of formic acid in the SPA detection step should be kept at or below 20 mM: higher concentrations of formic acid reduced the signal-to-noise ratio. The reduced concentration used for lysis removed the requirement for dilution of the lysate going into the detection step. Conveniently, with respect to automation, we found that it was not necessary to wash the cells after labeling with ³H-Ins. The bulk of this label is removed when the assay medium is replaced with formic acid for lysis.

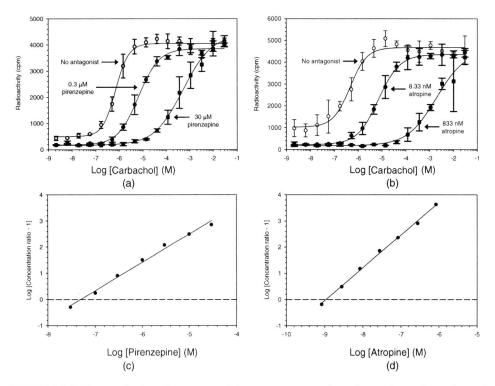

FIGURE 21.4 Characterization of two muscarinic receptor antagonists, pirenzepine and atropine, using the IP-SPA method in a 384-well, semiautomated format. M1-T24 cells were seeded into 384-well tissue culture plates (Corning 3701) at a density of 3×10^4 cells per well, and cultured for 1 day. Cells were then labeled with ^{3}H-Ins using 0.25 µCi of radiolabel in 20 µL IF-DMEM + 0.3% bovine serum albumin (BSA) per well, for approximately 16 h as described in [10]. Cells were not washed before proceeding to the assay steps. Five microliters of IF-DMEM + 0.3% BSA containing LiCl and antagonist (pirenzepine or atropine) was added to each well, and the cells were incubated for 10 min at 37°C. A further 5 µL of the same medium containing carbachol was added to each well, and the cells were incubated for 30 min at 37°C. The concentrations indicated are final, after the addition of carbachol. Extracts were prepared by aspirating the medium from the wells and adding 40 µL of 10 mM formic acid per well and incubating at room temperature for 20 min. Twenty-five microliters of each extract was mixed with 0.25 mg of YSi SPA beads in white 384-well plates in a final volume of 35 µL. The beads were allowed to settle for 2 h before measurement of radioactivity using a Topcount-384 (Packard). The values reported are means ± s.d. (n = 3). (a) Cells were treated with varying concentrations of carbachol in the presence of 10 m*M* LiCl and 0, 0.03, 0.1, 0.3, 1, 3, 10, or 30 µ*M* pirenzepine. For clarity, only the data for the no antagonist (○), 0.3 µM pirenzepine (●), and 30 µM pirenzepine (■) carbachol dose–response curves are shown. (b) Cells were treated with varying concentrations of carbachol in the presence of 10 mM LiCl and 0, 0.83, 2.77, 8.33, 27.7, 83.3, 277, or 833 nM atropine. For clarity, only the data for the no antagonist (○), 8.33 n*M* atropine (●), and 833 nM atropine (■) carbachol dose-response curves are shown. EC_{50} values were estimated from the dose response curves and used to construct Schild plots for pirenzepine (c) and atropine (d). This analysis yielded pA_2 values of 7.3 and 9.0 for pirenzepine and atropine, respectively. (From Brandish, PE et al. *Anal Biochem.* 2003; 313(2):311–318. With permission.)

The IP-SPA method is rather robust, and this meant that miniaturization from the 96-well format to the 384-well format did not require solving any significant technical problems. It really did work the first time. Once established in the 384-well format, we tested the signal *vs.* amount of beads added per well and the number of cells per well that gave a near confluent monolayer at the time of labeling. The former variable is decided based on a compromise between signal and cost of the beads. The first example, shown in Figure 21.4, is the characterization of the muscarinic acetyl-choline receptor antagonists pirenzepine and atropine as reported in [10]. The specific method used

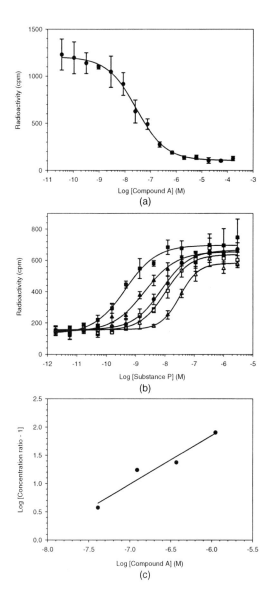

(a)

(b)

(c)

FIGURE 21.5 Functional characterization of an NK1 receptor antagonist using the IP-SPA method in a 384-well format. CHO cells expressing the human NK1 receptor growing in 384-well plates were washed once with HANKS buffer and incubated at 37°C for 18 h with 20 μL/well of ^{3}H-*myo*-inositol (10 μCi/mL) in inositol-free DMEM containing 0.2% BSA. Cells were then treated with LiCl (final concentration = 5 mM) and antagonist for 10 min at 37°C, followed by addition of agonist (substance P) and incubation for a further 20 min at 37°C. Cells were washed once with HANKS buffer before lysis with 10 mM formic acid for 20 min at room temperature. As for the M1-T24 cell line, the cell bodies were observed to remain attached to the plate following treatment with formic acid. A portion of resulting solution from each well was aspirated to a white 384-well plate, and 0.25 mg of YSi beads were added to each sample. The plate was counted in a scintillation plate reader after 2 h incubation at room temperature. (a) Cells were treated with 5 nM substance P (the concentration giving 70% of the maximal signal in the absence of antagonist) in the presence of the indicated concentrations of NK1 receptor antagonist, compound A. The EC$_{50}$ value of compound A was 17 nM. (b) Substance-P concentration–response curves were measured in the absence and presence of various concentrations of compound A: 0 ($\blacksquare$), 41 ($\blacktriangle$), 123 ($\bullet$), 370 ($\circ$), and 1100 ($\triangle$) nM. The concentration–response curves of substance P were parallel shifted to right in the presence of antagonist. (c) Schild analysis for compound A gave a pA$_2$ value of 8.13, equivalent to a K$_i$ value of 7.4 nM, with a slope of 0.86.

in this experiment is described in the legend to the figure. As noted in that report, the pA$_2$ values (which are an approximation to pK$_B$, the negative log of the dissociation constant for the receptor–antagonist complex) are in reasonable agreement with literature values of pK$_B$ of 7.8 to 8.5 and 9.0 to 9.7 for pirenzepine and atropine, respectively, at M1 receptors. This experiment served as validation of the technique. In the second example shown (Figure 21.5), we used the 384-well IP-SPA method to characterize a new neurokinin-1 (NK1) receptor antagonist, dubbed compound A. Characterization of a NK1 receptor antagonist using CHO cells expressing the human NK1 receptor is coincidentally the same system used for validation by Liu et al. [11]. Again, the method is described in the legend to the figure. The response to substance P was concentration-dependently inhibited by the antagonist (Figure 21.5a). Schild analysis revealed that this antagonist was competitive, with a pA$_2$ value of 8.13 (equivalent to a K$_i$ value of 7.41 nM) (Figure 21.5b and Figure 21.5c).

This general assay method has now been used by numerous experimenters on multiple programs around Merck's research labs, and experience has shown that there are no special tricks or knowhow associated with the assay beyond that described here and in the published papers. All parameters of the assay, including buffer constituents, incubation times, and orders of reagent addition, can be varied to suit the cell line in question and the laboratory equipment to hand. So the expectation is that with flexibility inherent in the procedure, this assay should be readily applicable and usable in support of discovery efforts for any G protein– or tyrosine kinase–linked receptor that can be coupled to PI turnover. In summary, this scintillation proximity assay for inositol phosphates is suitable for compound testing to support lead optimization in drug discovery and can be tailored to the needs of specific projects. In addition, it can be adapted into automated robotic assay format for the primary compound screening [7] and for confirmatory screening to complement more miniaturizable screening formats such as the beta-lactamase reporter assay.

REFERENCES

1. Rhee SG. Regulation of phosphoinositide-specific phospholipase C. *Annu Rev Biochem* 2001; 70:281–312.
2. Berridge MJ. Inositol trisphosphate and calcium signalling. *Nature* 1993; 361:315–325.
3. Downes CP, Macphee CH. *myo*-Inositol metabolites as cellular signals. *Eur J Biochem* 1990; 193:1–18.
4. Majerus PW. Inositol phosphate biochemistry. *Annu Rev Biochem* 1992; 61:225–250.
5. Barnaby RJ. Mass assay for inositol 1-phosphate in rat brain by high-performance liquid chromatography and pulsed amperometric detection. *Anal Biochem* 1991; 199:75–80.
6. Dean NM, Beaven MA. Methods for the analysis of inositol phosphates. *Anal Biochem* 1989; 183:199–209.
7. Zheng W, Brandish PE, Kolodin DG, Scolnick EM, Strulovici B. High-throughput cell-based screening using scintillation proximity assay for the discovery of inositol phosphatase inhibitors. *J Biomol Screen* 2004; 9:132–140.
8. Chengalvala M, Kostek B, Frail DE. A multi-well filtration assay for quantitation of inositol phosphates in biological samples. *J Biochem Biophys Meth* 1999; 38:163–170.
9. Tian Y, Wu L, Chung F. High throughput 96-well plate assay for receptor-mediated phosphatidyl-inositol turnover. *J Biomol Screen* 1997; 2:91–97.
10. Brandish PE, Hill LA, Zheng W, Scolnick EM. Scintillation proximity assay of inositol phosphates in cell extracts: high-throughput measurement of G-protein–coupled receptor activation. *Anal Biochem* 2003; 313(2):311–318.
11. Liu JJ, Hartman DS, Bostwick JR. An immobilized metal ion affinity adsorption and scintillation proximity assay for receptor-stimulated phosphoinositide hydrolysis. *Anal Biochem* 2003; 318:91–99.

22 Measuring Calcium Mobilization with G_q-Coupled GPCRs Using the Fluorometric Imaging Plate Reader (FLIPR)

John Dunlop, Yingxin Zhang, Robert Ring, and Dianne Kowal

CONTENTS

22.1 INTRODUCTION

An increase in cytosolic free calcium ion (Ca^{2+}) concentration represents a ubiquitous intracellular signaling mechanism influenced by a diverse array of extracellular signaling molecules, including hormones, neurotransmitters, and growth factors. Calcium signaling impacts on the complete repertoire of normal cellular functions, while dysregulation of calcium signaling and calcium overload are critical triggers for cell death. Consequently, intensive efforts have focused on the development and utilization of tools for monitoring changes in intracellular calcium concentration. In this regard, fluorescent dye indicators for intracellular calcium measurements have become widely used tools in biology. This has been particularly exploited in the area of drug discovery, more specifically in the field of G-protein–coupled receptor (GPCR) pharmacology and functional characterization. The use of fluorescent dye indicators in combination with the fluorometric imaging plate reader (FLIPR), a high-throughput platform for 96-well and 384-well cell-based assays, has been adopted universally in the pharmaceutical industry.

    One of the first calcium indicator dyes to receive widespread use, FURA-2 [1], reports increases in intracellular calcium by virtue of changes in its fluorescence signal upon calcium binding. In the case of FURA-2, the spectral shift following calcium binding allows for ratiometric measurement of calcium concentration with dual-wavelength ultraviolet (UV) excitation and single-wavelength emission. FURA-2 has largely been replaced by newer-generation calcium indicators such as FLUO-3 [2] and FLUO-4 [3], offering greater sensitivity with larger changes in fluorescence intensity following calcium binding, excitation with visible light sources, and lack of a significant spectral shift following calcium binding, limiting their use typically to assessment of qualitative changes in calcium. High-throughput screening demands for homogeneous assays have resulted in the introduction of a number of no-wash calcium assay kits specifically in support of the FLIPR platform. These reagents, the FLIPR Calcium, Calcium Plus, and Calcium 3 assay kits, utilize quenching technology [4] to eliminate background fluorescence, thereby providing a homogeneous no-wash assay format as an alternative to the more time consuming FLUO-3 and FLUO-4 wash protocols. Despite the significant advantage of a homogeneous assay, the FLIPR Calcium and Calcium Plus assay kits have not eliminated the use of FLUO-3 and FLUO-4, which are often more sensitive and yield better responses across a more diverse spectrum of biological targets. In the case of the FLIPR Calcium 3 assay kit, a 2003 newcomer to the calcium indicator dye reagent assortment, it remains to be fully evaluated.

Limitations of the current FLIPR dyes aside, the utility of the FLIPR in performing functional characterization of receptors is well validated, as evidenced by an increasing number of publications. In particular, the characterization of GPCRs signaling through the G_q-phospholipase C cascade, leading to increases in intracellular calcium, has been facilitated by the introduction of this high-throughput technology. To date, a number of G_q-coupled receptors have been profiled in some detail with respect to both agonist and antagonist pharmacology including the 5-HT_2 receptor subfamily [5,6], the orexin receptors [7–9], the histamine H1 subtype [10], bradykinin B1 and B2 receptors [11], the bombesin BB2 receptor [12], and the P2Y purinergic receptors [13]. In addition, the VR1 vanilloid receptor, a receptor-gated cation channel highly permeable to calcium ions, has also been extensively studied using this approach [14,15]. In addition to the G_q-coupled receptors, it has been possible to study G_s- and G_i-coupled receptors using the FLIPR assay platform by taking advantage of the promiscuous G protein, G15/16, or using chimeric G proteins comprising G_q with the C-terminal five amino acids replaced with those from G_s or one of the G_i family members [16]. By taking advantage of this approach, the G_s-coupled 5-HT_6 [17] and 5-HT_7 [18] receptors, and the G_i-coupled GABA-B [19], H3 [20], and mGlu 2 and 4 receptors [21] have been profiled pharmacologically.

In this chapter, we present a comparison of the various fluorescent calcium indicator dyes for the measurement of functional responses following activation of three different G_q-coupled GPCRs, the 5-HT_{2C} subtype of the serotonin receptors, the cholecystokinin-B neuropeptide receptor, and the metabotropic glutamate (mGlu) receptor subtype 5. In addition, important considerations for transition of assays from 96- to 384-well high-throughput screening (HTS) format are discussed.

22.2 MATERIALS AND METHODS

22.2.1 MATERIALS

1. Three "No Wash" protocol calcium indicator dyes: Calcium Assay Reagent Kit. Component A (Calcium 1), Calcium Plus, and Calcium 3 Bulk Assay Kits (Molecular Devices) are stored at −20°C.
2. Two "Wash" protocol calcium indicator dyes: Fluo-3/AM (acetoxymethyl ester) and Fluo-4/AM (Molecular Probes). Two-millimolar stocks are prepared by dissolving in dimethylsulfoxide (DMSO). Stock solutions may be aliquoted and stored at −20°C for subsequent use.

3. Hank's balanced salt solution (HBSS) buffer (Mediatech) supplemented with 20 mM HEPES (Invitrogen/Gibco) and 2.5 mM probenecid (see Materials, step 4), pH 7.4 (FLIPR buffer).

4. Probenecid (anion exchange inhibitor) (SIGMA); prepare a 250 mM probenecid stock solution as follows:

   • Probenecid (710 mg) is dissolved in 5 mL 1 N NaOH, add 5 ml HBSS/20 mM HEPES. A 1:100 dilution of the stock is supplemented into all wash and Dye Loading Buffers prepared on the day of the assay to help prevent dye efflux from the cells.

5. Pluronic acid F-127, 20% (w/v) in DMSO (Molecular Probes).

6. Fetal bovine serum, dialyzed and heat inactivated (SIGMA).

7. Glutamic pyruvic transaminase (GPT), porcine heart derived (Calzyme Labs Inc.).

8. Pyruvic acid (SIGMA).

9. Agonists used in this study: serotonin (5-HT) (SIGMA), cholecystokinin-8S (CCK-8S) (Peptide Products), and L-glutamate (SIGMA).

10. Costar 96-well black/clear bottom tissue culture-treated plates (Corning). These plates are used for 5-HT$_{2C}$ and CCK-B.CHO cell line setup.

11. Biocoat poly-D-lysine 96-well black/clear bottom plates (Becton Dickinson). These plates are used for mGluR5.HEK cell line setup. *Note*: Use of poly-D-lysine or other coating matrices is recommended in cases where cell adherence may be problematic for a plate-based assay. HEK cells are a good example of this as are many primary cell culture preparations. CHO cells, on the other hand, do not perform as well when plated onto a poly-D-lysine matrix.

12. Black disposable pipet tips (Robbins Scientific).

22.2.2 Dye Loading with Calcium 1, Calcium Plus, and Calcium 3: "No-Wash" Protocol

1. Cells are plated approximately 24 h prior to the FLIPR assay at the following cell densities in their appropriate black/clear bottom plates to ensure an adherent confluent monolayer of cells (37°C/5% CO$_2$): 5-HT$_{2C}$ and CCK-B.CHO: 50,000 cells/well; mGluR5.HEK: 80,000 cells/well.

2. On the day of the assay, prepare the Calcium 1, Calcium Plus, and Calcium 3 Dye Loading Buffers as follows:

   • Remove one vial of lyophilized FLIPR Assay Dye Reagent from the freezer and equilibrate at room temperature. Dissolve the contents of the vial with 200 mL HBSS containing 20 mM HEPES, pH 7.4.

   • Supplement with freshly prepared 2.5 mM probenecid (Materials, step 4) to the volume of Dye Loading Buffer to be used on the day of the experiment.

   • For mGluR5-expressing cells, 3 U/ml GPT and 3 mM pyruvic acid are also added to the Dye Loading Buffer to prevent receptor desensitization by intracellular glutamate released into the buffer during the loading process.

   • Aliquots of Dye Loading Buffer in the absence of probenecid, GPT, and pyruvic acid can be refrigerated or frozen for up to 5 days without loss of activity.

3. The culture medium is then manually aspirated and replaced with 180 μL/well of Dye Loading Buffer as prepared above. *Note*: An alternate approach is to add Dye Loading Buffer directly to the cell culture medium, adjusting the dye concentration in the loading buffer accordingly, eliminating the aspiration step. An important factor to consider here is any potential interference of medium/serum components with the assay.

4. The cells are incubated in a 37°C/5% CO$_2$ incubator for 1 h.

22.2.3 Dye Loading with Fluo-3/AM and Fluo-4/AM: "Wash" Protocol

1. Cells are plated as indicated above for the "No Wash" protocol.
2. The next day, prepare the "Wash" protocol Fluo-3/AM and Fluo-4/AM Dye Loading Buffers as follows:
   - For each 96-well assay plate, combine 12 mL FLIPR buffer with 120 μL FBS.
   - Add 25 μL Fluo-3/AM or Fluo-4/AM stock and 25 μL pluronic acid to achieve final concentrations of 4 μM and 0.04%, respectively.
   - In the case of mGluR5-expressing cells, see methods "No Wash" protocol, step 2c.
3. The culture medium is manually aspirated, and the cells are washed once or twice with 150 μL FLIPR buffer.
4. One hundred microliters of Dye Loading Buffer is added to each well, and the plates are returned to a 37°C/5% CO_2 incubator for 1 h.
5. The Loading Buffer is aspirated and the cells are washed once with 200 μL FLIPR buffer. Assay plates containing 5-HT$_{2C}$ and CCK-B.CHO cells receive a final volume of 180 μL FLIPR buffer, while mGluR5-expressing cells are provided FLIPR buffer supplemented with 3 U/mL GPT and pyruvic acid. The mGluR5-containing plates are returned to a 37°C/5% CO_2 incubator for at least an additional 20 min to allow the degradation of potential extracellular glutamate after the wash process.

22.2.4 Fluorescence Determination by FLIPR

1. Once the cells have completed the dye loading process, the plates are transferred to the FLIPR chamber and assayed one at a time.
2. Typical FLIPR system parameters are as follows:
   - Excitation wavelength — argon ion laser: 488 nM.
   - Emission wavelength: 525 nM.
   - Camera aperture: F Stop/2.
   - Camera exposure time: 0.4 sec.
   - Pipettor height: 150 μL.
   - Pipettor dispense speed: 40 μL/sec.
   - Laser power range: approximately 300 to 800 mW, depending on the results of the "signal test" (see step 3 below).
     Note: These settings are based on cells with good dye loading properties. In cases where poor dye loading is experienced, signals can be enhanced in a number of ways, including plating more cells (ensuring a confluent monolayer), increasing dye concentration in the loading step (limited by potential dye toxicity), increasing laser power and/or exposure time, and increasing camera aperture (or decreasing the F-stop).
3. A "signal test" is performed on each plate to determine the uniformity of the dye loading of the cells and to adjust the laser output needed to obtain a background fluorescence range of approximately 10,000 to 15000 and 7000 to 10,000 relative fluorescence units (RFUs) per well for "Wash" and "No Wash" protocols, respectively.
4. The "plate viewer" option may be used to reveal potential cell plating and/or dye loading differences among the wells in the event that a "signal test" indicates a variable distribution of background RFUs.
5. The "compound plate" is placed in the lower right-hand position of the FLIPR chamber. This conventional microtiter plate contains 10× concentrations of the agonists 5-HT, CCK-8S, and L-glutamate, prepared in FLIPR Buffer, over a range needed to determine EC$_{50}$ values at 5-HT$_{2C}$, CCK-B, and mGluR5, respectively.

6. The assay is initiated by the simultaneous addition of 20 μL of agonist from the "compound plate" by the FLIPR fluidics system into each well (180 μL volume) of dye-loaded cells to achieve a final volume of 200 μL. *Note*: Volumes can be varied depending on the specific application, e.g., a reduced volume for agonist addition can be utilized for expensive or limited reagents.

7. Agonist-stimulated calcium mobilization is measured by a time-dependent kinetic readout produced by the initial collection of 10 1-sec baseline counts followed by RFUs determined every second for 55 sec and five 6-sec readings. *Note*: The timing of data collection is dependent on the nature of the experiment and is controlled via the FLIPR software.

22.2.5 DATA EXAMPLES

For the purposes of comparison, data are presented for three different G_q-coupled GPCRs: two rhodopsin-like family A receptors, the 5-HT$_{2C}$ subtype of serotonin receptors and the CCK-B subtype of cholecystokinin receptors, activated by the neuropeptides CCK-8S and CCK-4, and the family C metabotropic glutamate receptor subtype 5. While the intention is to illustrate a few examples of data generated using the different dyes, no generalizations can or should be made, and it is recommended that investigators evaluate the various dye options using their own derived cell lines. Perhaps a good example of this is from our own observations with a CHO cell line expressing mGlu receptor 4, where we have found that results with the Calcium 3 dye have been highly variable (Kowal and Dunlop, unpublished observations), yet the same dye has performed well with many of the other cell lines we have examined.

An example of the raw data output obtained from a 96-well plate run on FLIPR is shown for the 5-HT stimulated calcium signaling in 5-HT$_{2C}$ receptor-expressing cells (Figure 22.1), in this case using the Calcium 3 dye. A graphical display of the time sequence for the change in relative fluorescence units (RFUs), a direct index of change in intracellular calcium concentration, is captured for each well. Average temporal profiles for the calcium responses elicited following addition of maximally effective concentrations of 5-HT, CCK-8S, or L-glutamate to 5-HT$_{2C}$, CCK-B, or mGlu receptor 5-expressing cells are presented (Figure 22.2 through Figure 22.4). Visual inspection of the data immediately provides a qualitative comparison of the different sensitivities associated with each dye. A calculation of the maximum minus minimum RFUs, achieved within the FLIPR software, is typically used to provide a quantitative assessment of the maximum change in signal in response to agonist, correspondingly the dynamic range of the assay. The greater the change in fluorescence observed, the better the signal:noise ratio associated with the assay. In the examples shown here, use of Calcium 3 yields the maximum sensitivity for each of the receptor systems examined. In the case of the 5-HT$_{2C}$ receptor responses, the performance of Fluo-4, Fluo-3, and Calcium Plus are comparable with Calcium 1 clearly inferior. For the CCK-B receptor responses, Fluo-4 performs slightly better than Calcium Plus, which is superior to Fluo-3, and again Calcium 1 is inferior. Lastly, the Calcium Plus dye is most sensitive after Calcium 3 for the mGlu receptor 5 responses, Fluo-3 and Calcium 1 are comparable, and Fluo-4 performs least well in this instance. Again, it is important to emphasize that any of these cell lines generated in the hands of other investigators might well perform differently when comparing dye responses than described here.

Data in the form shown in Figure 22.1 can be transformed to generate log-concentration response curves for agonist-stimulated calcium mobilization defining pharmacology for functional response. This is achieved by expressing each of the maximum minus minimum data point calculations at each agonist concentration as a percentage of the response observed with the maximally effective concentration of agonist (the defined 100%). Log-concentration response curves for 5-HT stimulation of 5-HT$_{2C}$ receptors (Figure 22.5), CCK-8S stimulation of CCK-B receptors (Figure 22.6), and L-glutamate stimulation of mGlu receptor 5 (Figure 22.7) are illustrated.

Some important points are evident here. Although the dye sensitivities differed quite substantially when evaluating the maximum window of fluorescence change for the 5-HT–stimulated

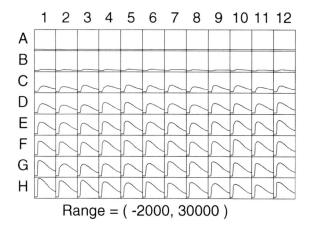

FIGURE 22.1 Typical calcium traces observed in 5-HT$_{2C}$-expressing cells stimulated with increasing log-concentrations of 5-HT (–11 to –5 in Rows B–H, respectively) using the Calcium 3 dye in a 96-well format. Row A represents the effect of the addition of FLIPR buffer alone.

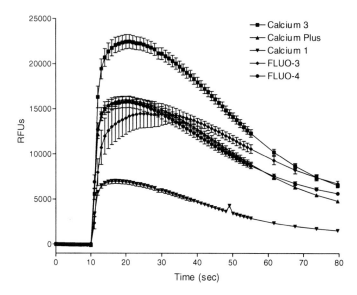

FIGURE 22.2 Comparison of time sequence kinetic data of agonist-stimulated calcium signaling in 5-HT$_{2C}$-expressing CHO cells using five calcium indicator dyes. A maximally effective concentration of 5-HT (10 μM) was added after 10 1-sec baseline counts, and the resulting changes in fluorescence over time were measured. Data measured from a representative experiment are shown as mean values ± SEM from 12 replicates for each time point.

response (Figure 22.2), the derived EC$_{50}$ values for 5-HT were identical, as is evident from the essentially overlapping log-concentration response curves (Figure 22.5; Table 22.1). Use of different dyes in this case has no impact on the functional pharmacology, despite the different dynamic ranges associated with each. In contrast, use of the various dyes gave rise to substantially different EC$_{50}$ values for the stimulation of the CCK-B receptor by CCK-8S (Figure 22.6; Table 22.1). While in this case Calcium 3 offered the greatest sensitivity with respect to dynamic range of the assay, Fluo-3 and Fluo-4 afforded better sensitivity in terms of the estimated agonist potency. Notably, the functional pharmacology determined using Fluo-3 and Fluo-4 is similar to that determined previously by ourselves and others using different functional endpoints. This highlights the importance of not only considering the dynamic range of an assay but also ensuring that receptor

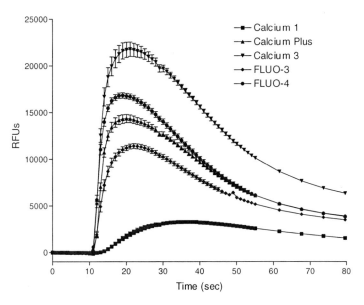

FIGURE 22.3 Comparison of time sequence kinetic data of agonist-stimulated calcium signaling in CCK-B-expressing CHO cells using five calcium indicator dyes. A maximally effective concentration of CCK-8S (10 μ*M*) was added after 10 1-sec baseline counts, and the resulting changes in fluorescence over time were measured. Data measured from a representative experiment are shown as mean values ± SEM from 12 replicates for each time point.

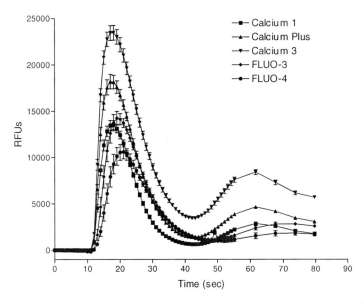

FIGURE 22.4 Comparison of time sequence kinetic data of agonist-stimulated calcium signaling in mGluR5-expressing HEK cells using five calcium indicator dyes. A maximally effective concentration of L-glutamate (100 μ*M*) was added after 10 1-sec baseline counts, and the resulting changes in fluorescence over time were measured. Data measured from a representative experiment are shown as mean values ± SEM from 12 replicates for each time point.

pharmacology is faithfully reproduced. Finally, in the example of the mGlu receptor 5–stimulated response, use of the different dyes, although offering different degrees of maximal sensitivity, had minimal effects on the determined EC_{50} values for L-glutamate (Figure 22.7, Table 22.1).

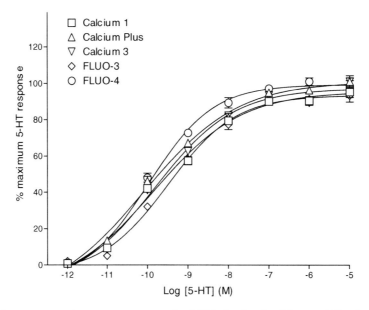

FIGURE 22.5 Log-concentration response curves for 5-HT-stimulated calcium mobilization in CHO cells expressing the 5-HT$_{2C}$ receptor using the five calcium dyes. Data are expressed as a percentage of the maximum response observed with 10 μM 5-HT and represent mean values ± SEM from 12 replicates for each concentration of 5-HT.

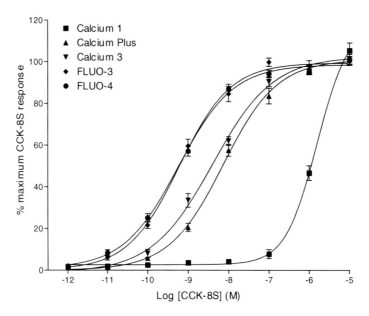

FIGURE 22.6 Log-concentration response curves for CCK-8S-stimulated calcium mobilization in CHO cells expressing the CCK-B receptor using the five calcium dyes. Data are expressed as a percentage of the maximum response observed with 10 μM CCK-8S and represent mean values ± SEM from 12 replicates for each concentration of CCK-8S.

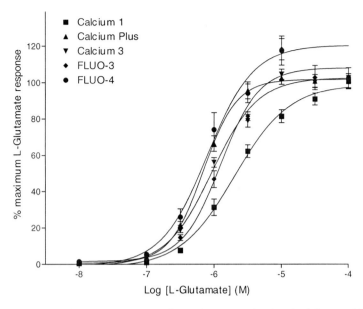

FIGURE 22.7 Log-concentration response curves for L-glutamate-stimulated calcium mobilization in HEK cells expressing mGluR5 using the five calcium dyes. Data are expressed as a percentage of the maximum response observed with 100 μM L-glutamate and represent mean values ± SEM from twelve replicates for each concentration of L-glutamate.

TABLE 22.1
EC_{50} Values (nM, 95% C.L.) of Reference Agonists Determined for $5\text{-}HT_{2C}$, CCK-B, and mGluR5 in a Comparison Study of Calcium Dye Indicators Using FLIPR

GPCR	Agonist	Calcium 1	Calcium Plus	Calcium 3	FLUO-3	FLUO-4
$5HT_{2C}$.CHO	5-HT	0.15	0.087	0.16	0.33	0.11
		(0.077–0.28)	(0.044–0.17)	(0.093–0.26)	(0.22–0.48)	(0.067–0.18)
CCK-B.CHO	CCK-8S	1560	7.3	4.0	0.61	0.58
		(1050–2318)	(5.5–9.8)	(2.9–5.5)	(0.45–0.81)	(0.45–0.75)
mGluR5.HEK	L-Glutamate	2020	691	935	1253	793
		(1595–2558)	(587–814)	(793–1102)	(995–1584)	(537–1171)

22.3   CONSIDERATIONS FOR TRANSITIONING G_q-COUPLED GPCR ASSAYS FROM 96- TO 384-WELL FORMAT

Miniaturization of 96-well formatted FLIPR assays to the 384-well microplate format is a necessary step toward achieving the high-throughput capacity required of screening large compound collections. To assist in the transition of validated G_q-coupled GPCR assays from 96- to 384-well format, we have highlighted here a number of experimental issues that should be considered.

22.3.1   BENCHMARKING OPTIMIZATION EFFORTS USING THE Z-FACTOR

Perhaps the most important consideration for transitioning assays from one format to another is accurately accounting for the relative success or failure of optimization/validation efforts employed during the transition process. Quality control statistics such as signal-to-noise ratios (S/N), signal-to-background ratios (S/B), and percent coefficient of variance (%CV) all provide valuable information regarding assay performance and are routinely used to identify issues with assay performance during the types of receptor characterization described above. "Hit identification," however, is the primary focus of the majority of 384-well based assays, and the Z-factor coefficient is the most commonly used statistical benchmark for evaluating the suitability of any assay for identifying active compounds in a HTS screening effort [22]. The Z-factor is a screening window coefficient that enables one to capture and assess variation in assay performance (e.g., signal measurements, dynamic range of signal) as a reflection of its suitability for process of hit identification. The use of a Z-factor score allows for simple categorization of assay quality, with high Z-factor values (1 > Z > 0.5) indicating excellent assays, and low values (0.5 > Z) indicating failure to achieve good separation between sample and control signal variation.

22.3.2 EYE TOWARD AUTOMATION

Most issues that require attention when transitioning 96- to 384-well format are related to an assay's tolerance of measures required to automate the preparation and handling of cells within the dimensions of a 384 microplate and the necessity for HTS to handle throughput. Manual "incubate-and-wash" protocols, such as those described earlier using FLUO-3 and FLUO-4, used frequently for 96-well assay formats, are commonly replaced with robotic dispensation and use of "mix-and-read" reagents in HTS. These no-wash dyes/protocols certainly help to reduce variability in data caused by cell detachment and assay performance with incubate-and-wash protocols in microplates. The exposure time of cells to these new dyes, however, can adversely affect cell health and performance, and thus needs to be carefully examined during any transition. This is a good example of an issue encountered as a function of the needs of HTS, specifically handling of large numbers of plates and longer dye exposure times, compared with the lower-throughput examples presented above where stricter adherence to an optimum dye loading time is possible. Tolerance of cells to compounds presented in the various solvent types (DMSO, MeOH, EtOH) used in large screening libraries can also be an issue to be defined during the transition. This is especially true for the G_q-coupled GPCRs, for dimethyl sulfoxide (DMSO) has known effects on intracellular calcium release in variety of cells [23].

22.4 CONCLUDING REMARKS

Assays measuring the function of G_q-coupled GPCRs using the FLIPR remain an integral component of the drug discovery process. Investigators have an assortment of fluorescent dye reagents at their disposal as they develop new assays. Although we have illustrated some examples of dye performance in this chapter, each new application merits an evaluation of various dyes to determine optimum performance.

REFERENCES

1. Grynkiewicz, G., Poenie, M., Tsien R.Y., 1985. A new generation of Ca^{2+} indicators with greatly improved fluorescence properties. *J. Biol. Chem.* 260, 3440–3450.
2. Minta, A., Kao, J.P., Tsien, R.Y., 1989. Fluorescent indicators for cytosolic calcium based on rhodamine and fluorescein chromophores. *J. Biol. Chem.* 264, 8171–8178.
3. Gee, K.R., Brown, K.A., Chen, W.N., Bishop-Stewart, J., Gray, D., Johnson, I., 2000. Chemical and physiological characterization of fluo-4 Ca^{2+}-indicator dyes. *Cell Calcium* 27, 97–106.
4. Krahn, T., Paffhausen, W., Schade, A., Bechen, M., Schmidt, D., 2002. Masking Background Fluorescence and Luminescence in Optical Analysis of Biomedical Assays, U.S. patent 6,420,183.

5. Porter, R.H.P., Benwell, K.R., Lamb, H., Malcolm, C.S., Allen, N.H., Revell, D.F., Adams, D.R., Sheardown, M.J., 1999. Functional characterization of agonists at recombinant human 5-HT$_{2A}$, 5-HT$_{2B}$, and 5-HT$_{2C}$ receptors in CHO-K1 cells. *Br. J. Pharmacol.* 128, 13–20.

6. Jerman, J.C., Brough, S.J., Gager, T., Wood, M., Coldwell, M.C., Smart, D., Middlemiss, D.N., 2001. Pharmacological characterization of human 5-HT2 receptor subtypes. *Eur. J. Pharmacol.* 414, 23–30.

7. Smart, D., Jerman, J.C., Brough, S.J., Rushton, S.L., Murdock, P.R., Jewitt, F., Elshourbagy, N.A., Ellis, C.E., Middlemiss, D.N., Brown, F., 1999. Characterization of recombinant human orexin receptor pharmacology in a Chinese hamster ovary cell line using FLIPR. *Br. J. Pharmacol.* 128, 1–3.

8. Smart, D., Jerman, J.C., Brough, S.J., Neville, W.A., Jewitt, F., Porter, R.A., 2000. The hypocretins are weak agonists at recombinant human orexin-1 and orexin-2 receptors. *Br. J. Pharmacol.* 129, 1289–1291.

9. Smart, D., Sabido-David, C., Brough, S.J., Jewitt, F., Johns, A., Porter, R.A., Jerman, J.C., 2001. SB-334867-A: the first selective orexin-1 receptor antagonist. *Br. J. Pharmocol.* 132, 1179–1182.

10. Miller, T.R., Witte, D.G., Ireland, L.M., Kang, C.H., Roch, J.M., Masters, J.N., Esbenshade, T.A., Hancock, A.A., 1999. Analysis of apparent noncompetitive responses to competitive H1 histamine receptor antagonists in fluorescent imaging plate reader–based calcium assays. *J. Biomol. Screen.* 4, 249–258.

11. Simpson, P.B., Woollacott, A.J., Hill, R.G., Seabrook, G.R., 2000. Functional characterization of bradykinin analogs on recombinant human bradykinin B-1 and B-2 receptors. *Eur. J. Pharmacol.* 392, 1–9.

12. Brough, S.J., Jerman, J.C., Jewitt, F., Smart, D., 2000. Characterization of an endogenous bombesin receptor in CHO/DG44 cells. *Eur. J. Pharmacol.* 409, 259–263.

13. Patel, K., Barnes, A., Camacho, J., Paterson, C., Boughtflower, R., Cousens, D., Marshall, F., 2001. Activity of diadenosine polyphosphates at P2Y receptors stably expressed in 1321N1 cells. *Eur. J. Pharmacol.* 430, 203–210.

14. Smart, D., Gunthorpe, M.J., Jerman, J.C., Nasir, S., Gray, J., Muir, A.I., Chambers, J.K., Randall, A.D., Davis, J.B., 2000. The endogenous lipid anandamide is a full agonist at the human vanilloid receptor (hVR1). *Br. J. Pharmacol.* 129, 227–230.

15. Smart, D., Jerman, J.C., Gunthorpe, M.J., Brough, S.J., Ranson, J., Cairns, W., Hayes, P.D., Randall, A.D., Davis, J.B., 2001. Characterization using FLIPR of human vanilloid receptor pharmacology. *Eur. J. Pharmacol.* 417, 51–58.

16. Coward, P., Chan, S.D.H., Wada, H.G., Humphries, G.M., Conklin, B.R., 1999. Chimeric G proteins allow a high-throughput signaling assay of Gi-coupled receptors. *Anal. Biochem.* 270, 242–248.

17. Zhang, Y., Nawoschik, S., Kowal, D., Smith, D., Ochalski, R., Schechter, L., Dunlop, J., 2001. Functional characterization of the 5-HT$_6$ receptor coupled to calcium signaling using an enabling chimeric G-protein. *Soc. Neurosci. Abstr.* Vol. 27, 265.6.

18. Wood, M., Chaubey, M., Atkinson, P., Thomas, D.R., 2000. Antagonist acivity of meta-chlorophenylpiperazine and partial agonist activity of 8-OH-DPAT at the 5-HT7 receptor. *Eur. J. Pharmacol.* 396, 1–8.

19. Wood, M.D., Murkitt, K.L., Rice, S.Q., Testa, T., Punia, P.K., Stammers, M., Jenkins, O., Elshourbagy, N.A., Shabo, U., Taylor, S.J., Gager, T.L., Minton, J., Hirst, W.D., Price, G.W., Pangalos, M., 2000. The human GABA(B1b) and GABA(B2) heterodimeric recombinant receptor shows low sensitivity to phaclofen and saclofen. *Br. J. Pharmacol.* 131, 1050–1054.

20. Uvegas, A.J., Kowal, D., Zhang, Y., Spangler, T.B., Dunlop, J., Semus, S., Jones, P.G., 2002. The role of transmembrane helix 5 in agonist binding to the human H3 receptor. *J. Pharm. Exp. Ther.* 301, 451–458.

21. Kowal, D., Nawoschik, S., Ochalski, R., Dunlop, J., 2003. Functional calcium coupling with the human metabotropic glutamate receptor subtypes 2 and 4 by stable coexpression with a calcium pathway facilitating G-protein chimera in Chinese hamster ovary cells. *Biochem. Pharmacol.* 66, 785–790.

22. Zhang, J.H., Chung, T.D., Oldenburg, K.R., 1999. A simple statistical parameter for use in evaluation and validation of high-throughput screening assays. *J Biomol. Screen.* 4, 67–73.

23. Morley, P., Whitfield, J.F., 1993. The differentiation inducer, dimethyl sulfoxide, transiently increases the intracellular calcium ion concentration in various cell types. *J Cell Physiol.* 156, 219–225.

23 Development of FLIPR-Based HTS Assay for G_i-Coupled GPCRs

Cailin Chen, Charles Smith, Lisa Minor, and Bruce Damiano

CONTENTS

23.1 INTRODUCTION

G-protein–coupled receptors (GPCRs) are a large family of seven-transmembrane spanning receptors [1,2]. GPCRs mediate signaling of stimuli as diverse as light, ions, small molecules, peptides, and proteins and include, among others, the adrenergic, dopaminergic, serotonergic, cholinergic, and histaminergic receptor types. Recent estimates suggest that approximately 1% of the proteins encoded by the human genome belong to this family. GPCRs have been the most successful family of protein targets for small-molecule drug discovery programs, such that agonist and antagonist ligands for these receptors comprise 45% of the sales of clinically effective drugs [3,4].

GPCRs consist of an extracellular *N*-terminus, seven transmembrane-spanning loops, and an intracellular *C*-terminal tail. After receptor activation by various ligands, GPCRs interact with heterotrimeric G proteins, which are composed of three subunits: α, β, and γ. G proteins then undergo conformational changes that lead to the exchange of GDP for GTP, which binds to the α-subunit. Consequently, the G_α and $G_{\beta\gamma}$ subunits stimulate effector molecules, such as adenylyl and guanylyl cyclases, phosphodiesterases, phospholipase A2 (PLA2), phospholipase C (PLC), and PI_3 kinase. This results in activation or inhibition of the production of a variety of second messengers such as cAMP, cGMP, diacylglycerol (DAG), and inositol-triphosphate, in addition to promoting increases in the intracellular concentration of Ca^{2+} and the opening or closing of a variety of ion channels. Changes in these second messengers result in a variety of gene expression and changes in biological responses.

The G-protein α-subunit is the key element mediating G-protein signaling. There are four major classes of G-proteins, defined by their α subunits ($G_{\alpha i}$, $G_{\alpha s}$, $G_{\alpha q}$, $G_{\alpha 12}$). Each couples to a distinct class of receptors and signals through a specific biochemical pathway. The G_s proteins primarily activate adenylyl cyclase, leading to increased levels of cAMP. G_q proteins primarily activate phospholipase C, which stimulates inositol-1,4,5-triphosphate (IP_3) formation and a subsequent increase in intracellular Ca^{2+} concentration. G_{12} proteins primarily regulate small GTP-binding proteins. The G_i family, which contains six members ($G_{\alpha i1, 2, 3}$, $G_{\alpha o1, 2}$, and $G_{\alpha z}$), mediates inhibition of adenylate cyclase leading to decreased cAMP [5].

Currently there are almost 2000 cloned GPCRs [3,6] of which more than 100 activate G_i-signaling pathways in response to activation [7]. G_i signaling is involved in a variety of physiologic processes, including chemotaxis, neurotransmission, proliferation, hormone secretion, analgesia, and regulation of platelet function [8–13]. Several methods exist for measuring signaling by G_i-coupled receptors. Incorporation of radiolabeled GTPγS into receptor-activated G proteins is a commonly used method. However this method requires handling of radioisotopes, which has safety concerns, as well as purification of membrane components, which may change the integrity of the receptor. In addition, this system lacks the intrinsic signal amplification that would occur in a whole cell [14]. Alternatively, G_i signaling has also been studied by measuring the inhibition of forskolin-stimulated cAMP accumulation [15]. The action of an antagonist is detected as a gain in signal generated following the coadministration of forskolin and agonist to the cellular assay. This assay requires several wash steps and changes of the assay buffer and suffers from a limited dynamic range due to low signal-to-noise ratio of stimulation. In addition, the assay is expensive, time-consuming, and not well suited for high-throughput screening ([7], our unpublished data).

Recently, the scientists from Molecular Devices Corporation have developed a fluorescence-based FLIPR (Fluorescent Imaging Plate Reader) assay for detecting changes in intracellular calcium [16]. This assay provides a high-throughput approach for agonist and antagonist screening in 96- or 384-well microplate format and is widely used in the study of both ligand-gated ion channels and GPCRs, where receptor activation changes intracellular Ca^{2+} levels.

In this chapter we present the development of the FLIPR assay for measuring the activation of G_i-coupled GPCRs.

23.2 PRINCIPLES OF FLIPR INTRACELLULAR CALCIUM ASSAY

The FLIPR system is a fluorescence-based assay system employing a fluorometric imaging plate reader to detect a fluorescent Ca^{2+}-sensitive dye [16]. Using a unique integration of optics, fluidics, and thermoregulation, the system is ideal for homogeneous, kinetic, cell-based fluorometric assays. The most extensive use has been in measuring the increases in intracellular calcium. The FLIPR system measures changes in cells stimulated with agonists in the presence of antagonists. An argon-ion laser excites the fluorescent indicator Ca^{2+}-sensitive dye, and the emitted light is detected using an optical detection system consisting of a highly sensitive, cooled CCD camera, which images the entire plate and integrates data signals over a time interval specified by the user. The assay includes three major steps: cell preparation, dye loading, and assay on the FLIPR.

23.2.1 CELL PREPARATION

The FLIPR calcium assay is suitable for both adherent and nonadherent cells. Cells are seeded in clear, flat-bottom black-wall, tissue-culture treated polystyrene 96- or 384-well plates. The flat bottom ensures that the cellular fluorescence is localized to a single horizontal plane. Adherent cells are seeded at least the day before the experiment while nonadherent cells can be plated on the same day. Depending on the individual cell line, some cells require more than 24-h seeding time to ensure cell attachment to the plate and stable protein expression. Use of attachment matrix coated plates, such as poly-D-lysine-coated plates, and precoating of specific attachment factors for some primary cells could improve cell adherence and minimize cellular "blow-off" during compound additions. It is necessary to optimize the cell seeding density so that a uniform, 90 to 100% confluent monolayer is formed on the day of the assay. Over-confluent cell monolayers may lead to reduced cellular response to the test compounds and increase in "blow-off."

23.2.2 DYE LOADING

In order to observe changes in intracellular Ca^{2+} levels, cells must be loaded with a calcium-sensitive fluorescent dye. The dye solution is added to cell plates before the assay and incubated using normal cell culture conditions. During incubation, the cells take up the dye. Upon agonist stimulation, GPCR activation induces a rapid intracellular influx of calcium, which binds to the fluorescent dye and greatly increases its fluorescence intensity. The most common calcium indicators used in the FLIPR system are Fluo-3, Fluo-4, and the FLIPR Calcium Assay Kits: the FLIPR Calcium Plus Assay Kit and the FLIPR Calcium 3 Assay Kit. The optimal dye loading time will depend on the cell type and the dye type. In general, cells can tolerate the calcium assay kits much longer than they can tolerate conventional dyes.

23.2.3 ASSAY ON THE FLIPR

23.2.3.1 Basal Fluorescence Signal Test and Adjustment

After the completion of dye loading, the plates are transferred to FLIPR one at a time or by a stacking system. A signal test to check the basal fluorescence signal of each cell plate is suggested before a data run. It is best to work with a basal fluorescence signal of 10,000 to 12,000 counts above background. Three parameters are used to adjust the basal fluorescence to the desirable range: laser power, exposure time, and camera f/stop. Laser power is normally started at 0.600 W for FLIPR^{384}. Laser power is increased if the basal fluorescence signal is too low or decreased if the basal fluorescence signal is too high. The exposure time is initially set at 0.4 sec but can be modified based on the basal fluorescence signal for optimum exposure. The camera f/stop of 2 is used for FLIPR^{384}. Increasing the camera f/stop decreases the aperture. Thus, if the basal fluorescence signal is too high, the f/stop can be increased from 2.0 to 2.8 or more. If the basal fluorescence signal is too low, an f/stop of 1.4 can be used. Other factors such as cell density and dye-loading time also affect the cell basal signal. A nearly confluent plate is necessary to ensure a sufficient basal signal. Increasing dye-loading time can increase the basal signal.

23.2.3.2 Pipette Height

Pipette height is a critical factor. If it is set too high, the solution is not well mixed. If it is set too low, the possibility of "blow-off" of cells will increase. In addition, for 96-well pipettes, a small air bubble is drawn into the tip of each pipette after picking up agonist/antagonist solution to ensure that fluid does not leak out. This bubble will be the first thing out of the pipette tip when fluid is dispensed. Thus, to avoid blowing bubbles in the wells (which can cause random light reflections and spurious signals), it is best to start dispensing with the tips above the liquid level in the well. It is also preferable to have the pipette tips submerged after the addition has been completed to ensure complete sample dispensing. Therefore, the pipette height should be set somewhere above the starting fluid volume in the wells but below the final volume after the addition. For example, if the wells contain 100 µL and the sample volume to be added is 50 µL, the pipette height should be set at 120 to 140.

23.2.3.3 Fluid Dispensing Speed

The default pipette dispensing speed is 50 µL/sec for a 96-well plate on FLIPR[384]. This value should be experimentally determined for each cell type, but it is generally preferable to dispense as fast as possible to enhance mixing of the compounds in the wells. However, the pipetting speed must not be so forceful as to dislodge cells from the well.

23.3 PRINCIPLES OF TRANSDUCING G_i SIGNALING THROUGH THE G_q PATHWAY

Conklin et al. have reported using an adapter G-protein, G_{qi5}, to transduce G_i signaling through the G_q pathway [17]. This is a valuable tool since the activation of G_i-coupled receptors would not normally alter the intracellular Ca^{2+} levels and provides a means to easily measure G_i activity. In G_{qi5}, the five C-terminal amino acids of the G-protein subunit α_q are replaced with the corresponding amino acids of α_{i2}. This creates an adapter protein that couples receptors normally linked to the stimulation of adenylyl cyclase to stimulation of phospholipase C, leading to a rapid intracellular calcium influx. This intracellular calcium change can be measured by FLIPR as shown in Figure 23.1. Using this technology, several G_i-coupled receptors have been studied pharmacologically [11–13].

23.4 MATERIALS AND METHODS

23.4.1 MATERIALS

- Cell culture flasks, canted neck, 225 cm^2 (Corning/Costar 3001)
- Cell culture medium (Life Technologies)
- Black-wall, clear, flat-bottom, tissue-culture treated, 96-well cell culture plates (Corning/Costar 3603)
- DMSO, Dimethylsulfoxide (Pierce 20684)
- Calcium Ionophore A23187, free acid (Sigma C-7522)
- HBSS, Hank's Balanced Salt Solution, 10× (Life Technologies 14065-056)
- HEPES Buffer Solution, 1 M (Life Technologies 15630-080)
- BSA, Bovine Albumin Fraction V Solution, 7.5% (Life Technologies 15260-037)
- 1 N NaOH
- FLIPR Calcium Assay Kit, for 100 plates (Molecular Devices R-8033)
- Probenecid, crystalline (Sigma P8761)
- Pipette tips, black, nonsterile, 200 µL (Robbins Scientific 1043-24-5)

Transducing G_i to G_q Signaling Pathway

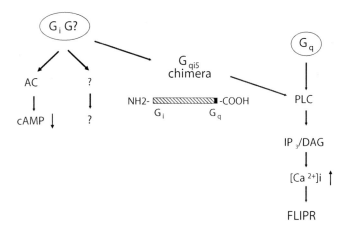

FIGURE 23.1 Schematic presentation of G_i and G_q signaling pathways: Stimulation of G_i protein mediates inhibition of adenylate cyclase leading to decreased cAMP. G_q protein primarily activates phospholipase C, which stimulates inositol-1,4,5-triphosphate (IP_3) formation and a subsequent increase in intracellular Ca^{2+} concentration. This intracellular calcium change can be measured by FLIPR. The chimera protein G_{qi5} creates an adapter protein that converts G_i signaling to the G_q pathway, hence enabling its measurement on FLIPR.

23.4.2 CELL CULTURE AND STABLE EXPRESSION OF P2Y12 RECEPTOR AND $G_{\alpha qi5}$

Human embryonic kidney (HEK) 293 cells were grown in Dulbecco's Modified Eagle Medium (DMEM) supplemented with 10% FBS, 2 mM L-glutamine, antibiotics (100 U of penicillin per mlL and 100 µg of streptomycin per mL) while Chinese Hamster Ovary (CHO) cells were cultured in F-12 medium. Cells were maintained at 37°C in an atmosphere of 5% CO_2. The cells were cotransfected with pcDNA3hygroP2Y12 containing a hygromycin resistance gene and pLEC1-G_{qi5}-HA containing a neomycin resistance gene (licensed from Molecular Devices, Sunnyvale, CA) with SuperFect (Qiagen, Valencia, CA). Cell clones were selected in the presence of 600-µg/mL hygromycin B (Life Technologies, Carlsbad, CA) and 1 mg/mL of G418 (Mediatech, Herndon, VA). Drug-resistant colonies were picked and screened by FLIPR, and the gene expression was further confirmed by RT PCR. Positive clones were maintained in growth medium containing 400 µg/mL of G418 and 200 µg/mL of hygromycin.

23.4.3 RT-PCR

The cells were washed with phosphate-buffered saline (PBS), and total RNA was isolated with Trizol Reagent (Invitrogen, CA). First-strand cDNA was prepared with an oligo (dT) primer and Superscript Preamplification System kit (Life Technologies, CA). After reverse transcription, the cDNA product was amplified by polymerase chain reaction (PCR) following a standard protocol. Briefly, approximately 25 ng of the cDNA was used in a 50 µL preparative 30-cycle PCR reaction using Advantage GC-polymerase (Clontech) with a 30 sec 94°C denaturation followed by a 30 sec 60°C annealing and a 60 sec elongation at 68°C. The forward primer 5′-GGCTCATGCACAATT-AGT-3′ and the reverse primer 5′- TCAGAAGAGGCCACAGTC-3′ produced a 756-bp fragment corresponding to the C-terminal sequences of the $G\alpha qi5$ chimera gene. The primer sequences used to amplify the housekeeping gene GAPGH were as follows: 5′-GGGGAGCCAAAAGGGTCAT-CATCT-3′ and 5′-GACGCCTGCTTCACCACCTTCTTG-3′. A primer pair specific for the P2Y12 receptor was used to determine the transcript of the receptor gene. The PCR products were separated and visualized in ethidium bromide–agarose gels.

23.4.4 FLIPR ASSAY

23.4.4.1 Plating the Cells

On the day prior to the assay, cells were washed once with PBS and trypsinized with 2 to 4mL of trypsin, which coated the bottom of the flask, incubated 2 to 3 min at RT, resuspended in medium, counted by hemocytometer, adjusted to a density of 2.0×10^5 cells/mL and dispensed into black 96-well plates, 100 μL/well. Cells were then incubated at 37°C, 5% CO_2.

23.4.4.2 Buffer Solution Preparation

- HTS buffer for compound preparation:
  - Stock 30% DMSO/50 mM HEPES was prepared as follows: combine 65 mL H_2O, 30 mL DMSO, and 5 mL 1 M HEPES; store at 4°C for up to 6 months.
  - 10% HTS buffer (3% DMSO/23 mM HEPES) working solution was made by 1:10 dilution with Complete Buffer.
- Calcium ionophore A23187 (10 mM aliquots):
  - Dissolve 1 mg of calcium ionophore A231987 (FW 523.6) in 191 μL DMSO to prepare a 10 mM solution. Store at –20°C.
- Buffer for agonist preparation:
  - Basic buffer

Component	[Stock]	[Final]	mL/Plate	mL/__ Plates
HBSS	10×	1×	1.6	
HEPES	1 M	20 mM	0.32	
H_2O	—	—	14	
Total volume	—	—	16	

Note: Adjust buffer to pH 7.4 using NaOH.

  - Complete buffer

Component	[Stock]	[Final]	mL/Plate	mL/__ Plates
Basic buffer	—	—	16	
BSA	7.5%	0.1%	0.21	
Probenecid	250 mM	2.5 mM	0.16	
Total volume	—	—	~16	

- Probenecid solution
  - Working solution is 250 mM (final concentration is 2.5 mM)

Component	Amount/Plate	Amount/__ Plates
Probenecid	29 mg	
1 N NaOH	0.2 mL	
Basic buffer	0.2 mL	
Total volume	0.4 mL	

23.4.4.3 Preparation of Dye Loading Solution

- Remove a vial of lyophilized FLIPR Calcium Assay dye from the freezer and equilibrate at room temperature. Dissolve the dye in 50 mL of basic buffer as prepared above. The Dye Loading Solution is stable for up to 8 h at room temperature. Any Dye Loading Solution not used on the day it is prepared can be stored at –20°C for up to 4 weeks without loss of activity.
- Complete dye loading solution preparation on the day of the assay as follows:

Component	[Stock]	[Working]	[Final]	mL/Plate	mL/__ Plates
Dye loading solution	—	—	—	5	
BSA	7.5%	0.3%	0.1%	0.2	
Probenecid	250 mM	7.5 mM	2.5 mM	0.15	
Total volume	—	—	5 mL/plate	~5	

23.4.4.4 Dye Loading

Fifty microliters of complete dye loading solution was added to each well containing 100 µL of culture medium. The cells were incubated for at least 30 min at 37°C in 5% CO_2 before initiating the assay on the FLIPR.

23.4.4.5 Measurement on the FLIPR

The FLIPR machine was warmed up for at least 30 to 60 min before using. The antagonist compounds were prepared in 10% HTS buffer at 4× and the agonists were diluted in complete buffer at 4× (for agonist addition only) or 5× (for antagonist followed by agonist additions) the concentration desired in the test wells and then aliquoted into the sample plates. The sample plates and the cell plate were placed in FLIPR's assay chamber. A signal test was taken and laser power adjusted to obtain a basal level of 10,000 to 12,000 fluorescence intensity units (FIU). The pipette height was set at 175 for the first addition and at 225 for the second addition. The cells were then excited at 488 nm using the FLIPR laser and fluorescence emission determined using the CCD camera with a band pass interface filter at 525. Fluorescence readings were taken at 1-sec intervals for 60 sec, and an additional 80 or 120 readings were taken at 3-sec intervals for a total of 5 (one addition only) or 7 (two additions) min. The first addition (50 µL) was made at the beginning while the second addition (50 µL) was made after 5 min of reading. Raw fluorescence data were exported for each well and tabulated vs. time within an ASCII file. Data were then imported into Excel, and the peak response over basal level was determined.

23.5 EXAMPLE OF ASSAY DEVELOPMENT

23.5.1 SELECTING PARENTAL CELL LINE, EXPRESSING RECEPTOR, AND CHIMERA PROTEINS

In order to transduce the G_i pathway to G_q signaling, a cell system must be developed that is suitable for a FLIPR assay and expresses both the receptor and the chimeric protein. A cell line expressing low levels of ADP receptors was chosen. Several cell lines were screened by measuring the calcium influx induced by ADP on FLIPR. In the examples shown in Figure 23.2, the EC_{50} for ADP is 0.1 M in CHO cells, while the EC_{50} is about 3 M in HEK293 cells. In addition to the expression of low levels of ADP receptors, HEK293 cells are very easy to work with, require no specific

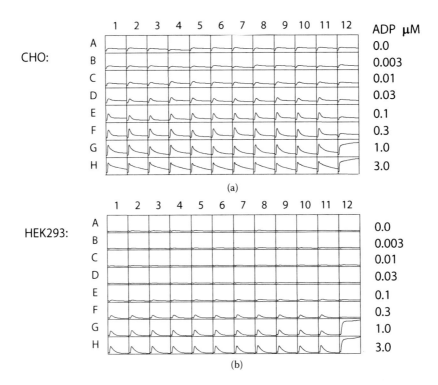

FIGURE 23.2 Choosing of a parental cell line: Typical calcium traces from the FLIPR software display the fluorescent increases in (a) CHO and (b) HEK293 cells stimulated by different concentrations of ADP using FLIPR Calcium Assay Kits in a 96-well format. Row A and B12 to F12 represent the effect with assay buffer alone. A generic calcium ionophore was used as a control and is shown in wells 12G and H.

growth medium, are easy to scale up, and adhere well on FLIPR assay plates. Therefore, the HEK293 cells were selected as the parental cell line. We then cotransfected the cells with both ADP P2Y12 receptor and G_{qi5} chimera plasmids or G_{qi5} plasmid alone. A hygromycin resistance gene was included in the ADP receptor construct, while a neomycin resistance gene was included in the G_{qi5} plamid. The double-transfected cells were selected for about 2 weeks in medium containing both hygromycin and neomycin. The positive cell clones were identified by a FLIPR assay, as shown in Figure 23.3. An ionophore was used as a control to ensure similar cell density and dye loading for all transfected clones (row B). As can be seen in column 1, 10 μM ADP induced only a slight change in intracellular calcium in HEK293 cells transfected with the G_{qi5} gene alone. A similar response was observed with clones 6, 7, 9, and 10 (columns 6, 7, 9, and 10), while strong ADP responses were observed for clones 1, 4, 5, 8, and 11 and moderate responses with clones 2 and 14. A total of 25 clones were tested, with nine being highly responsive to ADP stimulation. Clone 4 was used for further studies. RT-PCR was performed to further confirm the gene expression. As shown in Figure 23.4, both the P2Y12 receptor and the G_{qi5} chimera were highly expressed in clone 4 and as expected, only G_{qi5} was expressed in G_{qi5} control cells. Both genes were not detected in either nontransfected HEK293 cells or CHO cells.

23.5.2 CELL DENSITY AND DYE LOADING DURATION

To optimize the cell conditions, a series of cell densities were evaluated. Cells from 0.5 to 6.0×10^5 were plated in each well the day before the assay. The cells were incubated with calcium dye for 1 h. As can be seen in Figure 23.5a, a linear increase of fluorescence change was observed from 0.5 to 2.0×10^5 cells. Further increases were seen with increasing cell density. However, at the

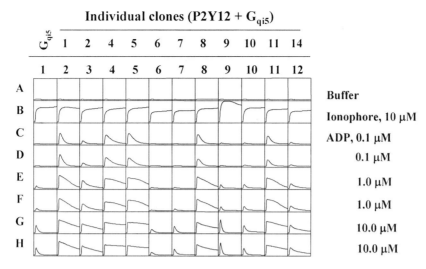

FIGURE 23.3 Testing the transfected cells in the FLIPR: Examples of calcium traces from FLIPR software showing the responsiveness to a dose response of ADP from individual clones stably transfected with both the P2Y12 receptor gene and the chimeric G_{qi5} gene. Row A represents the effect of assay buffer alone and Row B represents the effect of the ionophore.

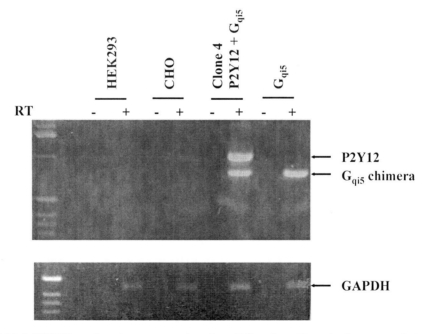

FIGURE 23.4 RT-PCR confirmed gene expression: An ethidium bromide–stained agarose gel showing RT-PCR analysis of P2Y12 receptor and G_{qi5} chimeric expression in HEK293 and CHO cells transfected with both P2Y12 and G_{qi5} (clone #4), or G_{qi5} alone. The housekeeping gene, GAPDH, was used to ensure that equal amounts of first-strand cDNA were used.

higher densities basal fluorescence levels were too high and the cells were overconfluent. The optimum cell density was 2×10^5/well, which provided $11,456 \pm 864$ basal fluorescence counts and cells that were near confluent. Next, we tested the duration of dye loading. As shown in Figure 23.5b, 2×10^5 cell/well were plated and incubated with calcium dye for different periods of time. Basal fluorescence was 6081 ± 755, 8968 ± 913, $10,377 \pm 1206$, $11,684 \pm 1643$, $12,532 \pm 1302$,

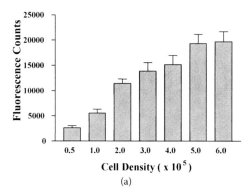
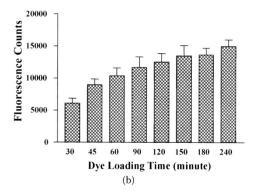

(a)

(b)

FIGURE 23.5 Optimization of assay conditions. (a) Different densities of cells were plated the day before the assay. The calcium FLIPR dye was incubated for 60 min prior to stimulation. (b) 2.5×10^5/mL of cells were plated. A time course of dye loading was tested.

$13,489 \pm 1599$, $13,651 \pm 1033$, and $14,961 \pm 1026$ at 30, 45, 60, 90, 120, 150, 180, and 240 min of incubation, respectively. The cells were intact for up to 4 h, the longest dye loading time tested. For compound library screening, we loaded five plates each time, started the first plate on FLIPR at 45 min, and completed assaying all plates 45 to 95 min post dye loading. Another group of five plates were loaded with dye after the second plate had been placed on FLIPR for assay.

23.5.3 RESPONSIVENESS TO AGONISTS AND ANTAGONIST

Next, we tested how the cells responded to agonists and antagonist. Two agonists were tested: ADP and an ADP analog 2-methylthioadenosinediphosphate (2MeSADP). As expected, parental HEK293 cells did not response to low levels of either agonist (Figure 23.6a and Figure 23.6b). Similar results were observed in the cells expressing the G_{qi5} chimera protein only. In contrast, stably transfected clone #4, expressing both P2Y12 and G_{qi5} proteins, responded to very low levels of the agonists. The EC_{50} values were 30 and 3 nM for ADP and 2MeSADP, respectively. On the other hand, 2-methylthioadenosinemonophosphate (2MeSAMP), a specific P2Y12 receptor antagonist, inhibited ADP-induced intracellular calcium increase with an IC_{50} of 0.3 μM (Figure 23.6c).

23.5.4 ASSAY VALIDATION FOR HTS AND DATA ANALYSIS

The assay was fairly easy to adapt for HTS in both 96- and 384-well format with a z' factor above 0.6. As an example (Figure 23.7), the "Z factor" plate was designed such that the top half of the plate received buffer alone and the bottom half received 30 nM of ADP. The plate format for our compound screen is shown in Figure 23.8. The percent inhibition for each compound was calculated using the average of the maximal fluorescent intensity of columns 1A to 1H and Columns 12 A, B, C, D.

23.6 SUMMARY

Prior to the development of a method to convert a G_i-coupled GPCR signaling to a G_q-coupled GPCR signaling, a significant barrier existed for the functional testing and screening of G_i-coupled GPCRs. One popular method, GTPγS migration, is slow, produces a small signal, and generates radioactive waste. It is a more difficult protocol to justify in higher throughput due to safety and radioactive disposal costs. The discovery that a small switch in the amino acid sequence in the G_i protein could convert it to a protein mediating a G_q-coupled response has led to the development of robust, high-throughput assays for this valuable receptor subtype. This chapter has detailed the

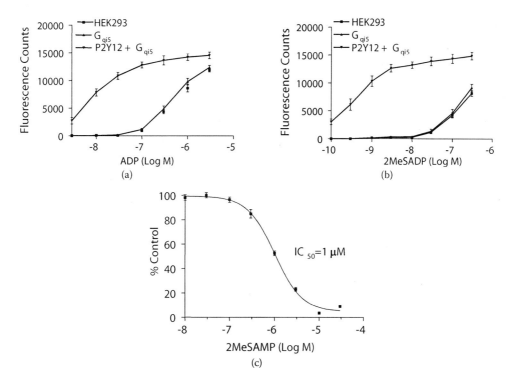

FIGURE 23.6 Responsiveness of cells to agonists and antagonist. Dose responses of (a) ADP or (b) 2MeSADP were determined in untransfected HEK293 cells and HEK293 transfected with G_{qi5} alone or with G_{qi5} and the P2Y12 receptor. Inhibition of (c) 2MeSAMP in HEK293 cells transfected with both G_{qi5} and P2Y12. Samples are the average of $12 \pm$ the standard deviation of the mean.

	1	2	3	4	5	6	7	8	9	10	11	12		
A	959	462	646	297	407	523	623	1521	484	458	402	1017	1435	Total SD
B	551	410	515	456	463	501	842	384	406	267	316	620	15443	Total mean
C	870	719	697	468	454	535	907	398	313	505	311	701	9	Total CV
D	471	794	966	534	439	657	949	465	486	362	520	1175	588	NSB mean
E	13747	13953	15621	16721	14279	14072	16723	14210	12178	16737	13438	16398	251	NSB SD
F	17119	18381	16055	14558	15618	14566	14431	16379	15240	15640	14294	15450	234	NSB CV
G	17999	16248	14311	15945	15938	15851	15337	15978	11757	13445	15391	17772	_0.66_	Z' factor
H	15496	16744	17176	17488	13639	16108	15939	16494	13287	16062	15901	15153	26	S:N
	Z factor = {1-[(3*totals SD)+(3*NSB SD)]}/(totals mean - NSB mean)													

FIGURE 23.7 Layout of a Z' factor plate: G_{qi5} and P2Y12 cells were stimulated with vehicle (rows A to D) or ADP at 30 nM (rows E to H). Total standard deviation (SD), total mean, total CV, total NSB (nonspecific background), NSB SD, NSB CV, Z' factor, and signal-to-noise (s:n) ratio were calculated.

strategy and basic methodology we have employed to both develop and validate such a screen. We have shown that the choice of cell is very important and that prescreening should be carried out to ensure that the receptor of choice is not already present in the cell. We have suggested that the presence of the receptor in the cotransfected cell line (G_{qi5} and P2Y12) be validated by RT-PCR and that its functional coupling to G_{qi5} be demonstrated using calcium mobilization. We have described a basic strategy and shown our results in validating the cell line and determining the

- 1° addition: 50 µl of compound in 3%DMSO in assay buffer

	1	2	3	4	5	6	7	8	9	10	11	12
A	♦	♣	♣	♣	♣	♣	♣	♣	♣	♣	♣	♥
B	♦	♣	♣	♣	♣	♣	♣	♣	♣	♣	♣	♥
C	♦	♣	♣	♣	♣	♣	♣	♣	♣	♣	♣	♦
D	♦	♣	♣	♣	♣	♣	♣	♣	♣	♣	♣	♦
E	♦	♣	♣	♣	♣	♣	♣	♣	♣	♣	♣	♦
F	♦	♣	♣	♣	♣	♣	♣	♣	♣	♣	♣	♦
G	♦	♣	♣	♣	♣	♣	♣	♣	♣	♣	♣	♦
H	♦	♣	♣	♣	♣	♣	♣	♣	♣	♣	♣	♦

♣ 4 x compounds ♦ Assay Buffer with 3% DMS O ♥ Specific antagonist

- 2° addition: 50 µl from agonist plate

	1	2	3	4	5	6	7	8	9	10	11	12
A	♣	♣	♣	♣	♣	♣	♣	♣	♣	♣	♣	♣
B	♣	♣	♣	♣	♣	♣	♣	♣	♣	♣	♣	♣
C	♣	♣	♣	♣	♣	♣	♣	♣	♣	♣	♣	♣
D	♣	♣	♣	♣	♣	♣	♣	♣	♣	♣	♣	♣
E	♣	♣	♣	♣	♣	♣	♣	♣	♣	♣	♣	♦
F	♣	♣	♣	♣	♣	♣	♣	♣	♣	♣	♣	♦
G	♣	♣	♣	♣	♣	♣	♣	♣	♣	♣	♣	♠
H	♣	♣	♣	♣	♣	♣	♣	♣	♣	♣	♣	♠

♣ 5 x agonist ♦ Assay Buffer ♠ 50µM (1 0µM final) C a^{2+} ionophore A231987

FIGURE 23.8 Layout of screening plate. To 200 µL of sample in the plate, 50 µL of compound is added first, followed by 50 µL of agonist. Samples are positioned as detailed in the figure.

optimal conditions for HTS on FLIPR. The strategy we have employed should be amenable to other G_i-coupled receptors and should serve as a guide for future assay development.

REFERENCES

1. Brady, A. E. and Limbird, L. E. (2002) G-protein–coupled receptor interacting proteins: emerging roles in localization and signal transduction. *Cell Signal* 14, 297–309.
2. Hur, E. M. and Kim, K. T. (2002) G-protein–coupled receptor signaling and cross talk: achieving rapidity and specificity. *Cell Signal* 14, 397–405.
3. Sautel, M. and Milligan, G. (2000) Molecular manipulation of G-protein–coupled receptors: a new avenue into drug discovery. *Curr Med Chem* 7, 889–896.
4. Marshall, F. H. (2001) Heterodimerization of G-protein–coupled receptors in the CNS. *Curr Opin Pharmacol* 1, 40–44.
5. Brink, C. B., Harvey, B. H., Bodenstein, J., Venter, D. P., and Oliver, D. W. (2004) Recent advances in drug action and therapeutics: relevance of novel concepts in G-protein–coupled receptor and signal transduction pharmacology. *Br J Clin Pharmacol* 57, 373–387.
6. Ji, T. H., Grossmann, M., and Ji, I. (1998) G-protein–coupled receptors. I. Diversity of receptor-ligand interactions. *J Biol Chem* 273, 17,299–17,302.
7. Coward, P., Chan, S. D., Wada, H. G., Humphries, G. M., and Conklin, B. R. (1999) Chimeric G proteins allow a high-throughput signaling assay of G_i- coupled receptors. *Anal Biochem* 270, 242–248.
8. Foster, C. J., Prosser, D. M., Agans, J. M., Zhai, Y., Smith, M. D., Lachowicz, J. E., Zhang, F. L., Gustafson, E., Monsma, F. J., Jr., Wiekowski, M. T., Abbondanzo, S. J., Cook, D. N., Bayne, M. L., Lira, S. A., and Chintala, M. S. (2001) Molecular identification and characterization of the platelet ADP receptor targeted by thienopyridine antithrombotic drugs. *J Clin Invest* 107, 1591–1598.

9. Zhang, F. L., Luo, L., Gustafson, E., Lachowicz, J., Smith, M., Qiao, X., Liu, Y. H., Chen, G., Pramanik, B., Laz, T. M., Palmer, K., Bayne, M., and Monsma, F. J., Jr. (2001) ADP is the cognate ligand for the orphan G-protein–coupled receptor SP1999. *J Biol Chem* 276, 8608–8615.

10. Hollopeter, G., Jantzen, H. M., Vincent, D., Li, G., England, L., Ramakrishnan, V., Yang, R. B., Nurden, P., Nurden, A., Julius, D., and Conley, P. B. (2001) Identification of the platelet ADP receptor targeted by antithrombotic drugs. *Nature* 409, 202–207.

11. Wood, M. D., Murkitt, K. L., Rice, S. Q., Testa, T., Punia, P. K., Stammers, M., Jenkins, O., Elshourbagy, N. A., Shabon, U., Taylor, S. J., Gager, T. L., Minton, J., Hirst, W. D., Price, G. W., and Pangalos, M. (2000) The human GABA(B1b) and GABA(B2) heterodimeric recombinant receptor shows low sensitivity to phaclofen and saclofen. *Br J Pharmacol* 131, 1050–1054.

12. Uveges, A. J., Kowal, D., Zhang, Y., Spangler, T. B., Dunlop, J., Semus, S., and Jones, P. G. (2002) The role of transmembrane helix 5 in agonist binding to the human H3 receptor. *J Pharmacol Exp Ther* 301, 451–458.

13. Kowal, D., Nawoschik, S., Ochalski, R., and Dunlop, J. (2003) Functional calcium coupling with the human metabotropic glutamate receptor subtypes 2 and 4 by stable coexpression with a calcium pathway facilitating G-protein chimera in Chinese hamster ovary cells. *Biochem Pharmacol* 66, 785–790.

14. Cerione, R. A. (1991) Reconstitution of receptor/GTP-binding protein interactions. *Biochim Biophys Acta* 1071, 473–501.

15. Wong, Y. H. (1994) G_i assays in transfected cells. *Methods Enzymol* 238, 81–94.

16. Schroeder, K. S. and Neagle, B. D. (1996) FLIPR: a new instrument for accurate, high-throughput optical screening. *J Biomol Screen* 2, 75–80.

17. Conklin, B. R., Farfel, Z., Lustig, K. D., Julius, D., and Bourne, H. R. (1993) Substitution of three amino acids switches receptor specificity of G_q alpha to that of G_i alpha. *Nature* 363, 274–276.

24 Aurora Assays

Priya Kunapuli

CONTENTS

24.1 INTRODUCTION

With the increasing popularity of functional assays to overcome some of the limitations of traditional ligand binding assays (inability to distinguish between agonists, antagonists, potentiators, allosteric modulators, inverse agonists, and partial agonists), reporter gene assays are making a significant impact on biological assays used routinely in academic and pharmaceutical laboratories. Reporter genes with readily measurable phenotypes have been used to measure the effects of signal transduction cascades on gene expression. Reporter gene assays offer high sensitivity, reliability, convenience, and adaptability to large-scale measurements [1]. Numerous developments in reporter gene technology have been made since the original chloramphenicol-acetyl-transferase (CAT) assay [2] was first used to measure gene transcription. Other commonly used reporter gene assays include beta-galactosidase [3,4], secreted alkaline phosphatase [5,6], luciferase [7,8], green fluorescent protein (GFP) [9,10], and beta-lactamase (BLA) [11]. Table 24.1 summarizes the key features of some popular reporter gene assays. Among these, the BLA reporter gene assay is colloquially referred to as "Aurora assay," as the first report on the use of the BLA enzyme as a reporter gene was described by scientists from Aurora Biosciences (now Vertex Pharmaceuticals) [11].

BLA is a bacterial enzyme encoded by the antibiotic resistance gene, TEM-1 BLA from *Escherichia coli*. The 29-kDa enzyme uses β-lactam-containing molecules (such as penicillins and cephalosporins) as substrates. BLA exhibits simple kinetics, has no known homologs in eukaryotes, and can function as a monomer or as a fusion protein (fused to the *N*- or *C*-terminus of heterologous proteins), thereby making it an attractive reporter gene [12]. The potential of BLA as a reporter gene was realized upon the development of a cell-permeable fluorogenic BLA substrate, CCF2/AM

TABLE 24.1
Reporter Gene Assays

Reporter Gene	Amplification	Miniaturizable	Ratiometric	FACS Compatible	Stable Cell Lines
BLA	Yes	Yes (3456-well)	Yes	Yes	Yes
Luciferase	Yes	Yes (384, 1536)	No	No	Yes
CAT	Yes	No	No	No	No
Secreated alkaline phosphatase	Yes	Yes (384)	No	No	Yes
GFP	No	Yes (384)	No	Yes	Yes
Beta-galactosidase	Yes	Yes (384, 1536?)	No	Yes	Yes

[11]. The CCF2 (or CCF4/AM developed later with improved solubility in aqueous solutions) consists of coumarin and fluorescein moieties connected by a β-lactam–containing cephalosporin core. The esterified CCF2/AM (or CCF4/AM) is lipophilic and readily enters cells through the plasma membrane. Once within cells, cytoplasmic esterases convert the substrate to CCF2 (or CCF4). Excitation of CCF2 (or 4) at 405 nm leads to fluorescence resonance energy transfer (FRET) from the coumarin moiety to the fluorescein derivative, resulting in green light emission at 530 nm. In the presence of BLA in the cell, the substrate is cleaved at the β-lactam ring, spatially separating the coumarin and fluorescein moieties, thereby disrupting FRET. Under these circumstances, excitation of coumarin at 405 nm results in blue fluorescence emission in the absence of FRET, detected at 460 nm (Figure 24.1). Thus, basal unstimulated cells appear green by fluorescence microscopy due to FRET while cells producing BLA appear blue (Figure 24.2). The BLA activity, reported as a ratiometric readout of 460 nm/530 nm, offers significant advantages (see subsequent sections).

This principle of the BLA reporter gene assay has been utilized to study a variety of molecular targets, including the popular class of cell surface G-protein–coupled receptors (GPCRs) [13]. Receptor signaling through intracellular second messengers can be linked to corresponding modulation of the BLA gene via appropriate promoters: Nuclear Factor of Activated T cells (NFAT), which is responsive to changes in intracellular Ca^{2+} [14], or the cAMP Response Element Binding Protein (CREBP), which is responsive to changes in intracellular cAMP [15,16]. In the case of GPCRs coupled to the Gq pathway, the BLA gene is engineered under the control of the NFAT promoter. In such a case, receptor activation and the subsequent increase in i[Ca^{2+}] results in the activation of the Ca^{2+}-dependent phosphatase, calcineurin. Activated calcineurin in turn dephosphorylates and activates the cytoplasmically located inactive NFAT. Dephosphorylated and activated NFAT translocates into the nucleus and initiates transcription of NFAT-regulated genes (i.e., the BLA gene in engineered cells) (Figure 24.3a) [14]. For GPCRs coupled to modulation of intracellular cAMP by stimulation or inhibition of adenylyl cyclase via the G proteins Gs or Gi, respectively, the BLA gene may be engineered under the control of the cAMP response element, CRE. Thus, Gs-coupled GPCRs increase intracellular cAMP, resulting in activation of the cAMP response element–binding protein (CRE-BP), increased CRE activity, and BLA gene activation (Figure 24.3b) [15,16], while Gi-coupled GPCRs would result in a decrease in intracellular cAMP, CRE activity, and BLA gene transcription. Similar to other measurements of Gi-coupled receptors (direct measurement of cAMP), inhibition of cellular cAMP is more easily observed by measuring the agonist-induced inhibition of forskolin-stimulated cAMP. Under these conditions, agonist-induced activation of Gi-coupled receptors can be observed with a decrease in forskolin-stimulated cAMP response, as indicated by BLA activity in cells engineered with the CRE-BLA reporter gene. The BLA reporter gene can also be used for receptor tyrosine kinases (RTKs) signaling via intracellular Ca^{2+} [17]. In this case, the NFAT-BLA promoter construct may be used to engineer

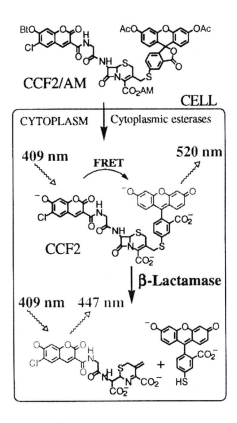

FIGURE 24.1 Schematic representation of CCF2 chemistry. The cell-permeable fluorogenic BLA substrate, CCF2/AM (or CCF4/AM), enters the cells and is converted to CCF2 (or CCF4) by cytoplasmic esterases. Upon excitation at 405 nm, the energy is transferred to the fluorescein moiety by FRET, thereby resulting in green emission at 530 nm. However, in the presence of BLA enzyme accumulated in the cells, the CCF2 (or CCF4) substrate is cleaved at the β-lactam ring by BLA. Under these conditions, excitation at 405 nm results in blue emission at 460 nm. (From Zlokarnik. G., Negulescu, P. A., Knapp, T. E., Mere, L., Burres, N., Feng, L., Whitney, M., Roemer, K., Tsein, R. Y. (1998) *Science* 279, 84–88. With permission.)

cells to study the downstream signaling effects of RTKs. Reagents for the cell-based BLA reporter gene assay are available through Invitrogen (Carlsbad, CA) as the GeneBLAzer system.

The BLA reporter gene may also be used to study nuclear receptors by several strategies. The BLA gene can be engineered as a C-terminal fusion of the target gene, creating a library of cells/clones with specific genes functionally tagged with BLA (Aurora's Genome Screen™ and/or CellSensor™) [18]. These clones can then be profiled for appropriate pharmacological response with known nuclear receptor ligands. Alternatively, the BLA gene can be introduced under the control of a response element known to be modulated by the nuclear hormone receptor of choice. Both these strategies can be used to study the ligand-dependent recruitment of coactivators, corepressors, and basal transcription machinery driving the nuclear hormone receptor activity, as reported by the BLA activity [19–21].

A key and unique feature of the BLA reporter gene assay is its unique compatibility with fluorescence activated cell sorting (FACS). The cell-permeable fluorescent substrate permits real-time live measurements during the cloning process, facilitating the isolation of *single* cell clones with the desirable signaling phenotype represented by the BLA response [11–13]. Such single cell clones can be isolated and expanded to create true clonal populations, which often exhibit a homogeneous response compared to the traditional cloning procedures, which tend to exhibit a

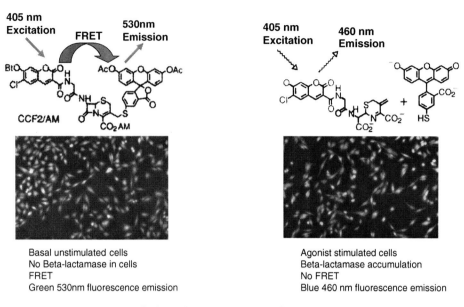

Ratiometric readout = 460 nm/530 nm

FIGURE 24.2 (See color insert following page 334.) BLA assay principle. Basal unstimulated cells emit green fluorescence at 530 nm and thereby appear green by fluorescent microscopy (left) due to FRET in the CCF2 (or CCF4) substrate. Cells expressing BLA exhibit blue fluorescence emission at 460 nm due to FRET disruption by BLA and thereby appear blue by fluorescence microscopy (right).

more heterogenous response. While homogeneous clones can also be obtained by the limited dilution procedure, this is a more tedious process spanning several months, requiring several hundred clones to be assayed, and it does not provide a direct measure of functionally active receptors during the cloning process. In comparison, FACS provides a more efficient and pure clonal population within a few weeks.

In addition to being used for cell-based functional measurements, the Aurora FRET technology has also been used for FRET-based biochemical measurements for kinases and phosphatases. The recently introduced Z-lyte technology (formerly known as PhosphoryLIGHT) is based on differential specificity of chymotrypsin for a phosphorylated and the corresponding nonphosphorylated peptide. This assay principle is applicable to both kinase and phosphatase reactions. In this assay, known peptide substrates containing a single phosphorylation site in proximity to the chymotrypsin consensus cleavage site are labeled with dual fluorophores, coumarin and fluorescein, one at each end. In the primary kinase reaction, the kinase phosphorylates a single residue of the FRET peptide substrate. In a secondary reaction, chymotrypsin recognizes and cleaves only the nonphosphorylated FRET peptides (the phosphorylated peptides are not recognized as substrates by chymotrypsin). Cleavage of the nonphosphorylated peptides by chymotrypsin disrupts FRET between the fluorophores, while uncleaved phosphorylated peptides undergo FRET (Figure 24.4). Similar to the BLA reporter gene assay, this FRET assay can be measured as a ratiometric readout. Excitation of coumarin in the uncleaved FRET peptide at 400 nm results in FRET to fluorescein, emitting at 520 nm. Chymotrypsin cleavage disrupts FRET, and hence excitation of coumarin now yields high-intensity coumarin fluorescence [22]. Reagents for the Z-lyte can be obtained from Panvera (Invitrogen).

Aurora assays also include the FRET-based voltage-sensing assay technology for measuring changes in membrane potential for ion channel applications. The voltage sensor probe (VSP) detects

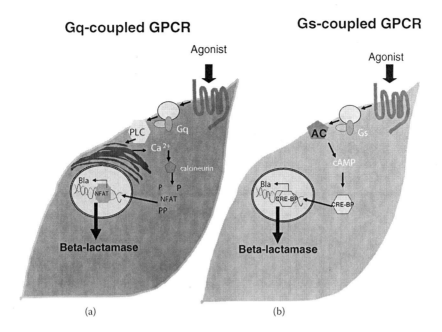

FIGURE 24.3 (See color insert following page 334.) Schematic representation of BLA reporter gene assay for GPCRs. (a) Gq-coupled GPCR. Agonist stimulation results in activation of Gq and phospholipase C, resulting in release of intracellular Ca^{2+} from internal stores. Elevated intracellular Ca^{2+} activates the Ca^{2+}-sensitive phosphatase, calcineurin, which in turn dephosphorylates and activates the transcription factor, NFAT. Activated NFAT translocates to the nucleus and activates transcription of the downstream BLA gene, resulting in BLA accumulation in the cytoplasm. (b) Gs-coupled GPCR: Receptor activation results in activation of Gs and adenylyl cyclase, increasing intracellular cAMP, and thereby activating the cAMP Response Element Binding Protein, which translocates to the nucleus and initiates transcription of the BLA gene under the control of the CRE promoter.

rapid dynamic changes in membrane voltage of live cells, reported as a FRET signal. This technology revolutionized the field of ion channels functional assays, enabling high-throughput screening (HTS) of ion channel targets in contrast to the extremely low-throughput, laborious electrophysiological measurements. VSP contains a FRET dye pair consisting of a mobile, fluorescent, voltage-sensitive acceptor oxanol (DiSBAC) and a fluorescent, membrane-bound coumarin phospholipid FRET donor, CC2-DMPE. When cells are loaded with the VSP dyes in resting condition, CC2-DMPE is inserted into the outer leaflet of the plasma membrane while DiSBAC is associated with the outer surface of the plasma membrane. Excitation of the CC2-DMPE at 400 nm results in FRET transfer to the oxanol and emission at 580 nm (strong red fluorescence). However, when the cells are depolarized, the intracellular membrane potential becomes relatively positive, which translocates the negatively charged oxanol rapidly into the inner surface of the cell membrane (Figure 24.5), resulting in disruption of FRET. Under these circumstances, excitation of the CC2-DMPE at 400 nm results in emission at 460 nm. As with the BLA technology, the FRET-based VSP assay can be represented by a ratiometric measurement with high sensitivity. This technology is primarily used for voltage-gated channels, including sodium channels, potassium channels, and chloride channels, particularly for HTS [23]. Reagents for the VSP technology can be obtained from Panvera/Invitrogen.

Thus, Aurora assays span three major biological areas: reporter gene BLA assays, biochemical Z-lyte assays, and ion channel VSP (Table 24.2). Among these, BLA reporter gene assays have been explored the most and will be described in detail in the following sections.

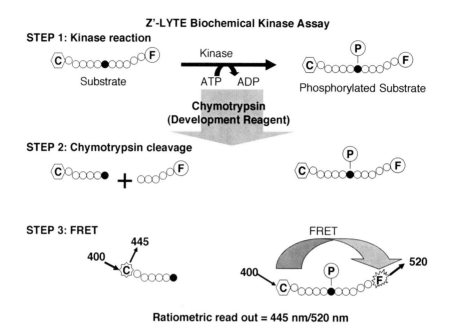

FIGURE 24.4 Schematic representation of the Z-lyte biochemical kinase assay. Kinase reaction on the FRET peptide substrate results in phosphorylation of the FRET peptide. When subjected to cleavage by chymotrypsin, only the native peptide is cleaved, while the phosphorylated peptide remains intact. Excitation of the coumarin at 400 nm in the phosphorylated peptide results in FRET to the fluorescein, resulting in 520 nm emission. In the cleaved peptide, excitation of the coumarin at 400 nm results in 445 nm emission. The assay can be represented as a ratiometric readout of 445 nm/520 nm emission.

24.2 DEVELOPMENT AND UTILIZATION OF A REPORTER GENE BETA-LACTAMASE ASSAY FOR GPCRS

The development of a BLA reporter gene assay for GPCRs in stably transfected mammalian cells requires the coexpression of the receptor of choice and the BLA reporter gene under the control of an appropriate promoter. The first step, therefore, is to pose the question: does your target GPCR signal via Ca^{2+} (Gq-coupled GPCR) or via cAMP (Gs-, Gi-coupled GPCR)? Accordingly, the BLA reporter gene should be constructed under the control of a NFAT or CRE promoter, respectively. If the signaling pathway is unknown, or for a more generic cell line, the BLA reporter gene may be constructed under the control of a dual CRE/NFAT promoter. Thus, different reporter gene constructs exist for GPCRs: NFAT-Bla and CRE-Bla for Gq-coupled and Gs-/Gi-coupled receptors, respectively. Alternatively, the more recently developed CellSensor™ cell lines, commercialized by Invitrogen as the GeneBLAzer system, in which the BLA gene is under the control of a variety of different promoters (including NFAT and CRE), may be used to study GPCR activation. These reporter gene constructs, cell lines, and the cell-permeable FRET substrate, CCF2- (or- 4)/AM, originally developed by Aurora Biosciences (now Vertex Pharmaceuticals), are now available through Invitrogen/PanVera. The appropriate BLA reporter gene construct may be cotransfected with the GPCR cDNA (in expression vectors like pcDNA3) directly into mammalian cells such as CHO-K1 or HEK293. Alternatively, CHO cells expressing the BLA gene under the control of the appropriate promoter may be used as the host cell line for transfection of the GPCR cDNA of interest.

An example of the generation and optimization of stable CHO cells expressing the NFAT-Bla reporter gene downstream to a Gq-coupled receptor is described below.

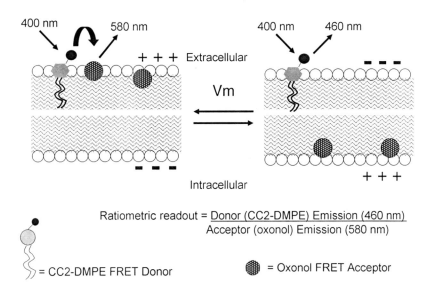

FIGURE 24.5 Schematic representation of VSP ion channel technology. In resting cells (left), the FRET dye pair, DiSBAC and CC2-DMPE, is located on the outer surface of the plasma membrane. Excitation of CC2-DMPE at 400 nm results in FRET to DiSBAC and subsequent 580 nm emission. Changes in electric potential of the cell membrane result in the translocation of the negatively charged oxonal (DiSBAC) to the inner surface of the plasma membrane, disrupting FRET. In the depolarized state, excitation of CC2-DMPE at 400 nm results in emission at 460 nm. The assay results are represented as a ratiometric readout of 460 nm/580 nm emission.

TABLE 24.2
Aurora Assays

Assay	Dye	Application
BLA reporter gene assay	CCF4/AM	GPCRs, NHRs, RTKs
Z-Lyte assay	Chymotrypsin	Kinases, phosphatases
VSP assay	DiSBAC and CC2-DMPE	Ion channels

Note: GPCRs: G-protein–coupled receptors; NHRs: nuclear hormone receptors; RTKs: receptor tyrosine kinases.

24.2.1 DEVELOPMENT OF STABLE CELL LINES

The development of stable BLA reporter gene cell lines for cell surface receptors is described in this section, with an example of the development of a cell line for a Gq-coupled GPCR in CHO cells.

CHO cells expressing the BLA reporter gene under the control of the 3X NFAT promoter may be used as the host cells for transfection of Gq-coupled GPCR in pcDNA3 by the Lipofectamine method. Alternatively, CHO-K1 cells may be cotransfected with the 3XNFAT-BLA reporter gene and the GPCR cDNA in pcDNA3. After transfection, cells should be maintained under antibiotic selection for the GPCR and BLA (zeocin selection) for 7 to 10 days. Cells may then be pooled to assay for BLA activity upon agonist stimulation to confirm the presence of functional GPCR. Figure 24.6 shows an example of BLA response from a transfected pool of cells in response to agonist

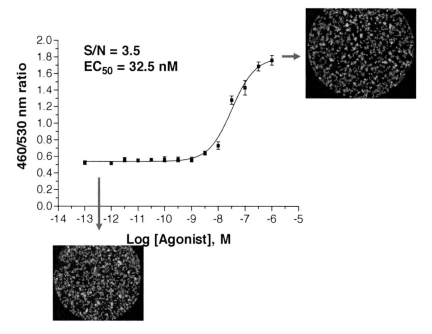

FIGURE 24.6 (See color insert following page 334.) Agonist dose response in heterogenous transfected pool of cells. CHO cells stably expressing the 3XNFAT-BLA reporter gene construct were transfected with a Gq-coupled receptor and maintained under antibiotic selection for 1 week. Cells were dissociated and assayed for agonist-induced BLA activity. Briefly, cells were plated into a 384-well black, clear bottom assay plate (Costar #3712) at a density of 4000 cells/well in growth medium and incubated at 37°C, 5% CO_2 for 18 h. On the day of the assay, growth medium was replaced with 40 μL of serum-free IMDM containing 25 mM Hepes and 0.1% BSA (assay buffer). Cells were stimulated with 10 μL (5× concentration) of agonist in assay buffer for 4 h at 37°C, 5% CO_2. BLA activity was measured by addition of 10 μL of the CCF4/AM substrate/dye mix (preparation: 12 μL of 1 mM CCF4/AM, 60 μL of 100 mg/mL pluronic F127, 925 μL of 24% PEG 400 with 12% Enhances Substrate Solution (ESS, from PanVera), and 75 μL of 15 mM probenecid in 200 mM NaOH) and incubation at room temperature for 1 h [12,13]. Assay plates were read in Tecan Spectrofluor plus (bottom read; excitation: 405 nm; emission: 460 nm, 530 nm).

stimulation. This heterogenous population of cells exhibits a ~3.5-fold increase in agonist-induced BLA activity, as observed by the FRET assay, and an EC_{50} of ~35 nM for the agonist, thus indicating the presence of functional receptors in this pool of cells. Fluorescence microscopy reveals that the basal unstimulated cells exhibit primarily green fluorescence emission due to FRET upon ultraviolet excitation, while at maximal agonist stimulation (1 $\mu$$M$), ~50% of the cells emit blue or blue/green fluorescence.

The next step is to isolate the cells transfected with functionally responsive receptors (preferably to isolate cells exhibiting the most sensitive response to agonist stimulation). This can be accomplished by fluorescence activated cell sorting (FACS) [12,13,18], detailed in Figure 24.7. First, in order to identify and eliminate cells exhibiting leaky, agonist-independent BLA activity, a "negative sort" should be accomplished in the absence of receptor agonist, as described. Briefly, cells in a T-75 flask should be washed with phosphate buffered saline (PBS) (without Ca^{2+} or Mg^{2+}), harvested by trypsinization, pelleted by centrifugation, resuspended at a concentration of 1×10^6 cells/mL in FACS buffer (PBS, 25 mM Hepes, 0.1% bovine serum albumin [BSA]) and loaded with the cell-permeable substrate, CCF2/AM (2 $\mu$$M$ CCF2/AM in dry dimethylsulfoxide [DMSO], 1 mg/mL pluronic-F127 in DMSO containing 0.1% acetic acid, 1 mL of 24% PEG 400), in the dark with mild shaking for 90 min. FACS may be accomplished on a Becton Dickinson FACS Vantage cell sorter as described previously [13,18]. During FACS, cells with different emission characteristics

Negative Sort

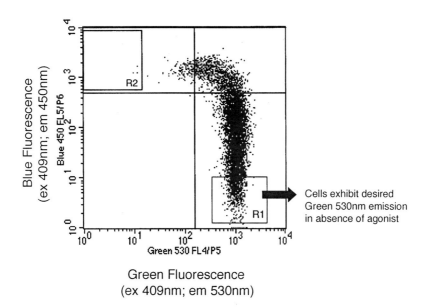

FIGURE 24.7 Clonal selection by FACS: Negative sort. FACS on transfected cells in absence of agonist stimulation to eliminate endogenously signaling cells. Cells were harvested, resuspended at a density of 1×10^6 cells/mL in FACS buffer (PBS, 25 mM Hepes, 0.1% BSA) and loaded with the CCF4/AM substrate for 90 min with gentle shaking in the dark. Cells exhibiting green 530 nm emission (bottom right quadrant of histogram) upon excitation at 405 nm were collected in bulk using a Becton Dickinson FACS Vantage cell sorter using HQ460/50m (blue) and HQ535/40m (green) bandpass filters. R2 box represents cells with endogenously signaling (blue fluorescence emission). R1 box represents cells exhibiting desired green fluorescence emission for bulk collection.

can be collected with HQ460/50m (blue) and HQ535/40m (green) bandpass emission filters separated by a 490 long-pass dichroic mirror. Cells responding green in color (i.e., emitting light at 530 nm due to FRET) should be directly collected into 10 mL of fresh complete medium containing 25 mM Hepes and Pen/Strep (prewarmed to 37°C) in a T-25 flask and allowed to grow for 1 week, thus eliminating cells exhibiting endogenous blue fluorescence at 460 nm from this population of cells. These cells should then be subjected to a second round of FACS (positive selection), this time after stimulation with agonist for 4 h (typically, the EC$_{50}$ concentration of agonist). This time around, the goal would be to identify and isolate single cells exhibiting agonist-induced BLA activity (representing cells with functional GPCRs stimulating the BLA reporter gene). Accordingly, upon excitation at 405 nm, cells exhibiting intense blue fluorescence (460 nm emission) due to the absence of FRET (Figure 24.8) should be single-cell sorted into 200 μL of complete medium with 25 mM Hepes and Pen/Strep in a 96-well plate (Becton Dickinson FACS Vantage) and expanded over several weeks into regular tissue culture flasks.

24.2.2 ASSAY OPTIMIZATION

Upon the isolation and expansion of several single cell clones from the 96-well plates into T-flasks, a few (five to 10) should be analyzed for agonist dose response in the BLA assay to identify the clone with the most robust response (assay window, Signal/Basal or S/B) and the highest sensitivity (lowest agonist EC$_{50}$). Figure 24.9 shows an example of five individual clones, all grown up from single-cell FACS, responding with different sensitivities and robustness in the BLA assay. Such an

Positive sort

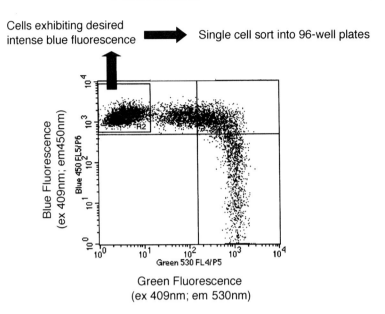

FIGURE 24.8 Clonal selection by FACS: Positive sort. Isolation of pure single cell clones by FACS on transfected cells after agonist stimulation. Cells were stimulated with EC_{50} concentration of agonist for 4 h at 37°C, 5% CO_2. Cells were then harvested, resuspended at a density of 1×10^6 cells/mL in FACS buffer (PBS, 25 mM Hepes, 0.1% BSA) and loaded with the CCF4/AM substrate for 90 min with gentle shaking in the dark. Cells exhibiting intense blue (460 nm) emission (top left quadrant of histogram) upon excitation at 405 nm were sorted as single cells into a 96-well plate containing 200 μL growth medium with 25 mM Hepes using a Becton Dickinson FACS Vantage cell sorter as described above.

experiment is useful for the selection of the final clone exhibiting a good S/B and a sensitive, reproducible EC_{50} for the agonist. It is important to select the final clone based not only on the assay window (S/B) but also on the sensitivity (low EC_{50}). In the example shown in Figure 24.9, clone K has the largest assay window, yet it exhibits a significantly higher EC_{50} for the agonist than some of the other clones do and hence is not the most sensitive one. The BLA assay using these stable cells may be further optimized as described below.

24.2.2.1 Optimal Assay Protocol

The BLA assay can be conducted on adherent mammalian cells (plated ~16 to 24 h before assay) or with suspension-format cells (dispersed cells, albeit normally adherent cells, for example, CHO, dissociated from flasks on the morning of the assay, resuspended at appropriate density, and directly plated for the assay). It is important to bear in mind that some clones respond well only in adherent cell format assays, while other clones may respond equally well in both adherent and suspension mode assays. Therefore, based on the goals of the project, clonal selection and assay development/optimization should be conducted with the appropriate protocol (suspension or adherent). For example, the "suspension cell" assay is often preferred for use in automated HTS assays with 24-h robotic protocols, etc. The suspension cell protocol provides the additional advantage of conducting the assay in serum-free conditions, as may be important in some cases for HTS, to prevent compounds from binding to serum. It is prudent to use nonenzymatic dissociation procedures when working with suspension cell assays. A schematic representation of the adherent and suspension cell assay protocols is presented in Table 24.3.

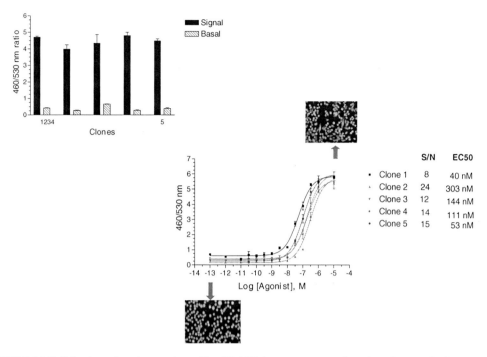

FIGURE 24.9 Selection of optimum clone. The BLABLA assay was conducted on "suspension" cells in a 384-well format using 30,000 cells/well in 40 μL assay medium. Cells were stimulated with 10 μL of 5× agonist for 4 h at 37°C, 5% CO_2, and loaded with 10 μL of 6× CCF4/AM substrate/dye for 1 h at room temperature. Assay plates were read as described in Figure 24.6.

TABLE 24.3
Beta-Lactamase Assay Protocol

Adherent Cell Assay Protocol	Suspension Cell Assay Protocol
Dissociate cells from flask	Dissociate cells from flask
Plate cells at 3–10 K cells/well in 20 μl medium	Plate 20–50 K cells/well in 20 μl medium
↓ Incubate ~20 hours 37°C, 5% CO_2	↓
Add 5 μl of 5× agonist	Add 5 μl of 5× agonist
↓ Incubate 3–5 hours 37°C, 5% CO_2	↓ Incubate 3–5 hours 37°C, 5% CO_2
Add 5 μl of CCF2/4-AM substrate/dye mix	Add 5 μl of CCF2/4-AM substrate/dye mix
↓ Incubate 1–2 hours RT	↓ Incubate 1–2 hours RT
Read plates in Fluorescent plate reader (bottom read; Excitation: 405 nm; Emission: 460 nm, 530 nm)	Read plates in Fluorescent plate reader (bottom read; Excitation: 405 nm; Emission: 460 nm, 530 nm)

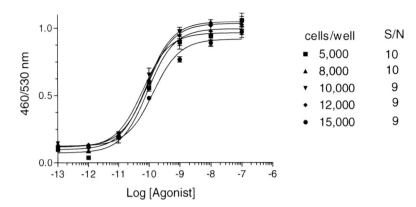

FIGURE 24.10 Optimum cell number per well. The BLA assay was conducted on adherent cells in a 96-well format. Cells were stimulated with 10 μL of 10× agonist for 4 h at 37°C, 5% CO_2, and loaded with 20 μL of 6× CCF4/AM substrate/dye for 1 h at room temperature. Assay plates were read as described in Figure 24.6. (From Kunapuli, P., Ransom, R., Murphy, K., Pettibone, D., Kerby, J., Grimwood, S., Zuck, P., Hodder, P., Lacson, R., Hoffman, I., Inglese, J., Strulovici, B. (2003) *Anal. Biochem.* 314, 16–29. With permission.)

24.2.2.2   Optimum Number of Cells per Well

An advantage of using a fluorogenic substrate and FRET as the readout is the ability to express the BLA activity as a ratio of intensities at the two emission wavelengths (460 nm/530 nm). The ratiometric readout is less susceptible to variations in cell size, cell number, or substrate concentrations. Variations in cell density from well to well or from plate to plate may occur due to unevenness in cell doublings or plating. Such variations in cell number do not adversely affect the assay results [13]. Typically, for the adherent cell assay protocol in 96-well assay plates, optimum assay windows may be achieved using 5000 to 15,000 cells/well for a given assay (Figure 24.10). Comparatively, in 384-well plates, optimum assay windows may be achieved using 1500 to 8000 cells/well. This large variation in cell number per well is not typically tolerated in other cell-based assays. Another source of variation in cellular response from well to well may be the distribution of the cells within the well. Uneven distribution or redistribution of cells within the well may occur at the time of dispensing the cells into the assay plates or later during addition of other assay reagents to the wells. Such unevenly distributed cells are well accommodated in a ratiometric readout due to the normalization of the signal by the number of live cells in the well (dead cells will not take up the CCF2/4-AM dye). For example, Figure 24.11 shows a well with evenly distributed cells (Figure 24.11a; ideal condition) and a well with an uneven distribution of cells (Figure 24.11b); both wells have comparable numbers of cells. When analyzed by a fluorescent plate reader, the well on the left with even cell distribution would register significantly higher green emission (inverted triangle, 530 nm) and slightly higher blue emission (square, 460 nm) than the well on the right. However, when represented as a ratiometric readout of 460 nm/530 nm (represented by circular, red data points in Figure 24.11), these wells do not differ significantly in their response.

24.2.2.3   Optimum Time of Stimulation

The optimal time of agonist stimulation varies in different types of assays and also among different reporter gene assays. Compared to CAT and luciferase assays, BLA assays require relatively shorter time periods for gene transcription and translation [13]. While this may vary somewhat from one cell line to another and should be determined empirically for each cell line, optimal BLA gene

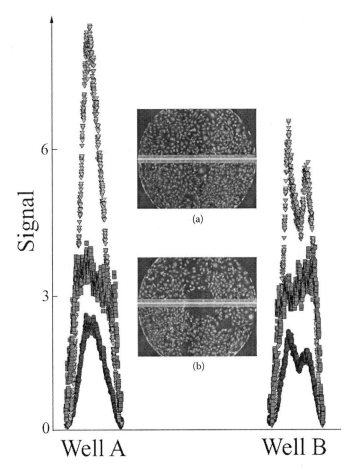

FIGURE 24.11 (See color insert following page 334.) Advantages of ratiometric readout. Ratiometric readout provides a method of compensation for uneven distribution of fluorescence in assay wells. Well A has a normal (even) distribution of cells. Well B exhibits uneven cell distribution. Blue circles represent the 460 nm emission, inverted green triangles represent the 530 nm emission, and red squares represent the green-to-blue ratio. The similarity of green-to-blue ratios for well A and well B demonstrates the effectiveness of the ratiometric method. (From Kornienko, O., Lacson, R., Kunapuli, P., Schneeweis, J., Hoffman, I., Smith, T., Alberts, M., Inglese, J., and Strulovici, B. (2004) *J. Biomol. Screen.* 9, 186–195. With permission.)

response is often achieved within 4 to 6 h of agonist activation (Figure 24.12). Similarly, the optimal time for substrate/dye loading also varies between 1 and 2 h and should also be verified empirically for each cell line.

24.2.2.4 DMSO Tolerance

Since most compounds are dissolved in DMSO, the DMSO tolerance of the cell lines should be analyzed. Agonist dose–response curves in the presence of varying concentrations of DMSO should be analyzed for each cell line (Figure 24.13). Different CHO-BLA cell lines can exhibit different sensitivity to DMSO, as judged by the EC_{50} of the agonist response and the S/B. DMSO sensitivity of cell lines also depends on the length of exposure (assay time). In the BLA assay, tolerance of ~0.75% DMSO is common, while higher concentrations of DMSO are detrimental to the cells [13]. However, this parameter should be determined empirically for each cell line.

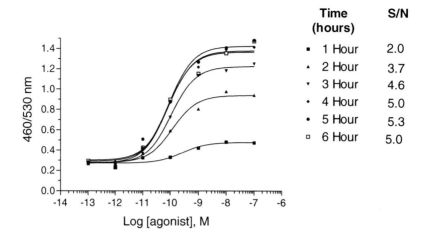

FIGURE 24.12 Time course of agonist stimulation. The BLA assay was conducted on adherent cells in a 96-well format. Cells were stimulated with 10 μL of 10× agonist for varying times at 37°C, 5% CO_2, and loaded with 20 μL of 6× CCF4/AM substrate/dye for 1 h at room temperature. Assay plates were read as described in Figure 24.6. (From Kunapuli, P., Ransom, R., Murphy, K., Pettibone, D., Kerby, J., Grimwood, S., Zuck, P., Hodder, P., Lacson, R., Hoffman, I., Inglese, J., Strulovici, B. (2003) *Anal. Biochem.* 314, 16–29. With permission.)

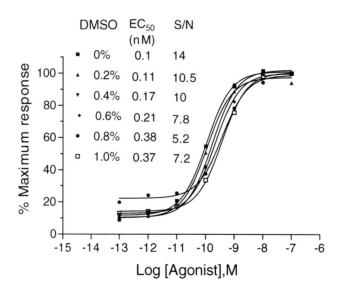

FIGURE 24.13 DMSO tolerance. The BLA assay was conducted on adherent cells in a 96-well format. Cells were stimulated with 10 μL of 10× agonist in buffer containing varying concentrations of DMSO for 4 h at 37°C, 5% CO_2, and loaded with 20 μL of 6× CCF4/AM substrate/dye for 1 h at room temperature. Assay plates were read as described in Figure 24.6. (From Kunapuli, P., Ransom, R., Murphy, K., Pettibone, D., Kerby, J., Grimwood, S., Zuck, P., Hodder, P., Lacson, R., Hoffman, I., Inglese, J., Strulovici, B. (2003) *Anal. Biochem.* 314, 16–29. With permission.)

24.2.3 Beta-Lactamase Assays for HTS

The BLA reporter gene assays are well suited for high throughput screening due to:

- Ability to clonally select a pure population of functionally responsive cells
- Homogeneous assay, compatible with a large variety of liquid dispensers and detectors
- Ratiometric readout, allows for high tolerance for variations in cell number from well to well
- Relatively short duration, minimizing compound toxicity issues
- Ease of miniaturization into 3456-well formats for 2-μL cell-based assays compatible with ultra high-throughput screening [24]

As shown in Figure 24.14, the BLA assay scales excellently from a 96-well assay plates (100 μL total assay volume) to 384-well plates (25 to 50 μL total assay volume), 1536-well plates (8 to 10 μL total assay volume) and 3456-well formats (2 μL total assay volume). The BLA assay in each of these plate formats can be performed using the adherent or suspension cell protocols. The assay scales proportionally to the plate density, while maintaining the same incubation times. Example protocols for 1536- and 3456-well formats are shown in Table 24.4. The 1536-well assays can be performed with standard 1536-compatible liquid dispensers (for example, CyBio, Cartesian) and fluorescence detectors capable of bottom reading (Tecan's Spectrofluor plus, Saphire, Ultra Evolution, Molecular Devices's LJL Aquest, PerkinElmer's Vicotr V). The 3456-well assays, on the other hand, require specialized instruments from Aurora Discovery Inc., such as the Flying Reagent Dispenser (FRD) for liquid dispensation into the 3456-well assay plates and the topology-compensating plate reader (tcPR), which appropriately corrects plate warping, for detection [24]. Principles, specifications, and use of these instruments for "Aurora assays" are detailed by Kornienko et al. [24].

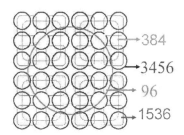

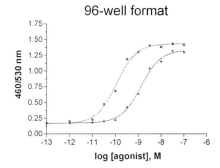
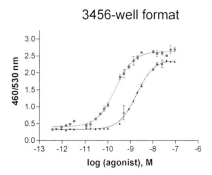

FIGURE 24.14 Assay miniaturization. A schematic representation of the well area of 96-, 384-, 1536-, and 3456-well plates is shown on the top. Comparison of agonist dose–response curves between a 96-well and a 3456-well assay is shown at the bottom for two different agonists. Both agonists exhibit comparable S/B and EC$_{50}$ in the two plate formats.

TABLE 24.4
Miniaturized Assay Protocol

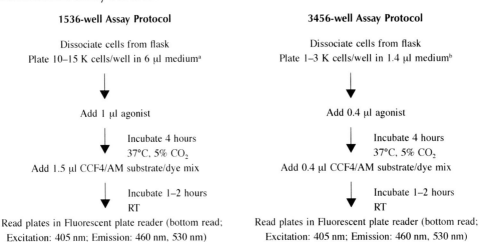

1536-well Assay Protocol	3456-well Assay Protocol
Dissociate cells from flask	Dissociate cells from flask
Plate 10–15 K cells/well in 6 µl medium[a]	Plate 1–3 K cells/well in 1.4 µl medium[b]
↓	↓
Add 1 µl agonist	Add 0.4 µl agonist
↓ Incubate 4 hours 37°C, 5% CO_2	↓ Incubate 4 hours 37°C, 5% CO_2
Add 1.5 µl CCF4/AM substrate/dye mix	Add 0.4 µl CCF4/AM substrate/dye mix
↓ Incubate 1–2 hours RT	↓ Incubate 1–2 hours RT
Read plates in Fluorescent plate reader (bottom read; Excitation: 405 nm; Emission: 460 nm, 530 nm)	Read plates in Fluorescent plate reader (bottom read; Excitation: 405 nm; Emission: 460 nm, 530 nm)

[a] For 1536 well adherent cell assays, cells should be plated at 2–7 K cells/well in 6 µl medium and incubated at 37°C, 5% CO_2 for 16–24 hours before assay.
[b] For 3456-well adherent cell assays, cells should be plated at a density of 500–1500 cells/well in 1.4 µl medium and incubated at 37°C, 5% CO_2 for 16–24 hours before assay.

24.2.4 TROUBLESHOOTING

This section describes problems that may be encountered during the course of developing and using the BLA reporter gene assay, and some strategies to overcome these problems:

- *Cells exhibit a high degree of basal 460 nm blue emission without agonist stimulation in the BLA assay*: Such a condition could occur due to leaky transcription/translation, constitutive activity of the receptor, and/or from some degree of receptor stimulation from components of the cell growth medium (for example, for metabotropic glutamine receptors, it is important to use glutamine-free medium for cell culture; other examples include receptors that respond to components in serum present in the complete growth medium). Such issues can be resolved as follows:
  - Choose the cell culture components carefully, mindful of the physiology of the recombinant target of interest; eliminate unnecessary components from cell culture medium.
  - To minimize basal FRET due to leaky transcription/translation, cell sorting (FACS) in the absence of agonist must be conducted. During this step, cells exhibiting 460 nm blue fluorescence emission should be eliminated and cells exhibiting 530-nm green fluorescence emission should be collected (bulk collection of cells is sufficient) for further growth and subsequent rounds of FACS, as necessary.
  - Use gentle cell dissociation procedures with minimal cell agitations to avoid excessive stress (stress-induced responses may result in BLA gene activation).
  - Basal blue fluorescence may be further decreased by the addition of clavulanic acid (Zuck and Kunapuli, personal communication). However, this needs to be titrated carefully for each assay, as excess clavulanic acid (an inhibitor of BLA enzyme) will inhibit the overall agonist-induced signal as well.

- Serum starvation of cells in the flask 16 to 24 h prior to dissociation and use in the BLA assay may further help in reducing the basal unstimulated activity (arising from serum components). This strategy could help increase the assay window (S/B).

- *Assay sensitivity not sufficient*: In order to maximize the assay sensitivity, the most sensitive cells should be selected at the time of clonal selection. This can be accomplished by using an EC_{50} concentration of agonist during FACS (instead of saturating concentration of agonist), selecting for intensely blue cells (single cell sort). This strategy has improved chances of ensuring the most sensitive clone for the assay.

- *Loss of response with cell passaging*: Although the the BLA assay works with stably transfected cells, instability in the response may occasionally be observed with cell passaging. The stability of the BLA response with cell passage needs to be empirically determined for each cell line. Inherent instability may be overcome by subcloning the promoter–reporter construct into a different expression vector with antibiotic selection. If the stable cell line displays acceptable response for a few passages (three to 10), an alternate strategy that works effectively is the generation of a large cell bank of frozen cells of early passage number. These frozen vials of cells may be thawed as needed and used for a few passages, depending on stability. Even for HTS purposes lasting several weeks, new vials of cells may be thawed every week without significant changes in assay results between the different thaws and the different passage numbers. A more dramatic version of the same strategy involves thawing cells and plating directly into assay plates for the assay, a strategy that works well with some cell lines even for HTS [40,41]. This strategy could even be adapted for a large batch of a transiently transfected heterogenous population of cells.

- *Good response during clonal selection by FACS, but suboptimal response in subsequent suspension cell assays*: It is important to avoid trypsin as the agent for dissociating cells from the tissue culture flasks when using the suspension cell format for the BLA assay. Since the cells are assayed within a few hours, receptor integrity at the cell surface is imperative. Therefore, nonenzymatic cell dissociation procedures such as enzyme-free (EDTA-based) solutions or accutase are strongly recommended. Another strategy to increase the assay window is to serum starve the cells for 16 to 24 h before cell dissociation. This could potentially decrease the basal 460 nm blue emission from unstimulated cells, resulting in a larger assay window upon agonist stimulation.

- *Issues with Gi-coupled receptors*: For this subclass of GPCRs, development of functional assays (reporter gene or direct second messenger assays) is rather challenging. The following strategies are recommended for the development of a BLA reporter gene assay for Gi-coupled GPCRs:
  - Transfection of receptor cDNA into cell expressing CRE-BLA. In this strategy, receptor activation results in a decrease in intracellular cAMP. In order to visualize this decrease, assays for Gi-coupled GPCRs (cAMP assay and reporter gene assays) are often conducted by stimulating the cells with forskolin, in addition to receptor agonist. This pharmacological agent causes an increase in intracellular cAMP by directly activating adenylyl cyclase by a nonreceptor mediated mechanism. Therefore, in the presence of forskolin, activation of the Gi-coupled receptor would result in a decrease in forskolin-stimulated cAMP [25], CRE, and subsequent BLA response.
  - Transfection of receptor cDNA into cells expressing Gqi5-NFAT-BLA. The identification of receptor-G protein interaction sites by Liu et al. [26] led to the development of chimeric G proteins. In the case of the Gqi5 chimera, the last five amino acids of Gi are substituted at the *C*-terminus of Gq, thereby redirecting Gi-coupled inhibitory

signals to the Gq-signaling pathway, providing an improved method for examining Gi-coupled GPCR activation with a positive readout (gain in signal assay) [27].

- *Need to increase assay robustness*: Assay robustness (S/B) may also be increased by prolonging the second messenger signaling. As demonstrated for several GPCRs [28,29], mutations in the carboxyl terminal tail (point mutations in Ser or Thr residues, or *C*-terminal truncations) may increase or prolong agonist-induced second messenger responses. Such mutations in the receptor may also help in increasing the assay window of the BLA assay and should be determined empirically for each receptor.

24.2.5 INTERFERENCES IN BLA ASSAYS

As with all assay technologies, the BLA reporter gene technology is also prone to interference. It is important to know the source of these interferences and to design appropriate strategies to overcome these assay-related artifacts. Similar to most reporter gene assays, which are significantly downstream from the target receptor, the BLA assay is sensitive to compounds that may modulate various components of the signal transduction pathway and hence appear to be a "hit" or "active" in the assay. Figure 24.15 depicts the potential targets in the signal transduction pathway downstream to a Gq-coupled receptor, including true receptor antagonists and inhibitors of the BLA

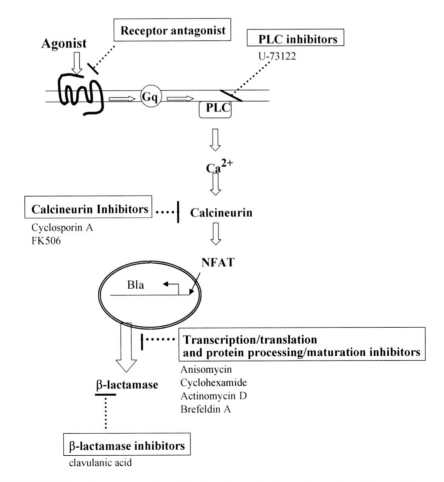

FIGURE 24.15 Schematic representation of the signal transduction pathway for a Gq-coupled receptor in a BLA reporter gene assay, indicating the types of nonreceptor targets in the assay.

enzyme. Probing the signal transduction pathway from receptor activation to BLA activity using well-characterized agents confirms that indeed the signal transduction pathway is vulnerable to a variety of interfering agents [13]. For example, as shown in Figure 24.16, receptor antagonists, pathway inhibitors, and BLA inhibitors all exhibit an "antagonist" or "inhibitor" phenotype, with a decrease in agonist-induced signal. However, this may in fact be due to inhibition of one of the several signal transduction components, such as phospholipase (PLC), calcineurin, or NFAT. Alternatively, some compounds may inhibit the transcription/translation machinery and thereby result in a decreased agonist-induced signal. However, it is important to note that such assay related nontarget positives are specific to the signal transduction pathway. Agents disrupting signal transduction components of other signaling pathways not related to the target receptor do not interfere with the BLA assay. For example, in Gq-coupled receptor signaling, inhibitors of rapamycin [30] or wortmannin, which target the TOR pathway and PI3K [31], respectively, do not have any effect on the agonist-induced BLA signal (see Figure 24.16a).

Other artifacts commonly seen in BLA assays include compound toxicity (similar to other cell-based assays) and fluorescence interference (similar to other fluorescence-based assays). As shown in Figure 24.16b, toxic compounds like oxalic acid result in cell death, as observed by a smaller number of live cells that take up the CCF dye. Such wells would give a low "blue" or agonist signal, mimicking an antagonist response. However, these wells would also exhibit low 530 nm emission (low green fluorescence) due to the lower number of cells in the well, and can therefore be identified by the analysis of individual emission channel values. In the case of fluorescence, in addition to fluorescent compounds, the BLA assay is sensitive to fluorescence introduced by lint (see Figure 24.16b) or dust particles, all of which fluoresce in the blue wavelength. Such fluorescence can also be identified by analysis of individual blue (460 nm) and green (530 nm) emission values. This feature of the BLA assay, by virtue of the FRET substrate, provides a unique advantage over other reporter gene technologies and over other fluorescence-based assays.

Table 24.5 lists the commonly encountered false positives and false negatives in the BLA assay. Most of these may be identified by analysis of the individual 460 nm and 530 nm emission values. Since all assay technologies are inherently prone to artifacts of some kind, it is important to understand and design assays and/or strategies to eliminate such "nontarget" activities. The ideal method of choice would be to design another assay for the same target based on a different assay technology (such as a direct second messenger analysis and/or receptor binding).

24.2.6 NOVEL APPLICATIONS OF THE BETA-LACTAMASE REPORTER GENE TECHNOLOGY

In addition to the obvious applications of the BLA reporter gene assay for popular molecular targets (discussed in previous sections), the BLA reporter gene technology has also been applied toward some unique assays, as reviewed below.

The development of a ubiquitin tag for heterologously expressed proteins in mammalian cells forms the basis of a versatile system for producing proteins with short half-life in mammalian cells [32]. This principle was shown to be applicable to cellular proteins and heterologous reporter proteins, and led to the development of a novel protease assay for the NS2/3 protease from the hepatitis C virus. In this protease assay, the Ns2/3 protease-BLA fusion was placed upstream of a ubiquitin destabilization domain. In stable cells, cleavage of the Ns2/3-BLA fusion protein by the Ns2/3 protease resulted in differential stability between the cleaved and uncleaved BLA fusion reporter. In this assay, the BLA activity was thus a measure of the Ns2/3 protease activity [33].

The versatility of the BLA reporter is evident in the ability of the enzyme to be expressed as secreted, intracellular, and membrane-bound forms, each with distinct advantages as a reporter system [34]. In addition, the enzymatic activity was shown to be proportional to transfected DNA and the level of mRNA expression, forming the basis for the development of a system for selecting human secreted and membrane proteins in *E. coli* by Tan et al. [35]. In order to isolate or enrich

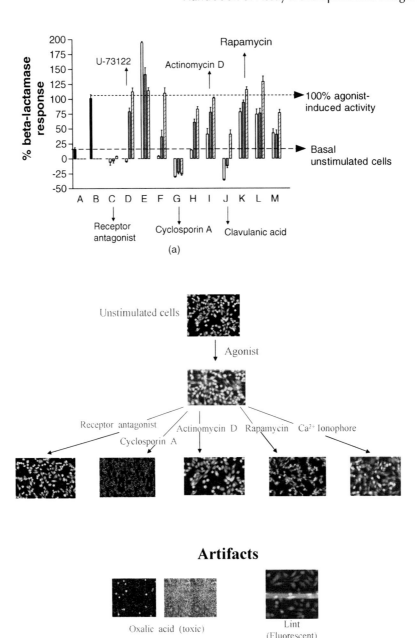

Artifacts

(b)

FIGURE 24.16 (See color insert following page 334.) Effect of signal transduction inhibitors in the BLA assay. (a) Cells were plated at a density of 4000 cells/well in a 384-well assay plate and stimulated with varying concentrations (0.5 μM, striped bars; 1.5 μM, filled bars; or 5 μM, open bars) of compounds in DMSO (final DMSO conc = 0.25%) for 20 min at 37°C, followed by the addition of 10 μL of 5× agonist at EC_{70} concentration and incubation at 37°C for 4 h. The 6× CCF substrate loading buffer was loaded, and plates were analyzed in Tecan Spectrofluor Plus as described earlier. A: unstimulated cells; B: EC_{70} agonist; C: receptor antagonist; D: U-73122; E: calcium ionophores; F: brefeldin A; G: cyclosporin A; H: cyclohexamide; I: actinomycin D; J: clavulanic acid; K: rapamycin; L: wortmannin; M: oxalic acid. (b) Fluorescence microscopy pictures of random fields of some representative wells at 200× magnification. (From Kunapuli, P., Ransom, R., Murphy, K., Pettibone, D., Kerby, J., Grimwood, S., Zuck, P., Hodder, P., Lacson, R., Hoffman, I., Inglese, J., Strulovici, B. (2003) *Anal. Biochem.* 314, 16–29. With permission.)

TABLE 24.5
Interferences in Beta-Lactamase Assays

Possible Reason	Agonist Assays		Antagonist Assays	
	False Positive	False Negative	False Positive	False Negative
Blue cpd fluorescence	X			X
Green cpd fluorescence		X	X	
Lint[a]	X			X
Dust[a]	X			X
Nontarget activity	X		X	
Toxicity	X	X	X	X

[a] Not reproducible.

for genes encoding secreted enzymes, growth factors and/or receptors, the cDNAs were fused to a leaderless BLA reporter gene to isolate clones encoding signal peptides of human genes. In this study, BLA fusion proteins with a eukaryotic signal peptide at the *N*-terminus were exported into the periplasm of *E. coli*.

The BLA reporter gene assay in mammalian cells can be directly applied as a genome-wide functional assay to identify patterns of regulation associated with specific genes [18]. For this application, a promoterless BLA reporter gene was used to create a library of reporter-tagged clones by transfection into mammalian cells. When loaded with the cell-permeable CCF2/4-AM substrate, this cell library reported the expression of a large number of endogenous genes. Such an assay could be used to monitor the regulation of specific genes or pathways with pharmacological agents or stimuli of interest. Along similar lines, BLA has been used as a marker to monitor gene expression *in vivo* during embryogenesis in live zebrafish [36]. The BLA reporter gene was found to be a suitable marker for monitoring spatially restricted patterns of gene expression by the injection of the CCF2 fluorescent substrate.

More recently, BLA has been used as a reporter for gene splicing activity in live cells [37]. In this report, the reporter–ribozyme construct consisted of a self-splicing intron ribozyme inserted into the BLA mRNA. The splicing activity in individual cells was visualized and quantitated to show that individual cells exhibit considerable heterogeneity in ribozyme activity. This BLA reporter assay allowed for the selection of ribozyme variants exhibiting improved activity.

The BLA reporter gene technology has also been used to evaluate cytotoxicity and antitumor potential in a high-throughput screening assay [38], examining the effects of platinum complexes and other agents like cisplatin on the BLA reporter gene expression in cancer cells.

Yet another unique example of the utilization of the BLA reporter gene technology is its use to study viral replication of hepatitis C virus [39]. In this assay, viral replication was measured using reporter genes such as luciferase or BLA to detect and quantitate viral replication, represented by viral replicons (subgenomic viral RNAs encoding elements required for autonomous replication). In this system, the BLA reporter gene had the additional advantage of providing a measure of both the number of cells harboring viral replicons and the number of replicons per cell, by virtue of the FRET assay for BLA detection, which allows single live cell quantification. In comparison, the luciferase reporter gene assay for viral replication only provided a measure of the amount of replicon RNA in a sample/well (aggregate measurement), but does not provide a measure on the number of cells in a heterogenous sample supporting replication.

Thus, the BLA reporter gene technology can be applied to a large variety of cell-based assays using the FRET detection system.

24.3 CONCLUSION

Among reporter gene assays, the BLA reporter gene technology has some unique features, which make this assay an attractive choice, particularly for clonal selection and high throughput screening. The BLA reporter gene assay offers a sensitive method for the selection of a functional clone, thereby significantly improving the quality of the assay. Since there are no endogenous homologs in mammalian cells, this assay has an inherently low background. When combined with the cell-permeable FRET substrate, this assay becomes homogeneous and does not require cell lysis. In addition, the assay is relatively rapid and is applicable to a large variety of molecular targets (GPCRs, nuclear hormone receptors, proteases, kinases, etc.). The homogeneous assay with the FRET substrate enables assay miniaturization to a 2-μL total assay volume, compatible with the 3456-well plate format. While the BLA assay (as with any other assay) appears to be vulnerable to certain pathway and detection artifacts, it is important to take advantage of the strengths of this assay and design strategies to overcome the limitations.

ACKNOWLEDGMENTS

I would like to thank Berta Strulovici for her continued support and review of this chapter, Wei Zhang for his critical review of the VSP technology, and Oleg Kornienko for his help with the miniaturized BLA assays.

REFERENCES

1. Naylor, L. H. (1999) Reporter gene technology: the future looks bright. *Biochem. Pharmacol.* 58, 749–757.
2. Gorman, C. M., Moffatt, L. F., and Howard, B. H. (1982) Recombinant genomes which express chloramphenicol acetyl transferase in mammalian cells. *Mol. Cell. Biol.* 2, 1044–1051.
3. Jain, V. K. and Magrath, I. T. (1991) A chemiluminescent assay for quantitation of β-galactosidase in femtogram range: application to quantitation of β-galactosidase in lacZ-transfected cells. *Anal. Biochem.* 199, 119–124.
4. Chen, W., Shields, T. S., Stork, P. J. S., and Cone, R. D. (1995) A calorimetric assay for measuring activation of Gs- and Gq-coupled signaling pathways. *Anal. Biochem.* 226, 349–354.
5. Berger, J., Hauber, J., Hauber, R., Geiger, R., and Cullen, B. R. (1988) Secreted placental alkaline phosphatase: a powerful new quantitative indicator of gene expression in eukaryotic cells. *Gene* 66, 1–10.
6. Durocher, Y., Perret, S., Thibaudeau, E., Gaumond, M. H., Kamen, A., Stucco, R., and Abramovitz, M. (2000) A reporter gene assay for high throughput screening of G-protein–coupled receptors stably or transiently expressed in HEK293 EBNA cells grown in suspension culture. *Anal. Biochem.* 284, 316–326.
7. Williams, T. M., Burlein, J. E., Ogden, S., Kricka, L. J., and Kant, J. A. (1989) Advantages of firefly luciferase as a reporter gene: application to the interleukin-2 gene promoter. *Anal. Biochem.* 176, 28–32.
8. Fitzgerald, L. R., Mannan, I. J., Dytko, G. M., Wu, H. L., and Nambi, P. (1999) Measurement of responses from Gi-, Gs-, or Gq-coupled receptors by a multiple response element/cAMP response element-directed reporter assay. *Anal. Biochem.* 275, 54–61.
9. Kain, S. R. (1999) Green fluorescent protein (GFP): applications in cell-based assays for drug discovery. *Drug Discovery Today* 4, 304–312.
10. Oakley, R. H., Hudson, C. C., Cruickshank, R. D., Meyers, D. M., Payne, R. E., Jr., Rhem, S. M., and Loomis, C. R. (2002) The cellular distribution of fluorescently tagged arrestin provides a robust, sensitive, and universal assay for screening G-protein–coupled receptors. *Assay Drug Devel. Tech.* 1, 21–30.

11. Zlokarnik. G., Negulescu, P. A., Knapp, T. E., Mere, L., Burres, N., Feng, L., Whitney, M., Roemer, K., and Tsein, R. Y. (1998) Quantitation of transcription and clonal selection of single living cells with β-lactamase as reporter. *Science* 279, 84–88.

12. Zlokarnik, G. (2000) Fusions to β-lactamase as a reporter for gene expression in live mammalian cells. *Meth. Enzymol.* 326, 221–241.

13. Kunapuli, P., Ransom, R., Murphy, K., Pettibone, D., Kerby, J., Grimwood, S., Zuck, P., Hodder, P., Lacson, R., Hoffman, I., Inglese, J., and Strulovici, B. (2003) Development of an intact cell reporter gene beta-lactamase assay for G-protein–coupled receptors for high-throughput screening. *Anal. Biochem.* 314, 16–29.

14. Rao, A., Luo, C., Hogan, P. G. (1997) Transcrioption factors of the NFAT family: regulation and function. *Annu. Rev. Immunol.* 15, 707–747.

15. Fink, J. S., Verhave, M., Kasper, S., Tsukada, T., Mandel, G., and Goodman, R. H. (1998) The CGTCA sequence motif is essential for biological activity of the vasoactive intestinal peptide gene cAMP-regulated enhancer. *Proc. Natl. Acad. Sci. USA* 85, 6662–6666.

16. Gonzales, G. A. and Montminy, M. R. (1989) Cyclic AMP stimulates somatostatin gene transcription by phosphorylation of CREB at serine 133. *Cell* 59, 675–680.

17. Chao, M. V. (2003) Neurotrophins and their receptors: a convergence point for many signaling pathways. *Nature Rev.* 4, 299–309.

18. Whitney, M., Rockenstein, E., Cantin, G., Knapp, T., Zlokarnik, G., Sanders, P., Durick, K., Craig, F. F., and Negulescu, P. A. (1998) A genome-wide functional assay of signal transduction in living mammalian cells. *Nature Biotech.* 16, 1329–1333.

19. Chin, J., Adams, A. D., Bouffard, A., Green, A., Lacson, R., Smith, T., Fischer, P. A., Menke, J. G., Sparrow, C. P., and Mitnaul, L. J. (2003). Miniaturization of cell-based beta-lactamase-dependent FRET assays to ultra-high throughput formats to identify agonists of human liver X receptors. *Assay Drug Devel. Tech.* 1(6), 777–787.

20. Peekhaus, N. T., Ferrer, M., Chang, T., Kornienko, O., Schneeweis, J. E., Smith, T. S., Hoffman, I., Mitnaul, L. J., Chin, J., Fischer, P. A., Blizzard, T. A., Birzin, E. T., Chan, W., Inglese, J., Strulovici, B., Rohrer, S. P., and Schaeffer, J. M. (2003). A beta-lactamase-dependent Gal4-estrogen receptor b transactivation assay for the ultra high-throughput screening of estrogen receptor b agonists in a 3456-well format. *Assay Drug Devel. Tech.* 1(6), 789–800.

21. Chin, J., Adams, A., Bouffard, A., Green, A., Lacson, R. G., Smith, T., Fischer, P. A., Menke, J. G., Sparrow, C. P., and Mitnaul, L. J. (2003) Miniaturization of cell-based beta-lactamase-dependent FRET assays to ultra-high throughput formats to identify agonists of human liver X receptors. *Assay Drug Devel. Tech.* 1(6), 777–787.

22. Rodems, S. M., Hamman, B. D, Lin, C., Zhao, J., Shah, S., Heidary, D., Makings, L., Stack, J. H., and Pollok, B. A. (2002) A FRET-based assay platform for ultra-high throughput screening of protein kinases and phosphatases. *Assay Drug Devel. Tech.* 1, 9–19.

23. Zheng, W. and Kiss, L. (2003) Screening technologies for ion channels targets in drug discovery. *Am. Pharm. Rev.* 6(4), 85–92.

24. Kornienko, O., Lacson, R., Kunapuli, P., Schneeweis, J., Hoffman, I., Smith, T., Alberts, M., Inglese, J., and Strulovici, B. (2004) Miniaturization of whole live cell-based GPCR assays using microdispensing and detection systems. *J. Biomol. Screen.* 9, 186–195.

25. Fitzgerald, L. R., Mamman, I. J., Dytko, G. M., Wu, H. L., and Nambi, P. (1999) Measurement of responses from Gi-, Gs-, or Gq-coupled receptors by a multiple response element/cAMP response element-directed reporter assay. *Anal. Biochem.* 275, 54–61.

26. Liu, J., Conklin, B. R., Blin, N., Yun, J., and Wess, J. (1995) Identification of a receptor/G-protein contact site critical for signaling specificity and G-protein activation. *Proc. Natl. Acad. Sci. USA* 92, 11,642–11,646.

27. Xing, H., Tran, H. C., Knapp, T. E., Negulescu, P. A., and Pollok, B. A. (2000) A fluorescent reporter assay for the detection of ligands acting through Gi protein–coupled receptors. *J. Recept. Signal Transduct. Res.* 20, 189–210.

28. Arai, H., Monteclaro, F. S., Tsou, C. L., Franci, C., and Charo, I. F. (1997) Dissociation of chemotaxix from agonist-induced receptor internalization in a lymphocyte cell line transfected with CCR2B. *J. Biol. Chem.* 272, 25,037–25,042.

29. Krafft, K., Olbrich, H., Majoul, I., Mack, M., Proudfoot, A., and Oppermann, M. (2001) Character-ization of sequence determinants within carboxyl-terminal domain of chemokine receptor CCR5 that regulate signaling and receptor internalization. *J. Biol. Chem.* 276, 34,408–34,418.

30. Heitman, J., Movva, N. R., and Hall, M. N. (1991) Targets for cell cycle arrest by the immunosup-pressant rapamycin in yeast. *Science* 253, 905–909.

31. Toker, A. (2000) Protein kinase as mediators of phosphoinositide 3-kinase signaling. *Mol. Pharm.* 57, 652–658.

32. Stack, J. H., Whitney, M., Rodems, S. M., and Pollok, B. A. (2000) A ubiquitin-based tagging system for controlled modulation of protein stability. *Nature Biotechnol.* 18, 1298–1302.

33. Whitney, M., Stack, J. H., Darke, P. L., Zheng, W., Terzo, J., Inglese, J., Strulovici, B., Kuo, L., and Pollok, B. A. (2002) A collaborative screening program for the discovery of inhibitors of HCV NS2/3 *cis*-cleaving protease activity. *J. Biomol. Screen.* 7, 149–154.

34. Moore, J. T., Davis, S. T., and Devv, I. K. (1997) The development of beta-lactamase as a highly versatile genetic reporter for eukaryotic cells. *Anal. Biochem.* 247, 203–209.

35. Tan, R., Jiang, X., Jackson, A., Jin, P., Yang, J., Lee, E., Duggan, B., Stuve, L. L., and Fu, G. K. (2003) *E. coli* selection of human genes encoding secreted and membrane proteins based on cDNA fusions to a leaderless beta-lactamase reporter. *Genome Res.* 13, 1938–1943.

36. Raz, E., Zlokarnik, G., Tsien, R., and Driever, W. (1998) Beta-lactamase as a marker for gene expression in live zebrafish embryos. *Devel. Biol.* 203, 290–294.

37. Hasegawa, S., Jackson, W. C., Tsein, R. Y., and Rao, J. (2003) Imaging *Tetrahymena* ribozyme splicing activity in single live mammalian cells. *Proc. Natl. Acad. Sci. USA* 100, 14,892–14,896.

38. Sandman, K. E., Marla, S. S., Zlokarnik, G., and Lippard, S. J. (1999) Rapid fluorescence-based reporter-gene assays to evaluate the cytotoxicity and antitumor drug potential of platinum complexes. *Chem. Biol.* 6, 541–551.

39. Murray, E. M., Grobler, J. A., Markel, E. J., Pagnoni, M. F., Paonessa, G., Simon, A. J., and Flores, O. A. (2003) Persistent replication of hepatitis C virus replicons expressing the beta-lactamase reporter in subpopulations of highly permissive Huh7 cells. *J. Virol.* 77, 2928–2935.

40. Fursov, N., Cong, M., Federici, M., Platchek, M., Haytko, P., tacke, R., Livelli, T., Zhong, Z. (2005) Improving consistency of cell-based assays using division-arrested cells. *Assay Drug Devel. Tech.* 30(1), 7–15.

41. Kunapuli, P., Zheng, W., Weber, M., Solly, K., Mull, R., Platchek, M., Cong, M., Zhong, Z., Strulovici, B. (2005) Application of division-arrest technology to cell-based HTS: comparison with fresh and frozen cells. *Assay Drug Devel. Tech.* 3(1), 17–26.

25 Membrane Potential Based Assays for Ion Channels and Electrogenic Transporters

Qiang Lü, Stephen Lin, and John Dunlop

CONTENTS

25.1 INTRODUCTION

The plasma membrane lipid bilayer separates internal and external components of a cell. Lipids, with their intrinsic properties of chemical inertness and electrical insulation, provide a protected environment, which is vital for some of the most crucial cellular functions. Just as important, proteins spanning across the membrane provide a means of communication and transport between the external environment and internal cytoplasm as well as between cells. Ion channels are a diverse family of proteins that lower the free energy required for ions to traverse across the plasma membrane. Ionic flux through ion channels provides the foundation for membrane excitability, which is essential for the proper functioning of cardiac cells, muscle cells, and neurons [1].

With a ~15% market share of the global 100 best selling drugs [2], ion channels are well-proven drug targets. Nevertheless, assay development, especially for high-throughput screening (HTS), has been below industry expectations and represents a significant bottleneck for ion channel drug development. Traditional binding assays, which are frequently used for HTS, are not suitable for screening ion channel targets due to the limited availability of high affinity ligands. Moreover, most ion channels have multiple conducting "states" based on different protein conformations. To maximize efficacy and minimize liabilities, small molecules should modulate ion channels in their physiological or pathological states. Thus functional assays are preferred for screening ion channel targets.

Since ion channels are membrane proteins, and since their function is characterized by the electrical signal that they generate under physiological conditions, it is necessary to incorporate in functional assays:

- An integrated biological membrane
- An appropriate electrical and chemical environment with which channels are configured

As permeation through ion channels involves ionic flux across the membrane, one direct assay for functional ion channel screening is flux measurement. For example, measurement of Rb^+ distribution, either by radioactive $^{86}Rb^+$ [3] or more recently, by atomic absorption [4,5], can be used to determine ion flux through potassium channels. Likewise, Li^+, which permeates sodium channels, can be used for sodium channel flux assays. However, for other channels, the lack of a detectable permeant ion precludes the use of flux measurements.

Under normal circumstances, a nonexcitable cell is at an electrically resting state, with no net charge movement across the membrane. All ionic movements, either via passive diffusion through ion channels or by active transport through pumps and electrogenic transporters, are at equilibrium. As charged particles, each ion is driven by both its concentration gradient and the electrical potential across the membrane. The Nernst equation [6] describes the *equilibrium potential* — which is the electrical potential at electrochemical equilibrium — for an arbitrary ion S with charge Z_s as:

$$E_s = (RT/Z_sF)\ln([S]_o/[S]_i) \qquad (25.1)$$

where E_s is the equilibrium potential, R is the gas constant, T is absolute temperature, F is Faraday's constant, and $[S]_o$ and $[S]_i$ are the respective concentrations of ion S outside and inside the cell.

Under physiological conditions, several ions contribute to the resting membrane potential, including Na^+, K^+, and Cl^-. The overall resting membrane potential is determined by the weighted sum of their contributions. This can be described by the Goldman–Hodgkin–Katz (GHK) voltage equation [7,8]:

$$E_{rev} = (RT/F)\ln((P_K[K^+]_o + P_{Na}[Na^+]_o + P_{Cl}[Cl^-]_i)/(P_K[K^+]_i + P_{Na}[Na^+]_i + P_{Cl}[Cl^-]_o)) \qquad (25.2)$$

Here, E_{rev} is called reversal potential, also known as zero-current potential, at which there is no net charge movement across the membrane. It is also called the membrane potential (V_m) and can be directly measured by electrophysiological recordings. P_K, P_{Na}, and P_{Cl} are the relative permeabilities of the respective ions across the membrane. $[K^+]_o$ and $[K^+]_i$ represent the K^+ concentration outside and inside of the cell, respectively, as do $[Na^+]_o$ and $[Na^+]_i$ and $[Cl^-]_o$ and $[Cl^-]_i$ for the sodium and chlorine ions, respectively.

There have been revisions of the GHK equation for membrane potential to account for contributions by ion pumps [9], by divalent ions such as Ca^{2+} [10], by ion transporters [11], and most recently, by osmotic stability [12]. Nonetheless, in most assays, the classic GHK voltage equation gives reasonable proximity in calculating membrane potentials. From the equation, it is obvious that the membrane potential is strongly dependent on ionic distribution as well as relative permeability. For ligand-gated channels, for example, ion permeability would be considerably changed by the presence of the gating ligand. As a result, the membrane potential can change significantly — given the right ionic conditions. Therefore, choosing an appropriate assay buffer and properly controlling channel gating states are the two most important considerations in optimizing assays for cell-based ion channel drug screening.

The progression of membrane potential and permeability theory occurred concurrently with advances in electrophysiological methods. Due to its fast (milliseconds) time resolution, rich information content (e.g., multistate test), and high signal–noise ratio, electrophysiological recording has been, and still is, the "gold standard" for all ion-channels assays. The ability to control either voltage or current across the membrane enables a detailed characterization of ion-channel activities and drug effects. Its biggest drawback, however, is throughput. Because of the rich content they provide, electrophysiological recordings are difficult to adapt to HTS. In fact, none of the electrophysiology systems currently available meet the standard of a true HTS (i.e., $>10^5$ data points per day). Instead, they serve primarily as secondary assays between HTS and hit validation.

To meet the demands of screening large compound libraries, the industry has adopted cell-based membrane potential assays compatible with HTS platforms. Instead of measuring ionic current from a particular channel, fluorescent dyes were developed to detect changes in membrane potential (for review, see [13]). Although it is not a direct measurement of ionic current through the channel target, membrane potential change is the ultimate outcome of all channel activities, including the targeted one. In some cases, it can provide information on the physiological roles for the channel target and the drug effects, especially if native cells are used in the assays. On the other hand, for detailed mechanistic studies, and for drugs with so-called "use" or "state" dependent properties, electrophysiological recordings are still irreplaceable. Therefore, rather than replacing electrophysiology, membrane potential assays serve as a complement, especially at the HTS stage, when the number of compounds is impossible to handle through traditional electrophysiology. Depending on the target, putting membrane potential assays and electrophysiological recordings at appropriate positions in the screening cascade is critical for success in ion channel drug development.

In this chapter, we will focus on the application of membrane potential dyes in developing functional ion channel assays. In addition to typical voltage-dependent ion channels, our targets include ligand-gated ion channels and electrogenic transporters — all of which involve ion flux in or out of cells, resulting in membrane potential changes. The source for these targets includes both heterologously expressed systems and native cells (e.g., neurons). We will cover two major platforms for fluorescence based membrane potential assays:

- A one-dye system that utilizes the FLIPR (fluorometric imaging plate reader) or FLEX station and membrane potential (FMP) dye [14]
- A two-dye system utilizing the VIPR (voltage-ion probe reader) for FRET (fluorescence resonance energy transfer) measurement and consisting of two dyes: coumarin-linked phospholipid CC2-DMPE, which acts as a donor, and an oxonol dye DiSBAC$_2$(3), which acts as an acceptor [15]

25.2 MATERIALS AND METHODS

25.2.1 Materials

- Standard assay buffer: 160 mM NaCl, 10 mM HEPES, 2 mM CaCl$_2$, 1 mM MgCl$_2$, 5.6 mM KCl, and 10 mM glucose, pH 7.4. The buffer was prepared from a 10× stock solution.
- FLIPR membrane potential dye, from Molecular Devices (Sunnyvale, CA) was dissolved as indicated by the manufacturer in standard assay buffer. Dye solution can be stored at −20°C for more than 6 months with multiple freeze–thaw cycles without a noticeable change in assay quality.

- VIPR dyes [CC2-DMPE and DiSBAC$_2$(3), from Invitrogen (Carlsbad, CA)] were dissolved in dimethylsulfoxide (DMSO) as 6-mM stock solution and stored at −20°C for subsequent use.
- Pluronic acid F-127, from Sigma (St. Louis, MO), was dissolved in DMSO as 200 mg/mL solution.
- VABSC-1 (voltage assay background suppression compound), from Invitrogen, was dissolved in water at 200 mM and adjusted pH to 7.4 with NaOH.
- All other chemicals were from Sigma. Wyeth compounds were dissolved in DMSO as 50 mM stock and diluted with assay buffer.
- All tissue culture reagents, media, and transfection reagents were from Invitrogen unless indicated otherwise.
- Costar 96-well black/clear bottom tissue culture-treated plates were from Corning (Corning, NY).

25.2.2 Cell Culture

- All stably transfected cells were cultured at 37°C, 5% CO$_2$ in Dulbecco's modified Eagle's medium (DMEM) supplemented with 10% fetal bovine serum and the following antibiotics as selection: 0.8 mg/mL zeomycin for LM(tk-) cells with Kv1.1 channel, 0.6 mg/mL G418 for CHO cells with another Kv channel (Figure 25.3). For HEK cells with glutamate transporter EAAT2 (Figure 25.6), a stable cell line was kept in the above media with both zeomycin and G418. One day prior to the assay, transporter expression was induced with 2 mg/mL Ponasterone A [16].
- One day prior to running the assay, cells were plated in poly-D-lysine–coated 96-well dark-wall plates at 30,000 to 50,000 cells/well.
- Kv-β1 subunit subcloned in pcDNA3 was transiently transfected into LM(tk-) cells using Lipofectamine 2000 reagent according to manufacturer's protocol. Transfection was performed to cells in 96-well plates, seeded at 30,000/well one day before. The assay was performed 24 h after transfection.
- Rat hippocampi were dissected from E18 embryos using methods modified from Furshpan and Potter [17]. Briefly, hippocampus was incubated at 37°C in 0.01% papain (Worthington, Freehold, NJ), 0.1% dispase (Roche Products, Hertforshire, U.K.), and 0.01% DNase (Sigma) dissolved in Hank's Balanced Salt Solution. Individual neurons were plated in 96-well format at a density of 100,000 cells/well. After a few hours, the cultures were fed with DMEM with the addition of fetal bovine serum and penicillin/streptomycin. Assays were performed 1.5 to 2 weeks after plating.

25.2.3 VIPR Assay

After washing once with assay buffer, cells were loaded with (100 μL/well) 6 μM CC2-DMPE (dissolved in assay buffer from stock with 0.1 mg/mL pluronic acid F-127) and incubated at room temperature for 40 min. Subsequently, cells were washed twice with assay buffer prior to being loaded with (100 μL/well) 3 μM DiSBAC$_2$(3) (dissolved in assay buffer from stock with 0.5 mM VABSC-1) for 40 min. Compounds, if any, were added with DiSBAC$_2$(3) at indicated concentrations — except in Figure 25.3a , where the compounds were added during the VIPR run, and in Figure 25.3b , where 50 μL compounds (in DiSBAC$_2$(3) solution) were added to 50 μL DiSBAC$_2$(3)-loaded cells 7 min before the VIPR run.

During a VIPR run, the instrument was set at column read mode with a sampling rate of 1 Hz. For each well, the first 10 points were averaged to obtain the R$_i$ [initial ratio of fluorescence intensity of CC2-DMPE to DiSBAC$_2$(3)], then 100 μL solution (160 mM KCl, except for Figure 25.3a, where compounds were added at 2× concentration in assay buffer) was added, and 25 more points were sampled. The average of the last 10 points was used to determine R_f [final ratio of fluorescence intensity of CC2-DMPE to DiSBAC$_2$(3)]. R_f/R_i was used as the final readout for VIPR.

25.2.4 FLIPR OR FLEX STATION ASSAY

After removing the culture medium, 180 μL (for FLIPR) or 200 μL (for FLEX station) FMP dye was added to each well, and the plates were incubated at room temperature for 1 h.

During FLIPR or FLEX station runs, 20 μL (for FLIPR) or 50 μL (for FLEX station) depolarizing solution was added to each assay plate after 15-sec reading of baseline. The fluorescence was measured every second (for FLIPR) or every other second (for FLEX station) for another 120 to 150 sec. The difference between the maximum and minimum was used to plot the FLIPR data, whereas the difference as a percentage of the baseline fluorescence was used to plot the FLEX station data.

25.2.5 WHOLE-CELL PATCH-CLAMP

Standard whole-cell patch-clamp was used for data collection with electrophysiological recording. Patch electrodes had a resistance of 2 to 5 MΩ with intracellular pipette solution containing (in mM): 140 KCl, 1 MgCl$_2$, 1.8 CaCl$_2$, 10 EGTA, 10 HEPES (pH to 7.4 with NaOH). The bath solution contained (in mM): 5 KCl, 150 NaCl, 2 CaCl$_2$, 1 MgCl$_2$, 10 HEPES, 5 glucose (pH to 7.4 with NaOH). A PC-controlled EPC-9 amplifier running HEKA PULSE v.8.0 (HEKA Elektronik, Lambrecht, Germany) was used for current and voltage recordings. Data were filtered with a built-in low-pass filter on a corner frequency of 1 kHz and sampled at 20 kHz.

25.3 DATA EXAMPLES

25.3.1 K$^+$-DEPENDENT DEPOLARIZATION

Kv1.1 transfected stable LM(tk-) cells were depolarized by 80 mM KCl; the results of fluorescent measurements are shown in Figure 25.1. For VIPR dyes, the sensitivity of the photomultiplier tubes is set so that the initial level of fluorescence signals [squares for CC2-DMPE and circles for DiSBAC$_2$(3)] is the same. Thus, their ratio (closed triangles in Figure 25.1) at prestimulation is 1.0. Upon addition of KCl, K$^+$ influx through Kv1.1 channels depolarizes the cells and decouples the FRET between the two dyes. The optical intensity of CC2-DMPE increases and that of the DiSBAC$_2$(3) decreases. As a result, the ratio increases from 1.0 to 1.5 to 1.8. The time constant of a single exponential for this change is on the order of 1 to 2 sec. With the FMP dye, the fluorescence level increases from a predefined baseline of 1.0 to 1.8 to 2.0, with a time constant on the order of 10 sec. This result is consistent with recently published data [18].

The different time constant between the two systems is highly attributable to their different mechanisms of action. VIPR dyes are coupled by FRET, with the range of energy transfer in the order of nanometers, which is shorter than the plasma membrane thickness. Upon depolarization, the acceptor dye, DiSBAC$_2$(3), which is negatively charged and membrane permeable, migrates inward toward the cytoplasm, uncoupling FRET. The initial loss of FRET due to cell depolarization can therefore occur in a subsecond time course, as shown in the increasing level of donor dye (CC2-DMPE, squares in Figure 25.1).

In contrast, the FMP dye, whose molecular identity has not been revealed, traverses the plasma membrane and accumulates inside the cells. It fluoresces by binding to specific protein components inside cells (David Donofrio, personal communication, 2003). Because of its negative charge, when cells are depolarized, more FMP dye will accumulate inside the cells resulting in the increase of fluorescence. In many cases, including the Kv1.1 channel exemplified here, the change of fluorescence of the FMP dye may be greater than that of VIPR dyes, yet occurring over a slower time

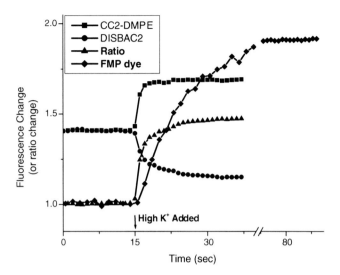

FIGURE 25.1 A typical kinetic response of VIPR dyes (squares and circles, ratio in triangles) and FMP dye (diamonds, running on FLEX station). LM(tk-) cells stably expressing Kv1.1 channel were depolarized by 80 mM KCl (final concentration) at the time point as indicated.

course because the dye has to traverse the plasma membrane before interacting with intracellular components to yield fluorescence.

For fast-inactivating channels (like most voltage-gated Na+ channels), faster VIPR dyes would appear to be superior to the FMP dye. However, depending on the background conductance of the host cell, the time course of membrane potential change does not have to match that of the change in current. It is only one of many parameters to consider in dye selection. It is also possible to prevent channel inactivation, either by mutagenesis or pharmacological modulation, to prolong the open channel state and produce a larger assay time window.

25.3.2 ASSAY FOR KV1.1 CHANNEL BLOCKERS

Using the two FRET-based dyes on the VIPR platform, we characterized the dose-dependent block of Kv1.1 channels stably expressed in LM(tk-) cells (Figure 25.2). A strong blocker (compound A) and a weaker blocker (4-aminopyridine, or 4-AP) dose-dependently blocked the depolarization produced by 80 mM KCl with IC_{50} values of 4.3 μM and 6.1 mM, respectively. Overall, the dose response curves of both blockers were well fit with a typical four-parameter logistic function.

It is worth noting that the IC_{50} values obtained through membrane potential assays are almost always higher (i.e., less potent) than those from electrophysiological recordings. For example, IC_{50} for compound A is 1.2 μM in whole-cell voltage clamp experiments (measured at 0 mV). One possible explanation is that cell conductance and membrane potential are not linearly related, i.e., inhibiting half of Kv1.1 current does not translate into inhibiting half of the change in membrane potential. In fact, due to background conductances, more than half of the Kv1.1 current needs to be blocked to depolarize the cells half way (Formula 2). Therefore, current measurements (as obtained through voltage clamp) usually show stronger blocker potencies than membrane potential measurements (as obtained through current clamp or via membrane potential dyes).

Another factor that contributes to the discrepancy between the two measurements of potency is the voltage- or state-dependent mechanism of channel blockers. This is likely the case for 4-AP, which has been reported to block K+ channels much more potently in their open state vs. the closed state [19–21], as much as 24-fold in terms of K_d [22]. In whole-cell patch-clamp experiments with Kv1.1 in the mostly open-channel state, 4-AP blocked the channel with an IC_{50} of 89 μM [23].

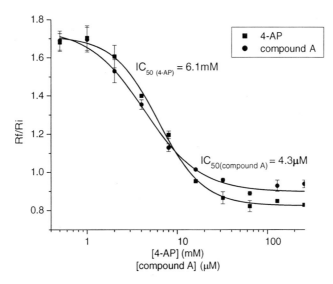

FIGURE 25.2 VIPR measurement of Kv channel blockers. LM(tk-) cells stably expressing Kv1.1 were dye loaded with various concentrations of compounds before being depolarized by 80 mM KCl. Ratio of fluorescent change was measured and plotted against the compound concentrations. Data represent mean value ± standard deviation of four wells in a 96-well format. The IC_{50} value was determined by nonlinear regression analysis.

This is 68 times more potent than the results with VIPR, where the channels are mostly closed. In addition to these factors, discrepancies between membrane potential assays and electrophysiology experiments could also be due to artifacts caused by compound interaction with membrane potential dyes, which is inevitable with all fluorescence-based assays.

Electrophysiological recordings provide the most detailed information about the actions of a compound on a channel, often under conditions that are most physiologically relevant. In order to take advantage of the high throughput while maintaining the quality of compound screening, it is important to develop membrane potential assays that maximally correlate with the electrophysiological results. So far, despite some optimization of conditions (e.g., host cell selection, dye loading, and depolarization conditions), this still remains a challenge.

25.3.3 Assays for Kv Channel Openers

Compounds that increase Kv channel activity (i.e., a Kv channel opener), increase efflux of K^+ ions and hyperpolarize the cell. With membrane potential assays, two paradigms can be applied to identify Kv channel openers:

- Measure cell hyperpolarization during compound addition
- Preincubate cells with compound, and measure depolarization upon addition of high K^+

In the first paradigm, with hyperpolarization, more FRET acceptor dye, $DiSBAC_2(3)$, moves to the external side of membrane and establishes FRET with CC2-DMPE. This causes a decrease in the final dye ratio (R_f), thus a decrease of the VIPR readout R_f/R_i. Unlike depolarization assays, where the loss of FRET is measured, this assay depends on the establishment of more FRET on the membrane surface. Due to the finite concentration of FRET donor and acceptor dyes in the membrane (depending on the initial membrane potential), the increase in FRET is limited. As shown in Figure 25.3a, with the addition of hyperpolarizing compounds, R_f/R_i decreases from the initial 1.0 (no potential change seen) to ~0.8 (the strongest hyperpolarization). This is not a significant assay window.

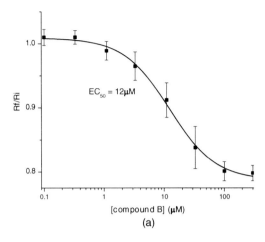

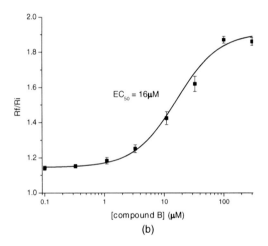

FIGURE 25.3 Comparison of two assay methods of Kv channel opener: (a) direct addition and (b) preincubation followed by high K+ depolarization. After dye loading CHO cells stably expressing a Kv channel, compound B was either directly added and change of fluorescence was measured as a result of *hyperpolarization* (panel a), or compound B was preincubated with cells for 7 min before 80 mM KCl was added to the cells for *depolarization* (panel b). The two methods gave similar EC_{50} values.

Alternatively, one may use assay paradigm 2 — measuring cell depolarization upon K+ addition following preincubation with the compound. In this case, after the compound hyperpolarizes the cell, the depolarization observed upon K+ addition is greater than in cells without compound. The amount of depolarization is dependent on the compound concentration (Figure 25.3b). In this example, using the same cell line and the same compound as in the first paradigm, R_f/R_i increases from ~1.1 (the intrinsic activity of Kv channel in this cell line) to ~1.8, a significantly larger assay window.

Due to its different dye mechanism, the FLIPR assay is able to produce a remarkable assay window with direct compound addition. For example, Whiteaker et al. [14] showed, in a K_{ATP} channel opener assay, that eight- to 10-fold signal/background ratio could be achieved. Therefore, at least in some cases, FLIPR is a better option for measuring hyperpolarization where direct addition of compound is preferred.

In addition to the assay window, many other factors should be considered in selecting an assay paradigm. For example, in cases where compounds open Kv channels very slowly, with a time

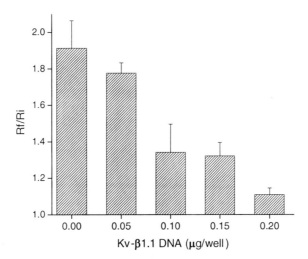

FIGURE 25.4 Kv beta subunit modulates Kv channel function detected by membrane potential assay. LM(tk) cells stably expressing a Kv channel were subject to transfection with various amount of DNA (0 to 0.2 µg/well in 96-well format) encoding Kv-beta1.1, which confers on the Kv channel fast inactivation. The total amount of DNA was kept at 0.20 µg/well with pcDNA3.1 vector. The result is reported as mean with standard deviation of eight wells.

course of more than 10 min, direct addition is suboptimal since FLIPR assays are typically run in a timeframe of 2 to 4 min/plate. For these compounds, preincubation followed by K+ depolarization is a better choice.

25.3.4 Assay for Kv1.1 Channel Kinetic Modulators

Besides increasing or decreasing current amplitude, ion channel modulators can also change activation or inactivation kinetics. For example, Kv-β1, a channel auxiliary subunit, is able to turn Kv1.1 current from delayed-rectifying into a fast-inactivating current [24] through a ball-and-chain mechanism. This significantly affects channel activity and membrane potential, thus providing a way of fine-tuning neuronal excitability. When transfected with Kv-β1, LM(tk-) cells expressing Kv1.1, which are normally very sensitive to KCl depolarization (Figure 25.1), now become much less sensitive (Figure 25.4). This effect is Kv-β1 DNA concentration dependent. As a tetramer, each Kv1.1 channel can have zero to four β1 subunits associated. While only one β1 subunit is sufficient to lead to fast inactivation, it has been shown that having more than one beta subunit brings even faster inactivation. This produces less charge entry and a smaller change in membrane potential upon KCl depolarization.

In most Kv channel drug discovery programs, modulators targeting the permeation pathway (e.g., the pore) are less favored because it is the most conserved region; thus, compounds are likely to be poorly selective. Secondly, the Kv channel pore is very efficient, making it unlikely to achieve increased permeability. Using this assay, it is possible to screen compounds that modulate the inactivation kinetics, such as disrupting the gating function of the Kv-β subunit, with satisfactory specificity.

25.3.5 FMP Dye-Based Assay for Ligand-Gated Channels in Primary Neurons

Embryonic rat hippocampal neurons contain an abundance of glutamate receptors, including NMDA receptors, which are cationic channels activated by extracellular NMDA or glutamate. NMDA receptors play a critical role in neurotransmission, and their antagonists have been suggested to

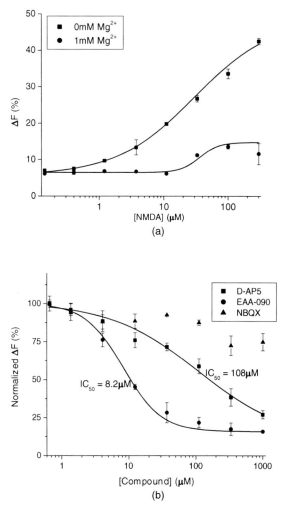

FIGURE 25.5 Ligand-gated ion channel assay using FLEX station: (a) NMDA receptors in rat primary hippocampal neurons. Neurons were dye loaded with (circles) or without (squares) 1 mM MgCl$_2$ and depolarized by various concentration of NMDA and (b) in the absence of MgCl$_2$, neurons were subject to depolarization by 100 μM NMDA after preincubation with various concentrations of D-AP5 (squares), EAA-090 (circles) or NBQX (triangles), and the change of fluorescence after NMDA addition was normalized against compound-free controls. Each data point represents mean and standard deviation of four wells in a 96-well plate.

treat a wide variety of neurological disorders including stroke [25], Alzheimer's disease [26], schizophrenia [27], epilepsy [28], and pain [29]. NMDA receptors are also important for forming and maintaining long-term potentiation, a process crucial for learning and memory [30]. Using membrane potential assays on FLEX station, and taking advantage of the FMP dye, robust NMDA-dependent depolarization can be observed in the absence of Mg^{2+} in assay buffer (Figure 25.5a). In the presence of 1 mM of Mg^{2+}, however, such depolarization is largely ablated, consistent with the fact that Mg^{2+} is a NMDA blocker [31]. The NMDA-dependent depolarization is also blocked, in a dose-dependent manner, by specific antagonists, such as D-2-amino-5-phosphonopentanoate (D-AP5) [32] and 2-[8,9-dioxo-2,6-diazabicyclo [5.2.0]non-1(7)-en2-yl]ethylphosphonic acid (EAA-090) [33], but not by non-NMDA glutamate receptor antagonists, such as 1,2,3,4-tetrahydro-6-nitro-2,3-dioxo-benzo[f]quinoxaline-7-sulfonamide (NBQX) (Figure 25.5b). As with a voltage-gated channel, blockers of ligand-gated channels demonstrate a slightly lower (three- to fivefold

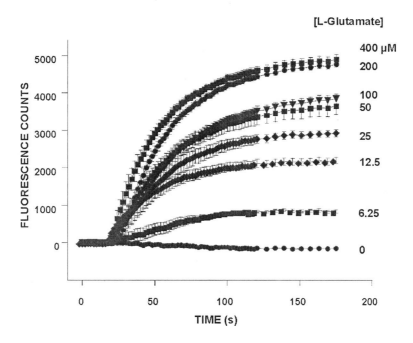

FIGURE 25.6 Electrogenic transporter assay using FLIPR: HEK cells stably expressing human EAAT2 were incubated with FMP dye for 45 min at room temperature in HBSS. During FLIPR assay, the indicated concentration of L-glutamate was added to each well, and the time course of fluorescent change was plotted. Fluorescent intensity was measured every 1.5 sec for the first 120 sec and every 6 sec for the last 60 sec. Data is mean ± standard deviation of n = 12.

in IC_{50}) potency using membrane potential assay than that obtained through current measurement using patch clamp.

The use of primary neurons enables compounds identified from HTS screens with heterologous expression systems to be tested on NMDA receptors in their native subunit composition, native cellular environment, and a more physiological state. Also, by using membrane potential as a readout, one is able to directly assess the compounds' effect on neuronal activity.

25.3.6 ASSAY FOR ELECTROGENIC TRANSPORTERS: EAAT2 AS AN EXAMPLE

EAAT2 (excitatory amino acid transporter 2) belongs to a family of transporters that actively transport glutamate or aspartate into cells. The transport of amino acids occurs concurrently with efflux of K^+ and influx of Na^+ and H^+, resulting in a net positive charge entry [34]. We have used FLIPR with the FMP dye to demonstrate a dose-dependent depolarization of EAAT2-expressing cells following glutamate application (Figure 25.6). Selective EAAT2 transporter blockers, such as DL-threo-β-benzyloxyaspartic acid (TBOA) or dihydrokainic acid (DHK), inhibit the depolarization by glutamate (data not shown).

It is worth noting that due to the small amplitude of charge entry, compared with a passively permeating ion channel, the time course of the depolarization in EAAT2-expressing cells (Figure 25.6) is slower than in a typical ion channel assay (Figure 25.1). In optimizing these assays, one should pick a host cell line with a relatively negative resting potential and low background conductance. This way, a modest amount of positive charge entry would be able to depolarize cells, and no endogenous channels would offset the transporter effect. Additionally, these assays benefit from selecting a cell line with high capacity for glutamate transport (V_{max} for accumulation of glutamate > 1 nmol/min/ng).

TABLE 25.1
Overall Comparison of Membrane Potential Assays: VIPR, FLIPR, and FLEX Station

	VIPR	FLIPR	FLEX Station
Dye base	FRET	Accumulation	
Time response (t$_{1/2}$)	<1 sec	<10 sec	
Wash needed?	Yes	No	
Throughput	Medium	High	Low
Typical Z′	0.50 to 0.85	0.45 to 0.70	>0.70
Cost	High	Medium	Low

25.4 CONCLUDING REMARKS

Ion channels are an important family of drug targets. Functional HTS for channels and transporters have just been emerging. Membrane potential–based functional assays using fluorescent dyes are among the first that have been adapted to HTS format. Currently, VIPR, FLIPR, and FLEX station assays are the most popular platforms for such assays. All of them are available in 384-well format. An overall comparison of the three platforms is shown in Table 25.1.

An overall winner combining throughput, quality, reliability, and to a lesser extent, cost, FLIPR has so far been the most popular choice for ion-channel HTS. On the other hand, because both VIPR dyes and alternative instruments for FRET measurement have become more widely available, VIPR has become a more serious choice now for more users. In addition, although not commercially available, the ability of adding field-stimuli to the assay plate in VIPR [35] makes the platform an attractive choice for some voltage-gated channels, where assays may be problematic in the normal cellular resting potential range. Finally, FLEX station is one of the most versatile instruments on the market: in addition to providing a broad spectrum of excitation and emission wavelengths, it can detect fluorescence and luminescence as well as time-resolved fluorescence, read dual wavelength, and do both top read and bottom read. Despite its drawback in throughput, FLEX station is an outstanding choice for assay development and for running secondary assays.

As the foundation for ion channel studies, electrophysiology still provides the gold standard for ion channel assays. No assay has yet to match the rich content and the high quality of data obtained through electrophysiological recordings. Very dynamic protein molecules that go through multiple conformations, often on a millisecond time scale, ion channels are best targeted during certain states that only electrophysiology can detect. Remarkable progress has been achieved recently in increasing the throughput for electrophysiological recordings — mainly through chip-based whole-cell patch-clamp. However, none of these comes close to a true HTS platform for the primary screening of a full compound library.

It is therefore important, in membrane potential–based assay development and compound prioritization, to maximize the similarity between results from the HTS assay and the electrophysiological assay. For example, in addition to traditional parameters for HTS assays, including assay window or variability, such considerations as state dependence of the compound activity should be taken rather early in the process — by choosing appropriate host cells and assay conditions.

ACKNOWLEDGMENTS

We are grateful to to Zhuangwei Lou and Susan Abulhawa for technical help in glutamate transporter and Kv1.1 channel experiments, respectively, and to Jia Xu of AVIVA for technical help on the VIPR instrument.

REFERENCES

1. Hille B. *Ion Channels of Excitable Membranes*, 2nd ed. Sunderland, MA: Sinauer Associates, 2001.
2. England PJ. Discovering ion channels modulators: making the electrophysiologists' life more interesting. *Drug Discovery Today* 1999; 5: 506–520.
3. Weir SW, Weston AH. The effects of BRL 34915 and nicorandil on electrical and mechanical activity and on 86Rb efflux in rat blood vessels. *Br. J. Pharmacol.* 1986; 88: 121–128.
4. Terstappen GC. Functional analysis of native and recombinant ion channels using a high-capacity nonradioactive rubidium efflux assay. *Anal. Biochem.* 1999; 272: 149–155.
5. Wang K, Mcilvain B, Tseng E, Kowal D, Jow F, Shen R, Zhang H, Shan QJ, He L, Chen D, Lu Q, Dunlop J. Validation of an atomic absorption Rb$^+$ efflux assay for KCNQ/M-channels using the Ion Channel Reader (ICR) 8000. *Assay Drug Dev. Tech.* 2004; 2: 525–534.
6. Nernst W. Zur Kinetit der in Lösung befindlichen Körper: Theorie der Diffusion. *Z. Phys. Chem.* 1888; 613–637.
7. Goldman DE. Potential, impedance, and rectification in membranes. *J. Gen. Physiol.* 1943; 27: 37–60.
8. Hodgkin AL, Katz B. The effect of sodium ions on the electrical activity of the giant axon of the squid. *J. Physiol. (London)* 1949; 108: 37–77.
9. Tosteson DC. In Regulation of cell volume by sodium and potassium transport, Hoffman JF, ed. *The Cellular Functions of Membrane Transport*. Englewood Cliffs, NJ: Prentice-Hall, 1964, pp. 3–22.
10. Spangler SG. Expansion of the constant field equation to include both divalent and monovalent ions. *Ala. J. Med. Sci.* 1972; 9: 218–223.
11. Lev VL, Freeman CJ, Ortiz OE, Bookchin RM. A mathematical model of the volume, pH, and ion content regulation in reticulocytes. Application to the pathophysiology of sickle cell dehydration. *J. Clin. Invest.* 1991; 87:100–112.
12. Armstrong CM. The Na/K pump, Cl ion, and osmotic stabilization of cells. *Proc. Natl. Acad. Sci. USA* 2003; 100: 6257–6262.
13. Xu J, Wang X, Ensign B, Li M, Wu L, Guia A, Xu J. Ion-channel assay technologies: *quo vadis*? *Drug Discovery Today* 2001; 6: 1278–1287.
14. Whiteaker KL, Gopalakrishnan SM, Groebe D, Shieh C, Warrior U, Burns DJ, Coghlan MJ, Scott VE, Gopalakrifhnan M. Validation of FLIPR membrane potential dye for high-throughput screening of potassium channel modulators. *J. Biomol. Screen.* 2001; 6: 305–312.
15. González JE and Tsien RY. Improved indicators of cell membrane potential that use fluorescence resonance energy transfer. *Chem. Biol.* 1997; 4: 269–277.
16. Dunlop J, Lou Z, Zhang Y, McIlvain HB. Inducible expression and pharmacology of the human excitatory amino acid transporter 2 subtype of L-glutamate transporter. *Br. J. Pharmacol.* 1999; 128: 1485–1490.
17. Furshpan EJ, Potter DD. Seizure-like activity and cellular damage in rat hippocampal neurons in cell culture. *Neuron* 1989; 3: 199–207.
18. Wolff C, Fuks B, Chatelain P. Comparative study of membrane potential–sensitive fluorescent probes and their use in ion channel screening assays. *J. Biomol. Screen.* 2003; 5: 533–543.
19. Bouchard R, Fedida D. Closed- and open-state binding of 4-aminopyridine to the cloned human potassium channel Kv1.5. *J. Pharmacol. Exp. Ther.* 1995; 275: 864–876.
20. Russell SN, Publicover NG, Hart PJ, Carl A, Hume JR, Sanders KM, Horowitz B. Block by 4-aminopyridine of a Kv1.2 delayed rectifier K+ current expressed in *Xenopus* oocytes. *J. Physiol. (London)* 1994; 481: 571–584.
21. Yao, JA, Tseng, GN. (1994) Modulation of 4-AP block of a mammalian A-type channel clone by channel gating and membrane voltage. *Biophys. J.* 67: 130–142.
22. Choquet D, Korn H. Mechanism of 4-aminopyridine action on voltage-gated potassium channels in lymphocytes. *J. Gen. Physiol.* 1992; 99: 217–240.
23. Castle NA, Fadous S, Logothetis DE, Wang GK. Aminopyridine block of Kv1.1 potassium channels expressed in mammalian cells and *Xenopus* oocytes. *Mol. Pharmacol.* 1994; 45: 1242–1252.
24. Rettig J, Heinemann SH, Wunder F, Lorra C, Parcej DN, Dolly JO, Pongs O. Inactivation properties of voltage-gated K+ channels altered by presence of beta-subunit. *Nature* 1994; 369: 289–294.
25. Hoyte L, Barber PA, Buchan AM, Hill MD. The rise and fall of NMDA antagonists for ischemic stroke. *Curr. Mol. Med.* 2004; 4:131–136.

26. Ferris SH. Evaluation of memantine for the treatment of Alzheimer's disease. *Expert Opin. Pharmacother.* 2003; 4: 2305–2313.

27. Coyle JT, Tsai G. NMDA receptor function, neuroplasticity, and the pathophysiology of schizophrenia. *Int. Rev. Neurobiol.* 2004; 59:491–515.

28. Malek R, Borowicz KK, Kimber-Trojnar Z, Sobieszek G, Piskorska B, Czuczwar SJ. Remacemide — a novel potential antiepileptic drug. *Pol. J. Pharmacol.* 2003; 55:691–698.

29. Planells-Cases R, Perez-Paya E, Messeguer A, Carreno C, Ferrer-Montiel A. Small molecules targeting the NMDA receptor complex as drugs for neuropathic pain. *Mini Rev. Med. Chem.* 2003; 3:749–756.

30. Roberts AC, Glanzman DL. Learning in *Aplysia*: looking at synaptic plasticity from both sides. *Trends Neurosci.* 2003; 26:662–670.

31. Coan EJ, Collingridge GL. Magnesium ions block an N-methyl-D-aspartate receptor-mediated component of synaptic transmission in rat hippocampus *Neurosci. Lett.* 1985; 53:21–26.

32. Raiteri M, Garrone B, Pittaluga A. N-methyl-D-aspartic acid (NMDA) and non-NMDA receptors regulating hippocampal norepinephrine release. II. Evidence for functional cooperation and for coexistence on the same axon terminal. *J. Pharmacol. Exp. Ther.* 1992; 260:238–242.

33. Sun L, Chiu D, Kowal D, Simon R, Smeyne M, Zukin RS, Olney J, Baudy R, Lin S. Characterization of two novel N-methyl-D-aspartate antagonists: EAA-090 (2-[8,9-dioxo-2,6-diazabicyclo [5.2.0]non-1(7)-en2-yl]ethylphosphonic acid) and EAB-318 (R-alpha-amino-5-chloro-1-(phosphonomethyl)-1H-benzimidazole-2-propanoic acid hydrochloride). *J. Pharmacol. Exp. Ther.* 2004; 310:563–570.

34. Arriza JL, Fairman WA, Wadiche JI, Murdoch GH, Kavanaugh MP, Amara SG. Functional comparisons of three glutamate transporter subtypes cloned from human motor cortex. *J. Neurosci.* 1994; 14:5559–5569.

35. Maher MP, González JE. Ion channel assay methods. U.S. Patent Application. 2002; 2002/0045159 A1.

26 Reporter Gene Assays for Drug Discovery

Keith A. Houck, Wayne P. Bocchinfuso, Michele S. Dowless, and Kristen M. Borchert

CONTENTS

26.1 INTRODUCTION

Reporter gene assays are versatile tools useful in a wide range of drug discovery applications. They consist of cells with introduced DNA sequences encoding a protein that can be detected directly by fluorescence or through enzymatic activity and whose level is regulated by heterologous gene promoter elements. Applications range from sensors for ligands of specific receptor targets, e.g., nuclear hormone receptors, to detecting modulation of entire signal transduction pathways, e.g., insulin pathway stimulators. A variety of reporter gene readouts provide formats that can be technically simple and inexpensive and allow miniaturized, ultra-high-throughput screening (uHTS). Because assays are conducted in intact cells, active compounds have a higher validation state compared to modulators detected in a strictly biochemical setting. However, active compounds discovered through reporter gene assays are subject to relatively high rates of both false positives and false negatives. Careful strategies must be developed to minimize false negatives and to eliminate false positives as efficiently as possible. This review will describe the types of reporter gene assays commonly employed in screening projects, methods of assay development, and strategies for dealing with the false positive/negative issue.

TABLE 26.1
Reporter Genes Used in Drug Discovery

Reporter Gene	Signal	HTS Advantages	HTS Disadvantages
β-Galactosidase	Colorimetric; fluorescence; luminescence	Sensitivity	False positives in antagonist mode; cell lysis required
Luciferase	Luminescence	Sensitivity; dynamic range	False positives in antagonist mode; cell lysis required
SEAP	Luminescence	No cell lysis; sensitivity	Media transfer required
GFP	Fluorescence	Clonal cell line generation	Sensitivity
Nitroreductase	Fluorescence	Live cell detection	Sensitivity
β-Lactamase	FRET	Live cell detection; internal control; clonal cell line generation	Fluorescent artifacts
Chloramphenicol acetyltransferase	Fluorescence; radioisotopic	Low background	Complicated assay

26.2 REPORTER GENE SYSTEMS

While reporter gene assays have been developed using a wide range of reporter molecules, this review will focus on those most commonly used in screening assays requiring medium- to ultra-high-throughput. Reporters fall into two major groups: proteins with intrinsic fluorescence and enzymes. Enzymes routinely used as reporter genes are summarized in Table 26.1. A common feature of these enzymes is either low or no endogenous expression in mammalian cells, yielding very low background levels of activity. Most of these enzymes require cell lysis and substrate addition before measuring the signal. Secreted alkaline phosphatase (SEAP) is an exception, as this enzyme contains a signal sequence allowing for its secretion into the cell culture medium. Assay readouts are colorimetric, fluorescent, or luminescent, depending on the reporter gene and substrate choice. Standard laboratory instrumentation such as absorbance or fluorescence plate readers, liquid scintillation counters, dedicated luminometers, and charge-coupled device (CCD) imagers is required for reading assay signals.

26.2.1 β-GALACTOSIDASE REPORTER

The β-galactosidase (*lacZ*) reporter consists of a bacterial gene product with little or no endogenous activity in mammalian cell lines [1]. Any background mammalian activity can be reduced by measuring at an acidic pH, much below the optimal pH of the mammalian enzyme [2]. It can be read using either colorimetric or fluorescent substrates with absorbance or fluorescence plate readers, respectively. Commonly used colorimetric substrates, such as *o*-nitrophenyl-β-D-galacto-side (ONPG), have limited dynamic range and sensitivity but can be used to stain intact cells. This provides the ability to measure individual cells expressing the reporter gene using light microscopy for needs such as measuring transfection efficiency [3]. Screening applications typically use chemoluminescent substrates derived from dioxetane chemistry, e.g.. the Galacto-Light™ and Galacto-Light Plus™ kits (Applied Biosystems). The chemoluminescent format yields three orders of magnitude greater sensitivity than the colorimetric version. For ease of HTS, reagents including cell lysis buffer, substrate, and signal enhancer are included in one solution.

26.2.2 LUCIFERASE REPORTERS

Probably the most commonly used reporter gene for HTS is the luciferase enzyme cloned from the firefly (*Photinus pyralis*) [4]. Through the ATP-dependent oxidative decarboxylation of the substrate luciferin, photons are generated that can be detected with dedicated luminometers, liquid scintillation counters, or CCD imagers. Since mammalian cells lack endogenous bioluminescence, background is nonexistent, leading to great sensitivity with detection of less than 0.05 attamole possible [5]. Equally important, the dynamic range is five to eight orders of magnitude, providing extremely broad versatility. Although early versions of the reporter system required injection of substrate and immediate read of the well to capture "flash" emission, stable or "glow" reagents are the current standard for HTS applications [6]. Through the use of emission stabilizers including dithiothreitol, coenzyme-A, AMP, and/or pyrophosphate, light emission can be stabilized over time scales of 30 min to several hours. Although some sensitivity is sacrificed, significant benefits are realized with regard to assay throughput. A variety of commercial kits are available, most consisting of one-step additions of lysis buffer and luciferase assay reagents (BD Biosciences Clonetech, Cambrex, PerkinElmer Life Sciences, Promega, Roche Molecular Biochemicals).

   Luciferase enzymes from other organisms have also been used for reporter genes, although less frequently. One luciferase gene, cloned from the sea pansy *Renilla reniformis*, has filled a niche as a second reporter in a dual-reporter assay system [7]. Such a dual-reporter system allows for internal normalization of gene expression to account for differences in growth rates, cell plating variations, transfection efficiency differences, cytotoxicity, and nonspecific gene regulatory effects. Two-step systems have been developed in which the firefly luciferase is read and a second reagent added that quenches the first signal and also contains the *Renilla* luciferase substrate, coelenterazine, to allow a second read (Dual-Glo™ Reporter Assay, Promega; Firelite™, PerkinElmer). A variation on the dual-reporter system was developed by modifying luciferase enzymes from the click beetle, resulting in reporters using the same substrate but emitting at different wavelengths (Chroma-Glo™ Luciferase Assay System, Promega) [8]. Thus the assay requires only a single addition step, although dual reads with different emission filters are necessary. Emissions are of the glow type, but different enzyme kinetics make timing a potentially critical factor. One additional dual reporter system is also available using both firefly luciferase and β-galactosidase. The Dual Light® Combined Reporter Gene Assay System from Applied Biosystems measures firefly luciferase in a glow reaction followed by a quench and subsequent chemoluminescent read of a β-galactosidase dioxetane substrate [9]. Although luciferase assays usually require cell lysis to measure activity, cell-permeable *Renilla* luciferase substrates are available. This permits use of *Renilla* measurements in live cells for kinetic reporter measurements. The EnduRen™ Live Cell Substrate from Promega provides reagents for this application.

26.2.3   GFP (Green Fluorescent Protein) Reporters

In addition to luciferase, another nonmammalian protein has been used as a reporter gene: green fluorescent protein (GFP), a bioluminescent protein cloned from the Pacific Northwest jellyfish *Aequorea victoria* [10]. GFP fluoresces upon ultraviolet (UV) irradiation and can be visualized in intact cells. Although GFP can be read in fluorescent plate readers, sensitivity as a reporter gene is low relative to luciferase since there is no enzymatic amplification of signal. However, the ease of visualizing individual cells expressing GFP using fluorescent microscopy or fluorescence-activated cell sorting (FACS) makes GFP an invaluable tool. Transfection efficiencies can be easily monitored through these techniques. By creating GFP fusions with proteins of interest, subcellular localization of specific proteins can be studied with fluorescent microscopy imaging systems [11]. A variety of GFPs are available showing enhanced fluorescence and different excitation/emission spectra [12]. BD Biosciences Clontech provides an interesting variant of GFP useful for situations where timing of promoter activity is of interest. The Living Colors™ Fluorescent Timer is a GFP that changes from green to red emission over time. Thus, green fluorescence indicates recent promoter activity while red reflects earlier promoter activity. Through use of a standard curve, timing of activation of a promoter of interest can be determined.

26.2.4 SEAP Reporter

Another enzyme that has utility as a reporter gene for HTS is SEAP [13]. The chief advantage of this system lies in the ability to measure reporter gene product secreted into the medium without cell lysis. Thus promoter activity can be measured over time from the same treated cell population. However, this procedure does require transfer of cell culture medium to a new plate. Signal can be read with either fluorescence (4-methylumbelliferyl phosphate substrate) or chemoluminescence (CSPD substrate) over a four-order-of-magnitude range of enzyme concentration [13,14]. Sensitivity is near that of luciferase, with SEAP as low as 100 fg detectable.

26.2.5 Nitroreductase and β-Lactamase Reporters

Recently, two reporter systems that can be used in live cell applications have become available, both utilizing bacterial enzymes and fluorescent substrates. The nitroreductase enzyme from *Escherichia coli* is used with a cell-permeable, quenched cyanine dye substrate, Cy5Q Quencher Dye [15]. Cleavage of the dye results in loss of fluorescent quench and is detected with a red-shifted excitation/emission (647 nm/667 nm) by a cellular imaging instrument, where response in individual cells can be measured. Due to the red-shifted response, the nitroreductase reporter can be multiplexed with a second reporter such as GFP. An additional advantage is the ability to sort transfected cells by FACS to facilitate selection of responding cells.

Another system, the bacterial enzyme β-lactamase, has been employed as an effective reporter gene using a novel cell-permeable substrate, CCF2-AM [16,17]. Before cleavage, the substrate generates green fluorescence (520-nm emission) due to fluorescence resonance energy transfer (FRET) from the coumarin half of the substrate to the fluorescein half; hydrolysis by β-lactamase eliminates FRET and the emission becomes blue (447-nm emission). A major advantage of this system is the internal control resulting from the ability to read both uncleaved and cleaved substrate. Results expressed as a ratio of blue/green normalize for such variables as well-to-well variation in cell number, differences in transfection efficiency, plate edge effects, and cytotoxicity. Therefore, assays can be miniaturized permitting screening in high-density plates, including 3456-well nanoplates [18]. This can be particularly useful when the desire is to find compounds that decrease reporter gene activity, an assay mode prone to high rates of false positives due to cytotoxicity. Again, with the advantage of detecting reporter activity in live cells by FACS, single-cell clones can be isolated, allowing rapid generation of stable cell lines with varying levels of reporter gene activity.

26.3 ASSAY DESIGN STRATEGIES

26.3.1 Cell Background Selection

As described above, a plethora of reporter gene systems exist with a variety of advantages and disadvantages. Selection of a particular system should be a function of:

- Desired throughput, e.g., HTS vs. uHTS
- Availability of readers, e.g., fluorimeter, luminometer, scintillation counter, CCD imager, cell imaging system
- Cost of reagents, including required reporter gene plasmids and reporter detection substrates
- Need for kinetic reads
- Need for live-cell readout
- Degree of sensitivity required
- Activation vs. inhibition of reporter gene readout

Once the reporter system is determined, the specifics of implementing the assay must be decided. One of the first considerations should be the type of cells to be used in the assay. If the reporter gene is being used to measure modulation of specific signal transduction pathways, then the cell phenotype should reflect, as accurately as possible, the pathological/physiological state being studied. This provides the best opportunity to translate *in vitro* findings into *in vivo* pharmacology. Although primary cells are desirable in this regard, practical considerations limit their utility. Introduction of plasmid DNA by transfection techniques into primary cells is very challenging, although viral introduction of DNA may be a suitable alternative [19]. In addition, since the cells are not immortalized, generating stable cell lines is not an option. While proliferative potential is limited, some cell types may have enough potential to provide sufficient numbers for moderate screening campaigns. As an alternative to primary cells, established cell lines are regularly used. Extensive characterization of the cell line should be conducted to ensure the signaling pathway of interest is functioning as expected. For example, modulation of the phosphatidylinositol 3-kinase (PI3K) pathway is greatly influenced by the presence or absence of the dual-specificity protein/lipid phosphatase PTEN [20], a gene mutated or deleted in many commonly used cell lines. In using a reporter gene to study the PI3K pathway, e.g., a p27^{Kip1} promoter driving a reporter gene, the status of PTEN should be determined so that modulators of the pathway can be understood in their proper context. In some cases, when endogenous signal transduction pathways are not critical, the choice of cell line is much less important. For example, ligand-sensing assays for nuclear hormone receptors require demonstration simply that the cell line is capable of responding to known agonists with appropriate EC$_{50}$ values. For such cases, robust cell lines easily transfectable and conducive to HTS applications are preferred choices.

26.3.2 Introduction of Reporter Gene

Thought must be given as to the method of introducing the reporter gene DNA. A number of methods exist, including transient plasmid DNA transfection, viral infection, BacMam transduction, or stable cell line generation. Generally the easiest and fastest route to building an assay is transient transfection of plasmid DNA [21]. Early transient transfection methods using carrier molecules, e.g., DEAE dextran and calcium phosphate, are effective but can be difficult to optimize for many cell types and are often associated with significant cytotoxicity. Cationic lipid-mediated transfection is generally more efficient but also usually accompanied by cytotoxicity. Newer lipid-based, non-liposomal methods such as FuGene™ from Roche Diagnostics [22] and Superfect™ from Qiagen [23] efficiently transfect a wide range of cell types with very good efficiency and little cytotoxicity.

Viral infection systems require creating a viral construct in a replication-deficient virus strain [24]. Viral stocks are created using helper cell lines. Infections are similar to transient transfections but more reproducible since the same viral stock can be used for many experiments. Disadvantages include the extra time required to generate the viral stock and potential safety issues in working with viruses.

BacMam is a viral delivery system that uses an insect viral vector backbone with the insect cell-specific polyhedrin promoter replaced by a mammalian promoter. This system has been used to transduce a wide variety of mammalian cell lines [25]. Safety is much less a concern than with other viral systems as the virus is incapable of replication in mammalian cells. Transduction efficiencies can be very high. Disadvantages include a requirement for experience with growing insect cells and the necessity of initially creating a viral stock.

Using the primary cells or cell line of interest, transfection, infection, or transduction efficiencies should be optimized with suitable reporter gene vectors, e.g., GFP, β-galactosidase, nitroreductase, or β-lactamase, controlled by constitutively active viral promoters to ensure expression. Because specific cells/cell lines vary in their response to these techniques, both transfection efficiency and cytotoxicity should be measured, if possible. For example, an experiment can be conducted by transfecting with a plasmid containing a constitutive viral promoter driving GFP expression followed by measuring the percentage of cells expressing GFP. To measure cytotoxicity, cell metabolic activity can determined using WST-1 reagent [26]. The reagent and conditions yielding the highest efficiency with little cytotoxicity for a particular cell line should then be selected.

Finally, one can also create stable clonal cell lines with an integrated reporter gene. This method, while time-consuming and often tedious to generate, offers several advantages. The first is a reduction of variability between experiments, as the cells used should have the reporter gene present in identical fashion. Second, since clones will vary in expression level depending on the number of copies of integrated reporter gene as well as the transcriptional activity of the locus of integration, testing many clones may identify specific ones with robust reporter response. The ability to measure reporter activity in viable cells can facilitate isolation of stable clones. For example, single-cell sorting by FACS of cells transfected with β-lactamase reporter can yield relatively rapid generation of stable reporter cell lines [27]. One should be aware that the response of reporter genes stably integrated vs. transiently transfected may differ both quantitatively, i.e., greater number of copies in transiently transfected, as well as qualitatively, i.e., integration into different chromatin regions that are subject to varied regulation by chromatin-modifying enzymes.

26.4 ASSAY OPTIMIZATION

26.4.1 TRANSFECTION OPTIMIZATION

Once the reporter gene system, cell background, and transfection mode are selected, assay optimization should be performed. Initially, it is useful to have an idea of the time course of the reporter gene response. The reporter gene with promoter of interest should be transfected using the optimal transfection conditions determined as described above. It is usually convenient to transfect and allow the cells to recover overnight before treatment with known modulators of the promoter followed by reading of the assay signal over a time course. Based on this response and factoring in the convenience factor, i.e., not having to harvest assays in the middle of the night, an assay endpoint should be selected. The next step is to vary the amount of reporter gene by holding the total DNA quantity constant at the determined optimum for transfection efficiency while varying the ratio of reporter gene to carrier plasmid DNA. Holding total DNA quantity constant will prevent confounding results between transfection efficiency and reporter gene response.

Optimal reporter response can be somewhat difficult to define. For assays designed to find compounds that increase reporter gene response, greatest sensitivity is usually found where background reporter activity is very low. However, sufficient signal must be elicited to ensure the entire response is within the dynamic range of the detection instrument. The gene expression literature commonly uses "fold-induction" as an indication of the level of modulation. This type of measurement suffers from artificially high results when background is very low. It is more appropriate to use raw data, e.g., "CPS," "Fluorescence Units," or "RLU," or data normalized as percentage of control modulator, when reporting data for reporter gene assays. Assay robustness can then be determined by Z'-factor calculations, a useful statistical measurement for HTS assays that is a function of signal window and standard deviations of the mean maximum and minimum signals [28]. For stable cell lines, it is useful to examine as many clones as possible covering a wide range of expression levels. Therefore, when FACS is used as a means of generating single cell clones, a range of responders should be saved. Again, testing for robustness of response using Z'-factor as an endpoint should identify the best clones. Cell number per well should also be determined empirically, optimizing for the best Z'-factor with the fewest number of cells.

26.4.2 OPTIMIZATION OF COTRANSFECTION ASSAYS

In circumstances where recombinant proteins are being introduced along with the reporter gene, i.e., cotransfections, it is advantageous to optimize the ratio of the introduced plasmids within the

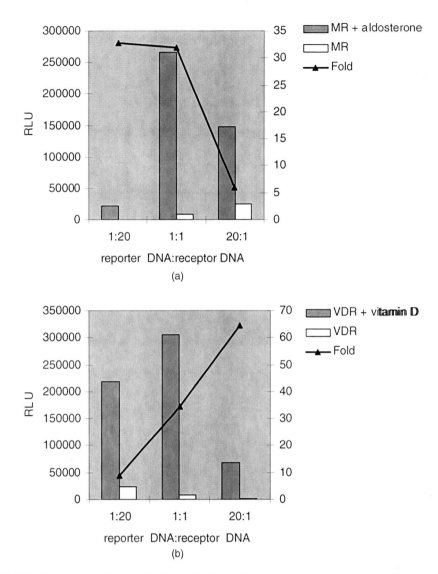

FIGURE 26.1 Comparison of cotransfection optimization for nuclear receptor reporter gene assays. HEK293 cells were transiently transfected under predetermined optimal conditions of 10 μg total DNA per 10⁶ cells with 10 μL Fugene 6 in six-well dishes. The reporter gene (pG5Luc) and receptor plasmid, (a) pGAL4-MR or (b) pGAL4-VDR, were mixed in the ratio indicated on the X-axis to make up the total of 10 μg DNA. After 24 h, cells were trypsinized, plated in 96-well plates at 10,000 cells/well, and treated with appropriate agonist (1,25 dihydroxyvitamin D3, 50 n*M*; or aldosterone, 100 n*M*) or control solvent. Luciferase activity was read 20 h later. Data are plotted as total RLUs for treated and untreated cells (left axis) and fold stimulation of treated cells relative to untreated cells (right axis).

limits of the total DNA determined to provide maximal transfection efficiency. Maximum levels of each plasmid do not always lead to the most robust reporter assay. For example, results with Gal4-nuclear receptor hybrid reporter systems are shown in Figure 26.1. The reporter gene pG5Luc, containing five Gal4 response elements, and an expression plasmid encoding a nuclear receptor ligand-binding domain fused to the Gal4-DNA binding domain were cotransfected into HEK293 cells [29]. The ratio of reporter gene to chimeric receptor plasmids was varied while keeping total transfected DNA constant at the optimum of 10 μg/10⁶ cells. Optimal activity based on considering fold stimulation with sufficient RLUs is found at a reporter:receptor ratio of 1:1

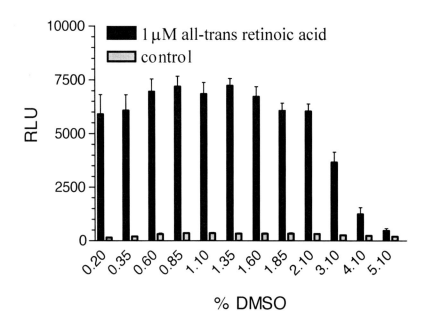

FIGURE 26.2 DMSO tolerance of transiently transfected HEK293 cells. Cells were transfected with pGAL4-RARα and pG5Luc using optimal conditions. After 24 h, cells were trypsinized, plated in 96-well plates at 10,000 cells/well, and treated with and without 1 μ*M* all-*trans* retinoic acid in the presence of the indicated concentration of DMSO. Luciferase signal was measured after 20 h.

for the mineralocorticoid receptor (MR) (Figure 26.1a), while for the vitamin D receptor (VDR) (Figure 26.1b), optimal activity is found at a ratio of 20:1. Such results are difficult to predict *a priori,* so empirical testing is the preferred method to establish optimal conditions.

26.4.3 DEALING WITH ASSAY VARIABILITY

As with any cell-based assay, dealing with assay variability in HTS is a significant issue. Day-to-day variability can be controlled using a stable cell line approach. However, care must be taken to ensure no change in signal response over time as passage number increases. Establishing limits to passages and then culturing fresh, early-passage cells limits this problem. A superior approach utilizes large batches of stable cells in the same growth state achieved by treatment with mitomycin C, causing division-arrested cells; these cells can then be frozen in aliquots providing cells for on-demand screening [30]. For transient transfections, a disadvantage is batch-to-batch variation and the necessity of transfecting each time the assay is performed. This can be largely overcome by doing large-scale, batch transfections followed by aliquoting and freezing the cells after 24 h. For screening, aliquots of transiently transfected, frozen cells are thawed and plated, ready for testing. Cell suspensions for plating should be as clump free as possible to ensure consistent plating numbers. Techniques to keep cells in suspension without mechanical damage to the cells during plating are also required. Treatment of cells with test samples, usually in dimethylsulfoxide (DMSO), can also be a source of variability. The tolerance of the assay system should be determined by testing in the range of 0.1 to 5%, as illustrated in Figure 26.2. As DMSO is toxic to most cells at concentrations above 2%, care must be used in treatment to make certain the liquid dispersion method does not result in locally high concentrations of DMSO on the cell monolayer.

Edge effects in microtiter-plate assays are a major source of variability in reporter assays, particularly for transiently transfected cells. Edge effects are apparent when wells on the perimeter of the microtiter-plate exhibit significantly different responses than wells on the interior. Reasons for this phenomenon are complex in nature and arise from multiple factors, including temperature

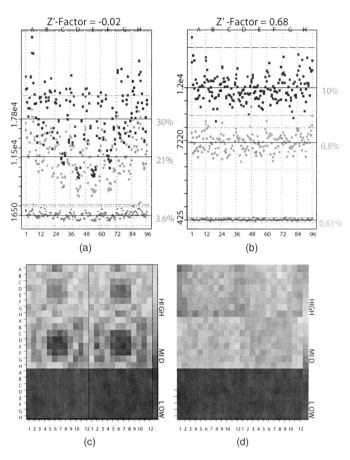

FIGURE 26.3 (See color insert following page 334.) Effect of cooling of plates before luciferase assay. HEK293 cells were cotransfected with pG5Luc and pGAL4-RARα using optimal conditions. After 24 h, cells were trypsinized, plated in 96-well plates at 10,000 cells/well, and treated with 1 μM 9-*cis*-retinoic acid (maximum signal), 5 n*M* 9-*cis*-retinoic acid (EC$_{50}$), or control solvent (0.25% DMSO). After 20 h, (a and c) plates were removed from the incubator and read with a glow luciferase reagent either immediately or (b and d) after 1 h incubation at room temperature. Edge effects are seen in the (a) scatter plot and the "bull's-eye" pattern in the (c) heat map of plates read directly from the incubator.

and gas gradients. One helpful procedure is to incubate newly seeded cells at ambient room conditions for 1 h before moving to the incubator; this ensures even distribution of seeded cells and a reduction in edge effect [31]. For enzymatic reporter assays, bringing the plates to room temperature before addition of lysis buffer and substrate can greatly reduce edge effects, as demonstrated in Figure 26.3.

Well-to-well variability during the read stage of the assay can also be a problem, especially for reporter assays requiring cell lysis. One-step read buffers contain detergent to cause cell lysis. The concentration of detergent used is a balance between efficient lysis of the cells and minimal interference with the reporter gene product activity. The effectiveness of lysis is affected by velocity and volume of lysis buffer addition, volume of cell culture medium, amount and type of mixing, and geometry of the well. All of these parameters should be examined, most easily with entire plates of maximum signal, to minimize variability. The time course for the duration of the reporter signal should also be determined, so that batch sizes can be planned appropriately. Internal controls provide an effective way to deal with much of the variability seen in cell-based assays. The dual luciferase system described earlier is one means of normalizing for well-to-well differences. The β-lactamase reporter is internally normalized as well, due to the ratiometric nature of the signal. The downsides of these approaches are the increased time required for two instrument reads as well as the increased cost where two enzyme substrates are required.

26.5 FALSE POSITIVES AND FALSE NEGATIVES

Reporter gene assays are well known for high rates of false positives and false negatives. False positives in promoter activation assays arise from a variety of mechanisms including inhibition of histone deacetylase (HDAC), membrane perturbations, genotoxic compounds, stress-inducing compounds, and inducers of apoptosis. HDAC inhibitors are frequent problems, due to their robust activation of a wide range of promoters [32]. It is useful to test the reporter system for sensitivity to HDAC inhibition with known inhibitors such as trichostatin A (TSA). Internal control reporter genes can be somewhat effective in dealing with this and other nonspecific modulators; however, because the genetic regulatory elements differ between the reporters, variation in response to such types of stimulation may result. Alternatively, a control system can be set up with a reporter gene promoter that either lacks the promoter response element being studied or contains a completely unrelated response element. If entire promoters are being studied, this is a challenge, as it can be difficult to rule out all potentially overlapping control elements. For stable cell lines with integrated reporter genes, the locus of integration may influence reporter gene responses. Hence, comparing two stable cell lines may reflect differences in integration sites, not inherent promoter control mechanisms. The Flp-In™ system from InVitrogen allows generation of integrated genes at the same chromosomal locus using a Flp recombinase target site in a stable cell line. Such isogenic cell lines would be ideal for comparing different reporter gene responses [33]. Further confounding comparison between two reporter responses is how to compare the magnitude of the induction in each. Should absolute increases in reporter activity be compared, as this is directly related to the amount of reporter gene present, or should fold-increase be used? While efficacy levels may differ between reporter systems, similar potencies may reflect related mechanisms of action in each. Studying the reporter system with as many compounds with known mechanisms of action as possible will provide some insight into recognizing false positives. False negatives are also common, primarily because concentration–response curves for reporter gene assays are often biphasic, i.e., activity is lost due to cytotoxicity or other effects at higher concentrations (Figure 26.4). Thus, in typical HTS campaigns, performed at one concentration for convenience, active compounds can be missed if screened too far above their maximal activity. Screening at multiple concentrations and/or retesting compounds with marginal activity in concentration–response format provides a means of avoiding this.

Often one is interested in modulation of targets, resulting in inhibition of reporter gene activity. Here reporters such as luciferase, SEAP, and β-galactosidase are very troublesome, as cytotoxicity results in false positives. Typical screening libraries produce unmanageable hit rates in this mode. Cell line controls consisting of the same cell type but with a reporter gene controlled by a constitutively active promoter can be run in parallel. Alternatively, use of an internal second reporter gene can help manage the hit rate, but cost, inconvenience, and the resulting potential for false negatives remain a concern.

26.6 SUMMARY

Reporter gene assays provide a cell-based screening platform that can be configured in a number of different ways for a diverse array of targets. There is much flexibility in the choice of reporter gene including extremely simple formats (GFP), exquisite sensitivity (luciferase, β-galactosidase), and no requirement for cell lysis (SEAP). More recently introduced reporter genes allow measurement of activity in intact cells using cell-permeable, fluorescent substrates (nitroreductase,

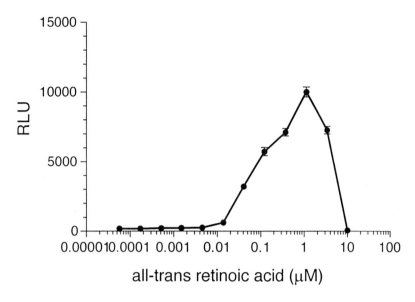

FIGURE 26.4 Example of biphasic concentration–response curve. Cells were transfected with pGAL4-RARα and pG5Luc using optimal conditions. After 24 h, cells were trypsinized, plated in 96-well plates at 10,000 cells/well, and treated with increasing concentrations of 9-*cis*-retinoic acid in a final concentration of 0.25% DMSO. Luciferase activity was measured 20 h later.

β-lactamase). Assay development is relatively rapid using transient transfection techniques, and this can be easily adapted to high-throughput screening using frozen stocks of transiently transfected cells. Alternatively, use of reporter genes in viable cells allows rapid isolation of clonal cells lines by FACS. Careful thought must be given to the design of the assays, however, in terms of appropriate cellular background and proper controls. Because hit rates for reporter assays can often be high, a method for elucidating true positives from the many nonspecific modulators must also be developed.

REFERENCES

1. An, G., Hidaka, K., and Siminovitch, L. 1982. Expression of bacterial beta-galactosidase in animal cells. *Mol. Cell. Biol.* 2:1628–1632.
2. Alam, J. and Cook, J.L. 1990. Reporter genes: application to the study of mammalian gene transcription. *Anal. Biochem.* 188:245–254.
3. Sanes, J.R., Rubenstein, J.L., and Nicolas, J.F. 1986. Use of a recombinant retrovirus to study postimplantation cell lineage in mouse embryos. *EMBO J.* 5:3133–3142.
4. De Wet, J.R., Wood, K.V., DeLuca, M., Helinski, D.R., and Subramani, S. 1987. Firefly luciferase gene: structure and expression in mammalian cells. *Mol. Cell. Biol.* 7:725–737.
5. Pazzagali, M., Devine, D.O., Peterson, D.O., and Baldwin, T.O. 1992. Use of bacterial and firefly luciferases as reporter genes in DEAE-dextran-mediated transfection of mammalian cells. *Anal. Biochem.* 204: 315–324.
6. Roelant, C.H., Burns, C.A., and Scheirer, W. 1996. Accelerating the pace of luciferase reporter gene assays. *BioTechniques* 20:914–917.
7. Lorenz, W.W., McCann, R.O., Longiaru, M., and Cormier M.J. 1991. Isolation and expression of a cDNA encoding *Renilla reniformis* luciferase. *Proc. Natl. Acad. Sci. USA* 88:4438–4442.
8. Wood, K.V., Lam, Y.A., and McElroy, W.D. 1989. Introduction to beetle luciferases and their applications. *J. Biolum. Chemilum.* 4:289–301.
9. Martin, C.S., Wight, P.A., Dobrestova, A., and Bronstein, I. 1996. Dual luminescence-based reporter gene assay for luciferase and β-galactosidase. *BioTechniques* 21:520–524.

10. Chalfie, M., Tu, Y., Euskirchen, G., Ward, W.W., and Prasher, D.C. 1994. Green fluorescent protein as a marker for gene expression. *Science* 263:802–805.

11. Kain, S.R., Adams, M., Kondepudi, A., Yang, T.T., Ward, W.W., and Kitts, P. 1995. Green fluorescent protein as a reporter of gene expression and protein localization. *BioTechniques* 19:650–655.

12. Hawley, T.S., Herbert, D.J., Eaker, S.S., and Hawley, R.G. 2004. Multiparameter flow cytometry of fluorescent protein reporters. *Meth. Mol. Biol.* 263:219–238.

13. Berger, J., Hauber, J., Hauber, R., Geiger, R., and Cullen, B.R. 1988. Secreted placental alkaline phosphatase: a powerful new quantitative indicator of gene expression in eukaryotic cells. *Gene* 66:1–10.

14. Kain, S.R. 1997. Use of secreted alkaline phosphatase as a reporter of gene expression in mammalian cells. *Meth. Mol. Biol.* 63:49–60.

15. Ismail, R., Cox, H., Kalinka, S., Williams, D., West, R., Millar, V., Michael, P., and Game, S. 2004. Nitroreductase: a new live cell gene reporter assay system. 10th Annual Conference of the Society of Biomolecular Screening, Orlando, FL, #P10055.

16. Zlokarnik, G., Negulescu, P.A., Knapp, T.E., Mere, L., Burres, N., Feng, L., Whitney, M., Roemer, K., and Tsien, R.Y. 1998. Quantitation of transcription and clonal selection of single living cells with β-lactamase as reporter. *Science* 279: 84–88.

17. Kunapuli, P., Ransom, R., Murphy, K.L., Pettibone, D., Kerby, J., Grimwood, S., Zuck, P., Hodder, P., Lacson, R., Hoffman, I., Inglesea, J., and Strulovici, B. 2003. Development of an intact cell reporter gene β-lactamase assay for G-protein–coupled receptors for high-throughput screening. *Anal. Bioch.* 314:16–29.

18. Chin, J., Adams, A.D., Bouffard, A., Green, A., Lacson, R.G., Smith, T., Fischer, P.A., Menke, J.G., Sparrow, C.P., and Mitnaul, L.J. 2003. Miniaturization of cell-based β-lactamase–dependent FRET assays to ultra-high throughput formats to identify agonists of human liver X receptors. *Assay Drug Dev. Tech.* 1:777–787.

19. Naldini, L., Blomer, U., Gallay, P., Ory, D., Mulligan, R., Gage, F.H., Verma, I.M., and Trono, D. 1996. *In vivo* gene delivery and stable transduction of nondividing cells by a lentiviral vector. *Science* 272:263–267.

20. Stambolic, V., Suzuki, A., de la Pompa, J.L., Brothers, G.M., Mirtsos, C., Sasak, T., Ruland, J., Penninger, J.M., Siderovski, D.P., and Mak, T.W. 1998. Negative regulation of PKB/Akt-dependent cell survival by the tumor suppressor PTEN. *Cell* 95:29–39.

21. Colosimo, A., Goncz, K.K., Holmes, A.R., Kunzelmann, K., Novelli, G., Malone, R.W., Bennett, M.J., and Gruenert, D.C. 2000. Transfer and expression of foreign genes in mammalian cells. *Biotechniques* 29:314–324.

22. Hellgren, I., Drvota, V., Pieper, R., Enoksson, S., Blomberg, P., Islam, K.B., and Sylven, C. 2000. Highly efficient cell-mediated gene transfer using nonviral vectors and FuGene6: *in vitro* and *in vivo* studies. *Cell. Mol. Life Sci.* 57:1326–1333.

23. Tang, M.X., Redemann, C.T., and Szoka, F.C., Jr. 1996. *In vitro* gene delivery by degraded polyamidoamine dendrimers. *Bioconjugate Chem.* 7:703–714.

24. Miller, A.D., Miller, D.G., Garcia, J.V., and Lynch, C.M. 1993. Use of retroviral vectors for gene transfer and expression. *Methods Enzymol.* 217:581–599.

25. Jenkinson, S., McCoy, D.C., Kerner, S.A., Ferris, R.G., Lawrence, W.K., Clay, W.C., Condreay, J.P., and Smith, C.D. 2003. Development of a novel high-throughput surrogate assay to measure HIV envelope/CCR5/CD4-mediated viral/cell fusion using BacMam baculovirus technology. *J. Biomol. Screen.* 8:463–470.

26. Scudiero, D.A., Shoemaker, R.H., Paull, K.D., Monks, A., Tierney, S., Nofziger, T.H., Currens, M.J., Seniff, D., and Boyd, M.R. 1988. Evaluation of a soluble tetrazolium/formazan assay for cell growth and drug sensitivity in culture using human and other tumor cell lines. *Cancer Res.* 48:4827–4833.

27. Peekhaus, N.T., Ferrer, M., Chang, T., Kornienko, O., Schneeweis, J.E., Smith, T.S., Hoffman, I., Mitnaul, L.J., Chin, J., Fischer, P.A., Blizzard, T.A., Birzin, E.T., Chan, W., Inglese, J., Strulovici, B., Rohrer, S.P., and Schaeffer, J.M. 2003. A β-lactamase-dependent Gal4-estrogen receptor transactivation assay for the ultra-high throughput screening of estrogen receptor agonists in a 3456-well format. *Assay Drug Dev. Tech.* 1:789–800.

28. Zhang, J.H., Chung, T.D., and Oldenburg, K.R. 1999. A simple statistical parameter for use in evaluation and validation of high-throughput screening assays. *J. Biomol. Screen.* 4:67–73.

29. Braselmann, S., Graninger, P., and Busslinger, M. 1993. A selective transcriptional induction system for mammalian cells based on Gal4-estrogen receptor fusion proteins. *Proc. Natl. Acad. Sci. USA* 90:1657–1661.

30. Fursov, N., Cong, M., Federici, M., Platchek, M., Haytko, P., Tacke, R., Livelli, T., and Zhong, Z. 2004. Improving consistency of cell-based assays by using division-arrested cells. *Assay Drug Dev. Tech.* 3:7–15.

31. Lundholt, B.K., Scudder, K.M., and Pagliaro, L. 2003. A simple technique for reducing edge effect in cell-based assays. *J. Biomol. Screen.* 8:566–569.

32. Hassig, C.A., Tong, J.K., Fleischer, T.C., Owa, T., Grable, P.G., Ayer, D.E., and Schreiber, S.L. 1998. A role for histone deacetylase activity in HDAC1-mediated transcriptional repression. *Proc. Natl. Acad. Sci. USA* 95:3519–3524.

33. Bethke, B.D. and Sauer, B. 2000. Rapid generation of isogenic mammalian cell lines expressing recombinant transgenes by use of Cre recombinase. *Methods Mol. Biol.* 133:75–84.

27 mRNA Detection from Cells Using Quantigene® Branched DNA Technology

Lisa Minor

CONTENTS

27.1 INTRODUCTION

Measuring changes in gene expression can be a useful tool to measure endogenous cellular responses to external events, such as receptor activation. The beauty of measuring expression of endogenous genes is twofold: it can avoid patent issues that exist with some reporter gene tools, and one is dealing with a physiological event. One could also measure the effect of a compound on that event, and if the appropriate message and cell type are selected, one could potentially use gene expression as a biomarker to measure activity of a compound in the clinic. However, measuring mRNA changes in any robust moderate-throughput manner requires that the sample handling be relatively straightforward and automatable. In addition, it would be preferable that the message did not have to be isolated and that it could be measured directly, not by message amplification.

Several methods exist for detecting message from samples. They include Northern blot analysis, ribonuclease protection, reverse transcription-polymerase chain reaction (RT-PCR), and Quanti-Gene® bDNA.[1,2] These methodologies differ in sensitivity, ease of sample preparation, radiolabel usage, and indirect or direct mRNA detection. In all of the methodologies except QuantiGene, RNA or mRNA is isolated from cells. These protocols typically require extraction of the RNA by

Trizol or some other reagent. In addition, since RNA is very sensitive to the action of ribonucleases, all reagents and tools that come into contact with the sample must be ribonuclease free.

Isolation of RNA can compromise throughput. The QuantiGene method requires no RNA isolation. In fact, the RNA is released from the cells by addition of a lysis reagent that can be added directly to the culture medium; hence, both suspension and adherent cultures can be used.

Northern blot analysis is performed by running the isolated RNA on a gel and detecting it with hybridization to a radiolabeled probe. Using this inexpensive and simple tool, it is possible to detect both the presence and size of a particular message. However, the use of gels is not a high-throughput method, and the use of radionucleotides is being discouraged within the pharmaceutical industry due to environmental risks as well as cost of disposal. Therefore, this method is not a viable one for moderate- to high-throughput message detection.

Ribonuclease protection is performed using solution hybridization to an antisense RNA. After hybridization, the unhybridized single-stranded material is digested by nucleases. The remaining product is run on a gel and message is quantified. The disadvantages for higher throughput are RNA isolation, using RNA as the probe, and gel quantification of the resulting fragment.

In RT-PCR, RNA is copied into DNA using retroviral reverse transcriptase. The cDNA is then amplified by PCR. Of all of the methods, RT-PCR is the most sensitive. Although it requires mRNA isolation, one can quantify minute amounts of message in a sample and detect multiple messages in a single tube. However, once again, RNA must be isolated, and the amplification occurs by copying and amplifying the original message.

QuantiGene bDNA methodology incorporates amplification of the signal not the message. Amplification is accomplished by bDNA. bDNA is a branched molecule of DNA. It was first described by Urdea[3,4] and has been used in message detection clinically as a tool to measure HIV load in patients.[5] Several versions of the bDNA amplifier have been generated to allow measurement of very low levels of viral load. The version in QuantiGene is such that each bDNA molecule has 15 branches, each capable of binding three molecules of DNA complexed to the enzyme alkaline phosphatase (Figure 27.1). This results in a 45-fold amplification of signal for each bDNA molecule that binds the message.

The bDNA assay itself is depicted in Figure 27.2. In this protocol, cells are plated in 96-well plates. Cells are treated with stimulus and then lysed in the presence of pools of capture and label probes (CE and LE respectively) within the lysis buffer. The lysis procedure does not require that the cellular supernatant be removed; hence, this protocol works for both suspension and adherent cultures. The probes within the lysis buffer (Figure 27.3) serve certain functions: CEs have complementary sequences to both the message and to a capture plate and so serve to adhere the message to the plate. The LE has sequences complementary to the message as well as to the bDNA and serves as a scaffold for binding the amplifying system. The BL are blocker probes and are synthesized complementary to sections of the message to which neither CE or LE can be designed. This serves to coat the message with probes in a contiguous manner. Following lysis of the cells, the samples are transferred to a capture plate. The capture plate is coated with an oligonucleotide that binds the CE thus capturing the message. After hybridization, the plate is washed and the amplifier is added (bDNA). After hybridization, the samples are washed and the label is added (DNA-coupled alkaline phosphatase). The samples are hybridized and washed and the luminescent substrate, dioxetene, is added; the samples are then incubated to allow the reaction to occur, and the plates are read on a luminometer.

This chapter describes the protocol for running the bDNA assay, general probe design, sample preparation, and assay validation.

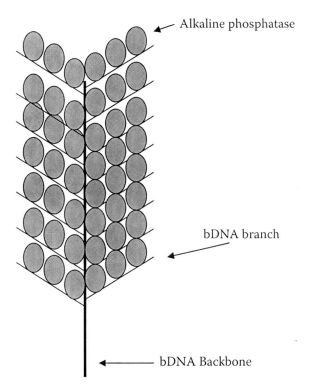

FIGURE 27.1 Branched DNA: A schematic representation of bDNA. bDNA consists of a trunk of DNA with 15 branches. Each branch can bind three molecules of alkaline phosphatase that are conjugated to DNA.

QuantiGene™ mRNA Assay

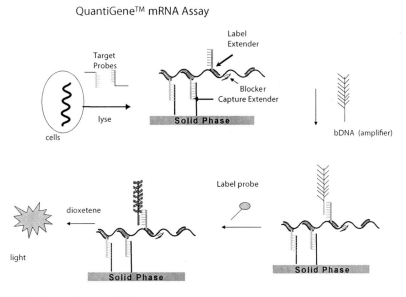

FIGURE 27.2 The QuantiGene mRNA assay.

Hybridization of Probes to the Target and Solid Phase

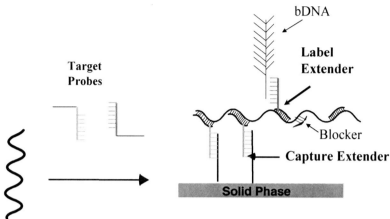

FIGURE 27.3 Hybridization of probes to the target and solid phase: A schematic of binding of LE and CE and blocker probes to target mRNA as well as to the capture plate.

27.2 GENERAL MATERIALS AND METHODS

- QuantiGene HV Lysis Mixture — Store at room temperature. *Note*: if chilled, the salts will precipitate. To dissolve, warm at 37°C until in solution.
- bDNA Solid White Capture Plates — Store at 4°C. *Note*: a partial plate can be used by keeping the unused section dry and then refrigerating the plate.
- QuantiGene HV bDNA Assay Amplifier/Label Diluent — Store at room temperature.
- QuantiGene HV bDNA Assay Amplifier — Store at 4°C.
- QuantiGene HV bDNA Assay Label — Store at 4°C.
- QuantiGene HV bDNA Assay Substrate — Store at 4°C.
- SSC Buffer (saline–sodium citrate buffer).
- LiLS (lithium laurel sulfate).
- Luminometer — We have used a Tropix reader and the LJL Analyst from Molecular Devices with no issues.
- Probe sets (CE, LE, blockers).

Probe sets can be designed using the Probe Designer™ software or can be purchased as a set from Genospectra (www.genospectra.com). The stocks of probes from Genospectra come prediluted into TE buffer as LE, CE, and blockers. They should be stored at 4°C. If you design your own probe sets, they should be handled as described in the next section.

27.3 GENERAL REAGENT PREPARATION

27.3.1 PROBES

- Concentrated stocks of individual probes:
  - Dilute each of your lyophylized probes (we purchase ours from InVitrogen) with TE buffer, pH 7.4. These will be marked CE, LE, or BL on the tube along with an oligo number. These numbers correspond to the numbers in the probe list you can retrieve from the Probe Designer software. Lyophilized oligos are diluted at 100 pmol/μL in

TE pH 7.4 and stored at –20°C. So for example, if the concentration on the bottle is 1.2 nmol, add 12 μL of buffer. Keep these probes separated and store at –20°C.

- Stock probes:
  - LE: Take 2 μL of each LE per mL (2 pmol/mL) of TE buffer and mix together (for example, for 6 mL, each LE would be 12 μL).
  - BL: Take 1 μL of each BL per mL of TE (1 pmol/mL).
  - CE: Take 0.5 μL of CE per mL of TE (0.5 pmol/mL).
  - We keep these stocks separate. These can be aliquoted into 1-mL tubes and frozen. Repeat freeze/thaw cycles do not seem to harm the probes. I have combined the LE and CE together and have had good success. This makes fewer reagents to store.

27.3.2 DAY 1 REAGENT PREPARATION

Probes at Day 1 of assay: Add 0.75 μL of pooled LE, CE, and BL stocks per 50 μL of lysis reagent used. You will add 50 μL of lysis reagent to each well of cells containing 100 μL of medium.

27.3.3 DAY 2 REAGENT PREPARATION

- Amplifier (1μL/mL in amplifier/label diluent). You will add 100 μL per well at the time of the assay, so calculate accordingly.
- Label probe (1μL/mL in amplilfier/label diluent). You will add 100 μL per well at the time of the assay, so calculate accordingly.
- Substrate: Add 50 μL of 20× wash solution per mL of substrate. You will add 100 μL per well at the time of the assay, so calculate accordingly. I would keep this in a drawer in the dark until you run the assay.
- Wash buffer
  - 20× solution (can be stored on the bench indefinitely)
    · SSC buffer (100 mL) per 1000 mL water.
    · LiLauryl sulfate (6 g) per 1000 mL water.
    · If this comes out of solution, put at 53°C for a short while.
  - Dilute 20× solution to 1×
    · (50 mL 20× plus 950 mL water).

27.4 RUNNING THE METHOD

Days 1 and 2 (Figure 27.4):

- Plate cells in 96-well plates
  - You should do a cell concentration curve to determine the best cell density and number of days in culture for your assay. Obviously, the more cells, the more viscous the sample could be when you lyse the cells. We have had problems when we have grown cells at 60K for 7 days. If you are not changing the medium, plate the cells in 100 μL of medium. The presence of serum has no effect on the bDNA mRNA detection assay itself.
- Start the assay. For the addition of an unknown compound in DMSO, we typically add up to 0.75% final concentration of DMSO followed by an immediate gentle mix. We have found that for most cell cultures, up to this amount of DMSO has minimal effects on the assay results.
- Incubate cell plate for time you want at 37°C or at room temperature (assay specific).
  - During incubation prepare Lysis buffer with probes:

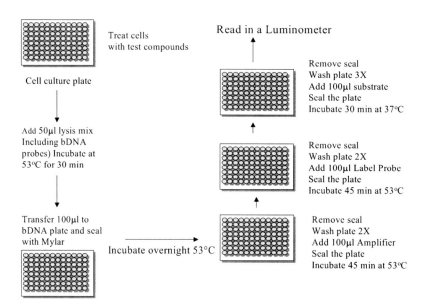

FIGURE 27.4 The bDNA assay protocol.

· Six milliliters of Lysis reagent +90 μL of each CE, LE, BL probes; 6-mL volume is good for one 96-well plate.

• After incubation is done, add 50 μL/well of Lysis reagent with probes to each well.

• Incubate cell plate at 37 or 53°C for 30 min to lyse the sample (or freeze at –20°C to hold the assay. If you freeze, be sure the sample is well sealed and when you thaw it, incubate at 37 or 53°C for 30 min after the sample is completely thawed.) This sample is very stable. We have inadvertently left the lysed sample in the cell culture incubator overnight and found that we had no degradation of the signal.

• After incubation, mix samples in well (triturate, i.e., pipet up and down) and aliquot 100 μL of each sample into solid white bDNA capture plate.

• Seal plate with foil seal and incubate at 53°C overnight. We find that this step is crucial. The seal must be complete, or the edges of the plate can dry out. Occasionally, an old seal is used, which compromises the integrity of the seal. Therefore, be sure that your seal is still good prior to running the assay.

• After overnight incubation, wash (warm) plate 2× with 1× SSC and LiLS wash solution. I typically use 400 μL per wash to ensure that I fill the well completely if washing by hand. I tend to use an eight-pin metal aspiration device instead of inverting the plate for this step. This ensures that no drops of reagent remain behind.

• Add 100 μL of Amplifier to each well (1 μL Amplifier stock per 1 mL of Amplifier/Label diluent stock).

• Seal plate (with plastic seal or just lid the plate, being sure not to get anything on the lid) and incubate capture plate for 45 min at 53°C.

• Wash (warm) plate 2× with 1× SSC and LiLS wash solution.

• Add 100 μL of Label to each well (1 μL Label stock per 1 mL of Amplifier/Label diluent stock).

• Seal plate and incubate capture plate for 45 min at 53°C.

• Wash (warm) plate 3× with 1× SSC and LiLS wash solution.

• Add 100 μL of Substrate to each well (50 μL concentrated wash buffer per 1 mL of Substrate; solution will be cloudy).

- Seal plate and incubate capture plate for 30 min at 37°C.
- Read plate on a luminometer. Can read on Tropix or Dynex luminometer with 1-msec read time. Alternatively, I have read on the LJL Analyst using the luminometer setting. Other systems should also work but would need to be validated.

27.5 PREPARING TO DESIGN PROBES

The first step in probe design is to acquire your message sequence. The sequence should be saved in FASTA format. You need to be sure that the file contains no odd spaces or returns since that can confuse the program. The next step is to BLAST search the gene of interest. We typically dissect the message into sections that are approximately 300 bases long and BLAST each one separately. The goal is to identify sections with homology to other sequences and eliminate them from consideration as capture probe sequences. We shoot for the capture sequences to have at least three base mismatches to homologous sequences and find that, in general, we can achieve specificity and selectivity of the probes.

27.6 PROBE DESIGN

The software package, ProbeDesigner, is provided by Genospectra. This is a relatively simple menu-driven software package that can take your FASTA sequence and generate probe lists.[6] Typically, you need about a 500-base sequence to generate enough probes. If 500 does not give you enough to work with, you can use a larger sequence. We have used up to 1000 bases successfully. The goal of the probe design is to generate enough capture probes to catch your message to the plate (we use between four and six) and as many label probes as possible since this is your signal amplification tool. Note that if you generate your probes and find that your signal is too high, you can always truncate your sequence to eliminate some of the label probes. When we do this, we always make sure that capture probes flank the chosen sequence.

The software itself is a menu-driven tool that walks you through the probe design in a stepwise fashion. After you import your target, you also import universals (all of the sequences contained within the bDNA assay itself) to be sure that your mRNA-specific sequence does not hybridize with any of these. If you did not have this safeguard, you could design a probe set whose LE sequence bound the capture oligo on the plate, resulting in a very high nonspecific background (Figure 27.5). In the end you have a very selective and specific probe set for your specific target. An example of a probe set is shown in Figure 27.6. Figure 27.7 shows a screen shot of one set of probes from the ProbeDesigner software.

A complete description of the software is beyond the scope of the chapter. However, Genospectra has a fairly complete technical bulletin describing the tool (www.genospectra.com, technical bulletin entitled "ProbeDesigner Software").

27.7 CHECKING PROBE SET SPECIFICITY

How do you know that the probes you have generated yield the expected results? There are several methods. One is to use purified target-specific mRNA. There are protocols to measure purified mRNA using bDNA (www.genospectra.com, technical bulletin entitled "QuantiGene Assays Using Purified RNA"). If you have purified mRNA to your target, you can use that as a control. An alternative is to measure the response of a cell to a stimulus, one that has been reported to generate mRNA changes in your specific target. Your data should mimic the literature values. An example is shown in Figure 27.8 using parathyroid hormone (PTH) or phorbol ester (PMA) stimulation of *cfos* in SAOS2 osteoblast cells. In these experiments, we showed that PTH and phorbol ester stimulated *cfos* in SAOS2 cells in a time-dependent manner measurable by bDNA technology.

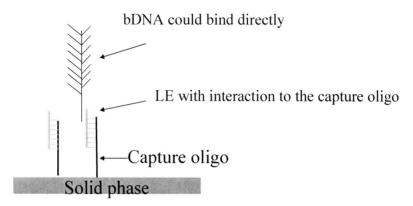

FIGURE 27.5 Background generated by nonspecific interaction between LE and the plate: If LE are generated that have sequence homology to the capture oligo bound to the plate, the LE can directly bind the plate without message and serve as a scaffold for bDNA binding resulting in a very large nonspecific background. The probe generation software has fail-safes in place to ensure that this does not occur.

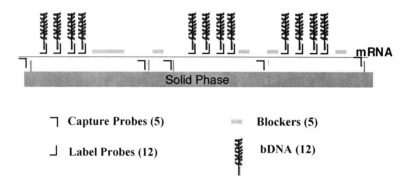

FIGURE 27.6 An example probe set.

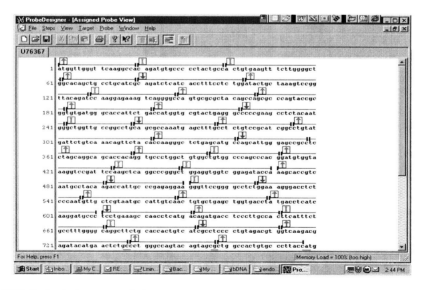

FIGURE 27.7 Probe design screen shot.

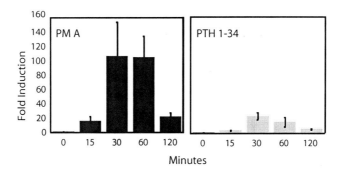

FIGURE 27.8 cfos mRNA Induction in SAOS-2 cells in response to PMA and PTH: SAOS-2 cells were plated at 60,000 cells per well in 96-well plates and allowed to grow for 3 days. Growth medium was removed, and 100 μL of fresh medium was added (0.1% BSA). Cells were stimulated for various times with either 10 μM PMA or 1 nM PTH. At the end of the experiment, 50 μL of lysis buffer was added, the samples were transferred to a bDNA plate, and the bDNA assay was performed. Results are the mean ± the standard deviation of four samples.

Another method is to validate the target first via RT-PCR. This could be particularly powerful if the induced messages have been identified by a gene chip analysis.

27.8 HOW SELECTIVE, SPECIFIC, AND SENSITIVE CAN PROBE SETS BE?

An example of selectivity, specificity, and sensitivity of probe sets was detailed by Zhou et al.[7] They looked at probes designed for uncoupling proteins 1, 2, and 3 (UCP1, UCP2, and UCP3 respectively).[8,9] These genes were selected because of their fairly high sequence homology (60 to 72%). The strategy for probe design was as described above. The probes that were generated are shown in Table 27.1. An experiment was done to look at selectivity/specificity for each probe set. In these experiments, probe sets to either UCP1, 2, or 3 were added to each well. Purified message to UCP1, 2, or 3 was added to all wells, the samples were hybridized, and the assay carried out using standard protocols. The data, in Figure 27.9, shows that the signal was generated only when the appropriate probe was matched to the corresponding mRNA. Zhou at al. also investigated sensitivity with his probe sets and found that there was a linear relationship between the amount of message added and the luminescent signal as shown in Figure 27.10. They were able to detect as little as 2×10^{-20} moles of message per well.

TABLE 27.1
Probe Types Generated for UCP1, 2, and 3 Probe Sets

Gene	CE	LE	Blockers
UCP1	7	24	9
UCP2	7	20	16
UCP3	7	22	20

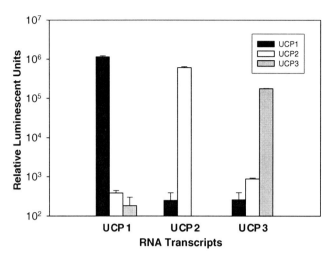

FIGURE 27.9 Specificity of bDNA assay for human UCP RNA: Approximately 5×10^{-17} mol of UCP RNA transcripts were added to the wells of a 96 well assay plate containing corresponding UCP primers. The bDNA assay was performed as described above. The data are the mean ± the standard deviation of triplicate samples. (From Zhou, L. et al. (2000) *Analytical Biochemistry* 282(1), 46–53. With permission.)

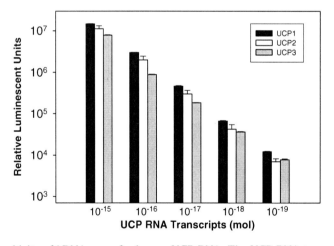

FIGURE 27.10 Sensitivity of bDNA assay for human UCP RNA: The UCP RNA transcripts were obtained by *in vitro* transcription, and the RNA was quantitated by absorbency at 260 nm. Known concentrations of RNA transcripts (1×10^{-15} to 1×10^{-19} mol) were added to the wells of a 96-well plate containing the corresponding UCP primers. The bDNA assay was performed as described above. The RLU represent the mean of triplicate samples. (From Zhou, L. et al. (2000) *Analytical Biochemistry* 282(1), 46–53. With permission.)

27.9 DETECTING MORE THAN ONE mRNA FROM A SINGLE SAMPLE

By far, the bDNA tool has been used mostly for measuring the expression changes of single mRNA from cell cultures treated with various things. However, if the expression levels of your genes of interest are adequate, one can generate more than a single data point per well. It is possible to split the sample and measure more than one message to, perhaps, simultaneously measure a control gene. To do this, the cells are stimulated and then lysed with lysis buffer (50 μL lysis buffer without probes per 100 μL of culture medium). The samples are incubated for approximately 15 min at

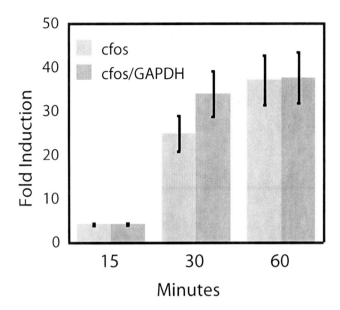

FIGURE 27.11 Fold induction of cfos mRNA by PMA with and without normalization to GAPDH in SAOS-2 cells: SAOS-2 cells were plated at 60,000 cells per well in 96-well plates and allowed to grow for 3 days. Growth medium was removed and 100 μL of fresh medium was added (0.1% BSA). Cells were stimulated for various times with 10 μM PMA. At the end of the experiment, 50 μL of lysis buffer was added without probes. The samples were split and transferred to two wells of a bDNA plate, and the bDNA assay was performed using probes for either GAPDH or cfos. Results are the mean ± the standard deviation of four samples.

53°C to be sure they are lysed, then the samples are pipetted up and down about three times, being careful not to generate bubbles, and aliquots are transferred to capture plates. The volume is brought up to 100 μL with either a 2:1 part media/lysis buffer or 2:1 part RNAse free water:lysis buffer mix. The probe sets are prepared by adding CE, LE, and blockers to lysis buffer and adding 10 μL of the appropriate probe set to each well. The remainder of the assay is performed normally. An example of this multiplexed assay is shown in Figure 27.11.

27.10 TROUBLESHOOTING

A few critical items need to be adhered to when running the assay:

- First, the overnight hybridization requires a tight seal. If the seal is compromised, the plates, particularly the edges, tend to dry out. I find that the foil that comes with the kits is adequate. To ensure the best seal, I recommend that the edges of the seal be rubbed with a Kimwipe. We have tested several automated sealers and have not had success. In general, the lysis solution loosens the glue and the film comes unsealed.
- Another issue is the wash step. If washed by hand, one can ensure that the wells have been completely evacuated. However, attempting to automate the wash program using an automated washer can raise issues. The biggest problem is aspirating the plate after the overnight hybridization. The solution is particularly viscous and does not suck out well. However, an addition of wash solution to the wells prior to their aspiration helped to thin the solution and make the washes amenable for automation. A wash program has been validated with the Biotek plate washer (Table 27.2). Program the washer with the settings for the dispense program D3 and the wash programs 34 and 35. Link the dispense

TABLE 27.2
Washer Protocol

Program	D3	34	35
Method			
Number of cycles		2	3
Soak/Shake		yes	yes
Soak Duration		10 sec	10 sec
Shake before soak		no	no
Prime		no	no
Prime volume			
Prime flow rate			
Dispense			
Dispense volume	290	395	395
Dispense flow rate	5	5	5
Dispense height	115	115	115
Horizontal dispense position	10	10	10
Bottom wash first	no	no	no
Bottom dispense volume			
Bottom flow rate			
Bottom dispense height			
Bottom horizontal position			
Prime	no	no	no
Prime volume			
Prime flow rate			
Aspiration			
Aspirate height		30	30
Horizontal aspirate position		–30	–30
Aspirate rate		5	5
Aspirate delay		2 sec	2 sec
Crosswise aspirate		no	no
Crosswise aspirate on			
Crosswise height			
Crosswise horizontal position			
Final aspirate		yes	yes
Final aspirate delay		2 sec	2 sec

program D3 to the wash programs 34 and 35 to yield Link 1 and 2, respectively. Link 1 is used to wash the capture plates after the overnight hybridization of the target with the gene-specific probe set and after the Amplifier hybridization. Link 2 is used to wash the capture plates after the Label Probe hybridization. We found that it was necessary to adhere to this program as well as be sure that the washer deck and head were level with each other. If this was not done, the assay tended to show gradient variations across the plate.

27.11 CONCLUSION

The bDNA assay is a relatively simple way to measure gene induction in a multiwell throughput. It requires no purification of message, making it amenable to automation. The probe design is straightforward, and one can design probes that are sensitive and selective. Cell-based target gene assays using bDNA can be a wonderful tool for measuring ligand activity. One can use this to measure the activation of nuclear receptors, ligand-activated transcription factors, and target gene induction/repression directly related to ligand binding. It is sensitive and specific, requires no RNA purification, and can avoid patent issues.

REFERENCES

1. Dvorak, Z. et al. (2003) Approaches to messenger RNA detection — comparison of methods. *Biomedical Papers* 147 (2), 131–135.
2. Wilber, J.C. and Urdea, M.S. (1995) Quantification of viral nucleic acids using branched DNA signal amplification. In *Molecular Methods Virus Detection*, Wiedbrauk, D.L. and Farkas, D.H. (Eds.), Academic, San Diego, CA, 131–145.
3. Urdea, M.S. et al. (1991) Branched DNA amplification multimers for the sensitive, direct detection of human hepatitis viruses. *Nucleic Acids Symposium Series* 24 (Synth. Oligonucleotides: Probl. Front. Pract. Appl.), 197–200.
4. Urdea, M.S. (1994) Branched DNA signal amplification: does bDNA represent post-PCR amplification technology? *Bio/Technology* 12(9), 926, 928.
5. Urdea, M.S. et al. (1993) Direct and quantitative detection of HIV-1 RNA in human plasma with a branched DNA signal amplification assay. *AIDS (London)* 7 (Suppl. 2), S11–S14.
6. Bushnell, S. et al. (1999) ProbeDesigner: for the design of probesets for branched DNA (bDNA) signal amplification assays. *Bioinformatics* 15(5), 348–355.
7. Zhou, L. et al. (2000) A branched DNA signal amplification assay to quantitate messenger RNA of human uncoupling proteins 1, 2, and 3. *Analytical Biochemistry* 282(1), 46–53.
8. Ricquier, D. and Bouillaud, F. (2000) The uncoupling protein homologues: UCP1, UCP2, UCP3, StUCP, and AtUCP. *Biochemical Journal* 345(2), 161–179.
9. Bouillaud, F. (2003) Uncoupling proteins: an overview. *Progress in Obesity Research* 9, 774–777.

28 Homogeneous Multiwell Assays for Measuring Cell Viability, Cytotoxicity, and Apoptosis

Terry L. Riss, Richard A. Moravec, Martha A. O'Brien, Erika M. Hawkins, and Andrew Niles

CONTENTS

28.1 INTRODUCTION

The use of cell-based assays is continuing to increase for high-throughput screening (HTS) in the drug discovery industry. Compared to biochemical assays, cell-based methods provide a more functional readout that reflects biology in a more physiological environment. In addition to direct cytotoxicity measurement, cell-based assays can provide information on parameters, such as cell permeability, that are impossible to obtain from biochemical assays.

The methods for doing cell-based assays have been improved to enable the homogeneous measurement of a variety of parameters used as markers for cell viability, cytotoxicity, and apoptosis. The introduction of simple add–mix–measure protocols in a miniaturized format has greatly increased the efficiency and reduced the expense of doing rapid screening of targets in multiwell plates and has contributed to the increased use of cell-based assays.

A major contribution has been the application of luminescent detection technology to improve sensitivity and enable the flexibility for measuring cell viability in 96-, 384-, and in some cases, 1536-well formats. The recent development of more robust homogeneous formats has facilitated the complete automation of cell-based assays for viability, cytotoxicity, and apoptosis.

There are a variety of markers available that can be used as indicators of the number of viable or dead cells remaining at the end of an experimental treatment. Choosing the most appropriate assay depends on the *in vitro* model system being studied. In some cases, cell-based assays are used simply to assess whether the cells are alive or dead after exposure to a test substance for a given length of time. In those situations, the method of measurement at the end of the experiment may not be critical. However, additional mechanistic information can be gathered by proper characterization of the kinetics of cell death in the *in vitro* model system and by the appropriate choice of which marker to measure and when to make the measurement.

To choose the most appropriate assay, you should first decide whether you want to determine the number of viable cells, determine the number of dead cells, or gather additional information about the mechanism of cell death. Although a variety of marker assays can be used for each option, only a few have been developed into a robust homogeneous format suitable for use in a high-throughput screening (HTS) environment. The choice among assays also must include an evaluation of the complete set of advantages and disadvantages including ease of use, cost, instrument availability, and the possibility of artifacts. Although more robust assays may help reduce artifacts, confirming results by using more than one type of measurement will help eliminate false hits.

Applying the correct strategy for multiplexing cell-based assays can increase the efficiency of the screening process. The use of cytotoxicity assay methods as a control to confirm observations from other multi-well cell-based assays is gaining in popularity [1]. For example, cytotoxicity assays can help distinguish between the specific effect of a drug treatment causing the down regulation of a reporter gene of interest or the nonspecific killing of host cells.

A variety of methods for measuring viable cell number, cytotoxicity, and apoptosis in a multiwell format are available. The most common method of detecting the presence of viable cells is to measure a marker of cell metabolism, such as the presence of ATP or the ability of the viable cells to reduce redox indicator dyes, including tetrazolium reagents [2,3] or resazurin [4]. Although the tetrazolium reduction assays including MTT, MTS, XTT, and WST-1 have historically been the most commonly used, the luminescent ATP assays are becoming widely accepted because of sensitivity, speed, and recent availability of a single-step homogeneous format. The characteristics of the ATP assay also make it ideal as a multiplex partner with other assay systems.

There are situations when it is desirable to specifically measure the appearance of a marker indicative of the number of dead cells at the end of an experiment rather than the disappearance of a marker of cell viability. The loss of cell membrane integrity is most commonly used to define cell death *in vitro* and is often measured by differential staining using dyes that are normally excluded from permeating and staining viable cells. Staining with vital dyes such as trypan blue, propidium iodide, or ethidium homodimer is useful for indicating the presence of nonviable cells; however, because of limitations caused by nonspecific staining and the need to perform wash steps, they have not been developed into homogeneous multiwell formats. Methods to measure the leakage of cytoplasmic components into the surrounding culture medium also can be used as markers of membrane integrity. Several enzymatic markers have been used for this purpose but many suffer from poor stability once released into culture medium. Enzymes such as lactate dehydrogenase (LDH), which exhibit prolonged stability after release into culture medium, have proven the most useful [5,6]. The assays typically use a coupled enzymatic procedure to measure LDH activity in an aliquot of culture medium removed from the experimental well. Reagent systems to measure LDH activity that do not damage viable cells have recently been developed [7]. The improved reagent design enabled development of a homogeneous method for measuring LDH activity released from dead cells in samples of culture medium containing both viable and nonviable cells.

Early cell-based methods to gather information on the mechanism of cell death and specifically whether or not it involves the process of apoptosis relied on multistep procedures to label DNA fragments in apoptotic nuclei using a TUNEL procedure or to measure biophysical changes in the membrane of apoptotic cells by labeling with tagged annexin V. Although these methods are valid, they are labor intensive and not easily adaptable to automation. The development of fluorescent homogeneous assay methods to measure caspase activity made possible efficient detection of apoptosis using cell-based assays in an HTS environment. More recently, the adaptation of luminescent protease assays to measure caspase activity has greatly improved assay sensitivity while eliminating the effects of interference by fluorescent compounds. With a few minor exceptions, activation of executioner caspase activity has been shown to be a nearly universal marker of the process of apoptosis and is becoming accepted as the method of choice for screening using cell-based assays in multiwell plates.

A thorough understanding of the *in vitro* model system being used will help with the choice for the most appropriate assay method and the interpretation of the results. Initial time course experiments to characterize the kinetics of cell death in the model system are necessary for determining when to make a measurement. An important consideration is that populations of cells in culture are heterogeneous, and even under optimal conditions *in vitro* cultures often contain a small percentage of cells undergoing spontaneous apoptosis. Because of this heterogeneity and considering that cells exist in different phases of the cell cycle, not all cells within a population will die at the same time when exposed to a toxin. Even established cell lines arising from an individual clone will exhibit differential response to a toxin. It is important to remember that measurements made using populations of cells in multiwell plates represent an average of the population, and interpretation of data may be much different than that obtained from analysis of individual cells from the same population using microscopy or flow cytometry.

A major challenge associated with using cell-based assays is the effort necessary to ensure a consistent source of cells used to prepare assay plates. The implementation of automated culture systems has helped with this problem in some large screening laboratories, but in most cases they are cost prohibitive. To obtain consistency of response from *in vitro* assays, the same starting material must be used each time. Establishing and following standard operating procedures to maintain stock cultures and prepare assay plates will decrease the variability among assays. Any environmental change that alters the physiology of the cells may affect the IC_{50} values for a toxic compound. Both the density of the stock culture used to set up assay plates and the density of cells in assay wells can affect the responsiveness of cells [8]. The challenge of consistency in handling and processing of experimental material is not unique to experiments examining cell viability and cell death. All cell-based assays have these same challenges. Assays that monitor cell viability can therefore be used to multiplex with other cell-based assays to reduce this variability.

Below are examples of recommended assay protocols for measuring the number of viable cells, the number of dead cells, and the number of apoptotic cells in multiwell plates. We also describe an example of multiplexing a cell viability assay and a reporter gene assay in the same sample well. All of the methods described are homogeneous, meaning removal of culture medium, cell washing, or multiple pipetting steps are not required. Although the homogeneous methods were designed for HTS, they have been widely accepted in smaller laboratories because of their ease of use for assaying just a few samples.

Each section will include a brief description, a detailed protocol, example assay data, and a discussion of the advantages and disadvantages of the assay. Also included will be a description of initial characterization of a model system to determine the appropriate timing for making a measurement.

28.2 ATP ASSAY TO DETECT VIABLE CELLS

28.2.1 DESCRIPTION

The cellular content of ATP has been established as a valid marker for cell viability [9–12]. Upon cell death, there is a rapid loss of the ability to synthesize ATP, and endogenous ATPases remove any remaining ATP that may be present. The ability of firefly luciferase to produce luminescence is dependent on the presence of ATP. This property has been harnessed to produce an extremely sensitive assay method to detect ATP and thus serve as an indicator of cell viability. Traditionally, the assay procedure has required a separate ATP extraction step using conditions to destroy endogenous ATPases, followed by steps to convert the sample to a condition compatible with detecting ATP with native firefly luciferase. Recent advances in directed evolution have produced highly stable mutant forms of luciferase [13] with properties that have facilitated the development of robust homogeneous assay methods for detecting ATP. The enabling feature is the stability of the mutant luciferase that maintains sufficient enzymatic activity to generate a luminescent signal that glows for several hours in the harsh conditions necessary to achieve cell lysis and inhibition to endogenous ATPases. The result led to the development of a rapid and sensitive homogeneous method of determining the number of viable cells in multiwell plates [14]. The assay reagent contains detergent to lyse viable cells and release ATP, supplies ATPase inhibitors to prevent destruction of ATP, and provides luciferase, luciferin, and the conditions necessary to generate a stable luminescent signal. After the ATP detection reagent is added to the sample of cells, the luminescent signal glows for several hours. Detection of ATP is currently the most rapid, most sensitive, and easiest method of measuring viability of cells in multiwell plates.

28.2.2 CELL VIABILITY ASSAY PROTOCOL

For more detailed information, see Promega Technical Bulletin #288 describing the CellTiter-Glo® Luminescent Cell Viability Assay Kit. The kit supplies a lyophilized Substrate component containing luciferase and luciferin and a Buffer component containing detergents to lyse cells and ATPase inhibitors to enable a long-lasting luminescent signal. The protocol is as follows:

- Thaw the CellTiter-Glo Buffer and equilibrate to room temperature prior to use. For convenience, the CellTiter-Glo Buffer may be thawed and stored at room temperature for up to 48 h prior to use.
- Equilibrate the lyophilized CellTiter-Glo Substrate to room temperature prior to use.
- Transfer the appropriate volume of CellTiter-Glo Buffer into the amber bottle containing CellTiter-Glo Substrate to reconstitute the lyophilized enzyme/substrate mixture. This forms the CellTiter-Glo Reagent.
- Mix gently to obtain a homogeneous solution.
- Prepare opaque-walled multiwell plates with mammalian cells in culture medium, 100 μL per well for 96-well plates or 25 μL per well for 384-well plates. (*Note*: Multiwell plates must be compatible with the luminometer used.)
- Prepare control wells containing medium without cells to obtain a value for background luminescence.
- Add the compound to be tested to experimental wells and incubate according to culture protocol (typically 2 days).

- Equilibrate the plate and its contents to room temperature for approximately 30 min. (*Note*: The optional equilibration period is to ensure that the plates are at a uniform temperature. Stacks of several plates will take longer to equilibrate. If wells are at different temperatures, the luciferase reaction rate and thus luminescence values will vary.)
- Add a volume of CellTiter-Glo Reagent equal to the volume of cell culture medium present in each well (e.g., add 100 μL of Reagent to 100 μL of medium containing cells for the 96-well plate format or add 25 μL of Reagent to 25 μL of medium containing cells for a 384-well plate).
- Mix contents for 2 min on an orbital shaker to induce cell lysis.
- Allow the plate to incubate at room temperature for approximately 10 min to stabilize luminescence signal.
- Record luminescence. (*Note*: Instrument settings depend on the manufacturer. An integration time of 0.25 to 1 sec per well should serve as a guideline.)

28.2.3 EXAMPLE ATP ASSAY DATA

The general performance of the luminescent ATP assay is shown in the first two figures. The ATP assay shows a long linear range of responsiveness over which the luminescent signal is directly proportional to the number of viable cells under most culture conditions (Figure 28.1). The 96-well plate experiment shown in Figure 28.1 demonstrates that the sensitivity of the assay is adequate for detecting small changes in cell number. Similar experiments done in opaque white 384-well plates (not shown) have demonstrated a similar limit of detection; however, for most applications, that level of sensitivity is not required.

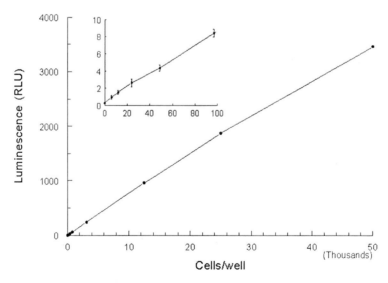

FIGURE 28.1 Cell number correlates with luminescent output. A direct relationship exists between luminescence measured with the CellTiter-GloAssay and the number of cells in culture, over three orders of magnitude. Serial twofold dilutions of Jurkat cells were made in a 96-well plate in RPMI1640 with 10% FBS. Luminescence was recorded at 10 min, using a Dynex MLX Microtiter plate luminometer. Values represent the mean ± SD of four replicates for each cell number. The luminescent signal from four Jurkat cells is greater than three standard deviations above the background signal resulting from serum supplemented medium without cells. There is a linear relationship (R^2 = 0.99) between the luminescent signal and the number of cells from 0 to 50,000 cells per well.

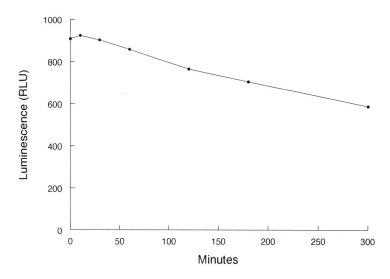

FIGURE 28.2 Extended luminescent half-life allows for high-throughput batch processing. The stability of the signal is shown here for the HepG2 cell line grown and assayed in MEM containing 10% FBS. Twenty-five thousand cells per well were added to a 96-well plate. After an equal volume of CellTiter-Glo Reagent was added, plates were shaken and luminescence monitored over time with the plates held at 22°C. The half-life of the luminescent signal was approximately 5.8 h.

After the ATP assay reagent is added to cells, the luminescent signal is relatively stable and exhibits a half-life generally greater than 5 h depending on the cell type and medium composition. The stability of the luminescent signal over time for HepG2 cells is illustrated in Figure 28.2. The long half-life of the luminescent signal eliminates the need to use instruments with reagent injectors, as are required for ATP assays emitting a flash of light from an uncontrolled luciferase reaction. The long signal half-life provides the flexibility for recording data if several plates are used in the same experiment.

An example of using the ATP assay to detect the effect of tamoxifen treatment on HepG2 cells is shown in Figure 28.3. This example was performed using 25,000 cells/well in a 96-well plate. The assay sensitivity has been demonstrated to be sufficient for use of as few as 1000 cells/well in a 384-well format, although low cell numbers result in greater variation among replicate samples [15]. Care must be taken to validate the reproducibility of dispensing small numbers of cells. In general, the variation among replicate samples does not result from the assay chemistry, but rather results from variation in the number of cells plated per well or the differential growth and responsiveness of cells to a particular toxin treatment.

Figure 28.3 illustrates the time-dependent cumulative effects of a toxin on the kinetics of cell death in the HepG2 model system. The data from 1 to 6 h exposure show that 30 μM tamoxifen has relatively little effect on toxicity. However, the 24-h data show that the same concentration of tamoxifen results in killing essentially all of the cells. This observation illustrates how a thorough understanding of the kinetics of cell death in the model system leads to selecting the most appropriate time to record the results of cytotoxicity assays.

28.2.4 ADVANTAGES AND DISADVANTAGES OF THE ATP ASSAY

The luminescent ATP assay is the fastest and most sensitive method currently available for measuring the number of viable cells in multiwell plates. The homogeneous add–mix–measure protocol helps to reduce error by reducing the number of pipetting and plate handling steps. Other homogeneous assays that measure cell metabolism (such as tetrazolium reduction or resazurin reduction assays) usually require returning the plates to an incubator for 1 to 4 h after addition of reagent to

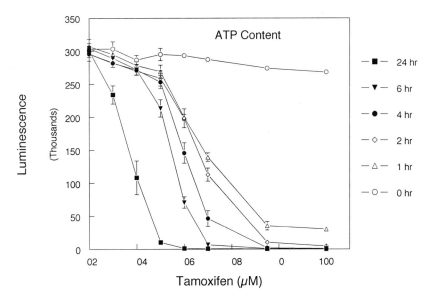

FIGURE 28.3 ATP content of HepG2 cells treated with 0 to 100 μM tamoxifen for 0 (○), 1 (△), 2 (◊), 4 (●), 6 (▼), and 24 h (■). HepG2 cells were cultured overnight at 25,000 cells/well in a 96-well plate. Tamoxifen stock solution in DMSO was further diluted into MEM and added at staggered times. All wells including 0 μM tamoxifen contained a final concentration of 0.2% DMSO. The CellTiter-Glo ATP assay reagent (100 μL/well) was added immediately after the time zero addition of tamoxifen and luminescence recorded after a 10-min equilibration period. Data shown represent the means ± SD (n = 3). The luminescence value for the control of culture medium without cells was 1316 ± 1439 RLU (not shown). (Reprinted from Riss, T. and Moravec R., *Assay and Drug Dev. Technol.*, 2004; 2(1):51–62.)

allow viable cells to convert the redox dye into a measurable signal. In contrast, the ATP assay measures the status of the cells at the point in time when the reagent is added to the sample and avoids potential changes that may occur during hours of incubation of viable cells with redox indicators.

The sensitivity of the ATP detection method allows lower cell numbers to be used and the development of miniaturized assays that help conserve test compounds. The assay can be scaled down and used in 96-, 384-, and 1536-well formats; however, care should be taken to ensure that the physiology and responsiveness of the cells is not affected by environmental factors such as evaporation of medium or edge effects that may occur in microscale plates. The luminescent signal from the ATP assay provides some flexibility in the type of instrument used to record data. The signal is most commonly recorded using a plate-reading luminometer or a charge-coupled device (CCD) sensor. Alternatively, although not as sensitive as a luminometer, some models of fluorometers can read luminescent signals successfully if the operator turns off the excitation beam and removes filters that restrict light entering the photomultiplier tube.

A potential disadvantage of the ATP assay is cost. Production of a lyophilized component containing luciferin and recombinant luciferase as ingredients is more expensive than creating solutions of redox dyes such as the tetrazolium reagents or resazurin. As a result, the ATP assays are generally more expensive when used in a 96-well format; however, the expense may be offset by the greater sensitivity that provides the ability to scale down the number of cells and volume of medium or test compound used per well.

There are two sources of artifacts that may be observed with the ATP assay. For any of the cell viability assay methods that measure metabolic markers, test compounds that alter cell metabolism without causing cell death may change the amount of signal generated per cell. For example, culture conditions that alter metabolic activity within cells such as a transition between active growth and reaching confluence may decrease the amount of ATP per cell. Alternatively, activation of resting lymphocytes will increase the amount of ATP per cell. These factors need to be considered when interpreting data from any assay that measures a marker indicating cellular metabolism.

Another potential artifact could result if the compound being tested interferes with the luciferase reaction, making it difficult to distinguish a cytotoxic event from a luciferase inhibitor. Interference with the assay chemistry can be confirmed or ruled out by testing the effect of the test compound in wells containing ATP in culture medium without cells. The results from screening of a library of 640 pharmaceutically active compounds indicated that inhibitors of luciferase are rare.

28.3 LDH-RELEASE ASSAY TO MEASURE MEMBRANE INTEGRITY

28.3.1 DESCRIPTION

Cell death is commonly defined by a loss of membrane integrity. The routine microscopic examination of cells in culture often includes a trypan blue staining assay to measure membrane integrity and assess percent viability. Trypan blue does not readily enter viable cells but can enter and stain cells that have lost membrane integrity. Dyes such as ethidium homodimer or propidium iodide that are not permeable to viable cells but enter cells with a compromised membrane and stain nucleic acids are examples of fluorescent indicators that have been used for measuring cell viability. The alternative approach of measuring membrane integrity is to assay for the leakage of components from the cytoplasm of dead cells into the culture medium. The *in vitro* LDH-release method was developed as a nonradioactive alternative for measuring the release of [^{51}Cr] from target cells in cell-mediated cytotoxicity assays [5,6]. The original format of the assay measured enzymatic activity of LDH in an aliquot of culture medium removed from the cells. The assay reagent provided the substrates and other ingredients needed to drive coupled enzymatic reactions involving LDH and diaphorase to generate an absorbance signal proportional to the number of dead cells. Recent improvements in the reagent formulation have enabled LDH release to be measured directly in the culture well in the presence of viable cells. This homogeneous format eliminates the need to remove an aliquot of medium to a separate plate [7]. Figure 28.4 shows the scheme of the coupled enzymatic reactions leading to a fluorescent signal proportional to the number of dead cells. The homogeneous procedure of the LDH-release assay is performed by adding reagent directly to culture wells, incubating for 10 min, and recording fluorescence.

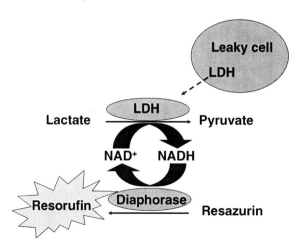

FIGURE 28.4 Release of LDH from a damaged cell is measured by supplying lactate, NAD$^+$, and resazurin as substrates in the presence of diaphorase. Generation of the fluorescent resorufin product is proportional to the amount of LDH.

28.3.2 CYTOTOXICITY ASSAY PROTOCOL

For more detailed information and additional considerations, see Promega Technical Bulletin #306 describing the CytoTox-ONE™ Homogeneous Membrane Integrity Assay. The kit includes a lyophilized Substrate component containing lactate, NAD⁺, and diaphorase, a Buffer component containing a physiologically balanced solution of resazurin, a Lysis Solution component consisting of 0.9% Triton® X-100, and a Stop Solution containing SDS to stop the reactions and stabilize the fluorescent signal. The protocol is as follows:

- Equilibrate Substrate Mix and Assay Buffer to 22°C and prepare CytoTox-ONE Reagent by adding 11 mL of Assay Buffer to each vial of Substrate Mix. Gently mix to dissolve the substrate. Protect the Reagent from direct light. Unused portions of the CytoTox-ONE Reagent may be stored tightly capped at −20°C for 6 to 8 weeks.
- Prepare 96-well (or 384-well) assay plates containing cells in culture medium. *Note*: Use opaque-walled (clear or solid bottom) tissue culture plates compatible with recording data using a fluorometer.
- Prepare a set of replicate wells containing culture medium without cells to serve as a control to determine background fluorescence (usually caused by trace amounts of LDH activity present in serum used to supplement culture medium).
- Prepare a set of replicate wells containing cells in culture medium to serve as a maximum LDH-release control.
- Add test compounds and vehicle controls to appropriate wells so the final volume is 100 μL in each well in the 96-well format (25 μL for a 384-well plate).
- Culture cells for desired test exposure period.
- Remove assay plates from 37°C incubator and equilibrate to 22°C, approximately 20 to 30 min. *Note*: This temperature equilibration is recommended to ensure a constant temperature among all the wells in the plate. Slight changes in temperature can affect enzymatic rate and thus the assay results.
- *Optional*: If Lysis Solution is used to generate a Maximum LDH Release Control, add 2 μL of Lysis Solution (per 100 μL original volume) to the positive control wells. If a larger pipetting volume is desired, use 10 μL of a 1:5 dilution of Lysis Solution.
- Add a volume of CytoTox-ONE Reagent equal to the volume of cell culture medium present in each well and mix or shake for 30 sec (e.g., add 100 μL of CytoTox-ONE Reagent to 100 μL of medium containing cells for the 96-well plate format or add 25 μL of CytoTox-ONE Reagent to 25 μL of medium containing cells for the 384-well format).
- Incubate at 22°C for 10 min. *Note*: If samples are assayed at 37°C, reduce incubation time to 5 min.
- Add 50 μL of Stop Solution (per 100 μL of CytoTox-ONE Reagent added) to each well. For the 384-well format (where 25 μL of CytoTox-ONE Reagent was added), add 12.5 μL of Stop Solution). This step is optional but recommended for consistency. It is recommended to add the Stop Solution to the wells in the same order that the CytoTox-ONE Reagent was added. This is especially important if manual addition of reagent takes a significant amount of time.
- Shake plate for 10 sec and record fluorescence of resorufin using an excitation wavelength of 560 nm and an emission wavelength of 590 nm. Protect plate from direct light and record data within 1 to 2 h to avoid increased background fluorescence.

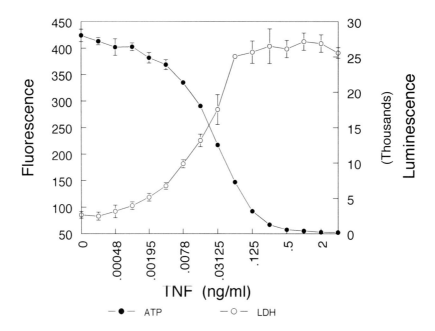

FIGURE 28.5 Half-maximal response values correlate well between the LDH-release assay of membrane integrity and the ATP assay for cell viability. Murine L929 cells were seeded at 2000 cells per well in a 384-well plate in serum-supplemented medium and cultured for 24 h; then, various amounts of TNFα (shown in log scale on the X-axis) were added and incubated overnight. The CytoTox-ONE Reagent used to measure LDH-release (open circles) or CellTiter-Glo Reagent used to measure ATP (solid circles) were added to parallel sets of wells, and fluorescence (560 nm excitation/590 nm emission) or luminescence values, respectively, were recorded. The values represent the mean ± SD of four replicate samples.

28.3.3 EXAMPLE LDH-RELEASE ASSAY DATA

Figure 28.5 shows the results of a cytotoxicity assay testing the effects of TNFα treatment on murine L929 fibroblasts. Increasing concentrations of TNFα resulted in greater fluorescence, indicating the release of LDH from damaged cells. For comparison purposes the results of an ATP assay from a parallel set of wells are shown. Concentrations of TNFα above 0.125 ng/mL result in the maximum release of LDH and almost no detectable ATP, demonstrating an inverse correlation between the markers of membrane integrity/cytotoxicity and cell viability. The data in Figure 28.5 indicate there is detectable LDH activity present in the sample of cells that were not treated with TNFα. There is a measurable contribution of LDH activity present in sera typically used to supplement cell culture media. Using serum-free medium can minimize the background fluorescence contribution from LDH.

28.3.4 ADVANTAGES AND DISADVANTAGES OF THE LDH-RELEASE ASSAY

One advantage of the LDH-release assay is that it provides direct evidence of damaged or dead cells present in assay wells. The data generated are consistent with the widely accepted definition of cell death being a loss of membrane integrity. There are many situations where the measurement of the number of damaged cells is more appropriate than the number of viable cells or the total number of cells present. This is especially true for short-term *in vitro* experiments investigating the induction of necrosis. There are multistep methods to measure membrane integrity using fluorescent dyes that are excluded from viable cells, but the choices have been limited for homogeneous *in vitro* assay methods.

The use of physiologically balanced solutions to create an LDH assay reagent that does not harm living cells has enabled the development of a homogeneous format. Addition of LDH assay reagent directly to cells in culture eliminates the need to transfer an aliquot of medium to a separate plate. The simplified procedure reduces errors introduced by multiple pipetting steps and saves time and the expense of an extra assay plate.

It should be noted, however, that there are applications where it is advantageous to determine LDH activity in a sample that has been removed from a culture well. Removing a small aliquot of culture supernatant leaves the original sample of cells unharmed and available for any other type of assay (e.g., gene reporter assay or cell imaging). This nonhomogeneous format of the LDH-release assay often serves as a cytotoxicity control to determine whether negative effects observed in other assays are the result of general cytotoxicity.

Lactate dehydrogenase is relatively stable in culture medium once it is released from cells with compromised membranes. Comparisons to other enzymes used as markers of membrane integrity, such as glucose-6-phosphate dehydrogenase, have demonstrated a rapid loss of enzymatic activity, making it more difficult to quantitate the number of dead cells (Riss and Moravec). The half-life of LDH released from cells is approximately 9 to 10 h. The loss of LDH activity over time becomes important to consider when interpreting data from experimental protocols with long exposure periods to agents that damage cells.

Although LDH-release is a valid marker of membrane damage, for experiments with long periods of exposure of cells to the test compounds, the changing levels of total LDH resulting from cell growth and the effects of stability of LDH once released from cells make it difficult to use this method as a precise quantitative measure. If all the cells are killed at the beginning of a 48-h exposure period, there will be little LDH activity remaining when the measurements are made. Including a 100% lysis control at both the beginning and end of the exposure period can help correct for loss of LDH activity and accumulation of LDH activity resulting from cell growth. If cells are actively growing during the drug exposure period, the total amount of LDH activity present at the end of the incubation period will be greater than when the cells are plated.

In addition to estimating the number of damaged cells, the measurement of LDH activity also can be used to determine the total cell number. The procedure involves addition of a lysis solution to rupture all cells, followed by measuring the total amount of LDH present [16].

The stability of LDH also results in a disadvantage for this assay method. The long-term stability of LDH in serum used to supplement culture medium adds to the background signal (shown in Figure 28.5) and ultimately limits the detection sensitivity of the assay. The LDH activity present in serum is the factor that limits sensitivity of this assay method; however, as few as 200 cells above background can be detected using the homogeneous method employing conversion of resazurin to the fluorescent resorufin product.

Another disadvantage of the coupled enzymatic assay approach to measure LDH-release is that culture medium supplemented with pyruvate will slow the rate of appearance of the fluorescent signal because it slows lactate-to-pyruvate conversion by LDH. An LDH assay that requires a 10-min incubation period in a medium that normally does not contain pyruvate (such as RPMI 1640) may require as long as 30 min to reach an optimal fluorescence signal in media that have been supplemented with 2 mM pyruvate.

Although measurement of LDH release has been used as a marker of necrosis, it cannot distinguish between primary necrosis resulting from a drastic insult to the cell and secondary necrosis that follows apoptotic cell death *in vitro*. A thorough understanding of the kinetics of cell death in the model system being used is critical to distinguish between apoptosis and necrosis as well as to decide when measurements should be taken.

28.4 CASPASE-3 ASSAY TO DETECT APOPTOSIS

28.4.1 DESCRIPTION

Many of the morphological and biochemical changes that occur in cells during apoptosis can be linked to the involvement of the executioner caspases. With few exceptions, the activity of caspase-3 or caspase-7 has been associated with the terminal phases of apoptotic cell death, and thus it has been accepted as a marker of apoptosis.

Many assays have been developed to serve as indicators of apoptosis, including the TUNEL assay to measure DNA fragmentation, the Annexin V binding assay to detect biophysical changes in the cell membrane, and immunocytochemical assays to measure the presence of neoepitopes generated by caspase activity. Unfortunately, none of those methods has led to the development of homogeneous cell-based assays for multiwell plates.

The identification of amino acid sequences that serve as selective substrates for the various caspases and the generation of aminomethyl coumarin (AMC)–labeled versions of those substrates have enabled the development of simplified cell-based assays for caspase-3/7. Caspase-3/7 cleavage between the *C*-terminal aspartic acid residue and the fluorophore of the Ac-DEVD-AMC substrate yields Ac-DEVD and free AMC and results in a red-shift in the fluorescence emission of AMC that is proportional to the amount of caspase-3/7 activity.

Rhodamine 110 (R110) has proven to be a better choice than AMC as a reporter fluorophore for protease substrates used in cell-based assays to screen for apoptosis. Rhodamine 110 has a larger quantum yield and does not require excitation in the ultraviolet region that can lead to interference by many other fluorescent compounds typically present in libraries of candidate drugs. The DEVD peptides covalently bound to R110 form a "profluorescent" caspase-3/7 substrate, meaning there is essentially no detectable fluorescence until the protease removes the peptides to liberate R110. Together, these properties result in an improved signal-to-background ratio and improved sensitivity.

The use of R110 as the fluorescent reporter for caspase substrates resulted in sufficient improvement in detection sensitivity to enable measurement of caspase-3/7 activity from the number of cells typically used in 96- and 384-well formats. Further improvement in the formulation to achieve cell lysis while maintaining the activity of caspase-3 led to the first widely accepted homogeneous "add–incubate–and–measure" method for detecting apoptosis in multiwell plates [17].

Further improvements in speed and sensitivity of homogeneous *in vitro* apoptosis assays have resulted from the application of luminescent detection technology to measure caspase-3/7 [18]. Figure 28.6 shows the general scheme of the luminescent protease assay. A caspase-3/7 substrate was designed to contain aminoluciferin instead of a fluorescent reporter molecule. Caspase-3/7 cleavage of the luminogenic Z-DEVD-aminoluciferin substrate liberates free aminoluciferin that can be utilized as a substrate by firefly luciferase to generate a luminescent signal. After an initial incubation period, a steady state is reached between the protease and luciferase enzymatic rates, which results in the generation of a luminescent signal proportional to the amount of active caspase. Because luciferase continually consumes the liberated aminoluciferin, and photons from the luciferase reaction dissipate rapidly, the signal represents the protease activity present at the time of reading. This is in contrast to the fluorogenic protease assays, which continue to accumulate fluorescent product as long as the protease remains active.

The luminescent assay kit provides a caspase-3/7 substrate, which contains the tetrapeptide sequence DEVD, in a reagent optimized to achieve cell lysis and maintain the enzymatic activity of caspase-3 and luciferase. The addition of a single reagent in an "add–mix–measure" format results in cell lysis, followed by caspase cleavage of the Z-DEVD-aminoluciferin substrate and generation of a "glow-type" luminescent signal, produced by luciferase (Figure 28.6). Cell washing, removal of medium, and multiple pipetting steps are not required. The luminescent protease assay relies on the properties of the same thermostable luciferase used for the ATP assay. The stability

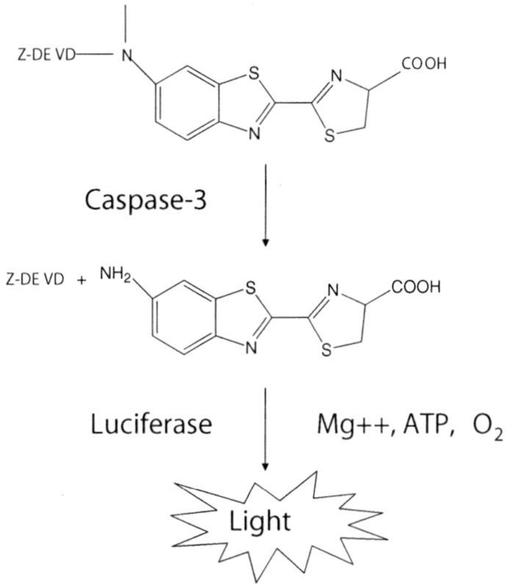

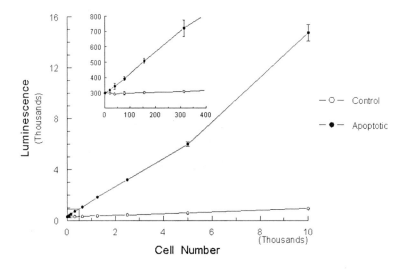

FIGURE 28.6 General concept of the luminescent protease assay. Caspase-3 cleavage of Z-DEVD-aminoluciferin results in release of aminoluciferin, which is used by luciferase to generate light.

FIGURE 28.7 Sensitivity of luminescent detection of caspase-3/7 activity in apoptotic Jurkat cells. Jurkat cells were treated with anti-Fas mAb for 4.5 h to induce apoptosis (●). An identical population of cells was left untreated (○). The Caspase-Glo 3/7 Reagent was added directly to cells in 96-well plates to a final volume of 200 μL per well. The plates were incubated at room temperature for 1 h before recording luminescence with a Dynex MLX luminometer. Each point represents the average of four wells. The "no-cell" blank control value has been subtracted from each point.

of the enzyme allows formulation of a reagent to generate a "glow-type" luminescent signal in conditions to lyse cells and liberate caspases. The caspase and luciferase enzyme activities reach steady state and generate a peak in the luminescent signal within 1 h. The luminescent signal is maintained as a glow for several hours with a marginal loss. Luminescence is directly proportional to the amount of caspase activity present (Figure 28.7).

28.4.2 APOPTOSIS ASSAY PROTOCOL

For more detailed information, see Promega Technical Bulletin # 323 describing the Caspase-Glo™ 3/7 Assay. The kit includes a lyophilized Substrate component containing Z-DEVD-aminoluciferin, luciferase, and ATP and a Buffer component containing detergent to lyse cells and other ingredients necessary to provide optimal conditions to maintain the enzymatic activities of caspase-3 and luciferase. The protocol is as follows:

- Equilibrate the Caspase-Glo 3/7 Buffer and lyophilized Caspase-Glo 3/7 Substrate to room temperature prior to use and transfer the contents of the Caspase-Glo 3/7 Buffer bottle into the amber bottle containing Caspase-Glo 3/7 Substrate. Mix by swirling or inverting the contents until the substrate is thoroughly dissolved to form the Caspase-Glo 3/7 Reagent. The reconstituted Caspase-Glo 3/7 Reagent may be stored at 4°C for up to 3 days with no loss of activity. Reconstituted reagent that has been refrozen and stored at –20°C for 1 week will have a 25% loss in activity.
- Prepare 96-well (or 384-well) assay plates containing cells in culture medium, typically fewer than 20,000 cells/well in the 96-well format. Use opaque-walled (clear- or solid-bottom) tissue culture plates compatible with recording data using a luminometer.
- Prepare a set of wells containing culture medium without cells to serve as a blank to determine luminescence from the combination of the cell culture system and the Caspase-Glo 3/7 Reagent. *Note*: The sensitivity of the luminescent assay is sufficient to detect DEVDase activity in serum used to supplement culture medium.
- Prepare a set of wells containing vehicle-treated cells in culture medium to serve as a negative control to determine the contribution of caspase activity from untreated cells.
- Add test compounds and vehicle controls to appropriate wells so the final volume is 100 μL in each well in the 96-well format (25 μL for a 384-well plate).
- Culture cells for desired test exposure period.
- Remove assay plates from 37°C incubator and equilibrate to 22°C, approximately 20 to 30 min.
- Add a volume of Caspase-Glo 3/7 Reagent equivalent to the volume of culture medium present to achieve a 1:1 ratio of culture medium to Reagent (e.g., add 100 μL of Reagent to 100 μL of medium containing cells for the 96-well plate format or add 25 μL of Reagent to 25 μL of medium containing cells for a 384-well plate).
- Cover the plate with a plate sealer or lid.
- Mix contents of wells using a plate shaker at 300 to 500 r/min for 30 sec.
- Incubate at room temperature for 30 min to 3 h. *Note*: Temperature affects the rate of enzymatic reactions and thus will impact the luminescent signal. The optimal incubation time will depend on the cell culture model system and should be determined empirically. In general, the luminescent signal remaining at 3 h is greater than 70% of peak luminescence.
- Measure the luminescence of each sample in a plate-reading luminometer.

28.4.3 EXAMPLE CASPASE-3 ASSAY DATA

The sensitivity of the luminescent assay is sufficient to detect caspase-3/7 activity in very low numbers of cells. Comparing serial dilutions of Jurkat cells in a 96-well plate that were either treated with anti-Fas monoclonal antibody to induce apoptosis or left untreated as a control demonstrate that the luminescent signal from as few as 20 cells per well was significantly greater than the average signal of the control population. Although this level of sensitivity is not required for all applications, it enables the assay to be scaled to smaller volumes for use in 384-well plates and to detect small increases in apoptosis.

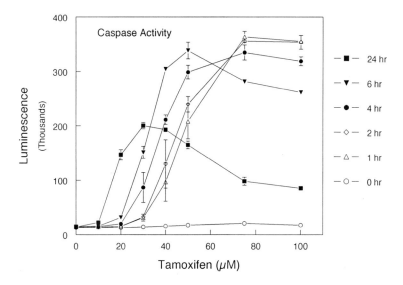

FIGURE 28.8 Caspase-3/7 activity measured from HepG2 cells treated with 0 to 100 μM tamoxifen for 0 (○), 1 (△), 2 (◊), 4 (●), 6 (▼), and 24 h (■). HepG2 cells were cultured overnight at 25,000 cells/well in a 96-well plate. Tamoxifen stock solution in DMSO was further diluted into MEM and added at staggered times. All wells including 0 μM tamoxifen contained a final concentration of 0.2% DMSO. The Caspase-Glo 3/7 Reagent (100 μL/well) was added immediately after the time zero addition of tamoxifen and the assay plates incubated at ambient temperature for 1 h prior to recording luminescence. The luminescence value for the control of culture medium without cells was 708 ± 68 (not shown). Data shown represent the means ± SD (n = 3). (Reprinted from Riss, T. and Moravec, R., *Assay and Drug Dev. Technol.*, 2004; 2(1):51–62.)

An example showing the measurement of caspase-3/7 as an indicator of tamoxifen-induced apoptosis in HepG2 cells is shown in Figure 28.8. Various concentrations of tamoxifen were added to the cells and incubated for different lengths of time before adding the luminogenic caspase-3/7 reagent. The Caspase-Glo 3/7 Reagent was added immediately after addition of tamoxifen for the 0-h control samples, and a 1-h ambient temperature incubation period was used to achieve a steady-state luminescent signal. All concentrations of tamoxifen had almost no effect on the 0-h samples, and there did not appear to be any significant induction of caspase-3/7 activity after cell lysis in the 0-h samples. The cells exposed to tamoxifen for 1 h prior to addition of assay reagent showed a dose-responsive induction of caspase-3/7 activity. Longer periods of incubation showed that induction of caspase-3/7 activity occurred at lower concentrations of tamoxifen. The IC$_{50}$ values were observed to decrease with increasing lengths of exposure to tamoxifen. Comparison of data from 100 μM tamoxifen show there was an increase in caspase activity from 0 to 1 h followed by a decrease in activity after 6 h of tamoxifen treatment. The maximum caspase-3/7 activity was recorded at 70 to 100 μM tamoxifen after 1 to 2 h exposure, whereas the peak shifted to 30 μM after 24 h exposure. The loss of caspase-3/7 activity from samples incubated with greater than 30 μM tamoxifen for 24 h probably resulted from destruction during secondary necrosis *in vitro*.

This example illustrates that the induction of apoptosis can be dependent on the dose of the test compound and the duration of exposure. Because the appearance of caspase-3/7 activity is a transient event, it is important to characterize the responsiveness of the model system used to measure apoptosis to determine the appropriate window of time to record data.

Appropriate characterization of the model system could be especially important in a screening environment where often only one concentration of test substance is assayed. The results from choosing too short or too long a treatment period could lead to vastly different interpretations. For example, if 20 μM were the only concentration of tamoxifen tested and the duration of exposure was less than 24 h, the data might be interpreted to suggest that tamoxifen does not induce apoptosis.

28.4.4 ADVANTAGES AND DISADVANTAGES OF USING THE LUMINESCENT ASSAY FOR CASPASE ACTIVITY

The luminescent apoptosis assay format is more sensitive and faster than corresponding fluorescent assays for measuring caspase activity. A major factor that contributes to the superior sensitivity is the low background luminescence. The signal-to-background ratios and thus the ultimate sensitivity of luminescent assays are greater than those of corresponding fluorescent formats, which can be more severely limited by contributions from background fluorescence. The greater sensitivity of the luminescent assay allows use of lower cell numbers and makes it easier to scale up to high-density plate formats. The assay can be easily adapted to different volumes, provided the 1:1 ratio of Caspase-Glo 3/7 Reagent to sample volume is used.

A maximum signal-to-background ratio is achieved faster with the luminescent assay format. The luminescent assay reaches peak luminescent output after about an hour of incubation with reagent when a steady state is achieved between the protease and luciferase activities. In contrast, fluorescent protease assays continue to accumulate signal as long as the protease reaction continues and there is sufficient substrate. We have found the fluorescent caspase-3/7 assay will continue to accumulate signal for up to 18 h. Several hours of incubation with the fluorescent assay reagents generally result in an increase in the signal-to-background ratio, but the limit of detection may not approach that of the luminescent assay at 1 h.

There is less chemical interference with the luminescent protease assay format. By screening a library of 640 pharmaceutically active compounds for caspase-3 inhibitors using both the luminescent (Caspase-Glo 3/7 Assay) and the fluorescent (Apo-ONE Assay) caspase-3/7 assays, a greater number of false hits was detected with the fluorescent format. The fluorescent properties of compounds in chemical libraries do not interfere with luminescence, and there is a greater separation between negative and positive control values in the luminescent assay format. Although there is the possibility that the chemical compounds being tested may interfere with the luciferase reaction, the rate of interference with luciferase was observed to be lower than the rate of interference with the fluorescent assay format. The occurrence of direct inhibition of the luciferase reaction can be confirmed or ruled out by testing control reactions in the presence of luciferin as a substrate.

The UltraGlow recombinant thermostable luciferase used for the luminescent assay kits is much less susceptible to inhibition than is native luciferase [13]. It is produced in large quantities with consistent properties. The successful development of the homogeneous luminescent caspase assay format was largely due to the ability of luciferase to maintain enzymatic activity for extended periods of time in the harsh conditions necessary to lyse cells and inhibit other destructive enzymes. The thermostable luciferase generates a signal with a long half-life that provides plenty of time to record data from several plates. As long as caspase-3/7 is generating aminoluciferin and the conditions are suitable for luciferase, a signal will be generated that is proportional to the activity of caspase.

Among the limitations of the luminescent assay format are that the signal must be recorded within a few hours after addition of reagent. The signal from a luminescent assay format is dependent on the presence of active luciferase to generate photons. After peak luminescence is achieved, the signal will begin to decrease gradually. This is in contrast to the fluorescent assay format, where liberated fluorophore continues to accumulate over time and the total accumulated fluorescence can be recorded for extended periods. The fluorescent assay format allows more flexibility for recording data because the assay can be stopped and the fluorescent signal can be stabilized and recorded at a later time.

28.5 MEASURING CELL VIABILITY TO NORMALIZE REPORTER GENE ASSAY DATA

28.5.1 DESCRIPTION

Cultured cells are used to develop assays to measure effects on a wide variety of targets. The biological complexity of cell-based assays dictates the need to develop appropriate controls to ensure the consistent behavior of the model system and the specificity of the effects measured for any target.

A cell viability assay can be used to normalize the effects of any experimental component that may affect the number of cells. Some common examples where normalization of cell number may be relevant include: correction for variability of cell plating, the effects of differential cell growth during the experimental period (especially within 384- or 1536-well plates), or the effects of transfection toxicity in transiently transfected cultures.

Cell viability assays also are frequently used as a control to determine whether treatments are specific for the target being measured or nonspecific due to cytotoxicity. For example, if the results from treatment with a test compound show a decrease in signal from a reporter gene assay, but the results from a cell viability assay do not indicate a decrease, it is likely the treatment is having a specific effect on expression of the reporter. On the other hand, if treatment with the test compound causes decreases in both the reporter gene function and cell viability, it is likely that the treatment is nonspecific and that the decrease in reporter gene activity is the result of cell death.

There are now examples where the measurement of cell viability can be multiplexed in the same sample well with assays to measure other targets. Reporter gene assays are among the most commonly used cell-based methods and are ideally suited for combining with cell viability determination as a normalizing control.

Combining more than one assay technology in a multiplex format requires an understanding of the chemistry underlying the assays. For example, some ATP assays contain harsh detergents that lyse all cells and inactivate endogenous ATPases. Those harsh lysis conditions also inactivate other cellular enzymes and compromise the researcher's ability to quantitate enzymatic reporters. Therefore, the reporter gene assay or other multiplex partner must be performed before the ATP assay. Understanding the characteristics of reporter gene assay also is important. For example, if the reporter gene assay contains a lysis step that results in the release of active cellular ATPases, the result will be consumption of ATP before it can be quantified as a marker of cell viability. These problems can be overcome by utilizing a live-cell luminescent reporter gene as the first component of a multiplex combination, followed by measurement of ATP to normalize for viable cell number.

A new technique to utilize *Renilla* luciferase as a live-cell reporter gene has recently been developed. The enabling technology was the invention of cell-permeable luciferase substrates that remain stable in culture medium for several hours. The substrates are added directly to culture medium and result in sustained luminescence that provides a real-time readout proportional to the reporter activity.

The *Renilla* Luciferase Live Cell Substrates are chemically modified coelentrazines that are relatively stable in culture medium, cell permeable, and biologically active. After the substrates enter cells, the chemical modifications are removed by endogenous esterases, and the native *Renilla* luciferase substrate (coelenterazine) is released within the cytoplasm (Figure 28.9). The most stable of these substrates, EnduRen™ Live Cell Substrate, can be added to culture medium at the time of plating cells because the luminescent signal generated from samples reflects *Renilla* luciferase content for at least 24 h.

After recording data from the reporter gene assay, the viable cell number can be determined by measuring ATP. The harsh conditions provided by addition of some ATP assay reagents result in reducing the luminescence from *Renilla* luciferase by at least 100,000-fold. The luminescence

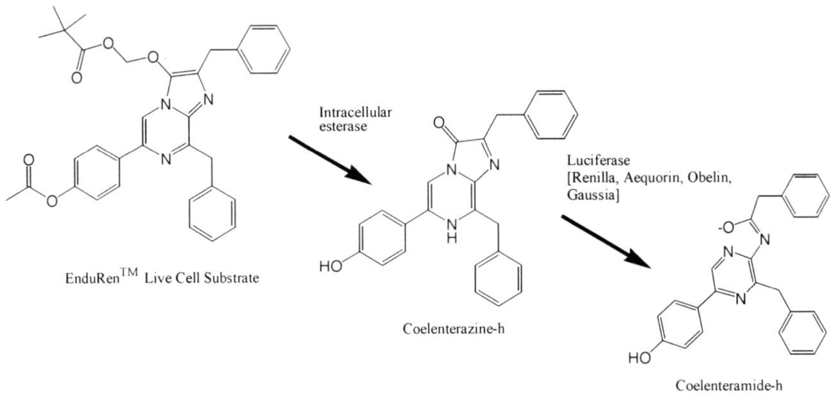

FIGURE 28.9 EnduRen Substrate is converted inside cells to coelenterazine-h, a substrate for *Renilla* luciferase.

from the *Renilla* luciferase reporter (which uses coelentrazine as a substrate) is turned off and simultaneously, the luminescence from the firefly luciferase (which utilizes luciferin and ATP as substrates) is turned on. Using this approach, no residual *Renilla* luminescence affects the quantitation of ATP. However, it should be noted that different ATP assays use different methods to inhibit ATPase activity, so if a method other than the CellTiter-Glo Assay is used to measure ATP, its effect on the luminescence generated by *Renilla* luciferase within cells should be examined.

28.5.2 MULTIPLEX GENE REPORTER AND CELL VIABILITY ASSAY PROTOCOL

For more detailed information, see Promega Technical Manual #244 and Promega Technical Bulletin #288 describing EnduRen Live Cell Substrate and CellTiter-Glo Luminescent Cell Viability Assay Kit, respectively. Each kit is used exactly as described in their respective technical literature. The EnduRen Substrate Kit provides a dried compound in a 1.5-mL tube. The CellTiter-Glo Assay provides a lyophilized Substrate component containing luciferase and luciferin and a Buffer component containing detergents to lyse cells and ATPase inhibitors to enable a long-lasting signal and to reduce the luminescence from the *Renilla* luciferase signal. The protocol is as follows:

- Prepare EnduRen Substrate by transferring an appropriate amount of dimethylsulfoxide (DMSO) into the tube containing the dried Substrate. Vortex until the Substrate is completely in solution. This can take up to 10 min depending on the amount of Substrate.
- Thaw the CellTiter-Glo Buffer and equilibrate to room temperature prior to use. For convenience, the CellTiter-Glo Buffer may be thawed and stored at room temperature for up to 48 h prior to use.
- Equilibrate the lyophilized CellTiter-Glo Substrate to room temperature prior to use.
- Transfer the appropriate volume of CellTiter-Glo Buffer into the amber bottle containing CellTiter-Glo Substrate to reconstitute the lyophilized enzyme/substrate mixture. This forms the CellTiter-Glo Reagent.
- Mix gently to obtain a homogeneous solution.
- Dilute EnduRen Substrate 1:1000 from the DMSO stock into cell culture medium covering cells. This dilution may be performed as a single step if the EnduRen Substrate is to be diluted into the cell suspension before plating and experimentation, or it may be performed as a serial dilution if Substrate is to be added to each well individually after experimental treatment.
- Return cells to incubator for at least 1.5 h.

- When appropriate, measure luminescence. Luminescence may be measured repeatedly from the same sample wells.
- At the end of the experiment, equilibrate the plate and its contents to room temperature for approximately 30 min. (*Note*: The optional equilibration period is to ensure that the plates are at a uniform temperature. Stacks of several plates will take longer to equilibrate. If wells are at different temperatures, the luciferase reaction and thus luminescence will vary.)
- Add a volume of CellTiter-Glo Reagent equal to the volume of cell culture medium present in each well.
- Mix contents for 2 min on an orbital shaker to induce cell lysis.
- Allow the plate to incubate at room temperature for approximately 10 min to stabilize luminescent signal.
- Measure luminescence.
- Normalize the luminescence from *Renilla* luciferase to that from the Cell Viability Assay.

28.5.3 EXAMPLE DATA

As stated above, reporter gene measurements can be normalized to cell number, reducing variability caused by transfection toxicity or inconsistent plating. In the example in Figure 28.10, *Renilla* luciferase was used as a model system for RNAi optimization. Normalization of the *Renilla* luciferase luminescence generated from these samples (Figure 28.10b) reduced the average relative standard deviation of the experiment by over twofold from the absolute luminescence (Figure 28.10a).

28.5.4 ADVANTAGES AND DISADVANTAGES OF NORMALIZING REPORTER GENE DATA USING CELL NUMBER

The advantages of normalizing reporter gene data to cell number, protein content, or a second reporter gene have been well documented and pertain to reducing artifacts and relative error in the data. Most scientists agree that if possible, it is best to normalize.

Both the *Renilla* luciferase assay and the ATP assay are very sensitive and are easily suited to scaling into 96-, 384-, and 1536-well formats. The speed and ease of luminescent measurements make the combination of *Renilla* luminescence measurements from live cells and ATP measurements from those same cells ideal for this normalization. The *Renilla* luciferase luminescence reflects the amount of luciferase present within the cells when they are measured. The signal is extremely long-lived, usually over 24 hours, and so a specific reagent delivery system is often not required for the *Renilla* luminescence measurement. Luminescent substrate can be added during cell plating. The only reagent-specific delivery system required is for the CellTiter-Glo Reagent. Other techniques for normalizing reporter gene data usually require equipment for two reagent deliveries, one for each assay to be performed.

There are two potential disadvantages to multiplexing the Live Cell *Renilla* Luciferase Substrate and the CellTiter-Glo Reagent. The first is cost. Although the combination of EnduRen Substrate and CellTiter-Glo Reagent is less expensive than many reagent systems intended for data normalization, it is more expensive than simply measuring a single reporter gene.

The second potential disadvantage is the *Renilla* luminescent signal's reflection of the current *in situ* content of *Renilla* luciferase. Unlike most reporter gene assays that are lytic end-point assays, luminescence generated by the *Renilla* Luciferase Live Cell Substrates reflect the *Renilla* luciferase content at the time of measurement. So, while this characteristic permits multiple measurements to be taken from the same sample for kinetic analysis of a phenomenon, it also precludes stopping the reaction for measurement of the samples at the scientist's leisure.

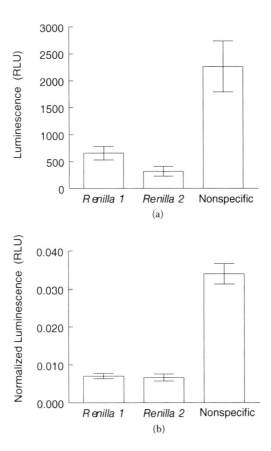

FIGURE 28.10 Relative error decreases when reporter gene values are normalized to cell number. The average relative standard deviation decreased from 26% for the raw data in (a) to 11% for normalized data in (b). HeLa cells, stably expressing *Renilla* luciferase, were transiently transfected with two different isolates of a vector (*Renilla* 1 and *Renilla* 2) expressing an siRNA against a *Renilla* luciferase target site or against a nonspecific sequence (Nonspecific). Forty-eight hours post-transfection, *Renilla* luciferase was quantitated by measuring luminescence after a 2 h exposure of cells to EnduRen Live Cell Substrate (a). The ATP assay was performed on the samples and the luminescence values were used to normalize for viable cell number (b). As expected, transfection with the siRNA against the *Renilla* luciferase gene decreased expression of *Renilla* luciferase compared to the nonspecific siRNA (four- to fivefold). Data are shown as averages of six replicates.

28.6 SUMMARY

The example protocols describe homogeneous assays to detect the number of viable cells, dead cells, and apoptotic cells in multiwell plates. Also included is an example of applying the ability to use cell viability in a multiplex format to normalize the results of a live cell reporter gene assay. The choice among available markers to measure cytotoxicity *in vitro* may depend on several factors including the ease of use, detection sensitivity, instrumentation availability, ability to multiplex, and cost. A comparison of factors to consider for the three assays described in this chapter is included in Table 28.1. The cell viability assay based on detection of ATP is the fastest to perform and can detect the smallest number of cells.

The type of cells used and the experimental design often limit the choice of available cytotoxicity markers. However, for some cytotoxicity applications, all that is necessary to measure is whether the population of cells is alive or dead at the end of the experiment. In those cases the marker endpoint chosen to measure cell death may not be as important, and other factors become more

TABLE 28.1
Summary of Factors to Consider When Choosing an Assay

Assay	Parameter Measured	Incubation Time[a]	Detection Sensitivity (Cells)[b]	Plate Format	Detection Instrument(s)
ATP quantitation	Viable cells	10 min	4	96, 384, 1536	Luminometer, CCD camera, or fluorometer[c]
LDH release	Necrosis	10 min	200	96, 384	Fluorometer
Caspase-3/7 activity	Apoptosis	0.5 to 5 h	20	96, 384	Luminometer

[a] Incubation time does not include optional time recommended to equilibrate assay plate to ambient temperature.

[b] Detection sensitivity varies among cell types and culture conditions.

[c] Fluorometer may be used for some luminescent assays by turning off excitation lamp and removing emission-side filters that restrict light from entering photomultiplier tube.

Source: Reprinted in modified form from Riss, T. and Moravec, R., *Assay and Drug Dev. Technol.*, 2004; 2(1):51–62.

influential. For example, for high-throughput screening applications, the ability to automate or multiplex with other assays strongly influences the choice.

Several factors can alter the outcome of *in vitro* toxicity measurements including the concentration of toxin, length of exposure, and responsiveness of the indicator cells. A thorough characterization of the cell model system and assays used will lead to increased confidence in the results from cytotoxicity screening. The data in Figure 28.3 and Figure 28.8 clearly indicate the effects of a cumulative exposure of a toxin over 24 h. A short period of exposure to a toxin may have no effect on cell viability, whereas the same concentration of toxin may kill all of the cells after longer periods of exposure. Those data highlight the value of understanding the kinetics of cell death in the *in vitro* model system before choosing a time to collect data. It is especially important to understand the kinetics of appearance of transient markers such as caspase activity that "appear and disappear" within a window of time. Performing a time-course experiment to establish when markers are present will help determine the window of time available to take measurements.

The apparent potency of cytotoxic compounds depends on the physiological state of the cells both during and immediately before the assay. Factors such as the density of cells in assay wells and the density of cells in parent stock cultures used to set up assay plates can affect cell physiology and thus responsiveness to toxins [8]. To overcome the many potential sources of variability and achieve consistent data from day to day, it is important to implement standard operating procedures for handling stock cultures and preparing assay plates.

Standardized procedures may be more important in high-throughput screening labs where there is a high demand for consistency across the entire screen. Additional benefits of implementing standard operating procedures include more control over temperature gradients, changes in nutrient content of culture medium, and changes in pH or other factors that may alter the responsiveness of cells.

REFERENCES

1. Riss T, Moravec R. Simplifying cytotoxicity screening: introducing the CellTiter-Blue™ cell viability assay. *Promega Notes* 2003; 83:10–13.
2. Mosmann T. Rapid colorimetric assay for cellular growth and survival: application to proliferation and cytotoxicity assays. *J Immunol Meth* 1983; 65:55–63.

3. Barltrop JA, Owen TC. 5-(3-carboxymethoxyphenyl)-2-(4,5-dimethylthiazolyl)-3-(4-sulfophenyl)tetrazolium, inner salt (MTS), and related analogs of 3-(4,5-dimethylthiazolyl)-2,5-diphenyltetrazolium bromide (MTT) reducing to purple water-soluble formazans as cell-viability indicators. *Bioorg Med Chem Lett* 1991; 1:611–614.

4. Ahmed SA, Gogal RM Jr, Walsh JE. A new rapid and simple nonradioactive assay to monitor and determine the proliferation of lymphocytes: an alternative to [³H]thymidine incorporation assays. *J Immunol Meth* 1994; 170:211–224.

5. Korzeniewski C, Callewaert DM. An enzyme-release assay for natural cytotoxicity. *J Immunol Meth* 1983; 64:313–320.

6. Decker T, Lohmann-Matthes ML. A quick and simple method for the quantitation of lactate dehydrogenase release in measurements of cellular cytotoxicity and tumor necrosis factor (TNF) activity. *J Immunol Meth* 1988; 115:61–69.

7. Riss T, Moravec R. Introducing the CytoTox-ONE™ Homogeneous Membrane Integrity Assay. *Promega Notes* 2002; 82:15–18.

8. Riss T, Moravec R. Use of multiple assay endpoints to investigate the effects of incubation time, dose of toxin, and plating density in cell-based cytotoxicity assays. *Assay Drug Dev Technol* 2004; 2(1):51–62.

9. Crouch SPM, Kozlowski R, Slater KJ, Fletcher J. The use of ATP bioluminescence as a measure of cell proliferation and cytotoxicity. *J Immunol Meth* 1993; 160:81–88.

10. Kangas L, Grönoos M, Nieminen A-L. Bioluminescence of cellular ATP: a new method for evaluating cytotoxic agents *in vitro*. Med Biol 1984; 62:338–343.

11. Lundin, A, Hasenson M, Persson J, and Pousette A. Estimation of biomass in growing cell lines by ATP assay. *Meth Enzymol* 1986; 133:27–42.

12. Sevin BU, Peng ZL, Perras JP, Ganjei P, Penalver M, Averette HE. Application of an ATP bioluminescence assay in human tumour chemosensitivity testing. *Gyn Oncol* 1988; 31:191–204.

13. Hall MP, Gruber MG, Hannah RR, Jennens-Clough ML, Wood KV. Stabilization of firefly luciferase using directed evolution. In: Roda A, Pazzagli M, Kricka LJ, Stanley PE, Eds. *Bioluminescence and Chemiluminescence – Perspectives for the 21st Century*. Chichester, U.K.: John Wiley & Sons, 1998:392–395.

14. Hannah R, Beck M, Moravec R, Riss T. CellTiter-Glo™ luminescent cell viability assay: a sensitive and rapid method for determining cell viability. *Cell Notes* 2001; 2:11–13.

15. Moravec R, Beck M, Hannah R. High-throughput screening with the CellTiter-Glo™ luminescent cell viability assay. *Cell Notes* 2001; 2:14–16.

16. Moravec R. Total cell quantitation using the CytoTox 96™ nonradioactive cytotoxicity assay. *Promega Notes* 1994; 45:11–12.

17. Humpal-Winter J, Niles A, d Bjerke M. Apo-ONE™ homogeneous caspase-3/7 assay: rapid apoptosis detection in high-throughput applications. *Promega Notes* 2001; 79:33–35.

18. O'Brien M, Moravec R, Riss T. Caspase-Glo™ 3/7 Assay: use fewer cells and spend less time with this homogeneous assay. *Cell Notes* 2003; 6:13–15.

29 Development of Image-Based Assays for Drug Discovery

Susan M. Catalano

CONTENTS

29.1 INTRODUCTION

Image-based assays consist of the analysis of digital images of fluorescently labeled cells taken with a optical magnification system. This analysis is performed by a variety of commercially available image processing programs and is colloquially referred to as "image processing" or "image analysis." Digital images are composed of pixels or squares of uniform gray values. These gray values commonly range from 0 (pure black) to 4095 (pure white) in the case of 12-bit digital cameras, and are sometimes referred to as intensity levels or values. Image analysis measures patterns of pixel gray values (two-dimensional arrangement of intensity values) contained in digital images. This analysis yields a large amount of unique information about the objects contained in the digital image. For example, changes in cell shape can provide information about toxicity or cell death, and the location of a particular protein within subcellular compartments can provide information about that protein's activity level (Figure 29.1).

Because a wide variety of assay technologies have readouts that can be detected with a digital camera, many assay formats can be configured as image-based assays. Due to their complexity and requirement for specialized equipment, these assays are most commonly found in secondary screening cascades, where the measurement of multiple parameters of cellular phenotype are a

407

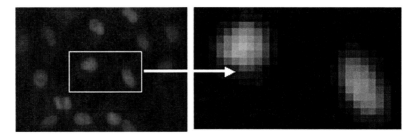

FIGURE 29.1 Digital images are composed of squares with a uniform gray value known as pixels. (©2003 Susan Catalano.)

valuable source of information about a compound's physiologically relevant effects. When developing assays that are surrogates of biological responses, temporal regulation of protein expression or modification can usually be configured as an "intensity readout." An example would be changes in levels of cyclin proteins driving the cell through the phases of the cell cycle to mitosis. Measurement of the intensity of a fluorescently labeled cyclin protein provides a measure of the amount of cyclin protein present and indirectly predicts the effects this would have on the cell cycle (usually confirmed in subsequent, separate assays. Note that this same assay configured as an image-based assay could potentially provide information about both the level of cyclin protein and the phase of the cell cycle in the population in a single assay). However, many biological responses are dependent on the spatial regulation of protein localization, and these responses require a "location readout." An example of this would be shifts in the subcellular localization of cyclin B1 that are required for its function in cell cycle regulation. During G2, cyclin B1 is localized in the cytoplasm, but it moves into the nucleus during prophase of mitosis. Image-based formats are currently the only format capable of distinguishing this shift in subcellular localization in nonengineered cells.

Location readouts are likely to play an increasing role in assay technology in the near future. As knowledge of cell biology increases, we are coming to see the interior of the cell as a highly spatially ordered environment where local protein concentrations frequently must increase 1000-fold in order to produce protein activation, and tight control over spatial organization of proteins is maintained. One recent estimation indicates that more than 22% of human proteins have a nuclear localization sequence [2]. Location is critical for entire classes of proteins such as transcription factors and cell cycle regulatory proteins. In addition, the topological organization of subcellular compartments and macromolecular structures such as DNA is a potential regulatory mechanism for events such as chromosome regulation and cellular development. Subcellular compartment organization reflects the physiological state of the cell and is stereotypically altered during cell division and in many phenotypes associated with toxic stimuli. A wide variety of structural proteins and their regulatory agents are required for neuronal long-term potentiation, a phenomenon thought to underlie learning and memory formation in the nervous system. The possibility of combining "intensity readouts" with "location readouts" in a single assay with the resulting cost and time savings is appealing to an increasing number of drug discovery investigators. Precisely because of this flexibility, image-based assays are slated to replace up to 50% of assay platforms in the high-throughput screening (HTS), flow cytometry, and ADMET fields in the near future.

29.2 IMAGE PROCESSING: GENERAL PRINCIPLES

29.2.1 OBJECT RECOGNITION

Image processing begins with object recognition. The digital image is divided into areas corresponding to objects of interest (cells growing *in vitro*) that the user desires to measure and background areas. The simplest example of object recognition is thresholding, illustrated in

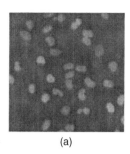
(a)

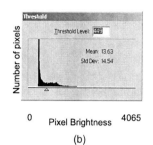
(b)

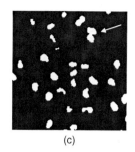
(c)

FIGURE 29.2 (a) Raw image of DAPI-stained nuclei. (b) Image histogram. The arrowhead at the bottom of the x axis is the threshold value. The user can change this value by sliding the arrowhead or typing in a specific gray value. (c) Resulting thresholded image. Arrow points to objects that are considered as a single object by the computer. (©2003 Susan Catalano.)

Figure 29.2. In order to manually set the threshold, the image histogram is sometimes displayed by the image processing program for the user to control. The image histogram is a display of the number of pixels that have a given gray value, arranged in order from black (usually equal to zero) to white (4095 in the case of images taken with 12-bit cameras). The user then chooses a gray level for setting the threshold, and a new image is then generated based on this threshold; pixels above the threshold gray value are changed to white, and pixels below this value are changed to black. As the user changes the threshold level, the objects in the thresholded image (Figure 29.2c) will appear to contract as the threshold is moved higher (whiter) and expand as the threshold is moved lower (blacker). By visually inspecting the thresholded image and comparing it to the original image, the user can determine which threshold value provides the most realistic representation of the actual boundaries of the objects in the original image. The computer can recognize contiguous white pixels in the thresholded image as discrete objects and can then count them and measure their features. Often the image processing program will not show the user the thresholded image but will instead superimpose the boundary drawn around the edge of the thresholded objects on the original raw image, allowing the user to assess the accuracy of this boundary in the appropriate context.

It is important that this procedure be done correctly and accurately. One of the important advantages of image processing is that surrogate measures of cell counting (for example, extrapolating cell number from the ATP content of a microtiter plate well) are not being used: an actual count of the objects present is being made, and statistically rigorous sampling of the assay can be performed as a result. However, this count is only as accurate as the image processing method used to produce it. Difficulties in the simple thresholding method described above arise in the case of two distinct objects that are close enough to be touching (arrow in Figure 29.2c). If not corrected, the computer will count these two objects as one, and inaccuracies in cell counting may propagate across the assay plate. The two touching cells can be easily separated by transforming the thresholded image with a variety of mathematical operations. A common procedure is known as a "watershed" transformation. In this procedure the center or most intensely white pixel of each nucleus is considered to be the peak of a mountain, and the location of the best boundary or watershed line between the two peaks is calculated. This is similar to the method the U.S. Geological Survey uses to produce topological maps of mountainous terrain. The pixels that constitute this boundary are changed from white to black in the image, separating the two touching nuclei.

This example illustrates several key themes that are central to image-based assays:

- Mathematical transformations of the raw image must be done to allow the computer to recognize features within it. These transformations are iterative and complex, and the results depend on the order in which they are done. The user may or may not be able to control the identity and order of these mathematical transformations.

- Regardless of the complexity of the mathematical transformations employed, the result is unlikely to be perfect registration between what a user considers an object and what the computer considers an object.
- The ultimate "gold standard" for determining the accuracy of the image processing method to identify and correctly measure cells *in vitro* is the visual system of a trained human observer. A person who is familiar with the phenotypic changes in cells that are likely to be encountered in the assay should assess the accuracy of the object recognition procedures and adjust them for maximal accuracy. An example would be a person who is familiar with the cellular morphology of nuclear fragmentation seen during apoptosis adjusting the parameters of the object recognition procedure until an acceptable compromise is reached between recognizing closely spaced nuclear fragments as separate objects vs. recognizing them as a single object.

Assay conditions do not usually permit cells to be cultured as isolated individuals with a great deal of separation between one cell and another (cell division in culture, survival). Nuclei are more likely to be separated from each other than cytoplasmic boundaries are, and so form the basis for object (cell) recognition in most image processing algorithms. The cytoplasmic compartment can be measured in a variety of ways. The pixel boundary of the nucleus can simply be dilated outwards by a fixed number of pixels that the user specifies. This new boundary now lies over the cytoplasm, and the user can measure attributes of this area in the same or a different color channel. Note that these boundaries demarcate specific areas of the image for measurement and can be superimposed on all color channels if desired. They are frequently referred to in image processing parlance as "masks" and can be combined in various ways. If the nuclear boundary or mask is subtracted from the cytoplasmic boundary or mask referred to above, then the resulting cytoplasmic mask will be shaped like a donut instead of a circle. Measuring the cytoplasmic brightness or intensity within the donut rather than the circle may increase the signal-to-noise ratio of the assay if the fluorescent label is absent from the nuclear compartment. Alternately, the nuclear boundary can be expanded outwards until it touches the borders of adjacent nuclear boundaries. These boundaries will tile the entire image if the cells are dense enough, and each boundary will encompass not only the cell cytoplasmic boundaries but the background dish plastic as well. Subtraction of all pixels of background intensity will leave just the cytoplasmic area to be measured. Lastly, the full extent of the cytoplasm can be measured if a cytoplasmic dye is used and the same thresholding procedure followed by watershed splitting described in the beginning of this section is employed on images within that color channel. Subcellular organelles stained with specific dyes or antibodies can be measured in the same way. The number of distinct subcellular compartments and individual proteins that can be measured depends on the shape, spatial overlap, and dye bleedthrough from one wavelength into another. It is important that the image processing program is able to measure specific regions within previously demarcated regions and keep them associated with a single object. Newer trainable classifier techniques operate on an object that is identified by traditional techniques and are therefore limited by the same considerations.

29.2.2 OBJECT MEASUREMENT

Once boundaries or masks associated with specific subcellular compartments of interest have been identified, various measurements of the pixels contained within these boundaries can be taken. These measurements consist of several types: intensity or absolute gray value (average intensity, total intensity), shape (roundness, irregular boundaries), or texture (standard deviation of intensity, randomness of intensity changes).

It is important to note that certain types of cell measurements are dramatically affected by specific aspects of image quality. Texture measurements are extremely sensitive to precision of focus of the image. Intensity measurements are sensitive to the evenness of the illumination across

TABLE 29.1
Image-Based Assay Categories

Assay Category	Example Assays
1. Intensity levels or protein interactions within specific subcellular compartments or complex contexts (tissue); measurements: color, intensity	NF-kB translocation GPCR dimerization (FRET) Phospho-H2 quantification Lineage dye quantification
2. Morphological or motility readouts seen during cell death, colony formation, B-cell capping, neurite outgrowth, or angiogenesis; measurements: texture, shape	Apoptosis AMES test Tumor cell clonogenicity *In vitro* angiogenesis Neurite outgrowth GPCR desensitization/beta-arrestin translocation
3. "Consolidation assays" — replacement of other assays for time/cost savings Flow cytometry assays — amount of labor and # of cells Plate based assays — expensive, inaccurate Difficult plates or cell types — Boyden chambers/Fluoroblock	PAD Assay — proliferation, apoptosis, and DNA content Haptotaxis

Source: ©2003 Susan Catalano.

the image. Comparison of absolute intensity measurements across instruments or assays requires precise calibration of the instrument with fluorescent standards.

29.3 IMAGE-BASED ASSAY CATEGORIES

Various categories of assays are suitable for configuration as image-based assays, such as cell-based disease models in primary or clonal cell lines or readouts of signal transduction targets in intact primary cells. However, because of their requirement for specialized equipment and expertise, and increased development and validation timelines, the most commonly utilized image-based assays are assays that cannot be measured with any other technology. Examples of these assay types are shown in Table 29.1.

The first category would be assays in which protein or compound amounts within specific subcellular compartments increase or decrease, usually reflected by a change in fluorescence intensity or color. Plate-based assays benefit from averaging across large numbers of cells and are well-suited to detect overall intensity or color changes rapidly and effectively. However, when the intensity changes are spatially restricted within the cell, the overall signal-to-noise ratio in the well may be too low to measure. High-magnification analysis with image processing may be the only technology capable of making these measurements. The same situation would apply to a mixed population of cells, in which only a subset consists of responders. Another example is morphological or motility changes that are seen during cellular responses such as B cell capping, neurite outgrowth, or GPCR internalization following desensitization. In these assays, overall intensity remains constant within the well, and the changes taking place are changes in shape or texture that can only be measured by image analysis.

Lastly, many types of assays can result in substantial time or cost savings if reconfigured as image-based assays ("consolidation assay"). Examples of this include flow cytometry assays of adherent cells. These assays are labor intensive and require large numbers of cells to achieve statistical significance. The same assay configured as an image-based assay allows the cells to remain *in situ* and can be five to 10 times faster to perform and require 10-fold fewer cells to reach the same level of performance. Another example includes plate-based assay formats that are expensive or inaccurate. BrDU assays are commonly used as a measure of antiproliferative compound effects. They measure incorporation of brominated uridine into DNA during the S phase of the cell cycle, which can be used as a surrogate of the number of cells present after compound treatment. However, compounds that induce arrest in the G2 phase of the cell cycle will appear similar to highly toxic compounds (both will yield zero signal). Preformatted BrDU assay–ready plates are also expensive. An image-based assay in which a DNA-binding dye is used to label nuclei can not only provide an accurate cell count following compound treatment but can also provide quantification of cell cycle arrest and/or nuclear fragmentation characteristic of late-stage apoptosis; it is also 50-fold less expensive. By shifting surrogate endpoints to morphology or quantitative dye measurements, "consolidation assays" capitalize on information that is already on the plate and lower the per-well cost of the assay (often eliminating other assays). This can free up cost and other color channels for more expensive reagents such as antibodies. However, this type of image-based assay is currently underutilized.

Many assays such as chemotaxis or haptotaxis assays require unusual plate formats in which a well insert is used that is a short distance above the bottom of the plate. Cells located on these inserts can be easily quantified with image analysis.

29.4 KEY CONSIDERATIONS FOR IMAGE-BASED ASSAYS

29.4.1 INFORMATION-HANDLING INFRASTRUCTURE

Image-based assays begin with an optical instrument that takes digital images and passes them on to an image processing program, which then quantifies the objects within the images according to the user's specifications. The resulting measurement data is passed on to a database. Some imaging instruments contain a database for images as well and maintain linkages between the images and their derived numerical measurements. This arrangement allows users to recall images and compare them to their derived numerical data side by side and makes it straightforward to curate and verify data from a particular screen. Because of the large size of the image sets and the numerical data, storage devices in the terabyte range are frequently seen. It is usually deemed important to keep this large amount of raw data separate from other corporate data and export only assay results to the company database.

None of the instruments currently available contain all of the features that would provide a complete, robust solution to image-based assay development, validation, and screening in one stand-alone package, although some come very close. It is important to carefully evaluate the analysis, databasing, and storage features of any potential imaging instrument purchase and plan integration with third-party software and hardware solutions accordingly. For this reason, long-term success of image-based assays often depends on close collaboration between scientists and in-house IT experts.

29.4.2 INSTRUMENTATION

Many manufacturers currently offer instrumentation suitable for measuring image-based assays. Instruments range from low resolution (in the millimeter range) to the highest possible optical resolution that is found in microscope-based imagers, 0.2 μM. (It should be noted that resolution is a function of both lens construction and detector sampling.) At the low end of resolution are familiar devices such as conventional microtiter plate readers configured for imaging. Figure 29.3a shows an image of cells in a well stained with a phosphor-specific fluorescently tagged antibody taken by LI-COR's infrared plate reader Odyssey. The overall intensity within the well can be clearly visualized and measured, but no individual cells are visible. In the intermediate range of

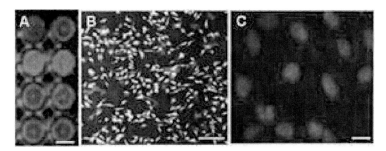

FIGURE 29.3 Optical detection systems. (a) Typical plate reader resolution. Scale bar = 1 mm. (b) Lower resolution imaging system. Scale bar = 200 μM. (c) Microscope-based imaging system. Scale bar = 5 μM. (©2003 Susan Catalano.)

resolution are low-magnification imaging instruments such as the Applied Biosystems 8200 Cellular Detection System (formerly FMAT 8100 HTS) and the Acumen Biosciences Explorer. Figure 29.3b shows an image of cells labeled with fluorescent ligand taken with the Applied Biosystems instrument. These instruments have the resolution to visualize individual cells, and in the case of certain cell types can even discriminate between nuclear and cytoplasmic compartments, but they do not have subcellular resolution capability.

At the upper limit of optical resolution are the many microscope-based instruments, including Cellomics' Arrayscan, and Molecular Devices' Discovery-1 and ImageExpress. Figure 29.3c shows an image of cellular nuclei stained with a DNA-binding dye taken with the Cellomics Arrayscan. In general, entire microtiter assay plates can be read faster with the lower resolution instruments. However, it is worth noting that many of the microscope-based instruments can usually accommodate both low and high magnification lenses, allowing them to easily take images similar to the ones shown in Figure 29.3a and Figure 29.3b. When operating in these low magnification modes, these instruments can exhibit the same throughput rates as conventional plate readers do.

The choice of instrumentation should take into account the many factors required for detecting the specific assay biology. Some imaging systems utilize laser illumination, which may help detect dim specimens but is expensive and needs to be properly aligned and maintained. Other imaging systems utilize arc lamp illumination, which gives a broad spectrum of illumination that can be used with a wide variety of fluorescent filters and is much less expensive than but not as bright as lasers. A subset of imaging instruments have confocal ability (though a true pinhole confocal instrument is not currently commercially available), which is useful for eliminating background fluorescence in an assay.

29.4.3 EXPERTISE

Despite their long history of use in science, optical magnification instruments such as microscopes are complicated precision instruments. Accurate data collection requires that users be trained in proper instrument operation and calibration. As discussed above, expertise in observing cellular phenotypes will also be a critical requirement for interpreting assay results. Lastly, some knowledge of image processing is critical for assessing and improving the accuracy of the analysis. Many manufacturers provide turn-key image processing methods that promise to provide one-button solutions for specific problems. However, it is extremely difficult to design one image processing routine that takes into account the entire range of biological variation that can be encountered in drug discovery, especially when there is a shift in cell types or assay conditions. The ability to perform basic customization of image processing methods is key to result accuracy.

29.5 ASSAY VALIDATION

29.5.1 Instrument Calibration

29.5.1.1 Flat Field Correction

Lenses are curved, and the light intensity that they transmit falls off toward their edges, a phenomenon termed "roll-off." Any image taken with a lens will exhibit an unevenness in light intensity across its extent (usually lower in the corners), and this will affect the quantification of brightness values within it. Misalignment of the light source (laser or lamp) will also affect the evenness of the illumination across the image. If the assay biology requires comparison of intensity values between wells or measures distribution of the population intensity values within a well, the images should be corrected for unevenness. Termed "flat field correction," this is usually accomplished by imaging a portion of the assay plate of uniform intensity that does not contain any features such as cells and calculating the unevenness across this image. The resulting offset is stored in a lookup table and is used to correct all of the subsequent images, usually automatically (provided the user specifies this procedure is to be part of the image analysis method). Due to variations in light source alignment and intensity, it is good practice to collect this set of flat field images before each analysis is performed.

29.5.1.2 Quantitative Intensity and Color Calibration

Less frequently, measurement of the absolute intensity or color of a particular biomarker is required. In these cases, images of standards of known intensity (Matech fluorescent plate standards) or color (Molecular Probes TetraSpeck beads) are used to calibrate images. Adjustments to illumination output or fluorescence filters may be required.

29.5.1.3 Computer Monitor Calibration

In situations where physicians must make diagnoses based on digital images, calibration standards for computer monitor output are used routinely, and image-based assays are no different. The intensity outputs and gray level display of different computer monitors is profoundly different and may result in substantial differences in the visibility of certain cellular details. The computer can distinguish among over 4000 gray levels with precision accuracy, but human eyes can detect only a fraction of this. Ambient lighting levels and the age of the observer can also affect what is visible. Assessment of image processing method accuracy against the gold standard of human observations requires that the conditions under which the human makes these assessments be as uniform and sensitive as possible. Although individual preferences may vary, it is good practice to use a bright monitor in a darkened room and allow your eyes to dark adapt for several minutes before beginning assessment work. Avoid frequent transitions between the darkened room and adjacent normally lit rooms.

29.5.2 Algorithm Validation

In addition to the optimization of the assay biology itself, time must be allowed for the validation of the image analysis method, sometimes colloquially referred to as an "algorithm." Typically, this begins with the development of several alternate algorithms and proceeds through the assessment of their performance on a pilot screen. Once an algorithm is chosen, the precise statistical parameters to be calculated should be determined, and the algorithm performance is then assessed on full screens. Frequently the assay biology and the analysis method are modified in an iterative manner: slight modifications to the assay biology are made to optimize algorithm accuracy, and the algorithm must then be slightly modified itself to adjust for the modified biology, etc. All of this adds time onto the already compressed time scales of assay development and should be taken into account when planning an image-based assay development campaign.

First assess the object identification algorithm's accuracy. An object count made by a human observer is compared to the object count provided by the algorithm. Does it separate cells that are touching? Does it recognize oddly shaped cells as a single object? Then assess the measured intensity values or textural values for accuracy. Finally, assess whether the measurements separate the population into realistic categories. Are 50% of the cells actually not responding to the treatment?

29.6 SUMMARY

The ultimate arbiter of assay technology utility is often stated as a question: "How does this give us better drugs faster?" There are several examples of image-based assays that are currently impacting small-molecule drug discovery. Most notably, image-based assays are enabling entirely new categories of targets to be accessed for small-molecule drug discovery and in some cases bringing new advantages to the exploration of familiar mechanisms of disease processes:

- New target classes: GPCR desensitization inhibitors
  - Xsira Pharmaceuticals Inc (formerly Norak Biosciences, Inc.) has developed a pro-prietary assay that measures GPCR desensitization by detecting fluorescent beta arrestin translocation from the cytoplasm to pits and vesicles associated with GPCR internalization [3]. They are currently in preclinical studies with small molecule desensitization inhibitors and nondesensitizing ligands.
- New targets with familiar mechanisms: cytoskeletal disruptors
  - Cytokinetics, Inc. is a pioneer in image-based assays and has utilized this as a core technology since its inception. Utilizing assays that imaged the process of cell division led directly to the discovery of SB-715992 and SB-743921, kinesin spindle protein disruptors that are currently in clinical trials.
  - Using an image-based assay for cell division, investigators at the Institute of Chemistry and Cell Biology discovered Monastrol, an inhibitor of the mitotic kinesin Eg5 [4].
- New target classes: translocation inhibitors
  - BioImage A/S has developed a series of proprietary assays that track the cellular redistribution of fluorescently labeled proteins within the cell. Interleaving these assays allows the responsiveness of entire pathways to be mapped. They have used these assays to discover small-molecule inhibitors of Crm1-dependent nuclear export of MK2 and nuclear export of FKHR, along with upstream inhibitors of the p38 and PI3K activation pathways and are currently progressing these through preclinical development [5,6].

There are many excellent sources of information on imaging available on the Web. For further information, readers are encouraged to participate in imaging community forums such as the ImageNet mailing list (http://imageanalysis.info).

REFERENCES

1. ©2003 Susan Catalano.
2. Cokol M, Nair R, Rost B. (2000) Finding nuclear localization signals. *EMBO Rep* 1(5):411–415.
3. Oakley RH, Hudson CC, Cruickshank RD, Meyers DM, Payne RE Jr, Rhem SM, Loomis CR. (2002) The cellular distribution of fluorescently labeled arrestins provides a robust, sensitive, and universal assay for screening G protein coupled receptors. *Assay Drug Dev Technol* 1:21–30.

4. Mayer TU, Kapoor TM, Haggarty SJ, King RW, Schreiber SL, Mitchison TJ. (1999) Small-molecule inhibitor of mitotic spindle bipolarity identified in a phenotype-based screen. *Science* 286:971–974.
5. Lundholt BK, Linde V, Loechel F, Pedersen HC, Moller S, Praestegaard M, Mikkelsen I, Scudder K, Bjorn SP, Heide M, Arkhammar PO, Terry R, Nielsen SJ. (2005) Identification of Akt pathway inhibitors using redistribution screening on the FLIPR and the IN Cell 3000 analyzer. *J Biomol Screen* 10:20–29.
6. Almholt DL, Loechel F, Nielsen SJ, Krog-Jensen C, Terry R, Bjorn SP, Pedersen HC, Praestegaard M, Moller S, Heide M, Pagliaro L, Mason AJ, Butcher S, Dahl SW. (2004) Nuclear export inhibitors and kinase inhibitors identified using a MAPK-activated protein kinase 2 redistribution screen. *Assay Drug Dev Technol* 2:7–20.

BIBLIOGRAPHY

Abraham V, Samson B, Lapets O, Haskins JR. (2004) Automated classification of individual cellular responses across multiple targets. *Preclinica* 2:349–355.

Abraham VC, Haskins JR. (2004). Utilizing automated classification of cellular responses to investigate apoptosis. *Biosci Tech* 29:8–10.

Arden, SR, Janardhan P, DiBiasio R, Arnold B, Ghosh RN. (2002) An automated quantitative high content screening assay for neurite outgrowth. *Chemistry Today* 20:64–66.

Bhawe, KM, Blake RA, Clary DO, Flanagan PM. (2004) An automated image capture and quantitation approach to identify proteins affecting tumor cell proliferation. *J Biomol Screen* 9:216–222.

Blumberg, H, Conklin D, Xu WF, Grossmann A, Brender T, Carollo S, Eagan M, Foster D, Haldeman BA, Hammond A, Haugen H, Jelinek L, Kelly JD, Madden K, Maurer MF, Parrish-Novak J, Prunkard D, Sexson S, Sprecher C, Waggie K, West J, Whitmore TE, Yao L, Kuechle MK, Dale BA, Chandrasekher YA. (2001) Interleukin 20: discovery, receptor identification, and role in epidermal function. *Cell* 104:9

Breider MA, Gough AW, Haskins JR, Sobocinski G, de la Iglesia FA. (1999) Troglitazone-induced heart and adipose tissue cell proliferation in mice. *Toxicol Pathol* 27:545–552.

Chen J, Zhong Q, Wang J, Cameron RS, Borke JL, Isales CM, Bollag RJ. (2001) Microarray analysis of Tbx2-directed gene expression: a possible role in osteogenesis. *Mol Cell Endocrinol* 177:43–54.

Conway BR, Minor LK, Xu JZ, D'Andrea MR, Ghosh RN, Demarest KT. (2001) Quantitative analysis of agonist-dependent parathyroid hormone receptor trafficking in whole cells using a functional green fluorescent protein conjugate. *J Cell Physiol* 189:341–355.

Conway BR, Minor LK, Xu JZ, Gunnet JW, DeBiasio R, D'Andrea MR, Rubin R, DeBiasio R, Giuliano K, DeBiasio L, Demarest KT. (1999) Quantification of G-protein–coupled receptor internalization using G-protein–coupled receptor green fluorescent protein conjugate with the ArrayScan High Content Screening System. *J Biomol Screen* 4:75–86.

Ding, GJF, Fischer PA, Boltz RC, Schmidt JA, Colaianne JJ, Gough A, RA Rubin RA, Miller DK. (1998) Characterization and quantitation of NF-kB nuclear translocation induced by interleukin-1 and tumor necrosis factor-α. *J Biol Chem* 273:28,897–28,905.

Gasparri F, Mariani M, Sola F, Galvani A. (2004) Quantification of the proliferation index of human dermal fibroblast cultures with the ArrayScan high content screening reader. *J Biomol Screen* 9:232–243.

Ghosh RN, Grove L, Lapets O. (2004) A quantitative cell-based high-content screening assay for the eperidermal growth factor receptor-specific activiation of mitogen-activated protein kinase. *Assay Drug Dev Tech* 2:473–481.

Ghosh, RN, Chen Y-T, DeBiasio R, DeBiasio RL, Conway BR, Minor LK, Demarest KT. (2000) Cell-based, high content screen for receptor internalization, recycling, and intracellular trafficking. *Biotechniques* 29:170–175.

Giuliano KA. (2003) High-content profiling of drug–drug interactions: cellular targets involved in the modulation of microtubule drug action by the antifungal ketoconazole. *J Biomol Screen* 8:125–135.

Giuliano KA, Haskins JR, Taylor DL. (2003) Advances in high-content screening for drug discovery. *Assay Drug Devel Technol* 1:565–577.

Giuliano KA, DeBiasio RL, Dunlay RT, Gough A, Volosky JM, Zock J, Pavlakis GN, Taylor DL. (1997) High content screening: a new approach to easing key bottlenecks in the drug discovery process. *J Biomol Screen* 2:249–259.

Guiliano K, Chan Y-T, Taylor DL. (2004) High-content screening with siRNA optimizes a cell biological approach to drug discovery: defining the role of p53 activation in the cellular response to anticancer drugs. *J Biomol Screen* 9:557–568.

Haskins JR, Rowse P, Rahbari R, de la Iglesia FA. (2001) Thiazolidinedione toxicity to isolated hepatocytes revealed by coherent multiprobe fluorescence microscopy and correlated with multiparameter flow cytometry of peripheral leukocytes. *Arch Toxicol* 75:425–438.

Honma K, Ochiyab T, Nagaharac S, Sanoc A, Yamamotob H, Hiraib K, Asoa Y, Teradab M. (2001) Atelocollagen-based gene transfer in cells allows high-throughput screening of gene functions. *Biochem Biophys Res Commun* 289:1075–1081.

Kain SR. (1999) Green fluorescent protein (GFP): applications in cell-based assays for drug discovery. *Drug Discov Today* 4:304–312.

Kapur R. (2002) Fluorescence imaging and engineered biosensors: functional and activity-based sensing using high-content screening. *Ann NY Acad Sci* 961:196–197.

Kawamura YI, Kawashima R, Shirai Y, Kato R, Hamabata T, Yamamoto M, Furukawa K, Fujihashi K, McGhee JR, Hayashi H, Dohi T. (2003) Cholera toxin activates dendritic cells through dependence on GM1-ganglioside which is mediated by NF-kB translocation. *Eur J Immunol* 33:3205–3212.

Li Z, Yan Y, Powers EA, Ying X, Janjua K, Garyantes T, Baron B. (2003) Identification of gap junction blockers using automated fluorescence microscopy imaging. *J Biomol Screen* 8:489–499.

Lovborg H, Nygren P, Larsson R. (2004) Multiparametric evaluation of apoptosis: effects of standard cytotoxic agents and the cyanoguanidine CHS 828. *Mol Cancer Ther* 3:521–526.

Mastyugin V, McWhinnie E, Labow M, Buxton F. (2004) A quantitative high-throughput endothelial call migration assay. *J Biomol Screen* 9:712–718.

Minguez JM, Giuliano KA, Balachandran R, Madiraju C, Curran DP, Day BW. (2002) Synthesis and high-content screening of simplified analogues of the potent microtubule stabilizer (+)-discodermolide. *Mol Cancer Ther* 1:1305–1313.

Minguez JM, Kim SY, Giuliano KA, Balachandran R, Madiraju C, Day BW, Curran DP. (2003) Synthesis and biological assessment of simplified analogues of the potent microtubule stabilizer (+)-discodermolide. *Bioorg Med Chem* 11:3335–3357.

Nam J-S, Ino Y, Kanai Y, Sakamoto M, Hirohashi S. (2004) 5-aza-2′-deoxycytidine restores the E-cadherin system in E-cadherin-silenced cancer cells and reduces cancer metastasis. *Clin Exp Metastasis* 21:49–56.

Richards GR, Millard RM, Leveridge M, Kerby J, Simpson PB. (2004) quantitative assays of chemotaxis and chemokinesis for human neural cells. *Assay Drug Dev Tech* 2: 465–472.

Schlag BD, Lou Z, Fennell M, Dunlop J. (2004) Ligand dependency of 5-HT2C receptor internalization. *J Pharmacol Exp Ther* 310:865–870.

Simpson PB, Bacha JI, Palfreyman EL, Woollacott AJ, McKernan RM, Kerby J. (2001) Retinoic acid–evoked differentiation of neuroblastoma cells predominates over growth factor stimulation: an automated image capture and quantitation approach to neuritogenesis. *Anal Biochem* 298:163–169.

Tencza SB, Sipe MA. (2004) Detection and classification of threat agents via high-content assays of mammalian cells. *J Appl Toxicol* 24:371–377.

Vakkila J, DeMarco RA, Lotze MT. (2004) Imaging analysis of STAT1 and NF-kB translocation in dendritic cells at the single cell level. *J Immunol Meth* 294:123–134.

Vogt A, Kalb EN, Lazo JS. (2004) A scalable high-content cytotoxicity assay insensitive to changes in mitochondrial metabolic activity. *Oncol Res* 14:305–314.

Vogt A, Cooley KA, Brisson M, Tarpley MG, Wipf P, Lazo JS. (2003) Cell-active dual specificity phosphatase inhibitors identified by high-content screening. *Chem Biol* 10:733-742.

Vogt A, Adachi T, Ducruet AP, Chesebrough J, Nemoto K, Carr BI, Lazo JS. (2001) Spatial analysis of key signaling proteins by high-content solid phase cytometry in Hep3B cells treated with an inhibitor of cdc2 dual-specificity phosphatases. *J Biol Chem* 276:20,544–20,550.

Wipf P, Reeves JT, Balachandran R, Giuliano KA, Hamel E, Day BW. (2000) Synthesis and biological evaluation of a focused mixture library of analogues of the antimitotic marine natural product curacin A. *J Am Chem Soc* 122:9391.

30 GPCR Internalization Measured by Image Analysis

Robert Graves, Michael J. Francis, Lynne Smith,
Jeffrey R. Cook, and Elaine J. Adie

CONTENTS

30.1 INTRODUCTION

G-protein–coupled receptors (GPCRs) continue to be a primary target for novel drug discovery [1,2]. They comprise seven trans-membrane-spanning helices linked by intracellular and extracellular loops with an extracellular *N*-terminus and an intracellular *C*-terminus [3]. Agonist stimulation of GPCRs activates a range of second messenger pathways via the family of heterotrimeric G-proteins. The agonist-mediated response usually decreases rapidly via a process known as desensitization and often involves GPCRs becoming phosphorylated by G-protein receptor kinases [4]. This allows the binding of an arrestin protein, leading to uncoupling of the GPCRs from their G-proteins and resulting in attenuation of second messenger signaling [5]. Agonist-mediated activation also promotes a rapid sequestration of GPCRs away from the plasma membrane into the endosomal pathway, from where they are sorted either into acidic perinuclear recycling endosomes or into lysosomes for proteolysis (Figure 30.1). Other cell surface–expressed receptors such as tyrosine kinases also internalize into the endosomal pathway after activation, where they are then sorted for recycling or future degradation [6].

The cellular distribution of GPCRs and their interactions with other proteins are readily monitored using a number of fluorescent techniques [7]. Incorporation of epitope tags into the extreme *C*- or *N*-terminal sequence of the receptors allows protein detection via immunoprecipitation or immunoblotting [8–11]. It also enables the application of immunocytochemical techniques to study GPCR distribution [12,13]. The recent discovery of a range of fluorescent proteins from a number of species has greatly enhanced the study of cellular protein function. Green fluorescent protein (GFP) from *Aequorea victoria* has been fused to the *C*-terminal tail of GPCRs, and following expression of such constructs in mammalian cells, agonist-induced internalization has been studied using standard microscopy techniques [14,15].

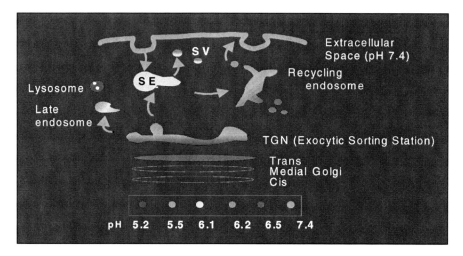

FIGURE 30.1 (See color insert following page 334.) Ligand-induced receptor-mediated endocytosis. A ligand binds to a GPCR, resulting in the ligand–receptor complex being internalized into the endosomal pathway. The receptor can either be recycled to the plasma membrane or degraded in lysosomes (adapted from www.adrenoreceptor.com/recycel.htm).

This chapter describes a technique to monitor the agonist-mediated internalization of GPCRs and other cell surface receptors into acidic endosomal vesicles. The assay uses the pH-sensitive cyanine dye, CypHer 5E (Figure 30.2) from GE Healthcare, which has an absorbance maximum of 644 nm and an emission maximum of 665 nm (Figure 30.3). The dye is minimally fluorescent at pH 7.4 and maximally fluorescent at pH 5.5 [16] and is therefore ideally suited to report the movement of a receptor from the cell surface into internal acidic endosomes [17,18].

The assay is suitable for live or fixed cell formats and has been performed both on automated epifluorescence and laser-based microscopes, such as the IN Cell Analyzer 1000 and 3000 from GE Healthcare, and manual research microscopes such as the Zeiss LSM410 confocal microscope. The assay can be readily adapted to other automated microscope platforms.

30.2 RECEPTOR INTERNALIZATION ASSAY DESIGN

The CypHer 5E assay relies on the intrinsic ability of a cell surface receptor to internalize in response to agonist-mediated stimulation. CypHer 5E labeled antibodies that recognize an exofacial epitope on the receptor can track the receptor's movement from the cell surface to intracellular endosomes via an intensity change and a translocation event. This exofacial epitope can be either a recombinant inserted tag, such as VSV-G and c-myc, or an epitope that is endogenous to the protein/receptor of interest. The CypHer 5E technology has been successfully applied to a broad range of cell lines, receptors (both GPCRS and receptor tyrosine kinases, RTKs), and labeled antibodies (Table 30.1).

In brief, the assay requires an antibody specific to the receptor of interest to be labeled with CypHer 5E. The labeled antibody is added to live cells expressing the receptor, which on agonist addition will internalize into the acidic environment of the cell, resulting in increased fluorescence. The assay results can be observed visually using lamp- or laser-based microscopy platforms, with subsequent analysis using commercially available software packages. To demonstrate the application of CypHer 5E in a high-throughput environment, the IN Cell Analyzer 1000 and 3000 automated microscope platforms have been used. Examples of images acquired and analyzed using the IN Cell Analyzer system with CypHer 5E are detailed.

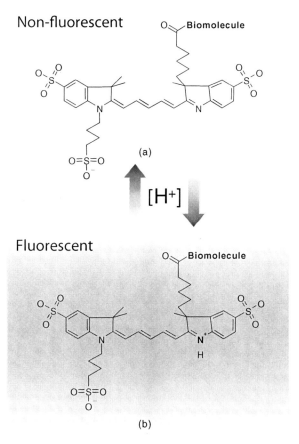

FIGURE 30.2 The structure of CypHer 5E in both pH (a) neutral and (b) acidic environments. The dye is shown conjugated to a biomolecule.

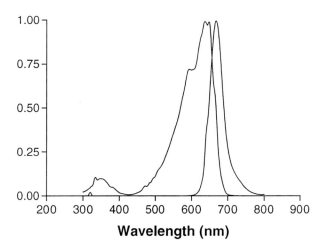

FIGURE 30.3 Absorption and emission profile of CypHer 5E mono NHS ester.

TABLE 30.1
Cell Lines and Antibodies That Have Been Implemented with the CypHer 5E Technology

Cell Line	Epitope Tag	Receptor Class	Receptor
CHO-K1	FLAG	Gs	IP (prostaglandin I) prostanoid receptor
HEK 293	VSV-G	Gs	IP (prostaglandin I) prostanoid receptor
CHO-K1	c-myc	Gi	δ-Opioid receptor
HEK 293	VSV-G	Gi	δ-Opioid receptor
CHO-K1	VSV-G	Gq	Thyrotropin releasing hormone receptor (TRHR)
HEK 293	c-myc	Gq	Thyrotropin releasing hormone receptor (TRHR)
HEK 293	VSV-G	Gq	Angiotensin II
HEK 293	VSV-G	Gs	β2-Adrenergic
HEK 293	c-myc	Gs	β2-Adrenergic
CHO-K1	VSV-G	Gq	Muscarinic M1
CHO-K1	VSV-G	RTK	Epidermal growth factor receptor (EGFR)

30.3 ASSAY — VSV-G-β2-ADRENERGIC RECEPTOR IN HEK 293 CELLS

The β2-adrenergic receptor was modified to incorporate a VSV-G tag at the *N*-terminus using a pCORON1000-VSV-G tag expression vector from GE Healthcare (25-8008-51). A clonal, stable HEK 293 cell line was established, which expressed this receptor at approximately 1.8 pmol/mg cell homogenate. Anti-VSV-G antibodies labeled with CypHer 5E were used to monitor agonist-mediated receptor internalization in these live cells. The assay was performed in the presence and absence of a specific agonist, isoproterenol. This protocol has been designed with a minimum number of steps to enable the assay to be run in high-throughput mode, amenable to automation and drug screening assay formats.

30.3.1 Labeling a Monoclonal Antibody with CypHer 5E Mono NHS Ester Dye

Described below is a general protocol for labeling a monoclonal antibody with CypHer 5E mono NHS ester dye. The materials and procedures have been optimized for one particular preparation of monoclonal IgG antibodies. Other antibodies and proteins may also be labeled; however, the choice of buffers and labeling technique to produce optimal results may vary:

- Antibody originating from mouse ascites fluid should be purified to a single immuno-globulin fraction by column chromatography, for example using a MAb Trap kit from GE Healthcare (17-1128-01), which contains a HiTrap Protein G column. Scan all fractions collected at 280 nm.
- Collect all fractions containing the purified antibody and dialyze in PBS at room temperature for 4 h with three changes of buffer or overnight at 4°C.
- Remove the antibody from the dialysis cassette and determine the protein concentration. It is recommended that the purified antibody be stored at 2 to 8°C for no longer than 1 week.
- Determine the antibody concentration by ultraviolet (UV) absorbance at 280 nm. Use the formula:

$$\frac{\text{Absorbance of antibody at 280 nm}}{\text{Molar extinction coefficient of antibody}} = \text{Molar concentration of antibody}$$

where UV absorbance at 280 nm ($E_{280}{}^{0.1\%} = 1.4$); molecular weight of IgG is assumed to be 150,000; and molar extinction coefficient ($1.4 \times 150,000$) of antibody = 210,000 $M^{-1}cm^{-1}$.

- Dissolve the 1-mg vial of CypHer 5E mono NHS ester dye (GE Healthcare, PA15401) in 100 μL sterile, freshly opened dimethylsulfoxide (DMSO; 500 μL DMSO for a 5-mg vial). Mix and sonicate for a few seconds, maintaining minimal exposure to light.
- Place 5 μL of reconstituted dye into 4 mL PBS in a glass vial and mix well. Scan for absorbance from 240 to 750 nm with 0.125% DMSO as a blank. There should be one peak at 480 nm. Calculate the concentration of dye using the formula:

$$\frac{\text{Absorbance of dye at 480 nm}}{\text{Extinction coefficient of dye}} \times \text{dilution factor (800)} = \text{Molar concentration of dye}$$

Extinction coefficient of CypHer 5E mono NHS ester at 480 nm = 40,000 $M^{-1}cm^{-1}$

- Dilute the antibody to 1 mg/mL in PBS/0.5 M sodium carbonate buffer pH 8.3 (9:1). Note that the rate of hydrolysis of the CypHer 5E mono NHS ester is pH dependent and the optimum pH for the labeling reaction has been carefully determined.
- Calculate the amount of dye to be added to the 1 mg/mL antibody solution using the following equation. A 20-molar excess of CypHer 5E mono NHS ester is recommended:

Molar concentration of antibody × volume of antibody solution × molar excess = Moles of CypHer 5E

Amount of dye to add:

$$\frac{\text{Moles of CypHer 5E}}{\text{Molar conc. CypHer 5E}} = \text{μL CypHer 5E mono NHS ester}$$

- Add the required amount of CypHer 5E mono NHS ester to the 1 mg/mL antibody solution and mix (care must be taken to prevent foaming of the protein mixture). Leave the solution to roll at room temperature in the dark for 1 h.
- Separate labeled antibody from unconjugated dye using overnight dialysis at room temperature. Dialysis at 4°C is not recommended because increased precipitation can occur.
- Remove the labeled protein solution and dilute a portion so that the maximum absorbance is 0.5 to 1.5 AU at 280 and 480 nm.
- Calculate the molar concentrations of dye and proteins as described below. The ratio of these values is the average number of dye molecules coupled to each protein molecule.
- Estimate the final dye/protein (D/P) ratio. The calculation is corrected for the absorbance of the CypHer 5E dye at 280 nm. The factor of 0.2 accounts for this absorption:

$$\text{D/P} = A_{DYE}\, \varepsilon_{PROTEIN} / (A_{280} - 0.2A_{DYE})\, \varepsilon_{DYE}$$

where A_{DYE} is the absorbance of the labeled dye at 480 nm at pH 7.4; A_{280} is the absorbance of the antibody at 280 nm; $\varepsilon_{PROTEIN}$ is the molar extinction coefficient of the antibody (210,000 $M^{-1}cm^{-1}$ at 280 nm); and ε_{DYE} is the molar extinction coefficient of the dye (40,000 $M^{-1}cm^{-1}$ at 480 nm).

- Dilute the labeled antibody to 0.5 mg/mL with PBS containing 0.1% IgG-free BSA (Sigma, w/v 1 mg/mL), and remove any precipitate by centrifugation at 20,000 r/min for 2 min. The presence of BSA enhances the freezing process and increases antibody stability. It is not recommended to add sodium azide to the storage buffer since it may be toxic to the cells under study.
- Dispense into suitable aliquots and freeze (−20°C). Avoid repeated freeze/thaw cycles, and protect from light.

Optimization of the CypHer 5E mono NHS ester dye:Ab ratio is important in order to avoid the presence of unwanted precipitate. A high background level of fluorescence (in the absence of agonist) may be observed if excess precipitate is present. When labeling an anti-VSV-G antibody with CypHer 5E, optimal assay results were obtained using anti-VSV-G antibody labeled with approximately seven to 12 dyes.

30.3.2 Cell Culture and Reagents

- HEK 293 cells expressing a VSV-G-tagged β2-adrenergic receptor are grown in MEM medium (Sigma, M2279) supplemented with 2 mM glutamine, penicillin and streptomycin, 10% fetal calf serum, and 0.2 mg/mL G-418 Geneticin.
- For HEK 293 cells it is recommended to coat the plates with poly-D-lysine prior to seeding cells. Dissolve 5 mg poly-D-lysine (Sigma, P6407) in 50 mL sterile PBS, add 50 μL per well to a 96-well Packard Viewplate, and leave for 45 min.
- Aspirate the coating solution and wash 4 times with 100 μL sterile PBS. Coated plates can be stored at 4°C for up to 1 week in PBS.
- Aspirate the PBS from the plate and add 100 μL per well of a 1×10^5 cells/mL solution (10,000 cells per well). Cultured cells can be added to the wells directly without first drying the plates. Incubate overnight at 37°C in a humidified atmosphere with 5% CO_2.
- The following day the cells should be 50 to 60% confluent for the assay.
- Reconstitute 250 μg lyophilized antimouse VSV-G CypHer 5E labeled antibody (GE Healthcare, PA 45407) with 0.5 mL deionized water (giving 0.5 mg antibody/mL). Mix thoroughly and centrifuge at 10,000 r/min for 2 min.
- Further dilute the reconstituted antibody to 5 μg/mL by adding 10 μL (0.5 mg/mL) to 990 μL serum-free and phenol red–free MEM medium (Invitrogen, 51200-038).
- Hoechst 33342 (Molecular Probes) can be added to the 5 μg/mL antibody solution in serum-free MEM to a final concentration of 5 μM. Alternatively, it can be added during the agonist addition as in assay example 1.
- Prepare 30 μM (3×) working solutions of isoproterenol (Sigma I5627) in serum-free MEM from a 10 mM isoproterenol stock solution in water. If a dose–response experiment is to be performed make an appropriate 3× concentrated dilution series.

30.3.3 Assay Procedure

The assay procedure is as follows:

- Aspirate the medium from the wells and add 100 μL antibody plus Hoechst solution to each well (from step 8). Incubate at room temperature for 15 min.
- Add 50 μL isoproterenol dilutions from step 9 giving 0 to 10 μM final in-well concentrations.
- Incubate for 30 min at 37°C in a humidified atmosphere with 5% CO_2.

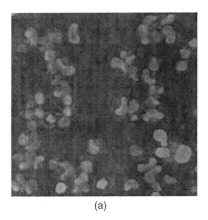

(a)   (b)

FIGURE 30.4 VSV-G β2-adrenergic receptor expressed in HEK 293 cells (a) before and (b) 30 min after stimulation with 10 μM isoproterenol agonist. (b) Receptor internalization is shown with CypHer 5E–labeled anti-VSV-G antibody by an increase in punctuate staining; cell nuclei are stained with 5 μM Hoechst 33342. Images were taken on the IN Cell Analyzer 3000 using a 40× objective lens.

- If required, fix the cells using a standard formaldehyde fixation protocol.
- Image the cells as described.

30.3.4 RESULTS AND IMAGE ANALYSIS

Translocation to the perinuclear recycling endosomes was observed in HEK 293 cells expressing a VSV-G tagged β2-adrenergic receptor after isoproterenol treatment, with a corresponding marked increase in CypHer 5E fluorescence (Figure 30.4). This relates to an agonist-specific internalization of CypHer 5E dye.

Numerous image analysis programs now exist with which to generate measurable data output defining cellular processes. An example of this is the analysis software that is used with the IN Cell Analyzer 1000 and 3000 platforms. This modular software enables multicellular processes to be defined. Figure 30.5 outlines the steps required to analyze images generated from a CypHer 5E assay using the IN Cell Analyzer 1000 granularity analysis module.

The granularity analysis module uses two channels, a marker channel and a signal channel. The marked objects are identified through a nuclear stain. In the signal or granule channel, granule definition is defined based on size and intensity. Following image analysis, colored bitmaps are automatically superimposed over acquired images to aid the visualization of the analysis results (Figure 30.6). The granularity analysis module reports population-averaged measurements for a range of parameters, including average granule area per cell, average granule count per cell, and number of nuclei per image.

The IN Cell Analyzer 3000 system enables both population and single cell information to be generated. The 3000 software also identifies each cell as an object by its nuclear stain. A boundary box is created around the dilated nucleus defining the individual cellular area. This box is searched for granules of a user-defined intensity and size. In this manner the number of granules associated with individual cells are calculated (Figure 30.7).

A sigmoidal dose response curve was obtained and is shown in Figure 30.8. To further demonstrate the capability of the CypHer 5E assay to generate pharmacologically relevant data, rank order potency experiments for various β2-adrenergic receptor agonists were performed. The calculated EC_{50} values and rank order, isoproterenol, 16 nM ≥ fenoterol, 15 nM < epinephrine, 35 nM < salbutamol, 85 nM < terbutaline, 280 nM < norepinephrine, 705 nM, are comparable to literature values.

Analysis Flowchart

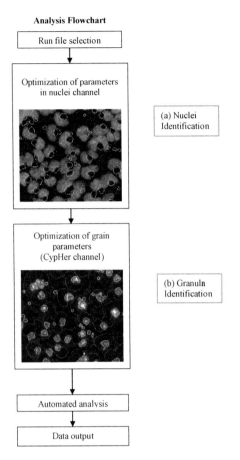

FIGURE 30.5 Flow diagram outlining the steps required to analyze images using the IN Cell Analyzer 1000 granularity analysis module. After acquisition, the run file is selected. (a) Nuclear parameters are defined to identify the nuclei. (b) Granule parameters are defined to identify grains. Automated analysis of the images provides data output.

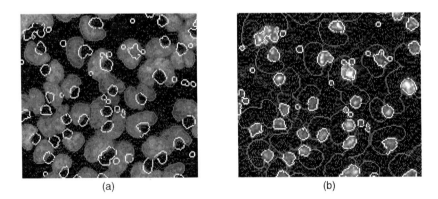

FIGURE 30.6 Postanalysis bitmap images demonstrating how the granularity algorithm analyzes CypHer 5E internalization. The larger outlines are for nuclei and the smaller outlines for granules. (a) The nuclei image channel and (b) granule image channel. The bitmaps identifying the nuclei and granules are shown in both channels.

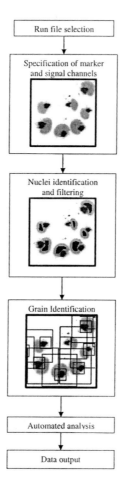

FIGURE 30.7 Flow diagram outlining the steps required to analyze images using the IN Cell Analyzer 3000 analysis module.

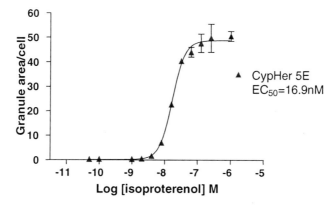

FIGURE 30.8 Quantification of the CypHer 5E dye (VSV-G-β2-adrenergic receptor) internalization using the granularity algorithm. Quantification of the agonist-mediated response was achieved using the IN Cell Analyzer 1000 granularity module.

30.4 TROUBLESHOOTING GUIDE

The following table describes possible problems with the assay, their causes, and remedies.

Problem	Possible Cause	Remedy
Low assay signal	Cell line passage number too high	Start a fresh batch of cells from an earlier passage number
	Incorrect incubation conditions	Ensure that recommended incubation conditions are maintained
	Antibody labeled with too little or too much CypHer 5E	Relabel antibody with the correct proportions of CypHer 5E
	Reagents were not stored properly or are out of date	Repeat assay with fresh reagents; titer reagents to determine optimal dilutions
	Analysis algorithm may not be optimized correctly	Check analysis algorithm parameters are correct
Low nuclear intensity	Counterstain concentration too low	Adjust counterstain concentration to recommended level
	Counterstain incubation time too short	Adjust counterstain incubation time to recommended length
Image is out of focus	Autofocus offset is chosen incorrectly	Change autofocus offset; check plate settings in software; check correction collar adjustment on objective lens

30.5 SUMMARY

CypHer 5E is a novel technology that enables study of receptor activation and internalization across a broad range of GPCRs as well as other receptors that are internalized via clathrin into acidic vesicles. CypHer 5E can be applied to both epifluorescence and laser-based microscopy platforms. The availability of high-throughput cellular imaging platforms, such as the IN Cell Analyzer 1000 and 3000, also enables CypHer 5E to be utilized as a generic method for high-throughput primary and secondary screening applications.

REFERENCES

1. Stadel J. M., S. Wilso, and D. J. Bergsma. (1997). Orphan G-protein–coupled receptors: a neglected opportunity for pioneer drug discovery. *TIPS*, 18: 430–436.
2. Wilson, S., D. J., Bergsma, J. K. Chambers, A. I. Muir, K. G. M. Fantom, C. Ellis, P. R. Murdock, N. C. Herrity, and J. M. Stadel. (1998). Orphan G-protein–coupled receptors: the next generation of drug targets? *Br. J. Pharmacol.* 125: 1387–1392.
3. Vaughan, M. (1998). G-Protein-coupled receptors minireview series. *J. Biol. Chem.* 273: 17,297–18,680.
4. Ferguson S. S. G. (2001). Evolving concepts in G-protein–coupled receptor endocytosis: the role in receptor desensitization and signalling. *Pharmacol. Rev.* 53: 1–24.
5. Perry, S. J. and R. J. Lefkowitz. (2002). Arresting developments in heptahelical receptor signalling and regulation. *Trends Cell Biol.* 12: 130–138.
6. Koenig, J. A. and J. M. Edwardson. (1997). Endocytosis and recycling of G-protein–coupled receptors. *TIPS* 18: 276–287.
7. Hovius R., P. Vallotton, T. Wohland, and H. Vogel. (2000). Fluorescence techniques: shedding light on ligand-receptor interactions. *TIPS* 21: 266–273.

8. Weill C., J-L. Galzi, S. Chasserot-Golaz, M. Goeldner, and B. Ilien. (1999). Functional characterization and potential applications for enhanced green fluorescent protein- and epitope-fused human M1 muscarinic receptors. *J. Neurochem.* 73: 791–801.

9. Peters D. M., E. K. Gies, C. R. Gelb, and R. A. Peterfreund. (1998). Agonist-induced desensitization of A2B adenosine receptors. *Biochem. Pharamcol.* 55: 873–882.

10. Roy S. F., S. A. Laporte, E. Escher, R. Leduc, and G. Guillemette. (1997). Epitope tagging and immunoreactivity of the human angiotensin II type 1 receptor. *Can. J. Physiol. Pharmacol.* 75: 690–695.

11. Gomes I., B. A. Jordan, A. Gupta, C. Rios, N. Trapaidze, and L. A. Devi. (2001). G-protein-coupled receptor dimerization: implications in modulating receptor function. *J. Mol. Med.* 79: 226–242.

12. McVey M., D. Ramsay, E. Kellett, S. Rees, S. Wilson, A. J. Pope, and G. Milligan. (2001). Monitoring receptor oligomerization using time-resolved fluorescence energy transfer and bioluminescence resonance energy transfer. *J. Biol. Chem.* 276: 14,092–14,099.

13. Oakley R. H., S. A. Laporte, J. A. Holt, L. S. Barak and M. G. Caron. (2001). Molecular determinants underlying the formation of stable intracellular G-protein-coupled receptor–arrestin complexes after receptor endocytosis. *J. Biol. Chem.* 276: 19,452–19,460.

14. Drmota T., G. W. Gould, and G. Milligan. (1998). Real-time visualization of agonist-mediated redistribution and internalization of a green fluorescent protein-tagged form of the thyrotropin-releasing hormone receptor. *J. Biol. Chem.* 273: 24,000–24,008.

15. Kallal L. and J. L. Benovic. (2000). Using green fluorescent proteins to study G-protein–coupled receptor localization and trafficking. *TIPS* 21: 175–180.

16. Briggs M. S., D. D. Burns, M. E. Cooper, and S. J. Gregory. (2000). A pH-sensitive fluorescent cyanine dye for biological applications, *J. Chem. Soc. Chem. Commun.* 23: 2323–2324.

17. Adie, E.J., S. Kalinka, L. Smith, M. J. Francis, A. Marenghi, M. E. Cooper, M. Briggs, N. P. Michael, G. Milligan, and S. Game. (2002). A pH-sensitive fluor, CypHer 5, used to monitor agonist-induced G-protein–coupled receptor internalization in live cells. *BioTechniques* 33: 1152–1157.

18. Adie, E. J., M. J. Francis, J. Davies, L. Smith, A. Marenghi, C. Hather, K. Hadingham, N. P. Michael, G. Milligan, and S. Game (2003). CypHer 5: a generic approach for measuring the activation and trafficking of G-protein–coupled receptors in live cells. *Assay Drug Devel. Technol.* 1: 251–259.

31 TRANSFLUOR® Provides a Universal Cell-Based Assay for Screening G-Protein–Coupled Receptors

Robert H. Oakley, Conrad L. Cowan, Christine C. Hudson, and Carson R. Loomis

CONTENTS

31.1 INTRODUCTION

G-protein–coupled receptors (GPCRs) make up a superfamily of membrane spanning proteins that mediate a large variety of physiological events throughout the body. Of the nearly 1000 members, approximately 200 have a known ligand and function [1,2]. The remaining members are termed orphan receptors because their natural ligands and physiology remain unknown. Ligand binding induces a conformational change in GPCRs that leads to coupling of the receptor with intracellular heterotrimeric guanine nucleotide binding proteins (G proteins). G protein binding leads to the exchange of guanosine triphosphate (GDP) for guanosine diphosphate (GTP) on the G_α subunit of the G protein. The GTP-primed G_α subunit dissociates both from its neighboring $G_{\beta\gamma}$ subunits and

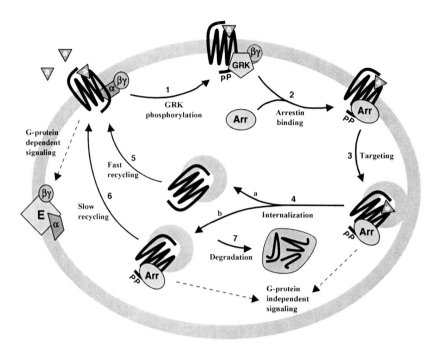

FIGURE 31.1 Model of GPCR desensitization and resensitization. Upon binding of agonist, GPCRs are phosphorylated (P) on serine/threonine residues in their third intracellular loop and/or carboxyl-terminal tail (*step 1*). Arrestins (Arr) are cytosolic proteins that translocate to and bind the agonist-occupied, GRK-phosphorylated receptor at the plasma membrane (*step 2*). By interacting with components of the internalization machinery, arrestins target the desensitized receptors to clathrin-coated pits for endocytosis (*step 3*). The receptors are internalized (*step 4*) and either recycle rapidly (*step 5*), recycle slowly (*step 6*), or are degraded in lysosomes (*step 7*). G-protein–dependent and G-protein–independent signaling pathways are indicated by dashed arrows.

from the receptor. Both the G_α and $G_{\beta\gamma}$ protein subunits interact with many intracellular effector proteins leading to the generation of second messengers, stimulation of multiple signal transduction pathways, and generation of receptor-specific cellular responses [3,4].

Nearly all GPCRs rapidly lose their ability to signal following sustained or repeated ligand activation. This waning or dampening of GPCR signaling is termed desensitization [5–7]. Desensitization is an adaptive mechanism that protects cells from acute and chronic overstimulation and is responsible for the decrease in drug effect with repeated dosing. This important regulatory feature of GPCRs occurs through a series of distinct cellular events shown schematically in Figure 31.1. After activating the G protein, the receptor is phosphorylated on serine and threonine residues located within the third intracellular loop and/or carboxyl-terminal tail (*step 1*). Phosphorylation is accomplished by a seven-member family of G protein-coupled receptor kinases (GRKs) [7–10]. This family can be divided into three subfamilies based on sequence and function. GRK1 and GRK7 form a distinct group that is only found in retinal cells. Of the nonvisual GRKs, GRK2 and GRK3 are cytosolic proteins recruited to the agonist-activated GPCR at the cell surface whereas GRK4, GRK5, and GRK6 comprise a unique subfamily that constitutively resides at the plasma membrane. Tissue distribution and levels of expression contribute to observed receptor specificity of the different GRKs.

Following GRK-mediated receptor phosphorylation, a cytosolic protein known as arrestin binds the receptor at the plasma membrane (*step 2*) [7]. Arrestins comprise a family of four proteins with visual arrestin (arrestin1) and cone arrestin (arrestin4) found only in the eye. The nonvisual arrestins, βarrestin1 (arrestin2) and βarrestin2 (arrestin3), are ubiquitously expressed throughout the body.

Binding of a single arrestin to the activated receptor sterically prevents further receptor–G-protein coupling, and thereby results in the arrest or termination of G-protein–dependent signaling (desensitization). By recruiting other proteins to the receptor, such as AP-2 and clathrin, arrestins target the desensitized receptors to clathrin-coated pits for endocytosis (steps 3 and 4) [11–13]. Once internalized, receptors are dephosphorylated and agonist is removed due to the acidic environment of endosomes [14–16]. The internalized receptors are then recycled back to the plasma membrane as fully competent GPCRs capable of responding again to agonist (steps 5 and 6). The rate of receptor recycling appears to be governed by the stability of the GPCR/arrestin interaction [17]. Not all GPCRs are recycled, however, as some are directed to lysosomes for degradation, leading to an overall reduction in receptor number (down-regulation) (step 7) [18,19].

Arrestins, by blocking signaling and controlling receptor internalization and recycling, play a critical role in the regulation of both the magnitude and duration of the receptor response. In addition to their role in terminating G protein–dependent signaling, arrestins have also been shown recently to function as signal transducers by recruiting other signaling molecules to the agonist-occupied GPCRs [20,21]. For example, arrestins have been shown to bind members of the Src tyrosine kinase family and to interact with components of the MAP kinase pathway. By recruiting these proteins to receptors, arrestins serve as scaffolding proteins that both facilitate and localize the receptor-mediated stimulation of MAP kinase activity and in this way initiate a second wave of G protein–independent signaling.

Multiple diseases and disorders have been linked to mutations and polymorphisms in GPCRs [22,23], and they are the targets of an increasingly large number of therapeutic agents. It has been estimated that 50% of all modern drugs and almost 25% of the top 200 best-selling drugs modulate GPCR activity [2,24]. GPCRs make excellent drug targets because their signaling induces highly specific and potent cellular changes, yet they are easily accessible to extracellular therapeutic agents [25]. Because of the success of this target class, the nonsensory group of orphan receptors (numbering approximately 160 of the 800 or so orphans) represents an untapped potential for new therapeutic discoveries [2,26]. Current drug discovery efforts for GPCRs are hampered by the lack of good structural information on these complex, membrane-bound proteins. Thus, drug discovery is based primarily on screening large diverse compound collections to identify novel chemical molecules followed by lead optimization. In contrast to other target classes, such as kinases and proteases, the lack of structural information on GPCRs increases the dependence of the discovery process on interpretation of screening data. Today, GPCR screening relies primarily on functional assays that measure a downstream signaling molecule (calcium, IP3, or cAMP) or reporter assays that measure changes in gene expression.

Transfluor is a novel GPCR screening platform that utilizes an arrestin–green fluorescent protein conjugate (arrestin-GFP) to detect ligand interactions with GPCRs. As described above and in Figure 31.1, arrestins reside in the cytoplasm and translocate to and bind agonist-activated GPCRs at the plasma membrane. By labeling the arrestin protein with GFP, the redistribution of arrestin to agonist-occupied receptors at the cell surface can be quantitated in live cells, thereby providing a proximal and sensitive readout for the activation and inactivation of GPCRs [27,28]. Consistent with the universality of desensitization, arrestin-GFP has been shown to translocate to nearly all GPCRs independent of the receptor signaling pathway, the receptor ligand, and the receptor sequence and structure (Figure 31.2). This feature of Transfluor has made it a particularly attractive platform for screening orphan GPCRs since no prior knowledge of the interacting G protein is required. Indeed, three orphan GPCRs in *Drosophila* have recently been deorphanized using Transfluor [29]. Transfluor is also less susceptible to false positives than classic signaling and reporter gene assays because arrestin binding is only two steps removed from ligand binding. Finally, Transfluor serves as a high-content screening (HCS) platform because image analysis of the translocation response allows additional information to be obtained on the nature and activity of the tested compounds. In this chapter, we describe the methodologies and optimization steps

Gs

A2a adenosine[45]
A2b adenosine[46-48]
β1-adrenergic[49]
β2-adrenergic[17,27,28,33]
CRF1 corticotropin rel. factor[50]
D1 dopamine[27,32,33]
D5 dopamine[50]
FSH follicle-stim. hormone[51]
Glucagon[50]
LH luteinizing hormone[52]
Octopamine(*Drosophila*)[50]
PTH1 parathyroid hormone[53-55]
E2 prostaglandin[30]
E4 prostaglandin[48,56]
Secretin[57]
VIP1 vasoactive intest. pept.[58]
V2 vasopressin[17,28,33]

Also

D6 chemokine[68]
Fz4 frizzled[86]

Gi/o

α2a-adrenergic[59]
α2b-adrenergic[59]
α2c-adrenergic[59]
A1 adenosine[60]
A3 adenosine[60]
Apelin[61,62]
C5a anaphylaxtoxin[63]
CCR5 chemokine[59]
CXCR1 chemokine[64]
CXCR2 chemokine[64]
CXCR4 chemokine[50]
D2 dopamine[65]
D3 dopamine[65]
D4 dopamine[50]
Edg1 endothelial diff. gene[66]
Edg2 endothelial diff. gene[50]
Edg3 endothelial diff. gene[50]
Edg5 endothelial diff. gene[50]
N-formyl peptide[67]
5HT1A hydroxytryptamine[50]
Leukotriene BLT1[68]
MCH1 melanin conc. hor.[61]
m2ACh muscarinic acetylcholine[50]
Neuropeptide FF[69]
δ-opioid[70]
μ-opioid[33,71]
E3 prostaglandin[50]
SST2a somatostatin[72]
SST3 somatostatin[72]
SST5 somatostatin[72]

Gq/11

α1b-adrenergic[33]
AT1A angiotensin II[32,33,35]
CCK-A cholecystokinin[50]
CCK-B cholecystokinin[50]
Cytomegalovirus US28[43]
ETA endothelin[32,33]
GnRH (type2) gonadotropin rel. hor.[73,74]
5HT2A hydroxytryptamine[75]
5HT2C hydroxytryptamine[42]
m1ACh muscarinic acetylcholine[30]
mGluR1 metabotropic glutamate[41,76]
NK1 neurokinin[33,35,77,78]
NK3 neurokinin[79]
NT1 neurotensin[32,33,35]
Orexin-1[61]
Oxytocin[35]
PAR2 proteinase-activated[80,81]
Platelet-activating factor[82]
TRHR-1 thyrotropin rel. hor.[33,83,84]
TRHR-2 thyrotropin rel. hor.[84]
V1a vasopressin[85]

FIGURE 31.2 GPCRs shown in the literature to translocate arrestin-GFP. Translocation of arrestin-GFP to activated GPCRs is independent of their interacting G protein (Gs, Gi/o, Gq/11), their interacting ligand (biogenic amines, peptides, glycoproteins, lipids, nucleotides), and their sequence and structure (rhodopsin-like subfamily, secretin-like subfamily, metabotropic-like subfamily). Coupling to a G protein has not been demonstrated for the seven transmembrane receptors listed under the "Also" heading.

necessary for the successful implementation of the Transfluor technology as a screening assay for GPCRs.

31.2 INITIAL VALIDATION STUDIES WITH TRANSFLUOR

Prior to screening a target GPCR with Transfluor, it is important both to verify that arrestin-GFP translocates to the receptor of interest and to evaluate the pattern and magnitude of the translocation response. These initial validation studies are most readily and easily accomplished in transiently transfected cells. While most receptors give a robust response in the assay and can be moved forward for development of stable cell lines and screening (see Section 31.3 and Section 31.4 below), a few receptors have been encountered that bind arrestin-GFP weakly or not at all. For these receptors, special technologies can be applied to improve the response. In the following section, we describe the steps for performing Transfluor in transiently transfected cells, the type of responses observed, and some of the special techniques employed to optimize and/or improve the response for both known and orphan GPCRs.

The arrestin-GFP conjugate we employ is generated by fusing the GFP moiety isolated from *Aequorea victorea* or *Renilla reniformis* to the carboxyl terminus of βarrestin2 (βarrestin2-GFP) (27,28). Fusion of the fluorescent moiety does not appear to alter the function of arrestin when compared to its wild-type counterpart [27]. Many different cell types have been shown to support Transfluor, including HEK-293, COS-7, CHO-K1, RBL-2H3, and U2OS cells. Each cell line has advantages and disadvantages. For example, HEK293 cells are easily transfected (transfection efficiency ~80%) but are small, grow in clumps, and require pretreatment of dishes with extracellular matrices for adherence to plastic and glass surfaces. In contrast, U2OS cells are large, flat, and

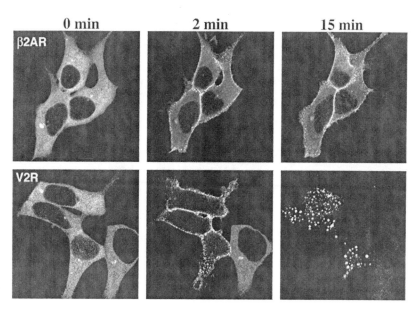

FIGURE 31.3 Translocation of βarrestin2-GFP to β2AR and V2R. HEK293 cells were transiently transfected with βarrestin2-GFP and either the β2AR or V2R. Photomicrographs show the distribution of βarrestin2-GFP in the same cells before (*0 min*) and after (*2 and 15 min*) treatment with the appropriate agonist. Agonists used were isoproterenol (10 μ*M*) for the β2AR and arginine vasopressin (100 n*M*) for the V2R. (From Oakley RH, Laporte SA, Holt JA, Barak LS, Caron MG. *J Biol Chem* 1999; 274:32,248–32,257. With permission.)

grow as a uniform monolayer, resulting in better visualization of the translocation response. U2OS cells also adhere very well to both plastic and glass surfaces but grow slow and transfect poorly (transfection efficiency ~20%).

Once a cell type is chosen, we transiently transfect the cells with cDNAs encoding the βarrestin2-GFP fusion protein and the GPCR of interest. Overexpression of the receptor is necessary. Visualization and subsequent quantitation of the translocation response depend ultimately on the ability to distinguish the population of receptor-bound βarrestin2-GFP from the unbound βarrestin2-GFP remaining in the cytoplasm. This is best accomplished in cells expressing the receptor at higher levels than the βarrestin2-GFP. We have never observed βarrestin2-GFP translocation to a specific agonist-activated endogenous GPCR. While this event undoubtedly occurs [30], the expression level of endogenous receptors is too low for the receptor-bound βarrestin2-GFP to be reliably distinguished from the unbound βarrestin2-GFP remaining in the cytoplasm. Cells expressing the βarrestin2-GFP and GPCR of interest are then plated on glass-bottom dishes and incubated overnight. One hour before analyzing the cells, the medium is removed and replaced with serum-free and phenol red–free medium to reduce autofluorescence. The distribution of the βarrestin2-GFP is then evaluated before and after agonist treatment at room temperature or 37°C using a 63× objective on a conventional fluorescent microscope.

The βarrestin2-GFP translocation response to two prototypical GPCRs, the β2-adrenergic receptor (β2AR) and vasopressin V2 receptor (V2R), is shown in Figure 31.3. HEK293 cells were transiently cotransfected with βarrestin2-GFP and either the β2AR or the V2R. In the absence of added agonist, βarrestin2-GFP is found evenly distributed throughout the cytoplasm of cells and excluded from the nucleus (*0 min* images). Upon addition of agonist, βarrestin2-GFP translocates within 1 to 2 min to both receptors at the plasma membrane (*2 min* images). The GFP fluorescence appears as distinct puncta or "pits" along the edge of the cells, reflecting the localization of the receptor/βarrestin2-GFP complex in clathrin-coated pits at the plasma membrane [13,17,31,32]. With longer agonist incubations, a striking difference is observed in the trafficking pattern of βarrestin2-GFP with these two receptors [17]. The βarrestin2-GFP remains localized in pits at the

Untreated   Pits   Vesicles

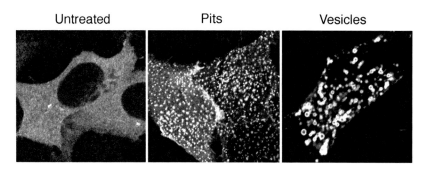

FIGURE 31.4 Prototypical "pit" and "vesicle" translocation responses observed in the Transfluor assay. HEK293 cells transiently overexpressing βarrestin2-GFP and GPCR were treated with or without agonist for 30 min. *Left panel* shows the distribution of βarrestin2-GFP in untreated cells. *Middle panel* shows the agonist-mediated redistribution of βarrestin2-GFP into clathrin-coated pits at the plasma membrane in cells expressing a class A GPCR. *Right panel* shows the agonist-mediated redistribution of βarrestin2-GFP into endocytic vesicles in cells expressing a class B GPCR. (From Oakley RH, Hudson CC, Cruickshank RD, Meyers DM, Payne RE, Rhem SM, Loomis CR. *Assay Drug Dev Technol* 2002; 1:21–30. With permission.)

plasma membrane in cells expressing β2AR but internalizes into endocytic vesicles in cells expressing V2R (*15 min* images). Colocalization studies monitoring both βarrestin2-GFP and the receptor in the same live cells have shown that βarrestin2-GFP dissociates from the β2AR and is excluded from receptor-containing intracellular vesicles [17]. In contrast, βarrestin2-GFP remains associated with the V2R throughout the endocytic process [17].

Of the approximately 100 GPCRs tested in Transfluor, all can be grouped into one of two classes based on the translocation response [33,34]. Class A or "pit-forming" receptors, such as the β2AR, form transient complexes with arrestin that dissociate at or near the plasma membrane after the receptor is targeted to clathrin-coated pits. Members of this family include dopamine receptor 1, mu opioid receptor, endothelial differentiation gene receptor 1, endothelin receptor A, and melanin-concentrating hormone receptor type 1 [34]. These receptors internalize without arrestin and appear to recycle rapidly back to the plasma membrane fully resensitized (Figure 31.1, *step 4a*). In the continued presence of agonist, the pit response is stable for several hours. Class B or "vesicle-forming" receptors, such as the V2R, form stable complexes with arrestin that internalize as a unit into intracellular vesicles (Figure 31.1, *step 4b*). Members of this family include the angiotensin II receptor type 1A, neurokinin receptor 1, thyrotropin-releasing hormone receptor 1, prostaglandin receptor E4, and *N*-formyl peptide receptor [34]. The internalized receptor/arrestin complexes can persist in cells for several hours (even after removal of agonist), and the receptors appear to recycle slowly or not at all. It appears that the majority of Gq-coupled receptors are vesicle-formers (class B) while the majority of Gi-coupled receptors are pit-formers (class A). Gs-coupled receptors are more evenly distributed between the two classes.

The pit and vesicle responses are shown at higher magnification in Figure 31.4. Two features should be noted:

• Vesicles are bigger than pits.
• Vesicles are brighter than pits.

As a result, vesicles provide a more robust response for quantitation by image analyzers [28]. Pits are also easily quantitated but require more precision in image acquisition. For example, the pit response is more sensitive to the plane of focus than is the vesicle response. Images of pit-forming receptors should be taken from the bottom of cells (as opposed to the middle of cells) to maximize the amount of plasma membrane and the number of observed pits (compare pit image taken from bottom of cell in Figure 31.4 to pit image taken from middle of cell in Figure 31.3). The focal plane for acquiring images of vesicles can be more variable because of their large size and high intensity.

The molecular motif mediating stable binding of arrestin to class B receptors has been identified and shown to be transferable [17,34,35]. Thus, a pit-forming GPCR can be converted into a vesicle-forming GPCR with no apparent alteration in pharmacology [17]. This technology is particularly useful for improving the Transfluor response when working with a receptor that gives either a very weak pit response or no translocation response at all. Belonging to the latter category is the β3-adrenergic receptor (β3AR), which has been shown in the literature not to be phosphorylated by GRKs, not to translocate βarrestin2-GFP, and not to desensitize [36,37]. We, however, have converted this receptor into a strong vesicle former by modifying the carboxyl terminal tail to include the motif mediating tight arrestin binding (R.H. Oakley and C.R. Loomis, manuscript in preparation). Following agonist addition, βarrestin2-GFP rapidly translocates to the modified β3AR and internalizes with it into endocytic vesicles.

For the small subset of GPCRs that bind βarrestin2-GFP very weakly, two alternative technologies can be used to improve the translocation response that avoid a genetic alteration of the receptor. One approach is to coexpress the GPCR of interest with the different nonvisual GRKs (GRK2, GRK3, GRK4, GRK5, or GRK6). Receptor selectivity has been shown to exist among the GRK family [38–40]; therefore, the endogenous complement of GRKs in the heterologous cell line may be insufficient to adequately phosphorylate the GPCR of interest. By overexpressing the appropriate GRK, agonist-dependent phosphorylation of the receptor and the subsequent translocation of βarrestin2-GFP can be restored. A second approach is to employ βarrestin1-GFP. Whereas class B or vesicle-forming GPCRs bind βarrestin1 and βarrestin2 with similar high affinities, class A or pit-forming receptors bind βarrestin2 with significantly higher affinity than βarrestin1 [33,34]. Therefore, we recommend βarrestin2-GFP for the initial validation studies. However, it is possible that some receptors will be found that selectively and/or preferentially bind βarrestin1 over βarrestin2 [41].

The validation studies discussed so far have focused on GPCRs that have a known agonist. What about orphan GPCRs that by definition have no known ligand? Before investing the time and money screening an orphan GPCR against a large library of compounds, it would be very beneficial to know whether βarrestin2-GFP will bind the orphan receptor of interest and whether the receptor will give a pit or vesicle response. We have recently developed a proprietary technology, termed the ligand-independent translocation assay or LITe™ assay, that allows such validation studies to be performed (R.H. Oakley and C.R. Loomis, manuscript in preparation). In short, this powerful technology employs a constitutively active GRK that will constitutively phosphorylate the receptor of interest and thereby induce constitutive βarrestin2-GFP translocation. For these studies, cells stably expressing βarrestin2-GFP are transiently transfected with the orphan receptor alone or the orphan receptor together with the constitutively active GRK. The GRK-mediated redistribution of βarrestin2-GFP into pits or vesicles indicates that the orphan receptor is expressed at the plasma membrane and can bind arrestin in a phosphorylation-dependent manner. Orphan receptors that give a translocation response in the LITe assay are moved forward for a Transfluor screening campaign.

31.3 DEVELOPMENT OF TRANSFLUOR-ENABLED STABLE CELL LINES

Target GPCRs can be screened with Transfluor using cells transiently or stably expressing the receptor of interest. Transient overexpression of receptor is achieved quickly but generally gives a more variable translocation response due to heterogeneity in cell-to-cell receptor expression levels and alterations in day-to-day transfection efficiencies. Stable overexpression is achieved slowly but generally results in less variability in the translocation response because individual clones can be chosen that express a uniform and optimum level of receptor. The combination of reduced variability and maximized signal window achieved with permanent cell lines leads to better signal-to-noise ratios and Z' values in a screen. Although we utilize both approaches, we typically employ transient expression system for smaller screens and stable expression for screening large compound libraries. In the following section we discuss the generation of Transfluor-enabled stable cell lines, highlighting the critical steps involved and some of the limitations encountered.

The cell type chosen for development of stables should be amenable to image analysis and high-throughput screening. For example, the pit and vesicle translocation responses are better visualized in large, flat cells that grow in a monolayer as opposed to small, round cells that aggregate. In addition, cells that adhere well to 96- and 384-microtiter plates are certainly preferred over those that tend to wash off during liquid handling. For these reasons, we favor U2OS cells. As discussed in the previous section, U2OS cells are large, flat, grow in a monolayer, and adhere well to plastic and glass, resulting in high-quality images. Other cell types exhibiting similar characteristics should also work well in a Transfluor screen. U2OS cells stably expressing both the target GPCR and βarrestin2-GFP can be developed by cotransfecting plain U2OS cells with both cDNAs. This approach, however, results in many cells that express either too much or too little βarrestin2-GFP. We recommend developing first a U2OS cell line permanently expressing a medium amount of the βarrestin2-GFP [28]. This backbone stable can then be transfected with the receptor cDNA of interest for subsequent development of the receptor/βarrestin2-GFP double stable [28]. This approach increases the probability of obtaining individual clones with optimum levels of the target receptor and βarrestin2-GFP.

Once the target GPCR is transfected into the βarrestin2-GFP stable, antibiotics are applied for the selection of receptor-expressing cells. The population of cells surviving the 1- to 2-week drug selection is referred to as the parent mix. The percentage of cells in the parent mix that express both the target GPCR and βarrestin2-GFP is important to determine because it foretells how easy or difficult it will be to generate the desired receptor stable. For this determination, we plate parent mix cells in a 35-mm glass-bottom dish and evaluate visually the distribution of βarrestin2-GFP before and after agonist addition using a fluorescent microscope, as described in the previous section. If a large portion of the parent mix cells (25% or more) displays a translocation response comparable in magnitude to that observed with transient overexpression, stable clones should be readily obtained. However, if the translocation response in these cells is significantly weaker than that observed with transient overexpression, then the resulting stable clones (though easily obtained) may not provide a sufficiently large signal window for screening. Finally, if only a small portion of the parent mix cells (less than 5%) displays a translocation response, then generation of a receptor stable may be problematic. The latter two scenarios (which occur infrequently) indicate that permanent expression of the receptor at high levels is toxic to the cells or inhibits the growth of the cells. For these receptors, Transfluor-enabled stables can sometimes be generated by removing activators of the receptor that are present in the serum, by using another cell type, or by employing an inducible expression system. Rather than pursue these alternative approaches, we typically turn to a transient overexpression system for screening such problematic receptors.

If the parent mix is found to have an acceptable amount of receptor-positive cells, we dilute the cells and plate them into 15-cm dishes to achieve individual clones. We prefer 15-cm dishes to 96-well plates for this process because the U2OS cells grow more quickly in the presence of neighboring cells secreting growth factors. The 15-cm dishes are incubated for 1 to 2 weeks until individual colonies have reached a size of approximately 100 cells. The colonies are then picked using cloning cylinders to minimize cross contamination of clones and expanded in 24-well plates. When the cells reach 90% confluency, they are split into a 96-well glass-bottom plate for evaluation of the translocation response in the absence and presence of a known ligand using a fluorescent microscope.

In the absence of added ligand, most receptor-positive clones will exhibit little or no basal translocation. The βarrestin2-GFP should be diffusely distributed in the cytoplasm of the cells and excluded from the nucleus. Occasionally, we have encountered receptors that give rise to a significant degree of basal translocation. If this basal response is due to an endogenous agonist in the serum, we can reduce or eliminate the response by incubating the cells for 1 to 2 h in serum-free medium or by culturing the cells in serum that has been charcoal stripped (which removes lipids, hormones, and growth factors) or dialyzed (which removes small molecules such as amino acids and nucleotides). Sometimes, however, the basal translocation response is due to a receptor that is constitutively active (activated in the absence of ligand), as has been reported recently for the serotonin 5HT2c receptor and the viral US28 receptor [42,43]. In the case of the 5HT2c receptor, the constitutive translocation response is blocked by treatment with inverse agonists, suggesting that receptors that demonstrate natural constitutive activity can be screened in the Transfluor assay for inverse agonists [42]. Finally, we also evaluate the morphology of the untreated cells. The cells should have a flat, epithelial-like shape. Clones with cells that are overly elongated, highly vacuolated, or rounded up are avoided.

Individual clones are treated with a saturating concentration of the receptor agonist for approximately 30 min. Positive clones that can be used for screening are selected based on the uniformity and magnitude of the pit or vesicle response [28]. We look for clones in which 100% of the cells show the translocation response and the magnitude of the response (be it pits or vesicles) is the same in all the cells. Clones that are mixed with unresponsive cells or cells that respond poorly are avoided because the subpopulation of cells expressing little or no receptor will frequently outgrow the desired population of cells expressing the receptor at higher levels. In deciding which positive clones have a translocation response that will quantitate well on an image analysis system, we look for clones in which the contrast between the fluorescence localized in pits or vesicles and the fluorescence remaining in the cytoplasm is greatest [28]. Ideally, all the cytoplasmic βarrestin2-GFP will be localized in pits or vesicles. Clones with a weak translocation response have less fluorescence localized in pits or vesicles and more remaining in the cytoplasm and consequently are more difficult for imaging platforms to detect and quantitate.

A clone of U2OS cells stably expressing the angiotensin II type 1A receptor (AT1AR) is shown in Figure 31.5 to illustrate several of the key features we look for when evaluating clones. In the absence of agonist (0 min image), βarrestin2-GFP is uniformly distributed in the cytoplasm and excluded from the nucleus of the cells. In addition, no basal translocation is observed. After agonist addition, the βarrestin2-GFP rapidly localizes in pits at the plasma membrane in all the cells (1 min image). The pits get bigger and brighter over the next several minutes as the βarrestin2-GFP continues to accumulate in these regions along the plasma membrane (2 and 5 min images). Eventually, βarrestin2-GFP localizes in vesicles inside the cells (15 and 30 min images). The large difference in contrast between the βarrestin2-GFP fluorescence in vesicles and the βarrestin2-GFP fluorescence remaining in the cytoplasm indicates that the response in this clone will be easily detected and quantitated by available image analysis systems. Moreover, the clone is not mixed with nonresponding cells or cells responding poorly to agonist because all the cells respond with similar kinetics and to a similar extent. Thus, these cells express a uniform level of receptor.

We have generated Transfluor-enabled cell lines not only for many known GPCRs but also for many orphan GPCRs. Orphan GPCRs present a unique challenge because there is no known ligand to utilize for evaluating the translocation response in surviving clones. We have circumvented this limitation by employing the LITe assay technology described above in Section 31.2. This technology allows us to induce βarrestin2-GFP translocation to the orphan receptor of interest by coexpression of a constitutively active GRK (R.H. Oakley and C.R. Loomis, manuscript in preparation). Thus, the surviving clones are transiently transfected with either empty vector or the constitutively active GRK. Coexpression of the GRK in orphan receptor–expressing clones results in the redistribution of βarrestin2-GFP into either pits or vesicles. Clones positive for the translocation response are then immunostained and analyzed by FACS analysis to determine whether receptor expression is uniform in all the cells. Orphan receptor stable clones displaying a robust translocation response in the LITe assay and a single peak profile on the FACS analysis are moved forward for screening.

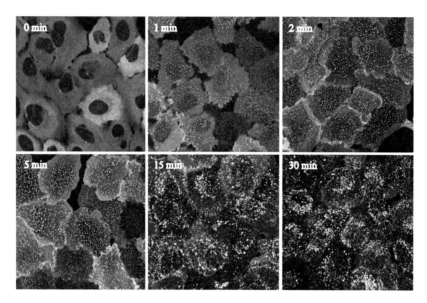

FIGURE 31.5 Translocation of βarrestin2-GFP to the AT1AR stably expressed in U2OS cells. U2OS cells stably expressing βarrestin2-GFP and the AT1AR were treated with or without 100 n*M* angiotensin II. Photomicrographs show the distribution of βarrestin2-GFP at indicated time points. The pit response is observed at 1-, 2-, and 5-min time points. The vesicle response is observed at the 15- and 30-min time points.

31.4 HIGH-THROUGHPUT SCREENING WITH TRANSFLUOR

Development of Transfluor into a robust, sensitive, and reproducible high-throughput screening platform was dependent upon the availability of instrumentation that could rapidly and accurately detect and quantitate the translocation response. A major challenge to the automation of the assay was that the number of arrestin-GFP molecules did not change upon receptor activation. Instead, as discussed above, the subcellular distribution of the arrestin-GFP changed from a diffuse cytoplasmic pattern to a pit and/or vesicular pattern. Without changes in overall fluorescent intensity, traditional detection instrumentation was unsuitable for Transfluor. Fortunately, interest in high-content screening (HCS) methods had initiated the recent manufacture of optical imaging instruments that could identify the subcellular location of fluorescently labeled molecules to follow intracellular physiology. The Transfluor technology has now been enabled on many of these image analysis systems, permitting tens of thousands of microplate wells to be imaged and quantitated per day, as demanded by today's drug discovery programs. In the following section, we discuss the available image analysis systems, the quantitation and validation experiments we perform for each receptor target, and the screening process itself.

31.4.1 Image Instrumentation

Some of the key characteristics differentiating the many imaging instruments on the market relate to confocal or nonconfocal imaging, focusing mechanisms, light source, breadth of applications, environmental control, and user interface. Since the Transfluor response involves a change in arrestin-GFP distribution within the cell rather than an increase or decrease in the overall level of fluorescence, systems capable of discriminating areas of fluorescence intensity within the cell are required. Typically, this is accomplished with either confocal or nonconfocal microscopy. Confocal-based image analysis systems such as Amersham Bioscience's INCell Analyzer 3000, Evotec's Opera, and Atto Bioscience's Pathway HT provide superior image quality by filtering out scattered light from the sample through a pinhole or slit. Removal of the scattered light improves clarity and resolution of the image but requires a greater intensity light source that typically is provided by a laser. Nonconfocal systems such as the Universal Imaging Corporation's Discovery-1, Cellomic's ArrayScan Vti, and Q3DM's EIDAQ 100 are still capable of detecting pit and vesicle responses with excellent sensitivity. An advantage of the nonconfocal systems is the ability to use white light, which reduces instrument complexity and cost while increasing the range of wavelengths of light provided. Recent advances in nonconfocal microscopy, such as the Apotome technology by Carl Zeiss, provide near-confocal image quality while still using a white light source. Most of these instruments come with computer-driven filter wheels that facilitate easy switching between excitation and emission wavelengths. This enables a wider range of assays to be run on these instruments.

In addition to microscope-based systems, Acumen Bioscience's Explorer is a PMT-based detection system that has had success in quantifying the Transfluor response. This system takes multiple readings throughout a sample well and then uses an algorithm to define the fluorescence intensity patterns throughout the well. To date only the vesicle response has been successfully quantitated on the Explorer. Systems such as these offer a great degree of flexibility for multiple assay formats.

The instruments on which Transfluor has been validated are given in Table 31.1. Each provides a different balance of image resolution, speed of acquisition, flexibility, light source, detectors, response analysis, and cost. The choice of a system obviously depends on how the instrument capabilities fit the overall goals of a program or department. When evaluating speed of acquisition, be aware that image size, number of images per well, and number of passes per well can dramatically affect plate read time. Additional important aspects in selecting an appropriate instrument are service support, ease of use, robustness, and user training. These machines tend to be very complex and require a fair degree of training and familiarity to use. Thus systems with intuitive interfaces that are both user-friendly and still provide flexibility for user-defined analysis are desirable. Given the complexity and technical nature of these machines, it is important that the manufacturer have skilled technicians who can respond quickly and effectively when problems arise.

31.4.2 QUANTITATION OF TRANSFLUOR RESPONSE

Imager manufacturers have tended to write their own algorithms to describe and quantitate the translocation of arrestin-GFP. We have worked closely with several of the imager manufacturers to develop algorithms that describe and relate the redistribution of arrestin-GFP to the known pharmacology of GPCRs [28,34]. At Norak, we have Amersham Bioscience's INCell Analyzer 3000. The algorithm for this instrument:

- Locates the nucleus of cells
- Sets a bounding box around the nucleus
- Expands the bounding box a user-defined amount to encompass a majority of the cell while minimizing overlap with neighboring cells
- Evaluates within this box the fluorescent intensity of pixel groups (βarrestin2-GFP localized in pits or vesicles) relative to the intensity of neighboring pixels (unbound βarrestin2-GFP remaining in the cytoplasm)

Pixel groups that have a fluorescence intensity greater than a user-defined threshold are marked as "grains." Three primary parameters are calculated based on the properties of these grains. Ngrains is the average number of grains per cell, Agrains is the average area of grains per cell, and Fgrains is the average fluorescence intensity of the grains. We have found that Fgrains correlates best with pharmacological responses measured by other methods such as binding and functional assays.

Quantitation of βarrestin2-GFP translocation to three different receptors, each stably expressed in U2OS cells and representing a different signal transduction pathway, is shown in Figure 31.6. For these experiments, the cells were plated in 96- or 384-well plates, treated with various

TABLE 31.1
Instrumentation for Quantitation of the Transfluor Response — Summary of Features[a]

Company	Amersham Biosciences[c]	Evotec	Cellomics	Axon Instruments	Q3DM	CompuCyte/ Olympus	Atto Bioscience	Amersham Biosciences[c]	Universal Imaging Corporation[TM,h]	TTP Labtech
Imager	IN Cell 3000	Opera	ArrayScan® Vti	Image Express® 5000A	EIDAQ™ 100[d]	iCyte™	Pathway HT™	IN Cell 1000	Discovery-1™	Acumen Explorer™
Standard objectives	40x objective	20x / NA .7 water immersion	5x, 10x, 10x high NA, 20x	10x, 0.5 NA S-fluor objective, 4x, 0.2 NA Plan Apo 20x, 0.45 NA Plan Fluor ELWD, 40x, 0.6 NA Plan Fluor ELWD, 6-position objective turret	2–40x magnification; 0.95 N.A.; plan/apo corrected	40x (others may be available)	20x / 0.75 NA Uapo/3400.17 mm working distance objective 1x, 4x, 10x, 40x, and 60x available	4x / 0.2, 10x / 0.45, 20x / 0.45 and 40x / 0.6 objectives	Fully automated 6-position objective turret (2x, 4x, 10x, 20x, 40x)	Non-image based
Confocal	Yes	Yes	No, optional ApoTome grating imager	No	No	No	CARV confocal module	No	No	Not Applicable
Light source	Krypton laser (647 nm), argon laser (364 nm and 488 nm), RGB LEDs (transmission mode)	Three laser sources (488 nm, 532 nm, 633 nm)	High Efficiency Metal-Halide illumination	300W xenon arc lamp	mercury arc lamp epi-fluorescence light source	Argon Ion (488 nM), Helium Neon (633 nM), Diode (405 nM)	Dual HBO 100 W mercury arc lamps (Xenon amps available)	100 W Xenon lamp	High intensity arc lamp for fluorescence excitation	Argon Ion laser (488 nM)
Excitation	647 nm, 364 nm, 488 nm	Filter sets for two color excitation and single color excitation	High speed 10-position excitation filter wheel	10-position excitation filter wheel	10-position excitation filter wheel	N/A	16 filter positions/5 dichroic mirror positions	Six-position computer controlled excitation and emission filter wheels	Optimized 10-position filter wheels (excitation, emission, and neutral density)	N/A

Emission	Selected wavelengths between 420 nm and 720 nm	N/A	User accessible Emission/ Dichroic Turret	10-position emission filter wheel	10-position emission filter wheel	445-485 nM, 515-545 nM, 565-585 nM, 600-635 nM, 650-700 nM, 750-800 nM	8 filter positions/5 dichroic mirror positions	Six-position computer controlled excitation and emission filter wheels	Optimized 10-position filter wheels (excitation, emission, and neutral density)	N/A
Detection	Three high-speed TE-cooled, 12-bit, CCD cameras	Two parallel detection channels 12 bit CCD camera	High resolution CCD camera	Axon Instruments 1280 × 1024 cooled CCD camera, with 12-bit readout	CCD camera: 10 bit per pixel (1024 gray levels per pixel); 9.9 um pixel size; 640 × 480 pixel frame size	4 multiplier tube fluorescence detectors	CCD camera	CCD camera 12 bit, 1392 × 1040 pixels, progressive scan with thermoelectric cooling to –30°C.	High-speed scientific grade CCD camera	4 PMT laser scanning system
Plate format	96- or 384-well format	96, 384, 1536, 2080	96, 384	96, 384	6 to 1536 well format, microscope slides	6-384-well, microscope slides, Petri dishes, chamber slides	96, 384, microscope slides	96, 384	Microplates formats up to 1536 wells	96, 384, 1536
Time to read 96-well	~ 2.5 min per 96-well plate, at 2.4 μm resolution	50,000 wells/hr 120,000+ wells/day	N/A	96-well plate, 1280 × 1024 images, 1 μm/pixel resolution, with 2 fluorescence channels/well, takes less than 4 minutes	30,000 wells per day	N/A	N/A	2000 images/h	N/A	
Time to read 384-well	~8.5 min / plate ~2500 wells/hr ~60,000 wells/24 hr	50,000 wells/hr 120,000+ wells/day	N/A	384-well plate, 1280 × 1024 images, 1 μm/pixel resolution, with 2 fluorescence channels/well, takes less than 13 minutes	30,000 wells per day	N/A	N/A	2000 images/h	N/A	

(continued)

TABLE 31.1 (CONTINUED)
Instrumentation for Quantitation of the Transfluor Response — Summary of Features[a]

Company	Amersham Biosciences[c]	Evotec	Cellomics	Axon Instruments	Q3DM	CompuCyte/ Olympus	Atto Bioscience	Amersham Biosciences[c]	Universal Imaging Corporation[TM,h]	TTP Labtech
Focusing	Continuous infrared laser autofocus	N/A	Image-based autofocus	High-speed, laser-based autofocus system	Autofocus in fluorescence, phase and brightfield based on image content	Fast laser-based autofocus	Automated focus and image acquisition	Confocal laser detection, Piezo positioner, focus time <400 ms	Laser auto-focus	Focus-free, area-based scanning
Automation	Compatible with externally scheduled automation packages	Fully automated system	Twister II robot	Zymark Twister plate-handling robot with bar code reader	N/A	Robotic specimen loader with 45-tray capacity	N/A	Designed for communication with all leading plate-handling robots	Thermo CRS CataLyst Express™, Hudson Control PlateCrane™, PlateExchange software, Generic robotics interface, Barcode reader	Twister, others
Temperature control	Ambient or 37°C, 70% humidity, 5% CO2	N/A	N/A	No	No	N/A	Integrated temperature and CO2 control	Live-cell, kinetic assays, ambient or 37°C	N/A	N/A
Transfluor optimized software[e]	Yes	Yes	Yes	Yes	Yes	Yes	Yes	No	Yes	No
Norak Validated[f]	Yes	Yes	Yes	Yes	Yes	Yes	Yes	No	Yes	Vesicles only

a All information copied directly from company websites / product literature as of June 2004.
b N/A = Not available.
c Now part of GE Healthcare.
d Acquired by Beckman Coulter in December 2003.
e Software developed for or compatible with quantification of Transfluor pit and vesicle responses.
f Norak has reviewed data generated on the instruments and confirms that the responses obtained conform to those observed by Norak.
g Rate based on experience at Norak Biosciences, Inc.
h A subsidiary of Molecular Devices.

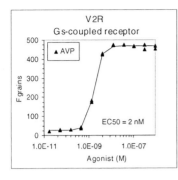 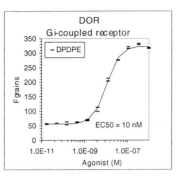 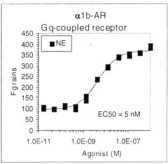

FIGURE 31.6 Quantitation of βarrestin2-GFP translocation to the V2R, delta opioid receptor (DOR), and alpha 1b adrenergic receptor (α1bAR). U2OS cells stably expressing βarrestin2-GFP and either the V2R, DOR, or α1bAR were treated with various concentrations of the appropriate agonists and analyzed on the INCell Analyzer 3000 (generation 1). Data are expressed as Fgrains.

concentrations of the receptor agonist for 30 to 45 min, fixed, and analyzed on the INCell Analyzer 3000 using the algorithm described above. The V2R is a vesicle former and therefore gives the greatest signal window with a signal-to-background ratio of about 18. The delta opioid receptor (DOR) and alpha-1b adrenergic receptor (α1bAR) are both pit formers and give a smaller signal-to-background ratio of about three to six. For all three receptors, the agonist EC_{50} values correlate well with the binding pKa values reported in the literature. The close correlation between Transfluor data and binding data has been observed with many different receptors for both agonists and antagonists and likely reflects the proximity of arrestin binding to ligand binding.

31.4.3 VALIDATION EXPERIMENTS

In preparation for a Transfluor screen, various parameters must be validated or optimized. These parameters are listed below with a brief description of the optimization criteria. With a modest amount of planning, many of these factors can be evaluated in parallel.

31.4.3.1 Plate Type

As with previous emerging high-throughput technologies, imaging instruments demand the highest level of optical clarity and plate consistency. For Transfluor, plate characteristics can have a substantial impact on the quality of the assay. Critical plate characteristics are clarity of the plate bottom material, flatness of the plate bottom, and cell viability. The material through which the samples are being imaged must be devoid of imperfections or material autofluorescence. It must reasonably resist scratches from common handling and should be able to be cleaned. It is good practice to clean the plate bottoms to remove fingerprints or other material that might interfere with the autofocus mechanism of the imager. Although the imaging systems have autofocus mechanisms, plate flatness is critical, as these systems have ranges outside of which they can no longer adjust the focus. In addition, substantial focusing adjustment can slow the screen rate. Finally the cells must adhere to, spread out, and thrive on the plate bottom. Different cell lines and different receptors can affect which plate works best or even which can or cannot pass quality control. Therefore, it is best to perform plate acceptance tests with the cell line and receptor combination that will be used in the screen.

31.4.3.2 Clone Selection

When stables are developed for a target GPCR, multiple clones are selected based on the translocation response observed under a fluorescent microscope (see Section 31.3). To identify the best clone for use during a screening campaign, concentration response curves are performed for each clone in the appropriate plate type. After quantitation on an image analysis system, the optimal clone is chosen based on a number of parameters including the signal-to-noise ratio, Z′ value [44], appropriate pharmacology, appropriate morphology, and growth rate.

31.4.3.3 Cell Number

The number of cells seeded per well in the microtiter plate is influenced by the area of the adherent cell, the rate of cell division, and the length of time the cells are plated prior to the experiment. The expressed receptor can impact each of these parameters, and therefore each new cell line requires a separate validation. Additionally, the plate type can impact the rate of growth or the degree of cell spreading. The goal is to have an even distribution of cells such that the cells are thriving but not so dense that the cells are confluent or overlapping. Having sufficient cell numbers to reliably quantitate the response can allow minimizing the image window and thus increase the rate of imaging. At Norak we use U2OS cells and typically target 4000 to 6000 cells per well, seeded 18 to 20 h prior to the screen in Becton Dickinson or Matrical 384-well plates.

31.4.3.4 Response Time and Temperature

The time needed for full development of the pit or vesicle response can vary with receptor, cell line, and temperature. At 37°C, pit responses develop rapidly within a few minutes and reach a maximal response within 5 to 10 min. The pit response is stable for 1 to 2 h. Vesicle responses develop more slowly than pit responses, with initial formation between 5 and 15 min and complete formation in about 30 to 60 min. The vesicle response can last up to 6 h. One of the advantages of Transfluor over other cell-based assays is that it can be run at room temperature. We have compared the responses of many receptors at 37°C and room temperature and not found any substantial difference in the translocation response at later time points. Consequently, we generally perform our screens at room temperature and employ compound and/or agonist incubation times of 30 to 40 min for either pit or vesicle formers. This minimizes our need to optimize between receptors of different response types. Both time and temperature are parameters that should always be validated for each new receptor target.

31.4.3.5 Signal-to-Background Ratio

The amount of βarrestin2-GFP recruited to the cell surface and how tightly it is bound to the receptor impact the amount of fluorescence accumulated in pits or vesicles and hence the magnitude of the detected response (see Figure 31.6 and [28,34]). This recruitment varies among receptors and is in part related to the receptor expression levels. In addition, the basal level of activity can vary due either to constitutive receptor activity or the presence of receptor agonist in the medium. In some cases removal of growth serum from the medium prior to the assay will reduce the basal signal. This is determined by following a time course of the basal activity after replacement of serum with non–serum-containing assay medium. Therefore, each new cell line must be validated to optimize the difference between basal and stimulated levels and to ensure an adequate response window for screening.

31.4.3.6 Cell Morphology

Another factor that can affect response magnitude is the morphology of the cell line. Cells that are thin and flat and do not overgrow provide the best images for Transfluor quantitation. Naive U2OS cells lay thin and flat. However, expression of a receptor of interest may change this morphology, which can then affect imaging of the translocation of βarrestin2-GFP. Furthermore, activation of

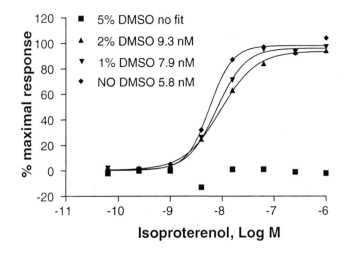

FIGURE 31.7 Effect of DMSO concentration on the translocation response. U2OS cells stably expressing the β2AR were treated with various concentrations of isoproterenol in the presence (1, 2, or 5%) or absence of DMSO. Data are expressed as a percentage of the maximal response to isoproterenol in wells containing no DMSO.

some receptors may also lead to morphologic changes, which can impact the ability to accurately quantitate the translocation response.

31.4.3.7 Dimethylsulfoxide (DMSO) Tolerance

As with all cell-based functional screens, there is a limit to the amount of DMSO that the cells can tolerate before the Transfluor response is affected. In U2OS cells, the redistribution of βarrestin2-GFP into pits and/or vesicles tolerates 1% DMSO, is slightly inhibited by 2% DMSO, and is completely prevented with 5% DMSO (Figure 31.7). While this data can offer guidance, each cell line and receptor combination should be individually checked.

31.4.3.8 Automated Handling

Many of the steps for Transfluor in our HTS lab are handled using bench-top automation. Therefore, we validate each new cell line for use in these systems. In particular, we verify that the cells can be plated using Multidrop dispensers (Titertek) and that we can achieve uniform plating. We confirm that the cells remain adherent during medium replacement using a Minitrak™ (PerkinElmer) liquid handling robot and during compound and fixative addition. We then use a Remp heat sealer to seal the plates without removal/replacement of the fixative. Other laboratories remove the fixative and replace with PBS. It is important to ensure that none of these procedures results in loss or morphological change of the cells.

31.4.3.9 Orphan Receptors

The validation steps for known and orphan GPCRs differ slightly. Clearly, there is no need to validate control reagents for orphan receptors, as these are by definition unavailable. We do employ a proprietary technique, the LITe assay described earlier in Section 31.2 and Section 31.3, which enables confirmation of receptor expression and coupling to the desensitization pathway. Using this technique, we can perform the above validation steps for orphan receptors to determine response magnitude and receptor suitability for automated handling.

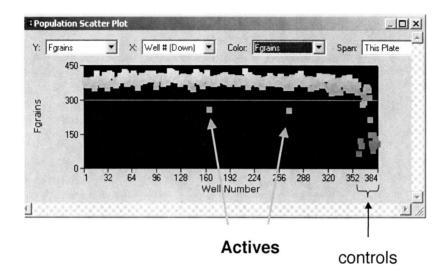

Actives   controls

FIGURE 31.8 Screen shot from INCell Analyzer 3000 (Raven software) showing the Fgrain values for each well of a representative 384-well plate from an antagonist screen of 350,000 compounds.

31.4.4 SCREENING

A typical screen at Norak involves multiples of 32 384-well plate runs using a PerkinElmer MiniTrak to perform the liquid handling steps in a timed manner. We remove the growth media and replace with phenol red– and serum-free medium. We then restack the plates prior to compound addition. If agonist addition is needed (for example, during an antagonist screen), the plates are restacked and agonist added. The final step, addition of PBS containing 4% formaldehyde and the nuclear stain Draq5, is conducted on the MiniTrak, Titan platewasher, or multidrop. Imaging on the INCell Analyzer 3000 can be performed at a rate of 120 384-well plates per day. A screen shot from the INCell Analyzer 3000 of a representative 384-well plate from an antagonist screen of over 350,000 compounds is shown in Figure 31.8.

We typically perform the assay and then fix the cells prior to imaging. This step is optional, however, as the technology is also amenable to being performed as a live cell assay. Thus in systems where timing can be controlled throughout the whole assay up to and including reading of the plates, one could remove the fixative step and image the plates with live cells. This has the added advantage of being able to perform time course experiments during assay development or when there is an interest in following the development of the response over time. Performing a live cell assay would also enable testing compounds sequentially for agonist and antagonist activity in the same experiment.

One of the advantages Transfluor provides as a high-content screening platform is the ability to evaluate undesired compound effects such as fluorescence and toxicity. Images from active wells are routinely visually inspected and assigned a score (active, toxic/active, fluorescent active, toxic, fluorescent, precipitate, etc.). We have worked with the software developers of the IN Cell Analyzer 3000 (generation 1) to provide readouts on well fluorescence (iCyte) and toxicity based on nuclear size and fluorescence intensity (iNuc). We have combined these readouts to automatically score the images. This greatly facilitates culling false positives from the primary screen, thereby increasing retest rates in this functional screen.

One of the challenges of HCS is the amount of data generated in the form of image files. These files can range from 400 to 2500 KB per image resulting in 150 MB to 1.0 GB per 384-well plate. Handling this amount of data requires additional infrastructure (high-capacity cables, network integration cards, routers, and servers) as well as retention policies. We have installed dedicated servers with 500-GB capacity for each of our INCell analyzers, connected by 1 MB network cards. This enables transfer of data from the INCell computer to a server, freeing capacity on the computer to continue operations. After completion of the screen and visual inspection for scoring, images from primary screening are deleted. Images from confirmation studies are saved on a near-line removable DVD system with 1-TB capacity.

In summary, the Transfluor assay has proven to be a facile and robust functional (cell-based) screening platform that consistently yields high quality ($Z' > 0.5$) and pharmacologically appropriate results. It has provided an assay format for GPCRs that is equally useful for screening Gs-, Gi-, or Gq-coupled receptors.

31.5 CONCLUSION

The Transfluor technology has been used successfully by a number of large and small pharmaceutical companies to find compounds active against target GPCRs. The technology has proven to be a stable and robust functional cell-based screening platform applicable to both known and orphan GPCRs. Transfluor has been employed by companies not only as a high-throughput assay to screen large compound libraries (>500,000) but also as a secondary screen to determine EC_{50} and IC_{50} values for active compounds. The assay readily distinguishes between full and partial agonists, inverse agonists, and neutral antagonists important in the identification of the best compounds to move forward into chemistry and development. With continued improvements in imaging technology, Transfluor represents an important addition to the family of assays required to realize the full promise of high-content screening.

REFERENCES

1. Bockaert J, Pin JP. Molecular tinkering of G-protein–coupled receptors: an evolutionary success. *EMBO J* 1999; 18:1723–1729.
2. Bailey WJ, Vanti WB, George SR, Blevins R, Swaminathan S, Bonini JA, Smith KE, Weinshank RL, O'Dowd BF. Patent status of the therapeutically important G-protein–coupled receptors. *Expert Opin Ther Patents* 2001; 11:1861–1887.
3. Neer EJ. Heterotrimeric G proteins: organizers of transmembrane signals. *Cell* 1995; 80:249–257.
4. Hall RA, Premont RT, Lefkowitz RJ. Heptahelical receptor signaling: beyond the G protein paradigm. *J Cell Biol* 1999; 145:927–932.
5. Hausdorff WP, Caron MG, Lefkowitz RJ. Turning off the signal:desensitization of β-adrenergic function. *FASEB J* 1990; 4:2881–2889.
6. Perry SJ, Lefkowitz RJ. Arresting developments in heptahelical receptor signaling and regulation. *Trends Cell Biol* 2002; 12:130–138.
7. Ferguson SS. Evolving concepts in G-protein–coupled receptor endocytosis: the role in receptor desensitization and signaling. *Pharmacol Rev* 2001; 53:1–24.
8. Benovic JL, Mayor F Jr, Staniszewski C, Lefkowitz RJ, Caron MG. Purification and characterization of the beta-adrenergic receptor kinase. *J Biol Chem* 1987; 262:9026–9032.
9. Chen CK, Burns ME, Spencer M, Niemi GA, Chen J, Hurley JB, Baylor DA, Simon MI. Abnormal photoresponses and light-induced apoptosis in rods lacking rhodopsin kinase. *Proc Natl Acad Sci USA* 1999; 96:3718–3722.
10. Pitcher JA, Freedman NJ, Lefkowitz RJ. G-protein–coupled receptor kinases. *Annu Rev Biochem* 1998; 67:653–692.
11. Ferguson SS, Downey WE, Colapietro AM, Barak LS, Menard L, Caron MG. Role of beta-arrestin in mediating agonist-promoted G-protein–coupled receptor internalization. *Science* 1996; 271:363–366.
12. Goodman OB Jr, Krupnick JG, Santini F, Gurevich VV, Penn RB, Gagnon AW, Keen JH, Benovic JL. Beta-arrestin acts as a clathrin adaptor in endocytosis of the beta2-adrenergic receptor. *Nature* 1996; 383:447–450.

13. Laporte SA, Oakley RH, Holt JA, Barak LS, Caron MG. The interaction of beta-arrestin with the AP-2 adaptor is required for the clustering of beta 2-adrenergic receptor into clathrin-coated pits. *J Biol Chem* 2000; 275:23,120–23,126.

14. Palczewski K, McDowell JH, Jakes S, Ingebritsen TS, Hargrave PA. Regulation of rhodopsin dephosphorylation by arrestin. *J Biol Chem* 1989; 264:15,770–15,773.

15. Krueger KM, Daaka Y, Pitcher JA, Lefkowitz RJ. The role of sequestration in G-protein–coupled receptor resensitization. Regulation of beta2-adrenergic receptor dephosphorylation by vesicular acidification. *J Biol Chem* 1997; 272:5–8.

16. Zhang J, Barak LS, Winkler KE, Caron MG, Feruson SS. A central role for beta-arrestins and clathrin-coated vesicle-mediated endocytosis in beta2-adrenergic receptor resensitization. Differential regulation of receptor resensitization in two distinct cell types. *J Biol Chem* 1997; 272:27,005–27,014.

17. Oakley RH, Laporte SA, Holt JA, Barak LS, Caron MG. Association of beta-arrestin with G-protein–coupled receptors during clathrin-mediated endocytosis dictates the profile of receptor resensitization. *J Biol Chem* 1999; 274:32,248–32,257.

18. Tsao P, Cao T, von Zastrow M. Role of endocytosis in mediating downregulation of G-protein–coupled receptors. *Trends Pharmacol Sci* 2001; 22:91–96.

19. Bohm SK, Grady EF, Bunnett NW. Regulatory mechanisms that modulate signaling by G-protein–coupled receptors. *Biochem J* 1997; 322:1–18.

20. Pierce KL, Lefkowitz RJ. Classical and new roles of beta-arrestins in the regulation of G-protein–coupled receptors. *Nat Rev Neurosci* 2001; 2:727–733.

21. Luttrell LM, Lefkowitz RJ. The role of beta-arrestins in the termination and transduction of G-protein–coupled receptor signals. *J Cell Sci* 2002; 115:455–465.

22. Spiegel AM. Defects in G protein–coupled signal transduction in human diseases. *Annu Rev Physiol* 1996; 58:143–170.

23. Rana BK, Shiina T, Insel PA. Genetic variations and polymorphisms of G-protein–coupled receptors: functional and therapeutic implications. *Annu Rev Pharmacol Toxicol* 2001; 41:593–624.

24. Howard AD. Orphan G-protein–coupled receptors and natural ligand discovery. *Trends Pharmacol Sci* 2001; 22:132–140.

25. Angrist M. GPCRs: Mining the richest vein in drug discovery. Cambridge Healthtech Institute Insights Report, LLC 2003.

26. Wise A, Jupe SC, Rees S. The identification of ligands at orphan G-protein–coupled receptors. *Annu Rev Pharmacol Toxicol* 2004; 44:43–66.

27. Barak LS, Ferguson SS, Zhang J, Caron MG. A beta-arrestin/green fluorescent protein biosensor for detecting G-protein–coupled receptor activation. *J Biol Chem* 1997; 272:27,497–27,500.

28. Oakley RH, Hudson CC, Cruickshank RD, Meyers DM, Payne RE, Rhem SM, Loomis CR. The cellular distribution of fluorescently labeled arrestins provides a robust, sensitive, and universal assay for screening G-protein–coupled receptors. *Assay Drug Dev Technol* 2002; 1:21–30.

29. Johnson EC, Bohn LM, Barak LS, Birse RT, Nassel DR, Caron MG, Taghert PH. Identification of *Drosophila* neuropeptide receptors by G-protein–coupled receptors/beta-arrestin2 interactions. *J Biol Chem* 2003; 278:52,172–52,178.

30. Mundell SJ, Benovic JL. Selective regulation of endogenous G-protein–coupled receptors by arrestins in HEK293 cells. *J Biol Chem* 2000; 275:12,900–12,908.

31. Laporte SA, Oakley RH, Zhang J, Holt JA, Ferguson SS, Caron MG, Barak LS. The beta2-adrenergic receptor/betaarrestin complex recruits the clathrin adaptor AP-2 during endocytosis. *Proc Natl Acad Sci USA* 1999; 96:3712–3717.

32. Zhang J, Barak LS, Anborgh PH, Laporte SA, Caron MG, Ferguson SS. Cellular trafficking of G-protein–coupled receptor/beta-arrestin endocytic complexes. *J Biol Chem* 1999; 274:10,999–11,006.

33. Oakley RH, Laporte SA, Holt JA, Caron MG, Barak LS. Differential affinities of visual arrestin, beta arrestin1, and beta arrestin2 for G protein-coupled receptors delineate two major classes of receptors. *J Biol Chem* 2000; 275:17,201–17,210.

34. Oakley RH, Barak LS, Caron MG. Real time imaging of GPCR-mediated arrestin translocation as a strategy to evaluate receptor–protein interactions. In: George SR, O'Dowd BF, eds. *G Protein–Coupled Receptor–Protein Interactions*. John Wiley & Sons, New York, 2005, 53–80.

35. Oakley RH, Laporte SA, Holt JA, Barak LS, Caron MG. Molecular determinants underlying the formation of stable intracellular G-protein–coupled receptor/beta-arrestin complexes after receptor endocytosis. *J Biol Chem* 2001; 276:19,452–19,460.

36. Liggett SB, Freedman NJ, Schwinn DA, Lefkowitz RJ. Structural basis for receptor subtype-specific regulation revealed by a chimeric beta 3/beta 2-adrenergic receptor. *Proc Natl Acad Sci USA* 1993; 90:3665–3669.

37. Cao W, Luttrell LM, Medvedev AV, Pierce KL, Daniel KW, Dixon TM, Lefkowitz RJ, Collins S. Direct binding of activated c-Src to the beta 3-adrenergic receptor is required for MAP kinase activation. *J Biol Chem* 2000; 275:38,131–38,134.

38. Jaber M, Koch WJ, Rockman H, Smith B, Bond RA, Sulik KK, Ross J Jr, Lefkowitz RJ, Caron MG, Giros B. Essential role of beta-adrenergic receptor kinase 1 in cardiac development and function. *Proc Natl Acad Sci USA* 1996; 93:12,974–12,979.

39. Peppel K, Boekhoff I, McDonald P, Breer H, Caron MG, Lefkowitz RJ. G-protein–coupled receptor kinase 3 (GRK3) gene disruption leads to loss of odorant receptor desensitization. *J Biol Chem* 1997; 272:25,425–25,428.

40. Gainetdinov RR, Bohn LM, Walker JK, Laporte SA, Macrae AD, Caron MG, Lefkowitz RJ, Premont RT. Muscarinic supersensitivity and impaired receptor desensitization in G-protein–coupled receptor kinase 5–deficient mice. *Neuron* 1999; 24:1029–1036.

41. Dale LB, Bhattacharya M, Seachrist JL, Anborgh PH, Ferguson SS. Agonist-stimulated and tonic internalization of metabotropic glutamate receptor 1a in human embryonic kidney 293 cells: agonist-stimulated endocytosis is beta-arrestin1 isoform-specific. *Mol Pharmacol* 2001; 60:1243–1253.

42. Marion S, Weiner DM, Caron MG. RNA editing induces variation in desensitization and trafficking of 5-hydroxytryptamine 2c receptor isoforms. *J Biol Chem* 2004; 279:2945–2954.

43. Miller WE, Houtz DA, Nelson CD, Kolattukudy PE, Lefkowitz RJ. G-protein–coupled receptor (GPCR) kinase phosphorylation and beta-arrestin recruitment regulate the constitutive signaling activity of the human cytomegalovirus US28 GPCR. *J Biol Chem* 2003; 278:21,663–21,671.

44. Zhang JH, Chung TD, Oldenburg KR. A simple statistical parameter for use in evaluation and validation of high throughput screening assays. *J Biomol Screen* 1999; 4:67–73.

45. Burgueno J, Blake DJ, Benson MA, Tinsley CL, Esapa CT, Canela EI, Penela P, Mallol J, Mayor F Jr, Lluis C, Franco R, Ciruela F. The adenosine A2A receptor interacts with the actin-binding protein alpha-actinin. *J Biol Chem* 2003; 278:37,545–37,552.

46. Mundell SJ, Matharu AL, Kelly E, Benovic JL. Arrestin isoforms dictate differential kinetics of A2B adenosine receptor trafficking. *Biochemistry* 2000; 39:12,828–12,836.

47. Matharu AL, Mundell SJ, Benovic JL, Kelly E. Rapid agonist-induced desensitization and internalization of the A(2B) adenosine receptor is mediated by a serine residue close to the COOH terminus. *J Biol Chem* 2001; 276:30,199–30,207.

48. Penn RB, Pascual RM, Kim YM, Mundell SJ, Krymskaya VP, Panettieri RA Jr, Benovic JL. Arrestin specificity for G-protein–coupled receptors in human airway smooth muscle. *J Biol Chem* 2001; 276:32,648–32,656.

49. Shiina T, Kawasaki A, Nagao T, Kurose H. Interaction with beta-arrestin determines the difference in internalization behavior between beta1- and beta2-adrenergic receptors. *J Biol Chem* 2000; 275:29,082–29,090.

50. Barak LS, Oakley RH, Shetzline MA. G-protein–coupled receptor desensitization as a measure of signaling: modeling of arrestin recruitment to activated CCK-B receptors. *Assay Drug Dev Technol* 2003; 1:409–424.

51. Kishi H, Krishnamurthy H, Galet C, Bhaskaran RS, Ascoli M. Identification of a short linear sequence present in the *C*-terminal tail of the rat follitropin receptor that modulates arrestin-3 binding in a phosphorylation-independent fashion. *J Biol Chem* 2002; 277:21,939–21,946.

52. Min L, Galet C, Ascoli M. The association of arrestin-3 with the human lutropin/choriogonadotropin receptor depends mostly on receptor activation rather than on receptor phosphorylation. *J Biol Chem* 2002; 277:702–710.

53. Ferrari SL, Behar V, Chorev M, Rosenblatt M, Bisello A. Endocytosis of ligand–human parathyroid hormone receptor 1 complexes is protein kinase C–dependent and involves beta-arrestin2. Real-time monitoring by fluorescence microscopy. *J Biol Chem* 1999; 274:29,968–29,975.

54. Vilardaga JP, Frank M, Krasel C, Dees C, Nissenson RA, Lohse MJ. Differential conformational requirements for activation of G proteins and the regulatory proteins arrestin and G-protein–coupled receptor kinase in the G-protein–coupled receptor for parathyroid hormone (PTH)/PTH-related protein. *J Biol Chem* 2001; 276:33,435–33,443.

55. Vilardaga JP, Krasel C, Chauvin S, Bambino T, Lohse MJ, Nissenson RA. Internalization determinants of the parathyroid hormone receptor differentially regulate beta-arrestin/receptor association. *J Biol Chem* 2002; 277:8121–8129.

56. Desai S, Ashby B. Agonist-induced internalization and mitogen-activated protein kinase activation of the human prostaglandin EP4 receptor. *FEBS Lett* 2002; 501:156–160.

57. Walker JK, Premont RT, Barak LS, Caron MG, Shetzline MA. Properties of secretin receptor internalization differ from those of the beta(2)-adrenergic receptor. *J Biol Chem* 1999; 274:31,515–31,523.

58. Shetzline MA, Walker JK, Valenzano KJ, Premont RT. Vasoactive intestinal polypeptide type-1 receptor regulation. Desensitization, phosphorylation, and sequestration. *J Biol Chem* 2002; 277:25,519–25,526.

59. Ferguson SS, Zhang J, Barak LS, Caron MG. Molecular mechanisms of G-protein–coupled receptor desensitization and resensitization. *Life Sci* 1998; 62:1561–1565.

60. Ferguson G, Watterson KR, Palmer TM. Subtype-specific regulation of receptor internalization and recycling by the carboxyl-terminal domains of the human A1 and rat A3 adenosine receptors: consequences for agonist-stimulated translocation of arrestin3. *Biochemistry* 2002; 41:14,748–14,761.

61. Evans NA, Groarke DA, Warrack J, Greenwood CJ, Dodgson K, Milligan G, Wilson S. Visualizing differences in ligand-induced beta-arrestin-GFP interactions and trafficking between three recently characterized G-protein–coupled receptors. *J Neurochem* 2001; 77:476–485.

62. Lee DK, Lanca AJ, Cheng R, Nguyen T, Ji XD, Gobeil F Jr, Chemtob S, George SR, O'Dowd BF. Agonist-independent nuclear localization of the Apelin, angiotensin AT1, and bradykinin B2 receptors. *J Biol Chem* 2004; 279:7901–7908.

63. Braun L, Christophe T, Boulay F. Phosphorylation of key serine residues is required for internalization of the complement 5a (C5a) anaphylatoxin receptor via a beta-arrestin, dynamin, and clathrin-dependent pathway. *J Biol Chem* 2003; 278:4277–4285.

64. Richardson RM, Marjoram RJ, Barak LS, Snyderman R. Role of the cytoplasmic tails of CXCR1 and CXCR2 in mediating leukocyte migration, activation, and regulation. *J Immunol* 2003; 170:2904–2911.

65. Kim KM, Valenzano KJ, Robinson SR, Yao WD, Barak LS, Caron MG. Differential regulation of the dopamine D2 and D3 receptors by G-protein–coupled receptor kinases and beta-arrestins. *J Biol Chem* 2001; 276:37,409–37,414.

66. Hobson JP, Rosenfeldt HM, Barak LS, Olivera A, Poulton S, Caron MG, Milstien S, Spiegel S. Role of the sphingosine-1-phosphate receptor EDG-1 in PDGF-induced cell motility. *Science* 2001; 291:1800–1803.

67. Key TA, Foutz TD, Gurevich VV, Sklar LA, Prossnitz ER. *N*-formyl peptide receptor phosphorylation domains differentially regulate arrestin and agonist affinity. *J Biol Chem* 2003; 278:4041–4047.

68. Galliera E, Jala VR, Trent JO, Bonecchi R, Signorelli P, Lefkowitz RJ, Mantovani A, Locati M, Haribabu B. beta-Arrestin–dependent constitutive internalization of the human chemokine decoy receptor D6. *J Biol Chem* 2004; 279:25,590–25,597.

69. Elshourbagy NA, Ames RS, Fitzgerald LR, Foley JJ, Chambers JK, Szekeres PG, Evans NA, Schmidt DB, Buckley PT, Dytko GM, Murdock PR, Milligan G, Groarke DA, Tan KB, Shabon U, Nuthulaganti P, Wang DY, Wilson S, Bergsma DJ, Sarau HM. Receptor for the pain modulatory neuropeptides FF and AF is an orphan G-protein–coupled receptor. *J Biol Chem* 2000; 275:25,965–25,971.

70. Zhang J, Ferguson SS, Law PY, Barak LS, Caron MG. Agonist-specific regulation of delta-opioid receptor trafficking by G-protein–coupled receptor kinase and beta-arrestin. *J Recept Signal Transduct Res* 1999; 19:301–313.

71. Zhang J, Ferguson SS, Barak LS, Bodduluri SR, Laporte SA, Law PY, Caron MG. Role for G-protein–coupled receptor kinase in agonist-specific regulation of mu-opioid receptor responsiveness. *Proc Natl Acad Sci USA* 1998; 95:7157–7162.

72. Tulipano G, Stumm R, Pfeiffer M, Kreienkamp HJ, Hollt V, Schulz S. Differential beta-arrestin trafficking and endosomal sorting of somatostatin receptor subtypes. *J Biol Chem* 2004; 279:21,374–21,382.

73. Heding A, Vrecl M, Hanyaloglu AC, Sellar R, Taylor PL, Eidne KA. The rat gonadotropin-releasing hormone receptor internalizes via a beta-arrestin-independent, but dynamin-dependent, pathway: addition of a carboxyl-terminal tail confers beta-arrestin dependency. *Endocrinology* 2000; 141:299–306.

74. McArdle CA, Franklin J, Green L, Hislop JN. Signalling, cycling and desensitisation of gonadotrophin-releasing hormone receptors. *J Endocrinol* 2002; 173:1–11.

75. Bhatnagar A, Willins DL, Gray JA, Woods J, Benovic JL, Roth BL. The dynamin-dependent, arrestin-independent internalization of 5-hydroxytryptamine 2A (5-HT2A) serotonin receptors reveals differential sorting of arrestins and 5-HT2A receptors during endocytosis. *J Biol Chem* 2001; 276:8269–8277.

76. Mundell SJ, Matharu AL, Pula G, Holman D, Roberts PJ, Kelly E. Metabotropic glutamate receptor 1 internalization induced by muscarinic acetylcholine receptor activation: differential dependency of internalization of splice variants on nonvisual arrestins. *Mol Pharmacol* 2002; 61:1114–1123.

77. McConalogue K, Dery O, Lovett M, Wong H, Walsh JH, Grady EF, Bunnett NW. Substance P-induced trafficking of beta-arrestins. The role of beta-arrestins in endocytosis of the neurokinin-1 receptor. *J Biol Chem* 1999; 274:16,257–16,268.

78. Barak LS, Warabi K, Feng X, Caron MG, Kwatra MM. Real-time visualization of the cellular redistribution of G-protein–coupled receptor kinase 2 and beta-arrestin 2 during homologous desensitization of the substance P receptor. *J Biol Chem* 1999; 274:7565–7569.

79. Schmidlin F, Dery O, Bunnett NW, Grady EF. Heterologous regulation of trafficking and signaling of G-protein–coupled receptors: beta-arrestin–dependent interactions between neurokinin receptors. *Proc Natl Acad Sci USA* 2002; 99:3324–3329.

80. Dery O, Thoma MS, Wong H, Grady EF, Bunnett NW. Trafficking of proteinase-activated receptor-2 and beta-arrestin-1 tagged with green fluorescent protein. beta-Arrestin–dependent endocytosis of a proteinase receptor. *J Biol Chem* 1999; 274:18,524–18,535.

81. DeFea KA, Zalevsky J, Thoma MS, Dery O, Mullins RD, Bunnett NW. beta-Arrestin–dependent endocytosis of proteinase-activated receptor 2 is required for intracellular targeting of activated ERK1/2. *J Cell Biol* 2000; 148:1267–1281.

82. Chen Z, Dupre DJ, Le Gouill C, Rola-Pleszczynski M, Stankova J. Agonist-induced internalization of the platelet-activating factor receptor is dependent on arrestins but independent of G-protein activation. Role of the C terminus and the (D/N)PXXY motif. *J Biol Chem* 2002; 277:7356–7362.

83. Groarke DA, Drmota T, Bahia DS, Evans NA, Wilson S, Milligan G. Analysis of the *C*-terminal tail of the rat thyrotropin-releasing hormone receptor-1 in interactions and cointernalization with beta-arrestin 1-green fluorescent protein. *Mol Pharmacol* 2001; 59:375–385.

84. Hanyaloglu AC, Seeber RM, Kohout TA, Lefkowitz RJ, Eidne KA. Homo- and hetero-oligomerization of thyrotropin-releasing hormone (TRH) receptor subtypes. Differential regulation of beta-arrestins 1 and 2. *J Biol Chem* 2002; 277:50,422–50,430.

85. Terrillon S, Barberis C, Bouvier M. Heterodimerization of V1a and V2 vasopressin receptors determines the interaction with beta-arrestin and their trafficking patterns. *Proc Natl Acad Sci USA* 2004; 101:1548–1553.

86. Chen W, ten Berge D, Brown J, Ahn S, Hu LA, Miller WE, Caron MG, Barak LS, Nusse R, Lefkowitz RJ. Dishevelled 2 recruits beta-arrestin 2 to mediate Wnt5A-stimulated endocytosis of Frizzled 4. *Science* 2003; 301:1391–1394.

Index

Printed in the United States
210721BV00006BA/3/P